FUNGAL ECOLOGY

FUNGAL ECOLOGY

Neville J. Dix
Department of Biological and Molecular Sciences
University of Stirling, UK

and

John Webster
Department of Biological Sciences
University of Exeter, UK

CHAPMAN & HALL
London · Glasgow · Weinheim · New York · Tokyo · Melbourne · Madras

Published by Chapman & Hall, 2–6 Boundary Row, London SE1 8HN, UK

Chapman & Hall, 2–6 Boundary Row, London SE1 8HN, UK

Blackie Academic & Professional, Wester Cleddens Road, Bishopbriggs, Glasgow G64 2NZ, UK

Chapman & Hall GmbH, Pappelallee 3, 69469 Weinheim, Germany

Chapman & Hall USA, One Penn Plaza, 41st Floor, New York NY 10119, USA

Chapman & Hall Japan, ITP-Japan, Kyowa Building, 3F, 2-2-1 Hirakawacho, Chiyoda-ku, Tokyo 102, Japan

Chapman & Hall Australia, Thomas Nelson Australia, 102 Dodds Street, South Melbourne, Victoria 3205, Australia

Chapman & Hall India, R. Seshadri, 32 Second Main Road, CIT East, Madras 600 035, India

First edition 1995

© 1995 Neville J. Dix and John Webster

Typeset in Times by Falcon Graphic Art Ltd, Wallington, Surrey

Printed in Great Britain at the University Press, Cambridge

ISBN 0 412 22960 9 (HB)   0 412 64130 5 (PB)

Apart from any fair dealing for the purposes of research or private study, or criticism or review, as permitted under the UK Copyright Designs and Patents Act, 1988, this publication may not be reproduced, stored, or transmitted, in any form or by any means, without the prior permission in writing of the publishers, or in the case of reprographic reproduction only in accordance with the terms of the licences issued by the Copyright Licensing Agency in the UK, or in accordance with the terms of licences issued by the appropriate Reproduction Rights Organization outside the UK. Enquiries concerning reproduction outside the terms stated here should be sent to the publishers at the London address printed on this page.

The publisher makes no representation, express or implied, with regard to the accuracy of the information contained in this book and cannot accept any legal responsibility or liability for any errors or omissions that may be made.

A catalogue record for this book is available from the British Library

Library of Congress Catalog Card Number: 94-72667

∞ Printed on permanent acid-free text paper, manufactured in accordance with ANSI/NISO Z39.48–1992 and ANSI/NISO Z39.48–1984 (Permanence of Paper).

To Heather, Katharine, Stephen and Eleanor,
and to Brom, Chris and Sarah.

CONTENTS

Preface ix

1 Introduction 1
  1.1 Life strategies of fungi 5

2 The mycelium and substrates for growth 12
  2.1 The mycelium 12
  2.2 Spores and other mycelial-derived structures 17
  2.3 Substrata and substrates for saprotrophs 26

3 Structure of fungal communities 39
  3.1 Introduction 39
  3.2 Development of fungal communities 45
  3.3 Successions 80

4 Colonization and decomposition of leaves 85
  4.1 Deposition 85
  4.2 Colonization of living leaves 88
  4.3 Factors affecting the development of fungal communities on leaf surfaces 99
  4.4 Fungal successions in leaf litter 114
  4.5 Decomposition of leaf litter 122

5 Development of fungal communities on herbaceous stems and grasses 128
  5.1 Distribution of fungal populations on *Dactylis glomerata* 129
  5.2 Interpretation of distribution patterns 135
  5.3 Distribution patterns on other plants 140

6 Colonization and decay of wood 145
  6.1 Wood as a resource 145
  6.2 Colonization of woody tissues 150
  6.3 Types of wood rot 161
  6.4 Water relations 169

7 Fungi of soil and rhizosphere — 172
7.1 Techniques for studying fungi of soil and roots — 172
7.2 Fungal distribution in soil — 175
7.3 Fungal activity in soil — 179
7.4 The rhizosphere and root colonization — 183

8 Coprophilous fungi — 203
8.1 Succession of coprophilous fungi — 207
8.2 Comparison of the fungal flora of different animal dungs — 222
8.3 Autecological studies — 224

9 Aquatic fungi — 225
9.1 Introduction — 225
9.2 Freshwater fungi — 227
9.3 Marine fungi — 265

10 Nematophagous fungi — 284
10.1 Techniques for studying nematophagous fungi — 290
10.2 Distribution and abundance — 291
10.3 Ecological characteristics — 294
10.4 Biological control of pathogenic nematodes — 300

11 Phoenicoid fungi — 302
11.1 Chemical, physical and biological changes in soil after burning — 303
11.2 Ecological characteristics and phenology — 305
11.3 Experimental studies — 313
11.4 Fruiting of phoenicoid fungi following volcanic eruptions — 319

12 Fungi of extreme environments — 322
12.1 Thermotolerant and psychrotolerant fungi — 322
12.2 Xerotolerant and osmotolerant fungi — 332

13 Terrestrial macrofungi — 341
13.1 Introduction — 341
13.2 Wood-decay macrofungi — 345
13.3 Litter decomposers — 365
13.4 Mycorrhizal macrofungi — 376

References — 398
Index — 499

PREFACE

Fungi play vital roles in all ecosystems, as decomposers, symbionts of animals and plants and as parasites. Thus their ecology is of great interest. It has been estimated that there may be as many as 1.5 million species of fungi, many of which are still undescribed. These interact in various ways with their hosts, with their substrates, with their competitors (including other fungi) and with abiotic variables of their environment. They show great variation in morphology, reproduction, life cycles and modes of dispersal. They grow in almost every conceivable habitat where organic carbon is available: on rock surfaces, in soil, the sea and in fresh water, at extremes of high and low temperature, on dry substrata and in concentrated solutions. Fungal ecology is therefore an enormous subject and its literature is voluminous. In view of this we have had to be selective in the material we have included in this book. We have chosen to concentrate on subjects in which we have some personal experience through either research or teaching. We preferred to tackle a few subjects in depth instead of attempting to cover a wider range of topics superficially. We are conscious of the extensive gaps in coverage: for example on the ecology of lichens, of fungal plant pathogens and of the complex interactions between fungi and animals. It is some justification that book-length treatments of these subjects are available elsewhere. We are equally conscious that many of the subjects which we have chosen to present are also very large and could themselves be expanded into books.

We are indebted to numerous publishers, colleagues and friends who have allowed us to make use of their published and unpublished figures. These are individually acknowledged. We are extremely grateful to Pat Brown, Khlayre Mullin and Joan Vaughan for competent typing. We also thank our wives for their continued patience and understanding.

Exeter and Stirling
September 1994.

1
INTRODUCTION

Fungi are heterotrophs, and because of this, they play several distinctive roles in ecosystems: as saprotrophs, as parasites of plants and animals, as mutualistic symbionts of many phototrophic organisms, e.g. cyanobacteria and algae in the form of lichens, and as mycorrhizal partners of vascular plants. Indeed the earliest vascular plants had mycorrhizal associations, and it has been claimed that mycorrhizal infection was the key to successful colonization of the land (Pirozynski and Malloch, 1975; Pirozynski and Hawksworth, 1988; Pirozynski and Dalpé, 1989.

There is an enormous number of fungi. Hawksworth (1991) has estimated that there may be as many as 1.5 million species. They exist in a wide range of habitats: in fresh water and the sea, in soil, litter, decaying remains of plants and animals, in dung, and in living plants and animals. In living plants, they are often present as symptomless endophytes (Chapters 4 and 6), sometimes as biotrophic or necrotrophic parasites. Some of the environments in which fungi grow are extreme, e.g. extremes of cold (as psychrophiles) or heat (as thermophiles and thermotolerants) or of low water content (as xerophiles) or of high osmotic content (as osmophiles). Some grow as oligotrophs and increase biomass on substrata extremely poor in nutrients, scavenging carbon dioxide and traces of organic carbon (Wainwright, 1988; R. Jones *et al.*, 1991).

Fungi are diverse in structure and vegetative organization, ranging from unicells (e.g. yeasts), holocarpic endoparasites of plants and animals, to filaments which may be single or aggregated into strands, cords or more highly differentiated rhizomorphs (section 2.2.3).

The reproductive cycles of fungi are very varied (Raper, 1966). Fungi reproduce by means of spores and some have cycles which are entirely sexual, whilst in others the reproductive cycle involves asexual and sexual phases. Such fungi are described as **pleomorphic**. The whole fungus is defined as the **holomorph**, which can be separated into the **anamorph** (asexual, imperfect or conidial state) and the **teleomorph** (sexual or perfect state). The two states are sometimes quite dissimilar and have often been given separate names. Whilst it would be desirable to use only one name for a single organism, in practice a dual nomenclature has developed, with separate names applied to the two states.

Spores have two primary functions: dispersal and survival. A distinction has been made (section 2.2.1) between **xenospores**, concerned with dispersal, and **memnospores**, concerned with survival. Asexual reproduction in many aquatic and some terrestrial fungi is by actively swimming **zoospores** which may be chemotropically attracted towards suitable substrata. In the Mucorales, the characteristic asexual spore is a non-motile sporangiospore (aplanospore), but in *Pilobolus*, which grows on herbivore dung, the whole sporangium is violently discharged and attaches to the herbage. In a few Mucorales, sporangiola containing relatively few sporangiospores may be the dispersal unit, whilst in others conidia are dispersed (Ingold, 1971). Violently discharged conidia are a feature of the Entomophthorales, many of which are insect pathogens. With relatively few exceptions, the conidia of ascomycetes, basidiomycetes and deuteromycetes are passively dispersed. They are variable in size, form and pigmentation. Variations in size and shape are correlated with habitat and mode of dispersal (see below).

Sexual reproduction (i.e. a process involving nuclear fusion and meiosis) often results in the formation of relatively resistant spores capable of survival, e.g. **oospores**, **zygospores**, **ascospores** and **basidiospores**. However, many fungi (Deuteromycotina) apparently have no sexual states (or these have not been demonstrated), and such fungi reproduce asexually by means of conidia. The absence of the sexual state does not prevent genetic recombination in Deuteromycotina, which may have other mechanisms of recombination, e.g. through a parasexual cycle.

A feature of almost all fungi is the copious production of spores, be they sexual or asexual. Coupled with prolific reproduction, the mechanisms and agents of dispersal are very effective in transporting fungal propagules. Spores may be dispersed in water, wind and water currents, by rain splash, by insects or other animal vectors. Wind dispersed spores are dry. Small spores are adapted to penetrate through vegetation and only impact effectively on narrow objects. Larger spores, such as those of leaf pathogens, impact with greater efficiency on wider objects (Gregory, 1973). Insect-dispersed spores are, in contrast, usually sticky and often associated with sweet and scented secretions (Ingold, 1971). Splash-dispersed spores are also sticky and usually aggregated into slimy masses often enclosed in mucilage, as in *Fusarium* and in the numerous pycnidial fungi which grow on the surface of twigs and herbaceous leaves and stems. The conidia of many freshwater fungi which grow saprotrophically on leaves and twigs in rapidly flowing streams are large, branched or sigmoid; these features are adaptations to attachment to underwater substrata (Webster, 1987) (Chapter 9). Marine ascomycetes form ascospores with elaborate appendages which aid in trapping (Chapter 9).

It is clear from these few examples that the form and structure of the fungus spore has a considerable bearing on the ecology of the fungus, in terms of substratum, habitat and vector.

Other means of dispersal may be within seeds or fruits, timber, in wind-blown leaves and soil, or on soil attached to plant-propagating material such as rootstocks, tubers, etc. (Hirst, 1965). Similarly, dispersal of coprophilous fungi may occur through the migration of herbivores carrying the spores in their guts. The dispersal of fungal pathogens of fish (e.g. *Saprolegnia*) may occur for many kilometres upstream as salmon migrate to spawning grounds, and fungal pathogens of insects are also dispersed by the flight of diseased hosts.

Although the majority of air-borne spores are deposited within 100 m of their source, clouds of spores carried up into the upper atmosphere may travel over thousands of kilometres, and a significant proportion of them may remain viable. For this reason, fungi are remarkably ubiquitous and cosmopolitan, and provided that a suitable host or substratum is exposed to infection, colonization usually proceeds quickly. Selective processes relating to the physical and chemical properties of the substratum, environmental variables such as temperature and water content, the physiology and enzymatic properties of the colonizing fungi, the availability and form of inoculum, competition between the colonizing fungi and other organisms which compete for the particular substratum, eventually result in a fungal community recognized as characteristic for that substratum (Chapter 3). The community takes time to establish itself. Pioneer organisms, which are often ubiquitous, rapidly growing and non-host-specific, capture available resources, but are then displaced by others. In this way a succession of fungi is involved in the exploitation of most substrata. Examples of fungal successions are given in sections 3.3 and 8.1 and Chapter 13.

Given the large number of fungi, the variation in their structure, reproduction, lifestyle, geographical distribution, habitat, host range or substrate preference, and that environmental variables may act directly on them and also indirectly through effects on their hosts and on their vectors, it should be apparent that a comprehensive account of fungal ecology is not possible. It is therefore necessary to be selective. In this book, we have, for the most part, excluded pathogenic fungi, especially biotrophs. We have also excluded lichens, despite the fact that much is known about lichen distribution and ecophysiology, especially in relation to pollution.

There are several ways of studying fungal ecology. The autecological approach focuses on a given species of fungus, and attempts to study its behaviour in relation to its changing substratum and to external abiotic and biotic variables. An example of this type of study is that of Frankland (1984) on *Mycena galopus*, a litter-decomposing basidiomycete (Section 13.3). Another approach is a study of communities characteristic of certain habitats, e.g. soil or marine and freshwater (Chapter 9), or communities tolerant of various kinds of environmental stress (Chapter 12). It is also possible to study the fungal communities which develop on particular substrata, e.g. herbivore dung (Chapter 8), leaves (Chapter 4), herbaceous stems (Chapter 5) or burnt ground (Chapter 11). Yet another approach is to study the fungal parasites of a group of animals, e.g. nematodes (Chapter 10). Emphasis may be placed on ecological phenomena such as antibiosis or mycoparasitism, and aspects of these subjects are discussed in various chapters.

Ecology can be defined as the study of organisms in relation to their environment. However, it is important also to define what aspect of this relationship is to be studied. The list below, whilst not exhaustive, gives some idea of the possibilities.

- Distribution on the microscopic, macroscopic and geographic scale.
- Presence or absence: form in which the fungus is present, e.g. as active mycelium, inactive spore, chlamydospore or sclerotium, reproductive state (anamorph or teleomorph).
- Relative abundance, e.g. in relation to stage of colonization, season, seral change or external variables.
- Amount of living biomass.
- Activity, viability, survival (and in what form?).
- Reproductive capacity.
- Competitiveness.

For each of these aspects of fungal ecology, different techniques may be required, and the techniques appropriate for one group of fungi may be inappropriate for another. For example, it is possible to estimate the concentration of the conidia of aquatic hyphomycetes in stream water (Section 9.2.3 (a)) by Millipore filtration, because the spores are sufficiently large and distinctive to be counted and identified, but this technique would be inappropriate for the sexually produced spores (ascospores and basidiospores) from the teleomorphs of the same group of fungi because these spores are smaller and less distinctive in form. It would, for the same reasons, be inappropriate for other groups of freshwater fungi. It is not always possible to identify fungi from their mycelial characters, so presence or abundance is often indexed by the counting of fruit-bodies (especially of the basidiomycetes), although it is obvious that the fungus must have been present in the mycelial form but possibly could not have been detected before fructifications developed. It is also likely that active mycelium persists after fructification has ceased. It may be necessary to isolate a fungus from its substratum and to induce it to sporulate in culture before it is possible to identify it. Whilst these problems are not unique to the study of fungal ecology, they present difficulties which can only be overcome by the choice of a technique appropriate to the organism and the facet of its biology that is being studied.

1.1 LIFE STRATEGIES OF FUNGI

Different fungi vary in their behaviour in relation to the substrata available to them, to environmental variables such as stress or disturbance, and in their relation to competitors. Even within the lifespan of a single fungal individual, its behaviour may change, e.g. from rapid exploitation during primary resource capture to a phase of assimilation and utilization, consolidation and defence. Attempts have been made to classify the types of fungal behaviour or life strategies which have evolved in response to selection pressures of various kinds. Before considering the life strategies of fungi, certain terms must be defined.

Stress has been defined as 'any form of continuously imposed environmental extreme which tends to restrict biomass production' (Cooke and Rayner, 1984). Stress may be caused by environmental features such as high or low temperature, unsuitable pH, water availability (too little or too much), or inadequacy or unsuitability of available nutrients or the presence of inhibitory compounds such as phenols in wood. Severe stress may prevent growth (i.e. biomass production) completely, but less severe stress may involve modification of the physiology and morphology of the fungus, i.e. in the production of structures which permit survival of the imposed stress, e.g. chlamydospores or sclerotia.

Disturbance implies disruption of the normal environment of a fungus. It has been defined as 'such dramatic alteration of the normal that (i) the biomass is reduced, or (ii) a new environment is superimposed on the existing one, and this results in a relative reduction in living fungal biomass' (Pugh, 1980). However, two types of disturbance have been distinguished:

- **Disruptive disturbance**, examples of which are fire, a landslip, the breaking off of a branch or the felling of a tree. In all these cases, the effect of the disturbance is to change one set of nutrient and environmental conditions and to substitute another set (Boddy, Watling and Lyon, 1988; Zak, 1992).
- **Enrichment disturbance**, where the habitat conditions are changed by the sudden addition, or increase in the rate of supply, of nutrients. Examples are the seasonal supply of leaf litter to the soil in deciduous forests or to streams flowing through forests, the deposition of dung on the surface of the soil or the death of a burrowing animal in the soil (Sagara, 1992). The effect is to remove nutrient limitation and to permit the establishment of a population of microorganisms capable of exploiting the newly available material (Boddy, Watling and Lyon, 1988).

Competition can be defined as 'an active demand by two or more individuals of the same or different species for the same resource' (Cooke and Rayner, 1984). This could imply active demand for nutrients, oxygen, water, or for space. Competition may take a variety of forms. Cooke and Rayner (1984) have distinguished between primary resource capture and combat whilst Lockwood (1992) and Wicklow (1992) refer to exploitation and interference competition.

Primary resource capture is the process by which a fungus gains access to an available resource. Garrett (1950, 1956) has introduced the term 'competitive saprophytic ability' to sum up the attributes which enable a fungus to colonize a virgin substratum. These might involve rapid spore germination and mycelial growth, versatile enzyme production and the production of antibiotics that limit the growth of other organisms.

Combat is the defence of resources which have already been captured, or the capture of resources which have previously been captured by other organisms, a process termed secondary resource capture. The actual mechanisms by which secondary resource capture is achieved may involve production of extracellular antibiotics, contact inhibition, e.g. hyphal interference (section 8.1.1(c)), and mycoparasitism (section 3.2.4(b)).

Some of the concepts and terminology which have been advanced as aids to understanding the ecology of other organisms have been extended to fungi. MacArthur and Wilson (1967) introduced the terms **K-selection** and **r-selection** strategies. The term K refers to the carrying capacity of the environment, i.e. the equilibrial population size attainable, whilst r is the intrinsic rate of increase, i.e. the per capita rate of increase of a population in a given environment. K-Selection favours a more efficient utilization of resources, whilst r-selection favours a high population growth rate and higher productivity, a form of selection advantageous during the colonizing episode, or in species which are frequently involved in colonizing episodes. Pianka (1970) views r–K as a continuum. At the r-end of the continuum which represents the quantitative extreme where there is no competition, 'the optimal strategy is to put all possible matter and energy into reproduction, with the smallest practicable amount into each individual offspring, and to produce as many total progeny as possible. The K-endpoint represents the qualitative extreme – density effects are maximal and the environment is saturated with organisms . . . The optimal strategy is to channel all available matter and energy into maintenance and the production of a few extremely fit offspring.'

Grime (1977, 1979) has suggested that during the evolution of plants, three fundamentally different forms of selection have occurred.

- **C-Selection**: Selection for highly competitive ability, depending on characteristics which maximize growth in productive, relatively undisturbed conditions (competitive strategy).

- **S-selection**: Selection for adaptations which allow endurance of conditions of environmental stress or resource depletion (stress-tolerant strategy).
- **R-selection**: Selection for short lifespan and high-speed production has evolved in disturbed but productive environments (ruderal strategy).

This concept has been accepted, modified and extended for fungi by Cooke and Rayner (1984) and Andrews (1992).

In relation to the ideas of r- and K-selection, the ruderal strategy corresponds with the extreme of r-selection.

Pugh and Boddy (1988) have stressed that the three primary ecological strategies should be used to define *behaviour* shown at a particular stage of its life cycle by a fungus, and should not be used to classify an individual fungus, because a fungus may switch its behaviour during different phases of its growth. Some examples of fungi apparently selected according to these differing primary strategies are given below.

1.1.1 R-Selected fungi (ruderals)

Many members of the Mucorales are regarded as ruderals. They include such genera as *Mucor* and *Rhizopus* which are abundant in soil. Their aplanospores survive in the soil and, when nutrients are provided, they germinate rapidly. Mycelial growth is rapid and they are able quickly to colonize (i.e. capture) a newly available resource. The enzymic equipment of most Mucorales is limited, and they are generally incapable of degrading cellulose or lignin, so that they are nutritionally dependent on soluble carbohydrates which are quickly depleted by microbial activity. However, some species have been isolated during late stages of fungal successions. Frankland (1966) showed that the frequency of Mucorales isolated from petioles of bracken (*Pteridium aquilinum*) laid on the soil surface reached a maximum after 5 years. The most probable reason was that they were obtaining soluble carbohydrates released during the breakdown of cellulose by associated organisms.

Members of the Mucorales reproduce asexually quickly and prolifically, often within 2 days. After available nutrients have been depleted, the mycelium undergoes lysis and survival is by aplanospores, chlamydospores and, in some species, by zygospores. Other soil-borne ruderals include *Pythium* spp., which are opportunistic non-obligate necrotrophs of seedling roots and hypocotyls capable of advancing rapidly through host tissues by means of pectolytic enzymes. Asexual reproduction occurs rapidly by zoosporangia which produce zoospores when soil water is plentiful. In *P. ultimum*, the sporangia germinate directly, i.e. by means of germ tubes, within 3–4 h when stimulated by exudates from *Phaseolus* seeds (Stanghellini and Hancock, 1971). Amongst coprophilous fungi, *Mucor* spp. fruit early on voided herbivore dung. The more specialized member of the Mucorales, *Pilaira anomala*, which is also amongst the first fungi to fruit on herbivore dung, has also been regarded as a ruderal. It makes good growth on glucose and fructose, but lacks the enzymic ability to break down polymers such as starch, cellulose, pectin or proteins (Wood and Cooke, 1987).

1.1.2 S-Selected fungi (stress-tolerant fungi)

Stress may be caused by a single factor, e.g. water limitation, nutrient limitation, anaerobiosis, high temperature, or by a combination of factors. Fungi characteristic of extreme environments are obvious examples of stress-tolerant forms which have adaptations to physical stress, and these are discussed more fully in Chapter 12. Fungi adapted to nutrient-imposed stress include oligotrophs (oligocarbotrophs – see Wainwright, 1988) which are capable of sparse hyphal growth in the laboratory on distilled or de-ionized water or carbon-free buffer solutions. Under such conditions *Trichoderma harzianum* produces sheets of fine hyphae ('gossamers') with numerous anastomoses and abundant chlamydospores. Common moulds such as species of *Mucor, Trichoderma, Fusarium, Gliocladium* and *Penicillium* have been grown on silica-gel media lacking organic carbon, and some have also been grown in distilled water on carefully cleaned glassware (Tribe and Mabadeje, 1972).

Figure 1.1 *Onygena equina*: ascomata on a cast sheep's horn. The specialized substratum of this fungus is related to its ability to utilize keratin. Scale bar = 1 cm.

Nutrient stress may be imposed by an inability to degrade polymers such as chitin, keratin and lignin, and fungi which are able to do so obviously have a selective advantage as compared with those that cannot, where these are the main carbon supplies in the resource. Specialized groups of fungi such as the Laboulbeniales have the capacity to penetrate insect cuticle and degrade chitin. *Onygena* spp. grow on keratin-rich substrata such as feathers, shed horns and hooves (Figure 1.1). The ability of some fungi, e.g. *Ctenomyces serratus*, to grow on feathers may also be correlated with the tolerance of or possibly the stimulation by substances present in feather fats (Pugh and Evans, 1970).

1.1.3 C-Selected fungi (combative fungi)

Combative fungi have the capacity for secondary resource capture and for defence of a captured domain, i.e. the taking-over of a domain previously captured by another organism (Rayner and Webber, 1984). Many good examples of combative or C-selected fungi are to be found in wood-decaying basidiomycetes (Chapters 6 and 13). Wood previously occupied by ruderals is later colonized by other fungi. From studies of the interactions between the participants in the struggle on the natural substratum or when they are opposed on agar, a hierarchy of combatants can be drawn up. For example, when the common fungi colonizing felled beech (*Fagus sylvatica*) logs were paired on agar (Chapela, Boddy and Rayner, 1988; Rayner and Boddy, 1988), they could be classified as follows:

- **Least combative**, being replaced by most other species: *Armillaria bulbosa, Lopadostoma turgidum* and *Xylaria hypoxylon*.
- **Better combatants**: *Coriolus versicolor* and *Stereum hirsutum*.
- **More combative still**: *Phallus impudicus, Phanerochaete velutina* and *Tricholomopsis platyphylla*.
- **Most combative**: *Hypholoma fasciculare, Lenzites betulina, Psathyrella hydrophilum* and *Sistotrema brinkmannii*.

The mechanisms by which combative fungi prevail over the fungi with which they interact are varied. They include the disabling of their opponents by diffusible antibiotics, as in the interaction between the necrotrophic mycoparasite *Hypomyces aurantius* and its susceptible host *Coriolus versicolor* (Kellock and Dix, 1984a,b), and hyphal interference, a form of contact inhibition (Ikediugwu and Webster, 1970a,b; Laing and Deacon, 1991) (section 8.1.1(c)). In some other cases, replacement of one fungus by another is the result of selective mycoparasitism. *Pseudotrametes gibbosa* is a white-rotting fungus especially common on stumps of beech (*Fagus sylvatica*) where it replaces populations of *Bjerkandera adusta*. In a similar way, *Lenzites betulina* can replace *Coriolus versicolor* by entwining around its hyphae and penetrating them (Rayner, Boddy and Dowson,

1987a,b). The result of this form of competition is that the successful combatant acquires domain previously occupied by its opponent and acquires nourishment from the fungal biomass.

1.1.4 Secondary strategies

It would be surprising if fungi (or the behaviour of fungi) could be classified into only one of three primary strategies, because the behaviour of a fungus will vary in relation to its stage of development, in relation to the state of decay of the substratum which it is colonizing, in relation to its competitors, and to abiotic variables. Even within the same mycelium, different kinds of behaviour can be recognized. For example, the tree root pathogen *Heterobasidion annosum* has a mycelium with great developmental versatility (Stenlid and Rayner, 1989), being able to switch from a rapidly growing, sparsely branched, mycelial growth to a slower-growing, densely branched form. It shows ruderal characteristics in rapid germination of basidiospores and conidia which enable it to colonize effectively 'disturbed' sites such as branch wounds and cut stump surfaces. It also shows features of K-selection in that it can persist for long periods in dead tree trunks and root systems. Although tolerant of the stress during survival, i.e. showing selection for R–S strategies, it is relatively intolerant of competition, and does not compete effectively with some other colonizers of the cut surfaces of stumps such as *Peniophora gigantea*. Indeed, *Peniophora* has been used as a very effective biological control agent against stump colonization by *Heterobasidion* (Rishbeth, 1963; Ikediugwu, Dennis and Webster, 1970).

Species of *Trichoderma* provide interesting examples of other combined strategies (Widden and Scattolin, 1988). *Trichoderma* spp. are common in soil and on wood lying in contact with soil. Most species grow very rapidly, produce abundant conidia and have a wide range of enzymes including cellulases (Eveleigh, 1985). They have the hallmarks of ruderals. At the same time, many species are combative, being highly antagonistic to other fungi by a variety of mechanisms including the production of soluble antibiotics, volatile antibiotics or by parasitism. Some species, for example, *T. harzianum*, as mentioned previously, are tolerant of stress imposed by nutrient scarcity (Wainwright, 1988). Others are tolerant of low temperature. In experiments designed to test the ability of five species of *Trichoderma* to colonize autoclaved spruce (*Picea abies*) needles competitively, over a range of temperatures it was found that all species were antagonistic towards each other, but temperature appeared to have little effect on their antagonistic activity (Widden, 1984; Widden and Hsu, 1987). However, *T. viride* and *T. polysporum* were displaced from the spruce needles at high temperatures by *T. hamatum* and *T. koningii*, whilst at low temperatures they survived better. This would enable *T. viride* and

T. polysporum to colonize newly available substrata such as freshly fallen spruce needles during the cooler months and remain in possession of these resources until replaced by *T. hamatum* and *T. koningii* in the spring and summer. Widden and Scattolin (1988) regard *T. hamatum* and *T. koningii* as better competitors (C-selected fungi). *Trichoderma viride* and *T. polysporum* are less aggressive and seem to have a stress-tolerant strategy, supported by the fact that, in the field, they are associated with colder soils, winter conditions and more acid soils.

It is therefore apparent that although the concepts of life strategies may be helpful in considering the role which a fungus may play in an ecosystem, its actual performance at any one time will be affected by many variables, and indeed its strategy (? tactics) may change during its development.

For further discussion of life strategies of fungi, see Chapter 3.

2
THE MYCELIUM AND SUBSTRATES FOR GROWTH

2.1 THE MYCELIUM

The vegetative body of most saprophytic fungi is the **mycelium**; only some members of the Chytridiomycetes and yeasts are non-mycelial. The unit of the mycelium is the **hypha**, which is typically filamentous except under abnormal physiological conditions. More diverse shapes and forms of hyphae, however, are to be found in vegetative survival structures and in macroscopic reproductive bodies. Hyphae are on average about 5–6 μm in diameter and grow by wall extension at the tip. The glucan polymers of the fibrillar components of the hyphal wall are elastic and are no hindrance to the stretching process that is an essential feature of forward growth. The actual region of growth is a few micrometres behind the tip, the very tip of the hypha being non-extendable and forming a resistant shield that enables the hypha to penetrate into solid substrata. With the exception of the Oomycetes and certain ascomycete yeasts, the hyphal walls of all other fungi contain small amounts of chitin. This is a hard polymer synthesized from N-acetyl glucosamine and is non-elastic but adds strength to the wall.

The mycelium is built up by the repeated branching of hyphae. In most systems branching is monopodial, by which the leading function is retained in the main branch. Branches are produced by young hyphae some way behind the tips at distances which vary according to species and conditions. Branching is normally not close to the tip so that the margin of the mycelium is made up of a peripheral zone of unbranched leaders which initiate exploratory forward growth. In the established mycelium the rate of forward extension is directly dependent upon the availability of exploitable resources in the peripheral growth zone since the rate of translocation from the distal parts is too slow. If conditions are favourable, the forward extension of hyphae is exponential.

Primary branches, such as those arising from the germ tube of a germinating spore, form at wide angles to the parent hyphae, but the angle of branching in subsequent orders of branching becomes increasingly narrower as hyphae seek to exploit uninvaded space. Spacing between hyphae in the expanding mycelium is probably regulated by positive chemotropic responses to oxygen or nutrient concentration gradients. The mycelium expands by the absorption of nutrients and extracellular enzymes are secreted for the purpose of solubilizing the substratum on which the hyphae are growing. Ecological adaptation is achieved through the range of enzymes produced and the multiple forms of individual enzymes (sections 2.3.2 and 2.3.6). The full range of these and hence the substrates that can be utilized depends upon species.

Extracellular enzymes are extremely stable glycoproteins that operate in the fluids of the substratum. Enzymes may diffuse through the substratum but, if pore sizes are limiting, enzymes will not be able to move into the substratum and reactions will then be restricted to interfaces between the substratum and the penetrating hyphae.

The manner in which the mycelium expands enables the fungus to exploit a large surface area of the substratum, capturing and holding resources for future use with a minimum expenditure of biomass. In culture the development of the mycelium from a point of inoculum is by radial extension of hyphae in three dimensions. In water or liquid culture the mycelium assumes a spherical form, but on solid substrata it is more convex with aerial hyphae rising above and submerged hyphae penetrating the medium below. Extracellular enzymes assist penetration into solid substrata but the depth of penetration is dependent on adequate aeration. Little is known about the shape of the mycelium when growing on solid substrata in nature.

During radial growth, the number of leading hyphae per unit area of the peripheral growth zone remains more or less constant. This is achieved by the accelerated growth of some laterals as the colony expands. Normally the rate of growth of laterals is subordinate to that of the leading hyphae, but if 'windows' appear in the zones of influence of the latter, laterals will increase their rate of growth and grow into them, an effect that can be reproduced if some leaders are excised. When nutritional conditions are poor in the peripheral growth zone branching is reduced and synthesized biomass is directed to the leading hyphae, enabling them to grow more rapidly. This device enables the fungus to grow over inhospitable substrata without unprofitable expenditure of biomass and is an obvious advantage in environments where exploitable resources are discontinuously dispersed.

Supporting cross-walls or **septa** are present in the vegetative hyphae of fungi in all taxonomic groups with mycelia, save those in the Oomycetes and the Zygomygotina, where they are mostly confined to the reproductive structures. The hyphae of the latter are **coenocytes**; in all others the presence of septa divides the hyphae into compartments which are analogous to cells. Septa are normally perforated and allow cytoplasmic continuity along hyphae and intercommunication between different parts of the mycelium. An important biological consequence of this arrangement is that free movement of suspended particles and the translocation of solutes is possible throughout the mycelium. Very active cytoplasmic streaming can be seen in hyphae growing under favourable conditions and this is evidence that there is movement of dissolved and suspended matter within hyphae. Transport is usually towards hyphal tips and serves to carry solutes, vesicles and other organelles to the actively growing parts. It is suspected that the driving force is the outflow of water from the distal ends of hyphae by transpiration. Another possibility is that movement is caused by the bulk flow of water due to internally generated hydrostatic pressure (Jennings, 1984).

In culture, the established mycelium extends by the advance of a slow-growing mycelial front of aerial and submerged hyphae that delimits the individual mycelium.

In nature, as parts of the established mycelium become older, the ageing hyphae become evacuated, the cytoplasm first becoming metabolically inactive, finally becoming solubilized by autolysis and moved forward to be recycled at the growing tips. After the withdrawal of the hyphal contents, the septal pores in the empty compartments are sealed off, blocking communication with the rest of the mycelium. Eventually the walls of emptied hyphae become lysed by extracellular chitinases and glucanases released by other microorganisms (Chesters and Bull, 1963; Mitchell and Alexander, 1963; Skujins, Potgieter and Alexander, 1965).

Autolysis will also follow if hyphae are starved (Ko and Lockwood, 1970) or if function is impaired by toxins and antibiotics produced by other microorganisms (D. Jones and Watson, 1969; Dennis and Webster, 1971a, b).

Lysis of hyphae can lead to the separation and independent development of daughter colonies. This is most obvious among certain 'fairy ring' forming toadstools (section 13.3). As time passes, the expanding mycelium of these species tends to fragment, leaving isolated portions of the mycelium to grow on to produce a number of genetically identical but scattered daughter colonies.

To be able to recycle biomass as the mycelium expands is not only economical but is also a mechanism of survival. By conserving accumulated resources, the fungus is able to continue to grow when carbohydrates and nutrients are in short supply, enabling it to survive for a time until conditions improve or it is able to grow to favourable conditions. As a survival mechanism it is seen in an extreme form when spores germinate under poor nutrient conditions on leaves. Here the endogenous food reserves of the spore can be continuously recycled to maintain life in a few anterior hyphal compartments. As the branching germ tube extends by this means, it leaves behind a series of empty ghost compartments, continuing in this fashion until conditions improve or the reserves become exhausted (Dickinson and Bottomley, 1980).

Anastomosis or **hyphal fusion** within the mycelium is a common feature of the hyphae of the established mycelium of higher fungi that distinguishes the mature mycelium functionally from the construction and exploration phase of hyphal growth in the young mycelium. Self-to-self fusion between hyphae in the same mycelium serves to increase intercommunication and functional efficiency. Fusion is always between young hyphae and is either by the tip-to-tip fusion of main hyphal branches or between hyphal tips and the tips of telemorphically produced short laterals (Figure 2.1).

Successful non-self fusion may also occur between the hyphae of compatible strains of the same species and if plasmogamy (the transfer and fusion of the two cytoplasms) follows hyphal wall fusion, nuclei and other organelles may be transferred and this may lead to establishment of a **heterokaryon**, where genetically different nuclei coexist in the same hyphal compartment. From such a heterokaryon a whole heterokaryotic mycelium can develop either by nuclear division and the migration of daughter nuclei throughout the mycelium, via septal pores, or by hyphal growth from the heterokaryotized portion. Rates of nuclear migration through the mycelium can be very fast and rates of movement as fast as 10 mm h^{-1} have been recorded. Movement of nuclei is possibly by the contraction of the cytoskeleton.

Genetic exchange and the establishment of a mycelium containing genetically different nuclei is very often an essential first stage in the cycle of sexual reproduction for many species in the Ascomycotina and Basidiomycotina, and in many species in these groups the formation of heterokaryons is restricted to sexually compatible partners.

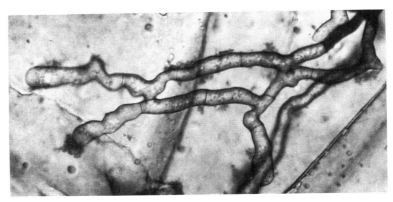

Figure 2.1 Anastomosing hyphae of *Sordaria fimicola*.

Little is known of the mechanisms that control anastomosis and the formation of heterokaryons. A successful union depends upon the recognition of a compatible hypha by another. Initial recognition is probably via a chemical signal which is produced by one hypha and results in a chemotactic growth response by another. Signals appear to be detectable over some distance and responses up to 250 μm away from the source have been observed (Ainsworth and Rayner, 1986). This phase appears to operate quite independently from plasmogamy and may occur between strains of the same species that are not compatible. Many hyphal unions take place but when the two cytoplasms make contact, a lethal reaction is precipitated some hours later.

Lethal reactions may be an important step in the evolution of new species (Kemp, 1975). They may prevent genetic recombination between evolutionary diverging strains, thus isolating them genetically so that they can evolve into new species. Paradoxically, it seems that when species have become sufficiently diverse from each other they may then fuse without initiating any lethal reaction, although this appears to happen only rarely.

Heterokaryons have many theoretical advantages. If genetically different haploid nuclei can coexist in the same mycelium, this allows the organism to maintain a large reservoir of variable genetic material in the population as in diploids but with additional advantages. In a heterokaryon, each nucleus retains its identity and each gene present is free to express itself in a way that is not possible in diploids. By selective division of appropriate nuclei, heterokaryons can respond to change without recourse to genetic recombination by the simple procedure of changing the proportions of the different nuclei present in the mycelium (Jinks, 1952).

Heterokaryons also introduce the possibility of genetic recombination outside sexual reproduction by means of the parasexual cycle. This may explain how many fungi without sexual reproduction in genera such as *Penicillium* are able to increase genetic variation (albeit at a rate that is not much faster than the rate of natural mutation) and be biologically successful in the absence of any genetic recombination through sexual reproduction.

Whilst recognizing the theoretical advantages of the heterokaryotic system, caution is needed in interpreting its effects on fungal biology. It should not be imagined that heterokaryons can form readily; in truth, somatically incompatible strains exist within species that prevent general heterokaryosis (Mylyk, 1975; Croft and Jinks, 1977). It may even be that in nature heterokaryon formation is even more restricted and they may only form when the nuclei of the participating strains possess a significant number of common loci on their chromosomes (Caten and Jinks, 1966).

In basidiomycetes, vegetative incompatibility mechanisms operate among heterokaryons (dikaryons) formed from interfertile populations of monokaryons that limit further fusions to those between genetically identical dikaryons. This is seen to delimit individuals within freely interbreeding populations (Todd and Rayner, 1980).

2.2 SPORES AND OTHER MYCELIAL-DERIVED STRUCTURES

2.2.1 Spores

Spores have several functions. They may be concerned with survival, migration, distribution of genetic variability and with the bringing together of compatible mating types for sexual reproduction in heterothallic species.

Spores whose function is primarily one of dispersal are called **xenospores**, survival spores are called **memnospores**; both can be produced either as a result of asexual or sexual reproduction. Xenospores are light with thin walls; memnospores are thick-walled and often contain oily food reserves. Mycelia may produce more than one type of xenospore and if these are dispersed by different methods, the chances of effective dissemination of the species is increased. Many species produce both xenospores and memnospores and the production of morphologically different spores on the same mycelium is known as pleomorphism.

a) Xenospores

These are the conidia, sporangiospores, basidiospores, etc. of fungi. They can be passively or actively released according to species, but widespread dispersal is usually by passive mechanisms. They are carried along by wind and animals and distributed by rain splash and water run-off. Arrival at suitable sites for colonization is therefore largely fortuitous. Dispersal by spores enables populations to build up rapidly on resources as they become available. Xenospores carry only limited food reserves and successful establishment depends on an adequate external supply of suitable carbohydrate and a source of nitrogen.

b) Memnospores

Memnospores and other survival structures have particular selective advantages for fungi, all of which can be affected by rapidly changing environmental conditions. They are especially advantageous for those fungi that occupy ephemeral substrata that are discontinuous in time and space.

A common form of memnospore is the chlamydospore. These usually form on the mycelium by the modification of a hyphal compartment.

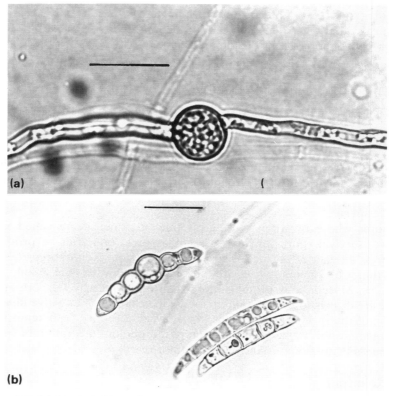

Figure 2.2 (a) Hyphal chlamydospore of *Trichoderma viride*. (b) Conidial chlamydospore of *Fusarium* species. Scale bar = 20 μm.

However in *Fusarium* species, *Mycocentrospora acerina* and other species they may form from conidia when conditions are unfavourable for germination (Figure 2.2). Certain sexually produced bodies such as oospores and zygospores also function as memnospores. These have to be mature before they will germinate and often need a period of dormancy, sometimes at low temperatures, to achieve this.

Wastage of spores is high because food reserves are often insufficient to overcome the effects of severe antibiosis. Many also fail to germinate due to the effects of severe competition for the uptake of exogenous nutrients (fungistasis). Ungerminated spores deplete food reserves by slow respiration, and as they age, membranes deteriorate and then become progressively more leaky and more sensitive to fungistasis (Dix and Christie, 1974). The loss of metabolites brings on starvation and ultimately autolysis sets in. As with hyphae, the process of deterioration is exacerbated by the activities of other microorganisms whose growth can create a nutrient sink

SPORES AND OTHER MYCELIAL-DERIVED STRUCTURES   19

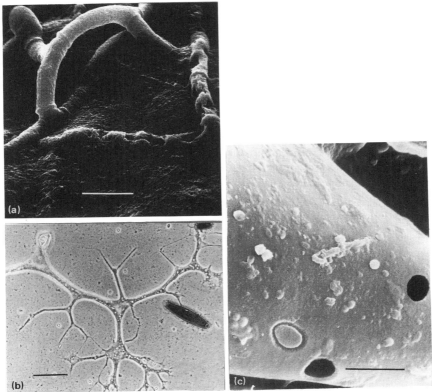

Figure 2.3 (a) Nematode damage to a hypha of *Mycogone perniciosa*. Scale bar = 10 μm. (Photograph by R.A.K. Szmidt.) (b) Vampyrelloid amoeba sending out a penetrating pseudopodium towards a conidium of *Cochliobolus sativus*. Scale bar = 50 μm. (Photograph by K.M. Old.) (c) Perforation holes in a *Cochliobolus sativus* conidium produced by a vampyrelloid amoeba. Scale bar = 5 μm. (Photograph by K.M. Old.)

around the spore into which metabolites and metabolic substrates from the spores drain. This condition has been created experimentally by Ko and Lockwood (1970), who artificially induced autolysis in spores by subjecting them to competition with *Streptomyces* species in buffered solutions. It is difficult to find any very precise information on the survival time of spores under natural conditions and for the majority the survival time may be very short. High wastage of spores is compensated for by the enormous numbers that are produced.

Consumption by animals is another factor contributing to high natural wastage of spores. Fungi act as secondary producers in food chains and form an important part of the diet of many fungivorous animals, including springtails (e.g. *Tomocerus minor*), various mites (e.g. *Creratoppia bipilis*) and nematodes, which inhabit plant litter and soil (Figure 2.3a). Amoeboid protozoans in the Rhizopoda feed on the contents of spores and other fungal bodies by lysing away the walls. Drechsler (1937) observed *Euglypha denticulata* and *Geococcus vulgaris*, two testaceous amoebae, feeding on the oospores of *Pythium* species in this manner by sucking out the contents through holes digested in the walls, while it seems that many of the larger types of hole (approximately 0.5 μm in diameter) which commonly perforate spores after a few weeks' incubation in soil are due to feeding by large vampyrelloid forms of amoebae in the Proteomyxida (Old, 1967; Anderson and Patrick, 1978; Chakraborty and Old, 1982). An intriguing account of the behaviour of one of these, *Arachnula impatiens*, has been related by Old and Patrick (1979). The spore is first engulfed by the animal which then penetrates and flows into the spore by the removal of a circular disc from the spore wall. Inside the spore, the pseudopodia digest the contents and reduce the spore to an empty shell (Figure 2.3b and c).

The trophozoite forms of these amoebae feed on a wide variety of spores and other fungal bodies. Other microorganisms may also cause wastage of spores. Certain small helical bacterial cells are frequently found colonizing the interior of spores recovered from soil, and Old and Wong (1976) suggested that these may be responsible for producing a smaller type of hole in the walls of lysed spores.

Mycoparasitic fungi and actinomycetes also destroy fungal spores and have often been observed invading spores in culture. Certain species or types of spore seem particularly vunerable. The oospores of *Phytophthora* and *Pythium* species when recovered from soil are often found to be dead and invaded by other fungi, actinomycetes and many species of soil bacteria (Sneh, Humble and Lockwood, 1977).

2.2.2 Sclerotia

A number of fungi in the Ascomycotina, Basidiomycotina and Deuteromycotina elaborate **sclerotia** on their mycelia. These form in various ways by localized profuse hyphal branching from a single hypha or by the intermingling of the branches of several hyphae (Cooke, 1983). They are hard, perennating structures that vary in shape and size according to species but are often elongated in shape. In some species these elongated forms can reach up to 1 cm in length. Sclerotia show degrees of differentiation. Some are protected by a pigmented rind covering several distinct zones or tissues within; others are rindless and show little differentiation. During maturation they become dehydrated and accumulate glycogen, trehalose and other food reserves. Under favourable conditions they germinate to produce a new mycelium or, in some species, reproductive structures. Sclerotia have the potential to survive for many years and longer than many types of spore. Sclerotia of *Verticillium* species have, for example, been recorded as surviving in soil in a viable state for about 14 years (Sussman, 1973). Low temperatures and damp conditions are generally unfavourable for survival but a major cause of loss under natural conditions is through colonization by fungi and actinomycetes. Significantly, survival is reduced in soils with a high organic content and the addition of vegetable matter to soil is one way of reducing the sclerotial populations of plant pathogens (Papavizas, 1977a,b).

Some sclerotia contain antibiotics, such as the unidentified pyrone with antibacterial properties discovered in the sclerotia of *Rhizoctonia tuliparum*. In this case, however, these substances seem to play little part in the long-term survival of sclerotia and the survival in soil of sclerotia in which the antibiotic had been destroyed by oven-drying at low temperatures compared favourably with those which were untreated or air-dried (Gladders and Coley-Smith, 1978).

Both chlamydospores and sclerotia form in unfavourable conditions, such as starvation, and low C:N ratios in the substratum (Park, 1965; Alexander *et al.*, 1966; Myers and Cook, 1972). Humpherson-Jones and Cooke (1977) increased sclerotial formation in *Sclerotium rolfsii* by applying acidic morphogens extracted from staled cultures of the fungus. Thus, these resting bodies seem to form under conditions which limit growth and when their potential for survival is most apposite.

Most sclerotia are pigmented, as are the spores and hyphae of many fungal species due to the deposition of melanin in spore and hyphal walls, and this appears to enhance survival; certainly when the survival time of comparable spores is measured, such as pigmented wild-type spores and hyaline mutants of the same species, the former survive for much longer periods (Old and Robertson, 1970). Enhanced survival of pigmented fungal structures is apparently due to a greater resistance to lysis produced by the formation of a chitin–melanin complex in the walls (Bull, 1970). Several experiments have shown that purified wall-degrading chitinases and glucanases readily lyse the hyphae of hyaline fungi (including the hyaline mutants of pigmented species) but have no obvious effect on the hyphae of pigmented forms (Potgieter and Alexander, 1966; Kuo and Alexander, 1967; Bull, 1970).

Sporopollenin, which forms from the oxidative polymerization of carotenoids, is another brown pigment resistant to microbiological attack. It is very common in pollen grain walls where it is thought to account for the very long-term survival of these bodies in soil. It has been found in the walls of the ascospores of *Neurospora crassa* and in the zygospores of *Mucor mucedo* (Gooday *et al.*, 1973, 1974) and it is possibly more widespread in fungal spores than is now appreciated.

22 THE MYCELIUM AND SUBSTRATES FOR GROWTH

2.2.3 Linear organs

Linear organs are produced by the mycelia of more than a few species of fungi in several diverse groups, but probably the best known are those associated with some members of the basidiomycetes. In this group they are produced by both saprotrophic and ectomycorrhizal species. The morphology, physiology, ecology and morphogenesis of these organs in the Basidiomycotina have been reviewed by Thompson (1984), Jennings (1984), Watkinson (1984) and Rayner *et al.* (1985). This account will only deal with their most salient features relevant to the ecology of those species that produce them. Examples of the simplest form of linear organ are those produced by *Agaricus* species and some non-basidiomycetes. These are no more than a federation of unmodified hyphae loosely connected together, the functional efficiency of which can be little more than that of their individual hyphae.

The more complex basidiomycete forms are called **cords** or **rhizomorphs** (Figure 2.4). These ramify through the substratum as interconnecting networks. In all those so far investigated these highly differentiated organs have a similar structure (Cairney, Jennings and Veltkamp, 1989). An outer tissue comprises the cortex, made up of a thick layer of narrow multiseptate hyphae that merges with a layer of loosely woven hyphae at the surface. Below the cortex is the medulla conspicuous, in which are the linearly arranged, greatly inflated, sparsely septate hyphae that form the so-called conducting vessels in which solutes are transported by bulk flow

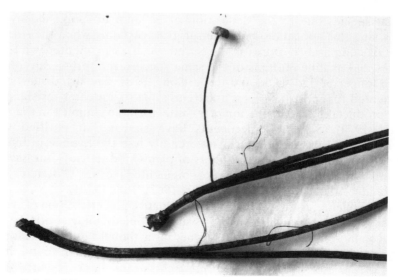

Figure 2.4 Basidiocarp of *Marasmius androsaceus* and rhizomorphs extending from one pine needle to another. Scale bar = 1 cm.

(Jennings, 1984; Eamus and Jennings, 1984, 1986). The driving force for this could be internally generated hydrostatic pressure.

Water is thought to move into the basal part of the cord due to the accumulation of solutes from the food base. This causes water to flow in and the pressure generated drives water and solutes from the base to the apex of the cord. In *Serpula lacrimans* and *Phanerochaete laevis* strengthening fibres (skeletal hyphae) with thick walls and small lumina also run longitudinally in the inner cortex. The whole is held together by adhesive glucan gums. In some, backwardly growing hyphae and anastomoses between hyphae give additional mechanical strength and ensure continuity in the system (Agerer, 1992). Details of cord structure are shown in Figure 2.5. Differentiation in cords is always towards the apex and hence the rate of development is only as fast as the forward growth of the leading hyphae. Cords show a varying capacity for co-ordinated growth and apical organization ranging from the brush-like diffuse form of growth at the apex shown by the cords of *Serpula lacrimans* to complete apical dominance in those of *Armillaria mellea* (Rayner et al., 1985). Whatever the mode of development, this appears to have little bearing on the degree of differentiation of tissues in the mature organ.

One of the chief biological functions of cords is to serve as organs of migration and exploration, transporting inocula from an established mycelium to new sites suitable for colonization. The more complex forms of cord and rhizomorph have several advantages over individual hyphae in this respect. They are more efficient in the deployment of biomass. They transport solutes more efficiently because there is less resistance to translocation in the medulla than in individual hyphae. This increases their inoculum potential, gives them a greater range and enables them to carry the inoculum further over non-supportive terrain. They can withstand competition better and what is especially interesting from an ecological point of view is that most form only in non-sterile conditions in culture. One of their specific functions seems to be to counteract any microbiologically generated antagonistic forces which, if present, would limit growth and prevent the migration of unmodified hyphae.

Cords also transport water and nutrients absorbed from the substratum. Water transport could be influential in establishing saprotrophic species on dry substrata.

Among saprotrophic basidiomycetes, cords are commonly produced by wood-rotting species. Here the evolution of cords is seen as an adaptation to overcome the problems of discontinuity in the distribution of suitable substrata for colonization. Significantly, they are less frequent among litter-decomposing basidiomycetes, few of which have to contend with the same problem. Litter-decomposing basidiomycetes tend to spread by the extension of mycelial sheets which cover a relatively small area on the forest floor; cord-formers on the other hand

24 THE MYCELIUM AND SUBSTRATES FOR GROWTH

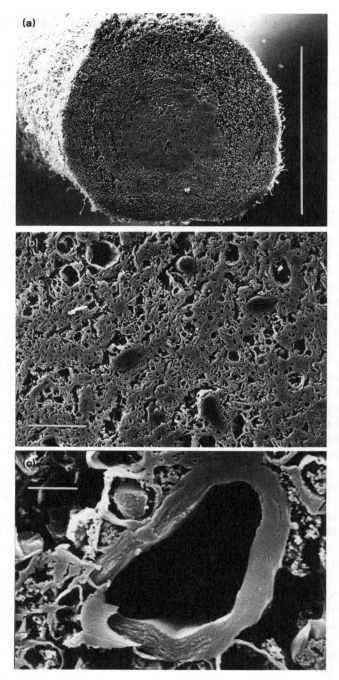

Figure 2.5 Scanning electron microscope sections of *Steccherinum fimbriatum* cords. (a) General view showing central medulla surrounded by the cortex with its outer covering of crystalline-encrusted interwoven hyphae. Scale bar = 1 mm. (b) Enlargement of medulla showing the large diameter hyphae, some with dense contents. Scale bar = 5 μm. (c) An empty conducting thick-walled large diameter hypha. Scale bar = 5 μm. (Photographs by J.W.G. Cairney.)

may have to extend quite long distances (100 m is not exceptional) in order to colonize new substrata. Consequently, in a defined area, genetically similar daughter populations of cord-forming wood-rotting species tend to become rather scattered, whereas those of non-cord-forming litter-decomposers tend to remain close together (Thompson and Rayner, 1982b; Thompson, 1984). For wood-rotting basidiomycetes, cords as agents of migration are likely to be more successful than spores in locating new substrata and establishing a new mycelium if only because they carry a much greater inoculum load. There is also some evidence that their efficiency may be additionally increased by virtue of the directional growth that the hyphae and cords of some species can make in response to soluble or volatile diffusates from wood (Mowe, King and Senn, 1983; Dowson, Rayner and Boddy, 1986). In some cases this reponse can be highly specific.

The concentration of external nutrients, particularly nitrogen, as reflected in the C:N ratio, appears to be a major influence on cord morphogenesis in some species (Watkinson, 1975). Low nitrogen concentrations and C:N ratios stimulate cord formation, whereas high external nitrogen concentrations cause loss of apical dominance, de-differentiation and reversion to diffuse hyphal growth. This could be an important biological adaptation to ensure that cords form where further mycelial expansion is impossible because of resource depletion and migration becomes necessary for survival. When a cord reaches a new site with higher nitrogen concentrations, de-differentiation would allow the fungus to spread out and take possession of the new resource. However, we are far from understanding all the factors that control cord morphogenesis and the mechanisms that cause switching from growth in one vegetative form to another. It seems likely that cord formation and other changes in vegetative growth form are triggered by independent endogenous mechanisms that control the rhythms of growth. In *Serpula lacrimans* cord initiation is associated with a switch to faster hyphal growth in a part of the mycelial front, leading to '**point growth**' (Coggins *et al.*, 1980). The acute branching of 'point growth' produces hyphae that lie nearly parallel to each other and become a focus for cord morphogenesis.

Cords are also often formed when competing mycelia come into opposition and are associated with replacement phenomena. Here their function can be seen as being essentially the same, namely to carry the fungus over a hostile environment and to capture and colonize new resources.

Cord formation in antagonistic interactions may be stimulated by physical contact, staling or low nutrient concentrations in contact zones. Bellotti and Couse (1980) successfully stimulated cord formation in *Calvatia sculpta* by using cellophane as a physical barrier. In some instances there is evidence that fungal volatiles may be involved (Rayner and Webber, 1984). The action of these stimuli may be specific. For example, *Collybia peronata* forms cords when replacing *Cladosporium* species but not when overgrowing *Aureobasidium pullulans* (Mohamed, 1986). Similarly, *Collybia peronata* forms cords when replacing *Clitocybe flaccida* but not when replacing *Collybia butyracea* (Rayner and Webber, 1984).

2.3 SUBSTRATA AND SUBSTRATES FOR SAPROTROPHS

In order to grow, fungi require sources of carbon and nitrogen, a supply of energy and certain essential nutrients such as potassium and phosphorus. Supplies of nitrogen may be obtained from proteins and other organic sources or from simple inorganic substances such as nitrates and ammonium salts, which are good alternatives as sole supplies of nitrogen for the majority of fungi. Supplies of energy and most of the carbon required for growth, however, must be obtained directly from living organisms or indirectly from their wastes or dead tissues. The latter two provide a great range of different substrata for the growth of saprophytic fungi and include such diverse resources as animal faeces, the cast-off skins, hooves, fur, feathers, nails and horns of vertebrates, the exoskeletons of arthropods, plant litter, the dead bodies of animals and even the mycelia and fruit-bodies of other fungi. These materials contain all the nutrients required for growth and substances rich in potential energy, including in some cases, supplies of readily metabolized simple organic substrates. In plant litters, for example, various amino acids and C_6 and C_5 sugars may be present. Eventually these simple substances become exhausted and further growth depends upon the dissimilation of some of the complex polymers present. In plant material the structural polysaccharides, the cellulose and hemicelluloses of the cell walls, lignins, starch and other food reserves, together with various other minor components, pectins, cutins and tannins are all potential sources of energy which can be degraded and respired by saprophytic fungi.

Animal tissues contain fats and proteins. The carbohydrate food reserve glycogen, which can be hydrolysed to glucose by fungi, is present in muscle. Keratin, a protein, is the principal component of fur, feathers, hooves and other skin structures. All of these may serve as sources of energy for saprotrophs. Arthropod exoskeletons are sources of protein and chitin. The latter is an energy supply for those fungi that can produce chitinases to hydrolyse it to its respirable monomer, *N*-acetyl glucosamine.

Details of the chemistry and distribution of some of the more abundant of these polymers and the way in which they are degraded by fungi are given below.

2.3.1 Hydrolysis of starch

Starch is the commonest food reserve of plants, and fungi, with the notable exception of most yeasts, produce **amylases** which catalyse starch hydrolysis. Chemically starch is made up of two polymers of glucose: amylose and amylopectin. These are present in varying proportions according to plant species but invariably amylopectin is in the greater amount, and is usually about 75–85% of most starches. Both polymers consist of chains of glucose molecules linked by $\alpha 1$–4 glucosic linkages but an important difference is that amylopectin is highly branched and carries side-chains which are linked to the main chain through $\alpha 1$–6 bonding.

The starch-hydrolysing enzymes and their distribution in microorganisms have been described by Fogarty and Kelly (1979). Alpha-amylase is the commonest starch-hydrolysing extracellular enzyme found in fungi. Many fungi also produce extracellular amyloglucosidase (glucoamylase), an enzyme which seems to be exclusive to fungi. Alpha-amylase hydrolyses both amylopectin and amylose to maltose and higher molecular weight fractions, bypassing $\alpha 1$–6 linkages and randomly cleaving chains in the fashion of an endoenzyme. Amyloglucosidase hydrolyses $\alpha 1$–4 and $\alpha 1$–6 glucose residues to glucose, working on the ends of chains in the manner of an exoenzyme and is also capable of hydrolysing amylopectin, amylose and glycogen almost completely to glucose. Since α-amylase cannot hydrolyse $\alpha 1$–6 linkages it cannot attack the branch points in amylopectin; thus in fungi which produce no amyloglucosidase, high molecular weight dextrins tend to accumulate when starch is hydrolysed.

All the maltose produced by the hydrolysis of starch is finally split into two glucose molecules by the catalytic action of intracellular α-glucosidase.

2.3.2 Hydrolysis of cellulose

Cellulose is the most abundant substance in plant litter and as a major constituent of all the layers of plant cell walls it forms about 30–40% of the dry weight of wood and can be as high as 45% of the dry weight of cereal straw.

Cellulose is a straight-chain $\beta 1$–4 glucan polymer containing as many as 10 000 glucose molecules linked together by the removal of water from two hydroxyl groups (Figure 2.6). Glucan chains join to form microfibrils, bundles of which run in the matrix of the plant cell wall as strengthening components. Each microfibril consists of about 40–100 glucan chains linked together by hydrogen bonding between adjacent hydroxyl groups.

In parts of the microfibril the glucan chains are regularly arranged in a parallel fashion forming cellulose with crystalline characteristics. Crystalline cellulose is the more resistant to decay, possibly because the close packing of the molecules prevents the penetration of microbial enzymes. It

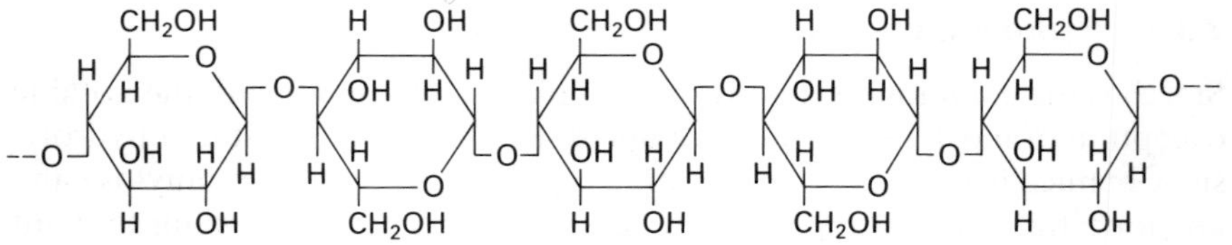

Figure 2.6 The structure of cellulose showing β1–4 linkages.

increases in proportion in the microfibrils as the tissues dry out and in mature tissues it usually constitutes the greater part of the cellulose present. However, in wood susceptibility to decay by cellulose hydrolysis is more likely to be dependent on the pore volume of the substratum.

Cellulolysis, or the hydrolysis of cellulose, is catalysed by an enzyme complex called **cellulase** that consists of a number of extracellular β1–4 glucanases, some of which are **endohydrolases** randomly disrupting linkages throughout β1–4 glucan chains, producing glucose, cellobiose and higher molecular weight fractions, while others are **exohydrolases** or **β1–4 cellobiohydrolases**, which act only on the ends of β1–4 glucan chains releasing the disaccharide cellobiose (Halliwell, 1979). **Glycohydrolases** that release single glucose units from glucan chains are also part of the cellulase complex of some microorganisms. Beta 1–4 glucan endohydrolase and β1–4 glucan cellobiohydrolase may correspond to the C_x and C_1 enzymes of Reese, Siu and Levinson (1950). Both of these components are essential for the rapid hydrolysis of cellulose with a high crystalline content. Many fungi do not produce a C_1 component, so while their cellulases hydrolyse amorphous cellulose and manufactured carboxymethylcellulose they have little effect on the crystalline cellulose of cotton fibres.

Certain experimental evidence indicates that the C_1 component may not be a cellobiohydrolase and many have now expressed the view that C_1 is neither an exo- nor an endoglucanase and that both the endoglucanase and cellobiohydrolase components are part of the C_x complex. They suggest that the function of C_1 is to separate cellulose molecules on the surface of the fibrils allowing C_x access (Sagar, 1988).

Beta 1–4 glucanases are thermostable glycoproteins which exist in multiple forms. Analysis of the β1–4 glucanase complex of *Sporotrichum pulverulentum* (*Phanerochaete chrysosporium*) showed that it is composed of five β1–4 endoglucanases of different molecular weights each working at its own optimum pH (Eriksson, Pettersson and Westermark, 1975) and *Trichoderma viride* produces at least four β1–4 cellobiohydrolases (Gum and Brown, 1977). Wood and McCrae (1986) measured an increase in the rate of hydrolysis when several of the different forms of cellobiohydrolase from *Penicillium pinophilum* were incubated with cellulose in the presence of an endoglucanase. These workers suggested that the multiple forms of cellobiohydrolase are complementary because they are stereospecific in action. It is also possible that each cellobiohydrolase works best with its own endoglucanase and this could be why the latter also exist in multiple forms. An interesting observation that has ecological significance is that the β1–4 glucan cellobiohydrolases produced by one fungus may act with the β1–4 glucan endoglucanases produced by another (Eriksson, 1981). The decomposition of cellulose is finally completed by the transformation of trisaccharides and disaccharides to glucose by the action of β1–4 glucosidases within the hyphae.

All wood-rotting fungi undoubtedly degrade cellulose as do apparently many microfungi from soil and litter as measured by their ability to hydrolyse carboxymethylcellulose and pure cellulose in the laboratory (Siu, 1951; Domsch and Gams, 1969; Flanagan, 1981). However, in nature cellulolytic activity depends upon a number of substratum-related factors, notably pH and mineral composition. This is reflected in differences in cellulolytic activity in different soils. It has been found that in extreme cases species that are very active in some soils show no activity in others (Gillespie, Latter and Widden, 1988). Where cellulose is embedded in a protective coat of lignin, as in xylem, cellulolysis tends to be limited to those fungi that can first modify or respire lignin. Thus the major decay of woody plant tissues is by specialized wood-rotting species (Chapter 6).

The ability to hydrolyse cellulose is very variable. Some fungi have very low rates of utilization and others are unable to degrade cellulose at all. Examples of the latter can be found in all the divisions of the fungi, but it appears to be commonest among the Zygomycetes and Oomycetes, and so far most species which have been tested in the genera *Rhizopus*, *Mucor* and *Mortierella* and the majority of *Pythium* species have been shown to be inactive (Siu, 1951; Reese and Levinson, 1952; Deacon, 1979; Flanagan, 1981). Notable exceptions in the genus *Pythium* include *Pythium fluminum* from fresh water (Park, 1975, 1980a) and several common terrestrial species, *Pythium graminicola*, *P. irregulare* and *P. intermedium* (Deacon, 1979). More have been found to hydrolyse cellulose when the fibres are in a swollen state (Taylor and Marsh, 1963). It must be said, however, that cellulolytic activity in *Pythium* species is usually rather weak. Few other genera in the Oomycetes have actually been tested but a few species in the Saprolegniales have also been shown to hydrolyse cellulose or carboxymethylcellulose (Thompstone and Dix, 1985). The ability to decompose cellulose (or other plant polymers) vigorously or not, as the case may be, has been used to classify fungi into several substrate-related ecological groups. However, low rates of hydrolysis of cellulose may not necessarily indicate poor adaptation to growth on cellulose for, as Garrett (1966) has pointed out, it is not the rate of cellulolysis which is important but the rate in relation to the production of enough respirable carbohydrate sufficient for the needs of a particular fungus. Successful growth on cellulose may not, therefore, necessarily be limited by the inherent rate at which the fungus hydrolyses cellulose. A measure of the potential for successful growth on cellulose is the cellulolysis adequacy index (CAI) of Garrett (1966), which relates rates of cellulolysis to units of growth on media containing simple sugars. Deacon (1979) suggested that this idea should be extended to provide a comparison of the degree of adaptation of fungi to growth on cellulose and could be used as a basis for the redefining of substrate groups.

A wide-ranging review of the ecology of microbial cellulose degradation has been produced by Ljungdahl and Eriksson (1985).

2.3.3 Hydrolysis of hemicellulose

Hemicelluloses are an integral part of all plant cell walls and of all the polysaccharides present in plants they are second in quantity only to cellulose. In herbaceous plants hemicelluloses form about 25% of the total dry weight but in woody species the figure can rise to about 40%. More hemicelluloses are usually present in angiosperm woods compared with gymnosperms, and the quality of hemicellulose present in the two types of wood is very different.

Hemicelluloses, which consist of chains of sugars, are non-fibrillar in organization and are linked to cellulose microfibrils by weak hydrogen bonding. They are usually heteropolymers of mixed C_5 and C_6 sugars such as xylose, arabinose, glucose, galactose and mannose with uronic acids. The commonest forms in plants are glucomannan, galactoglucomannan, arabinoglucuronoxylan and glucuronoxylan (Figure 2.7). All wood-rotting basidiomycetes and many ascomycetes and fungi imperfecti of soil, litter

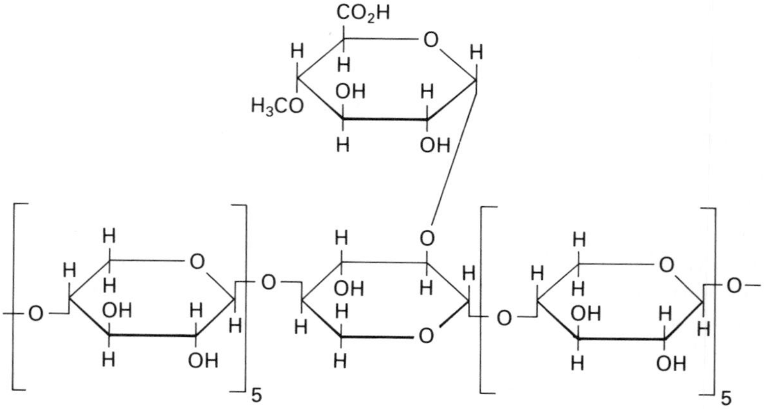

Figure 2.7 The structure of methyl glucuronoxylan.

and woody substrata also degrade a variety of hemicelluloses (Flannigan, 1970b; Flannigan and Sellars, 1972). Common microfungi may degrade xylan more actively than carboxymethylcellulose or pectin (Domsch and Gams, 1969). A list of fungi from soil with xylanase activity included 47 out of 69 ascomycetes and fungi imperfecti and, interestingly, all seven of the *Mucor*, *Mortierella* and *Rhizopus* species investigated (Flanagan, 1981). However, due to their chemical diversity, all hemicelluloses are not equally degraded by every fungus. They are thus very variable substrates for fungal growth and their variable distribution in plant tissues is thought to be ecologically important for the distribution of some fungi. This may be especially so for wood-rotting fungi (Chapter 6).

Xylanases, mannases and other glycanase complexes are produced by hemicellulose-degrading fungi (Dekker and Richards, 1976). These catalyse the disruption of specific hemicelluloses in a similar manner to the way that cellulases catalyse the hydrolysis of cellulose, but because of the complex structure of hemicellulose molecules more enzymes are needed. The basic sorts are **exohydrolases**, which liberate low molecular weight fractions from the ends of long polysaccharide chains, and **endohydrolases**, which produce higher and lower molecular weight fragments by the random disruption of linkages in the chains. The xylanase complex is known to consist of four endohydrolases, two capable of attacking branch points and branches, reducing the size of side-chains, and two that only reduce the size of the main chain (Reilly, 1981). Disaccharides and other low molecular weight fractions are acted on by specific glycosidases producing the corresponding sugar, e.g. xylosidase catalyses the hydrolysis of xylobiose to xylose.

During the decomposition of mixed substrates in culture it has been found that fungal hemicellulase production is initially much higher than that of cellulase (Lyr, 1960; Eriksson and Larsson, 1975). This sequence of enzyme production corresponds with the need to remove the hemicellulose from around the cellulose microfibrils first before cellulolysis can begin, and would seem to be a necessary adaptation for growth on cellulose in plant tissues. In some basidiomycetes a supply of sugars from hemicellulose hydrolysis may be a necessary first step in the utilization of lignin (Chapter 6).

2.3.4 Hydrolysis of chitin

Chitin is an amino-polysaccharide which is similar in structure and function to cellulose. It is found in low quantities as a strengthening substance in the bristles of annelids, in the hyphal walls of many species of fungi, in the cysts of protozoa and in the shells of molluscs. It is present in the greatest quantities in the exoskeleton of arthropods and this source makes up the bulk of chitin in the biosphere. The chitin monomer is

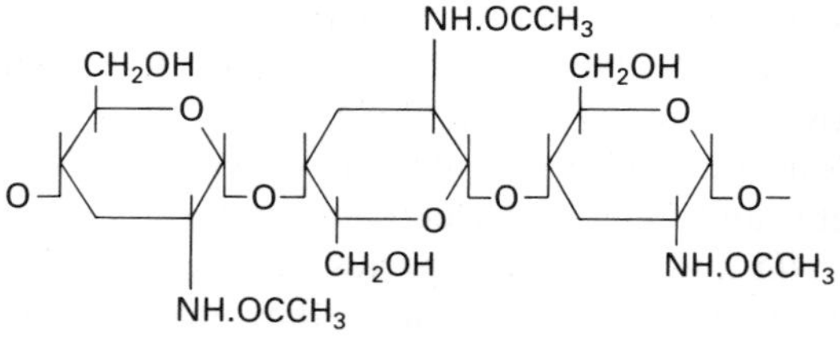

Figure 2.8 The structure of chitin.

N-acetylglucosamine, a glucose-like molecule in which one hydroxyl group of glucose is replaced by an acetyl amino group containing nitrogen (Figure 2.8). Chitin can serve, therefore, as a source of carbon and nitrogen for microorganisms and is particularly important as a source of energy, carbon and nitrogen in marine environments.

Chitin is relatively rapidly decomposed in most environments. In soil, actinomycetes generally dominate microbial communities growing on chitin, although many soil microfungi are undoubtedly also capable of decomposing it. The fungi most commonly listed as being chitinolytic are *Verticillium*, *Trichoderma*, *Penicillium*, *Paecilomyces* and *Mortierella* species. The degradation of chitin to respirable materials has not been thoroughly studied in fungi but, during decomposition, high molecular weight fractions arise together with the dimer diacetyl chitobiose, suggesting that an initial step involves the random hydrolysis of the N-acetylglucosamine chains by an **endochitinase**. Diacetyl chitibiose, which is equivalent to the disaccharides of cellulose and starch hydrolysis, is then presumed to be further degraded to produce N-acetylglucosamine through the action of a **chitobiase** (N-acetylglucosaminidase). During the metabolism of N-acetylglucosamine within the cell an amino group is removed and ammonia is liberated. Further evidence has indicated that chitinases may exist in forms which attack highly crystalline or amorphous forms of chitin corresponding to the C_1 and C_x enzymes of cellulose breakdown (Tiunova *et al.*, 1976). The ecology of chitin degradation has been extensively reviewed by Gooday (1990).

2.3.5 Dissimilation of aromatic polymers

A significant proportion of the carbon of plants is in the form of complex aromatic polymers, such as tannins, lignins and related phenolics. Lignin is most abundant in woody plants where it accounts for up to about 30% of the carbon content. Microorganisms in soil ultimately oxidize these compounds completely to carbon dioxide and water. The essence of this process is that in the final stages of degradation fission of the benzene ring must occur to produce straight-chain aliphatic substances which can be completely respired.

a) Degradation of lignin

Lignin is a complex three-dimensional branched polymer formed by the oxidative polymerization of substituted *p*-hydroxycinnamyl alcohols which form the basic phenyl-propane units. Derivatives of coniferyl, coumaryl and sinapyl alcohols are the common aromatic components (Figure 2.9) and these occur in a variety of hetero- or near homo-polymers, the proportions of which vary in different plant families. Lignins therefore tend to be highly characteristic for different plants, making them variable substrates for microbial growth and this, coupled with the fact that linkage between phenyl-propane units is random and very strong, makes lignin generally resistant to microbial degradation.

The importance of fungi in lignin degradation is well known, but the mechanisms whereby lignin-decomposing fungi dismantle the lignin polymer is still incompletely understood. The assimilation of the products of the dissimilated polymer in the final stages of lignin breakdown seems to follow similar metabolic pathways to those by which bacteria assimilate lignin phenolics (Dagley, 1971). The whole subject has been reviewed by several authors in Kirk, Higuchi and Chang (1980), Kirk (1983) and Leisola and Fiechter (1985). The dissimilation of lignin by fungi can be conveniently thought of as occurring by three mechanisms: (1) depolymerization by cleavage of bonds within the polymer; (2) removal and modification of side-chains with substitution on benzene rings; and (3) fission by ring-splitting enzymes to convert aromatic nuclei into respirable aliphatic compounds. The sequence of events does not appear to involve depolymerization as a necessary preliminary. About 15 separate enzymes are required for the complete oxidation of lignin polymers.

With the exception of some wood-rotting ascomycete and basidiomycete

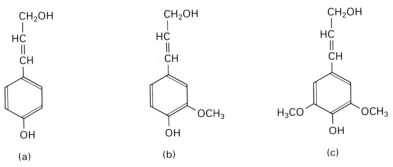

Figure 2.9 The lignin alcohols: (a) p-coumaryl alcohol; (b) coniferyl alcohol; and (c) sinapyl alcohol.

fungi known as white-rots and some litter-decomposing basidiomycetes, relatively few other fungi appear to be capable of producing all the necessary enzymes for the complete destruction of lignin. Although there have been no exhaustive surveys, the activities of most of the microfungi that abound in lignin-rich plant residues seem to be largely confined to the degradation of the low molecular weight fractions that have been released from the dissimilation of the polymer. Despite the fact that *Fusarium* and *Aspergillus* species grow well on modified and synthetic lignins (Iwahara, 1980; Higuchi, 1980; Betts, Dart and Ball, 1988), it is not thought likely that they can degrade plant lignins.

The use of radio-labelled model lignin compounds has established that cleavage of the polymer occurs through the actions of **ligninases** (peroxidases) that react with hydrogen peroxide, generating free radicals via an intermediate. The free radicals oxidize the polymer at specific linkages, cleaving the polymer at sites remote from the site of production of the enzymes (Buswell and Odier, 1987; Kirk and Farrell, 1987). Two aldehydes – veratraldehyde and benzaldehyde – have been identified as the products of cleavage of certain model lignin compounds.

Modifications of substituents of the benzene rings of the polymer, either before or after cleavage, lead to the formation of a variety of related phenolic acids, aldehydes and alcohols, increasing the solubility of the cleavage products. Vanillic acid, vanillin, ferulic acid and syringaldehyde are just a few of the detectable phenyl propanoid derivatives that form when white-rot fungi attack wood (Ishikawa, Schubert and Nord, 1963a,b). Ultimately, simple dihydroxyphenols are formed, on which **ring-splitting dioxygenases** work to produce respirable aliphatic compounds. Pathways that have so far been elucidated have shown that catechol and protocatechuic acid are two of the key substrates for these enzymes (Figure 2.10). Many soil microfungi can participate at this stage and play a role in the transformation of the cleavage products into respirable substrates (Henderson, 1960, 1961; Black and Dix, 1976a).

Many soil microfungi undoubtedly produce ring-splitting dioxygenases but relatively few have so far been investigated. Cain, Bilton and Darrah (1968) identified ring-splitting dioxygenases in a substantial number of soil microfungi that they investigated. Microfungi split benzene rings by orthofission. Metafission, which exists in bacteria, has so far not been demonstrated in fungi.

Many fungi possess **polyphenol oxidase** enzymes, which catalyse the hydroxylation of monophenols to diphenols (the **cresolase reaction**) and the oxidation of di- and trihydroxyphenols to quinones (the **catecholase reaction**).

Polyphenol oxidase is a general name given to two groups of copper-based enzymes, the Cu^{2+} laccases and the Cu^+ tryosinases that exist in multiple forms. The cresolase and catecholase reactions are catalysed by

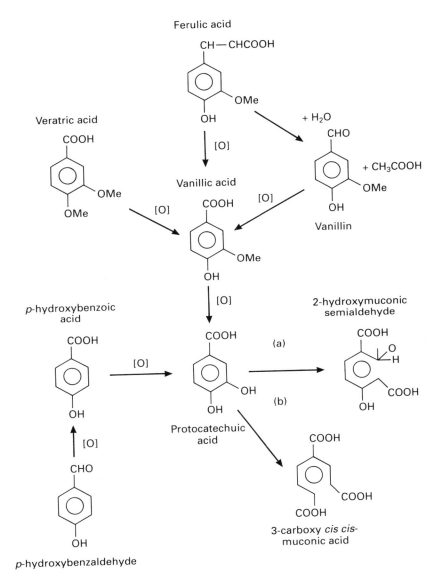

Figure 2.10 Enzymic interconversions of lignin-derived phenolic acids and aldehydes and mode of cleavage by dioxygenases (a) by metafission, (b) by orthofission.

both enzymes; each enzyme having two active sites. Some tyrosinases have strong cresolase activity and weak catecholase activity and vice versa, but the cresolase activity of all laccases appears to be weak. Much confusion has arisen over the specificity of these enzymes but it now seems that while tyrosinases will oxidize only o-diphenols, in the catecholase reaction, laccases have a much broader substrate specificity (Schanel and Esser, 1971). Tyrosinases appear to be exclusively intracellular and probably only laccases are involved in extracellular oxidations (Lindeberg and Holm, 1952).

What part these reactions play, if any, in the breakdown of lignin is still not completely understood. Extracellular polyphenol oxidase activity is associated with white-rot basidiomycetes that degrade lignin, and the formation of a brown coloration in the medium when phenols are being oxidized forms the basis of the **Bavendamm test** for ability to decompose lignin (Bavendamm, 1928a,b). This test is reasonably successful in distinguishing between white-rot fungi and brown-rot fungi which do not degrade lignin and produce no extracellular polyphenol oxidases (Davidson, Campbell and Blaisdell, 1938). This association appears to be more than coincidental and there is evidence to suggest that polyphenol oxidases could be involved in lignin degradation. Certainly it has been found that when *Pleurotus ostreatus* decays straw, the efficiency of the oxidation of lignin is reduced if the water-soluble phenols that induce laccase formation are removed (Haars, Chet and Huttermann, 1981; Platt, Hadar and Chet, 1983). Further, phenol oxidase-less mutants of *Phanerochaete chrysosporium* were found to be incapable of degrading lignin unless laccase was added to cultures (Ander and Eriksson, 1976).

Some researchers believe that laccases are involved in demethylation reactions and side-chain rupture in the polymer, leading to partial depolymerization and the formation of quinones (Ishihara, 1980). However, since the action of laccase would be restricted to phenol substrates with free hydroxyl groups, others think that this is unimportant and see a more likely role for these enzymes in the final 'mopping up' of some of the water-soluble products of depolymerization, transforming substances like vanillin to quinones.

The enzyme **cellobiose:quinone oxidoreductase**, which reduces quinones to catechol compounds, simultaneously utilizing cellobiose, provides a route whereby quinones can be transformed into substrates for ring-splitting dioxygenases. Significantly, cellobiose:quinone oxidoreductase has so far been found only in white-rot basidiomycetes (Ander and Eriksson, 1978). The function of polyphenol oxidases in lignin degradation is still considered to be somewhat controversial, but at least we know that their role cannot be all that crucial because there is poor correlation between polyphenol oxidase production and lignin degradation. Not all white-rot fungi produce laccase (Kirk and Kelman, 1965; Setliff and Endy, 1980) and many microfungi from soil and litter that cannot degrade lignin produce abundant laccase (Dix, 1979). The most convincing evidence supporting this interpretation comes from studies of phenol oxidase-less mutants of *Phanerochaete chrysosporium* where it was found that some mutants actually oxidized lignin more successfully than wild-types (Liwicki et al., 1985).

b) Degradation of tannins

Chemically, tannins are of two kinds: hydrolysable tannin and condensed tannin. **Hydrolysable tannins**, such as gallotannins, are polymers, in which the basic unit is 3,4,5-trihydroxybenzoic acid (gallic acid) attached to sugar molecules by ester linkages. Ellagitannins are chemically similar but release ellagic acid (a derivative of gallic acid) as well as gallic acid on hydrolysis. **Condensed tannins** are more complex and are formed by the oxidative polymerization of phenolic compounds only. Most frequently they form by the polymerization of flavan 3,4-diols (leucoanthocyanidins), but they also include flavan 3-ols (catechins), gallic acid esters and other flavonoids.

Tannins are particularly abundant in woody plants, where large deposits often occur in bark cells, complexing with or tanning the cell contents and imparting some resistance to fungal attack (section 3.2.2).

Tannins decompose slowly in soil, and among microfungi *Penicillium* species are especially important to their decomposition, while many other common soil fungi including *Mucor* and *Fusarium* species are inhibited by tannins in culture (Knudson, 1913; Cowley and Whittingham, 1961). Bark-colonizing yeasts and bacteria are also known to hydrolyse tannins (Deschamps and Leulliette, 1984). Hydrolysable tannins composed of sugar and simple phenolic substances are decomposed more readily in soil than the more complex condensed tannins. Little is known about the metabolic pathways whereby the latter are respired but, when gallotannins are decomposed in culture by *Penicillium* species, gallic acid and sugar are separated from the polymer. During respiration of these substances gallic acid is completely removed and no other phenolic materials accumulate in the medium, indicating that the metabolic steps involve ring fission (Lewis and Starkey, 1969). Communities of tannin-hydrolysing bacteria and fungi on bark may assist in the establishment of wood-decay fungi and increase the rate of wood destruction by removing inhibiting tannins.

2.3.6 Hydrolysis of proteins

Proteins are the most abundant nitrogen-containing constituent of living organisms. Soluble proteins, of about 30 amino acids or less in chain length, can pass through hyphal walls; insoluble proteins must be hydrolysed externally before they can be utilized by fungi.

Most fungi show extracellular proteolytic activity against most proteins over a range of environmental conditions. **Peptide endohydrolases (protei-**

nases) cleave internal peptide bonds, releasing soluble peptides. These can be taken into hyphae and degraded to their component amino acids by a large number of different peptidases.

There are four broad classes of proteinase that have been detected in fungal cultures. Each proteinase works over a range of pH, covering values between about pH 2 and pH 10. Serine proteinases usually have high pH optima, with some activity above pH 9. Aspartic proteinases have low optima, with activity below pH 4, while cysteine and metallo-proteinases operate in the range pH 4–8. The former work best at slightly acidic pH; the latter at neutral pH. Multiple forms of serine and aspartic proteinases appear to be the proteinases most widely produced by fungi (North 1982). Fungal proteinases have a low substrate specificity and are very durable under extreme environmental conditions.

Exceptions to the general availability of proteins as nitrogen sources for fungi are certain connective tissues and structural proteins of animals. Proteins such as collagen, elastin, keratin of horn, fur and feathers and the sclerotin of arthropod skeletons can only be degraded by fungi producing specific collagenases, keratinases, etc. The resistance of keratin to hydrolysis resides in the high level of disulphide cross-links in its structure and the resistance of collagen is due to its highly organized helical structure.

Fungi uniquely associated with the degradation of the more resistant proteins of animal remains are found in the Ascomycotina in the families Onygenaceae and Gymnoasaceae. In the former they include the genera *Aphanoascus* and *Onygena*; in the latter *Arthroderma* (*Trichophyton*) and *Nannizzia* (*Microsporum*). *Onygena equina* is illustrated in Figure 1.1. Not a great deal is known about the ecology of this group of fungi. They are undoubtedly keratinophilic but are probably not solely dependent upon keratin as a source of nitrogen. They are widespread in soils and can be isolated from soil by baiting with hair or wool (Griffin, 1960; Pugh and Mathison, 1962). Several species in the genera *Trichophyton* and *Microsporum* are important skin pathogens of man and animals, causing various ring-worm diseases.

3
STRUCTURE OF FUNGAL COMMUNITIES

3.1 INTRODUCTION

Saprophytic fungi have evolved to exploit every possible habitat on earth, wherever degradable organic matter exists. Fungi occupy terrestrial and aquatic environments (both marine and freshwater) in tropical, temperate and polar regions. There are fungi characteristic of forests, pastures, heaths and bogs, and within these habitats fungi form distinct but overlapping ecological groups associated with various resources. Some exploit the leanest of resources and grow in the most inhospitable of environments such as the microcolonial ascomycetes of rock surfaces in arid deserts (Staley, Palmer and Adams, 1982; Palmer *et al.*, 1987).

Many fungi within these ecological groupings have a restricted distribution and are confined to a particular type of resource, be it plant litter, wood, roots, animal dung, etc. Wood-decay fungi are good examples. They fruit only on wood and some, due to a variety of factors, are highly selective in their requirements. In the case of wood-rotting agarics, many produce their basidiocarps only on deciduous wood, while those of *Hypholoma capnoides* and *Tricholomopsis rutilans* appear only on that of conifers. Of the former group, *Tricholomopsis platyphylla, Coprinus disseminatus* and *Pluteus cervinus* occur on a range of deciduous woods but *Rhodotus palmatus* and *Oudemansiella mucida* are much more selective and are seldom found on wood other than elm and beech respectively. This more restricted type of distribution has been referred to as **taxon-selective** (Rayner, Watling and Frankland, 1985). Further distinction may be made between those species that grow on a particular part of a resource (e.g. a twig or a branch in the case of wood) and those that are not so restricted. These have been referred to by the above authors as **component** and **non-component restricted** species. A similar scheme can be set out for species inhabiting plant litter that distinguishes between those growing generally in litter, those confined to specific sorts of litter, e.g. deciduous or conifer, and within these resources species that are confined to specific components of the litter, e.g. *Xylaria carpophila* restricted to beech cupules and *Tubaria autochthona* to hawthorn berries.

Occasionally wood-decay fungi with strong selective tendencies may switch hosts as distribution changes, for example from north to south in Britain or vice versa. Sometimes this seems to be related to the limits of the distribution of the 'normal' host and sometimes it is not (Whalley and Watling, 1982). Good examples of this type of behaviour are *Daldinia concentrica* and *Fomes fomentarius*: *D. concentrica* is common in the south of Britain where *Fraxinus* is the preferred host; it is rarer in the north where it grows on *Betula*. *F. fomentarius* is the opposite; it is common in the north growing on *Betula*, rarer in the south where it grows on *Fagus*.

Distribution between resources can seldom be linked to lack of suitable substrates for catabolic enzymes, as witness the successful growth of fungi of restricted distribution on a variety of natural resources in culture; most natural resources, even wood, contain utilizable substrates. The main factors that appear to limit distribution are competition from better adapted species, chemical inhibitors present in the substratum and the physical environment. However, we should bear in mind, especially when referring to the distribution of larger fungi, that observations are nearly always based on the distribution of fruit-bodies, not mycelium, and fruiting may be stimulated only on certain resources.

Many other fungi are non-selective and are not restricted to a specific resource. They capitalize on the fact that certain substrates are widespread and present in abundance in many different resources. This point is well illustrated by the fungus *Cladosporium herbarum*, whose abundant airborne spores find conditions suitable for growth on a variety of resources. The commonest habitat of this fungus is the phylloplane of living and senescing leaves. However, it also occurs frequently on seed coats, wood, senescing plant stems, herbivore dung and litter and plant debris from soil.

The distribution patterns of fungi are naturally related closely to the distribution of resources in time and space. Component– or taxon–selective species will be non-randomly distributed, while non-selective species are ubiquitous and more randomly distributed. Non-selective species may be non-randomly distributed if their resource units are discontinuously spread, e.g. fallen logs.

The functional unit of decomposition of the resource is the **unit community** occupying a delimited volume of the resource or a specific unit of the resource, e.g. a single leaf or twig etc. (Swift, 1976). Analysis of the relative abundance of species in these communities nearly always follows a log-normal distribution pattern (Lussenhop, 1981), indicating that abundance is influenced by a number of different factors and suggesting that a fungal community is in reality composed of a number of small subunits, each of which exploits a particular feature of the resource.

Although the physical and chemical complexity of individual resource units might be expected to give rise to a variety of ecological niches, analysis of fungal communities for a number of different resources, such as tree stumps, small branches and roots, indicates that individual units of these resources are typically composed of a relatively low number of species. Taken as a whole, however, these communities are very diverse and share only some species with other communities on the same type of resource. Nevertheless, at the same time the unit community forms a recognizable recurring association of species for each type of resource.

Swift (1976) showed by mycelial sampling that the number of fungal species participating in communities on beech branches (*Fagus sylvatica*) was usually about 4, within the range 1–9. Of the total of 56 species isolated, the vast majority were found only once, with only 5 species occurring in more than 5 of the 32 branches sampled.

Community structures on French bean roots (*Phaseolis vulgaris*) are quantitatively similar (Dix, 1964). The majority of species arising from plated-out washed sections of young and old roots occurred in low frequency, with only two species present as dominants and as a near constant component of the community on roots of all ages (Table 3.1). These results would seem to suggest that while one or two fungal species may dominate a community, there are many species with overlapping strategies which allow them to compete for a number of different niches in the same resource unit.

To sum up, fungal communities are usually very diverse with recognizable dominant species. They are highly characteristic for different resources but at the same time may share some species with other dissimilar resources.

Table 3.1 **Frequency and distribution of fungi on French bean roots of different ages**

No. of species occurring at	Apices	Older roots	Decomposing roots[a]
2–20% frequency	10	16	3
21–40%	4	0	0
41–60%	1	1	1
61–80%	0	1	1
81–100%	0	1	1
Total	15	19	6

Total no. of different species 20.
[a]14 days after shoot removal.

The details of colonization dynamics as applied to fungi is increasingly attracting attention and such observations as are available indicate that the pattern of fungal colonization broadly conforms to the '**island theory**' of MacArthur and Wilson (1967). Isolated resource units resemble in many respects the geographically separated islands of plant biogeography. The island theory states that there is a dynamic equilibrium between species immigrations and extinctions such that at equilibrium new species can only be accommodated by the elimination of a corresponding number of existing occupants and also that the number of species in residence will be expected to increase as the area available for colonization increases. If this theory holds, the number of fungal species occupying a resource unit will increase as the resource unit increases in area and species in residence will vary, i.e. there will be species 'turnover', but numbers will remain constant when equilibrium is reached.

A number of these tenets have been found to hold for observations on time-course studies of colonization in both aquatic and terrestrial situations.

Bärlocher (1980) reported that the number of aquatic fungal species colonizing leaves increased until a plateau was reached. Thereafter numbers remained stable for several months. Sanders and Anderson (1979) showed that the numbers of fungi colonizing wood blocks in a stream increased as the block area increased. Details of this experiment are given in section 9.2.3(b).

Wildman (1987) followed the colonization of Cellophane squares of different area buried in soil and found that numbers of species present on each square reached a steady state after 2 days, but at equilibrium, while numbers remained relatively stable, the composition of the community constantly changed. He confirmed that the number of species supported is dependent upon area and increased as the area of the squares increased. Kinkel *et al.* (1987), studying the colonization of the phylloplane by filamentous fungi, obtained similar results and found that numbers reached equilibrium after about 3 weeks. Species 'turnover' occurred at equilibrium, but in this case it was found that there was no correlation between species numbers in occupation and leaf area. This may be because units of small area usually suffer from a shortage of unoccupied niches but in the case of leaves only a small proportion of available leaf area is ever actually occupied (Kinkel *et al.*, 1987). Here, other limiting factors appear to operate and any additional space provided by leaves of larger area seems to offer no extra advantage.

Fungal communities do not remain indefinitely in a steady state, however. Disturbance, such as that caused by invasion by animals, or increase in stress, caused by falling nutritional levels and antagonistic interactions between competing species, inevitably lead to a high rate of extinction that is not compensated for by recruitment, so that in time the unit of resource becomes dominated by one or two species that are usually highly antagonistic. Floristic descriptions of the development of fungal communities are called **fungal successions** (section 3.3).

In order to exploit the great variety of organic debris in many different environments as it becomes available, fungi have evolved a variety of different ecological lifestyles. Many common saprophytic moulds and microfungi are **'weed' species**, ubiquitous, non-resource-specific, utilizing a large but relatively restricted range of organic carbon sources. Some are non-cellulolytic and few, if any, can decompose lignin. Typically they grow quickly and their reproductive cycles are short because their substrates are transitory. Characteristically, they exploit disturbed environments, i.e. those changed by physical disruption and/or the incorporation of organic matter (e.g. leaf fall) where competition is typically low. A significant proportion of their biomass becomes invested in spores and also, in many cases, resting spores, sclerotia, etc., that remain as the units of survival. Their relatively limited reproductive capacity is compensated for by the large selection of resource units that are available to them for colonization. These have also been called **opportunistic decomposers** (Gochenaur, 1981).

Microfungal weed species frequently combine this type of lifestyle with a high tolerance of stress, whether it be due to low water potentials or to phenols in plant litter.

There are also some larger fungi that exploit disturbed environments. *Psilocybe* species, *Coprinus atramentarius*, *C. comatus* and *Lacrymaria velutina* are familiar toadstools that appear on soils disturbed by road construction and building operations. *Agrocybe praecox* and *Panaeolus subbalteatus* often appear on soils disturbed by cultivation. Such disturbances often incorporate a considerable amount of organic matter into soils that can be exploited by these fungi. Examples of other toadstools associated with disturbed soils are quoted by Watling (1988).

These toadstools are probably all cellulolytic. Of those that have been investigated experimentally *C. comatus* and *C. atramentarius* utilized straw lignin while *L. velutina* is a potential white-rot of wood. These have also been found to be poor competitors with toadstools from leaf litter and show very little tolerance of gallic acid, a typical leaf litter phenol (Mohamed and Dix, 1988).

'Non-weed' species include most of the larger fungi. The Basidiomycotina are typical examples. These have more restricted types of distribution and many are highly resource-specific. Typically they have a greater range of biochemical activity compared with weed species and metabolize a wide range of organic carbon sources, including in many cases lignin. They are slow to reproduce and devote a great proportion of their energy to the production of fruit-bodies and spores to ensure that at least some will reach sites suitable for colonization. Reproduction is usually seasonal and is triggered to coincide with conditions that

Table 3.2 Ecological adaptations of 'weed' and 'non-weed' fungi

	Exploitation of disturbance	Stress tolerance		Antagonistic propensity
		to low ψ	to plant phenols	
Ecological strategy (see text):	(R)	(S)	(S)	(C)
'Weed' species				
Mucor spp.	H	M	L	L
Penicillium spp.	H	H	H	M
Trichoderma spp.	H	M	M	H
Cladosporium spp.	H	M	L	L
Fusarium spp.	H	M	M	M
'Non-weed' species				
Wood-decay basidiomycetes	L	L	H	H
Litter-inhabiting agarics	L	L	H	H
Terricolous agarics (e.g. *Coprinus comatus*, *Lacrymaria velutina*)	H	L	L	L

Relative adaptation levels: L = typically low; M = typically moderate; H = typically high.
ψ = water potential

will maximize the successful establishment of progeny.

Non-weed species are characteristically highly competitive and antagonistic, with a propensity to replace other fungi in a community, taking over their territory and their accumulated resources. Hierarchies exist among them such that the most antagonistic species eventually become dominant. They grow on chemically stressed resources, e.g. wood and leaf litter with a high phenol content, but in temperate regions few are known that can tolerate dry conditions.

The behaviours of 'weed' and 'non-weed' species have been equated to the evolution of the R (ruderal), S (stress-tolerant) and C (competitive) strategies of higher plants (Pugh, 1980; Cooke and Rayner, 1984). However, in the case of fungi these strategies are often combined in overlapping forms of behaviour (Table 3.2). These terms should be also seen as being somewhat relative when applied to fungi. The concept of the R, S and C strategies is discussed in greater detail in Chapter 1.

Fungal growth is also influenced by pH, oxygen availability, carbon dioxide accumulation, and the presence of certain inhibitors or stimulators of plant origin. Fungi also show a range of adaptation to these factors.

3.2 DEVELOPMENT OF FUNGAL COMMUNITIES

Given that fungi are variously adapted, it is axiomatic that individuals will colonize under the ecological conditions that suit them best. The development of fungal communities will therefore depend upon a variety of environmental and biotic factors that can be identified. Among the most important of these are the nature of carbon supplies, the quality and quantity of nitrogen, the presence of specific inhibitors and stimulators, certain physical factors and interspecific competition.

3.2.1 Influence of C and N

Chemically simple sources of energy such as C6 and many C5 sugars are readily respired by nearly all fungi, but these usually form a very low proportion of the potential energy of most resources. Moreover, by virtue of the fact that they are also readily respired by many microorganisms, sugars tend to disappear soon after any resource becomes available for colonization. Colonization and the continuous occupation of many substrata therefore depend upon the ability of the fungus to produce the enzymes required to degrade some of the many complex polymers found in plant and animal tissues (section 2.3). To degrade all the diverse polymers found in nature would require a vast battery of catabolic enzymes, involving several for each polymer, with one enzyme being required for each step in the dissimilation process. While many fungi produce a considerable number of enzymes, no species is capable of producing a complete range such that it would be able to utilize all the potential substrates found in plant and animal tissues.

Some of the common *Mucor* and *Mortierella* species of soil and *Saprolegnia*, *Pythium* and *Aphanomyces* species of soil and water cannot utilize cellulose (Siu, 1951; Unestam, 1966; Deacon, 1979; Flanagan, 1981) and it is generally believed that the majority of the Zygomycotina and many of the saprophytic species in the Oomycetes are non-cellulolytic. These are the so-called '**sugar fungi**' of Burges (1958), supposedly solely dependent upon simple organic carbon compounds for energy and carbon supplies, although a number of *Mucor*, *Rhizopus* and *Mortierella* species have since been shown to hydrolyse xylan (Flanagan, 1981).

Most members of the Ascomycotina and Deuteromycotina produce a range of glycanases hydrolysing many hemicelluloses and cellulose, and the ability to hydrolyse xylan is particularly common (Siu, 1951; Reese and Levinson, 1952; Domsch and Gams, 1969; Flanagan, 1981). Few fungi in these subdivisions are capable of degrading lignin (some exceptions are certain Xylariaceae which decay wood, see section 6.3.3). All saprophytic basidiomycetes degrade cellulose (discounting basidiomycete yeasts), many also degrade lignin, and in general they produce a large range of enzymes which enable them to deploy a number of different strategies to obtain energy from plant polymers. The preference which these species show for a particular resource – wood as opposed to leaf litter, for example (some are also rather more specific) – appears to be largely non-enzymically based and is due to such factors as interspecific competition and the presence of specific chemical inhibitors (Chapter 6).

Enzyme production in relation to carbon supplies can therefore have an important role to play in influencing the distribution of species, partitioning species between resources, and even within resources if degradable polymers are heterogeneously distributed. The situation is much complicated, however, by the fact that there are fungi with restricted catabolic activities entering into commensal and mutualistic relationships with, and flourishing alongside species decomposing cellulose or lignin. A number of fungi growing on wood cannot attack the complex lignin polymer themselves, but are able to obtain some energy by the oxidation of the simple phenyl-propanoid-derived aldehydes, alcohols and organic acids released from the degradation of the polymer by true wood-decay fungi.

Similarly, where cellulose is being degraded, microbial communities build up containing fungi that produce no cellulases but which survive by utilizing the products of cellulose hydrolysis arising from the cellulolytic activities of true cellulose decomposers. Fungi which behave in this way have become known as **'secondary sugar' fungi** (Garrett, 1963), so called because they persist or appear in successions after the sugars of the resource have been respired.

Thermomyces lanuginosus is a good example of a secondary sugar fungus. This species is associated with *Chaetomium insolens, Humicola thermophile* and other cellulolytic fungi in composts. *T. lanuginosus* produces no glucanase other than β1–4 glucosidase, with which it hydrolyses the disaccharide cellobiose (Chang, 1967; Hedger and Hudson, 1974). Other examples include *Pythium oligandrum, Mucor hiemalis* and *M. oblongisporus*. *P. oligandrum* is a mycoparasite which grows at the expense of cellulolytic communities in soil (Tribe, 1966; Deacon, 1976), a characteristic that enables it to colonize substrata low in nutrients that are already occupied by other fungi (Foley and Deacon, 1985). *M. hiemalis* occurs in the sub-Antarctic in association with the cellulolytic fungus *Chrysosporium pannorum* in the tussocks of the grass *Poa flabellata* (Hurst and Pugh, 1983) and in cellulolytic communities of bracken petioles in the final stages of decomposition (Frankland, 1969). *M. oblongisporus* is a psychrophile from forest litter communities; it can hydrolyse neither sucrose nor cellulose (Hintikka, 1971a). For discussions of the ecology of 'sugar' and 'secondary sugar' fungi in successions see section 3.3.

Interspecific interactions such as these can increase rates of decay and it has been shown that co-culturing fungi that can directly degrade plant polymers with those that cannot, or that are only weakly active, synergistically increased the rate of decomposition of wood (Hulme and Shields, 1975; Blanchette and Shaw, 1978), leaves (Dix and Simpson, 1984), filter paper cellulose (Deacon, 1985) and the rate of dephenolization of industrial lignin (Sundman and Nase, 1972).

In these associations rates of cellulose hydrolysis are believed to increase because cellulases are induced enzymes whose production is regulated by glucose and other carbohydrates in feedback mechanisms. Cellulase production will cease if free glucose levels are too high but if the products of cellulose hydrolysis are respired by other microorganisms, intracellular glucose levels may not rise to critical levels and cellulase production may continue.

It has been demonstrated experimentally that the rate of cellulolysis may also increase in mixed fungal communities due to complementary interaction between the extracellular glucanases of the cellulase complex of different species. T.M. Wood (1969) demonstrated that when the cellobiohydrolases of *Trichoderma koningii* or *Fusarium solani* were mixed with the culture filtrates of a number of weakly cellulolytic fungi or those with endoglucanase activity only, this markedly increased the solubility (hydrolysis) of cotton fibres compared with culture filtrates acting alone. Synergism is even higher between cellobiohydrolases and endoglucanases of different origin when both organisms are capable of complete cellulose hydrolysis (T.M. Wood, 1980).

Moderately high levels of nitrogen are essential for good growth. The growth of fungi will cease at low nitrogen concentrations because the continuous exploitation of available resources necessitates the production of extracellular enzymes and new mycelial growth. If the nitrogen concentration is too low, insufficient nitrogen can be taken up, especially since, as nitrogen dilution increases, the network of mycelium that will be needed to collect it must necessarily increase proportionally.

Some fungi, e.g. wood-decay fungi, are adapted to grow at remarkably low nitrogen levels, while at the other end of the scale some coprophilous Mucorales have unusually high nitrogen requirements. Fungi adapted to grow at low nitrogen concentrations seem to be particularly resourceful at scavenging for nitrogen and are especially efficient in recycling amino acids from unneeded proteins and spent protoplasm. In these situations low nitrogen levels may be also compensated for, to some extent, by the diffusion of soluble nitrogen to hyphae down diffusion gradients. However, low nitrogen concentration is the usual rate-limiting factor of cellulolysis (Deacon, 1985).

The C:N ratio is an expression of the total amount of nitrogen present by weight in relation to carbon supplies, but for strict relevance to availability the concentration of nitrogen per unit volume is the only valid measurement (Park, 1976a). Fungi will grow over a range of C:N ratios but have optimum C:N ratios for maximum activity, and increases in the rate of decomposition do not necessarily follow increases in nitrogen levels (Park, 1976b). Optimum cellulolysis occurred in *Pythium* species, for example, when C:N ratios were near 400:1 but the rate declined to about a quarter of the optimum at supra-optimal C:N ratios of 40:1 (Park, 1976c).

The quantity of nitrogen available is undoubtedly an important factor influencing the distribution of fungi, and shifts in nitrogen concentration can affect the competitiveness of certain species. Analysis of fungal populations from soil or leaves developing on agar plates containing nitrogen levels in the range 106–1.06 mg l^{-1} (standard Czapek–Dox medium contains about 330 mg l^{-1}) showed that the composition of the mycoflora on the plate varied with the concentration of nitrogen supplied (Park, 1976a). A wide variety of common species, over 50% of the total isolated, were isolated at all nitrogen levels but some, such as *Penicillium spinulosum* and *Trichoderma koningii*, were only detected at higher nitrogen levels, while *T. harzianum* and *T. viride* only appeared when nitrogen levels were low.

The quality of nitrogen is not often important since most organic and inorganic forms of nitrogen are readily metabolized and satisfy all the nitrogen requirements of most fungi. Amino acids, ribonucleic acids and most proteins are rapidly utilized and ammonium salts and nitrates are acceptable inorganic substitutes. Important exceptions are some members of the Mucorales, Saprolegniales and a number of wood-decaying Hymenomycetes which cannot utilize nitrates. In addition, a number of fungi are heterotrophic for certain vitamins which contain nitrogen, particularly those of the B group such as pantothenic acid, pyridoxine, and vitamin H (biotin), concerned in the production of certain co-factors or co-enzymes. A more limited number of fungi are heterotrophic for certain amino acids. Although these vitamins and amino acids are vitally important for these organisms and supplies have to be ensured when they are cultured, in nature supplies are not thought to be limiting. Most are synthesized by plants, and are generally present in plant remains: alternatively fungi may obtain them from other microorganisms growing on the same substratum which release them at death or 'leak' them into the environment during life. Commensal and mutualistic relationships of this type are probably quite common in fungal communities.

3.2.2 Inhibitors and Stimulators

Specific and non-specific biologically active substances that have the potential to inhibit fungi occur commonly as constituents of plant tissues. Because of their heterogeneous distribution between different plants and within the different tissues of the same plant they are thought to have profound effects on the distribution of fungal species and the structure of communities. The colonization of certain types of resource may depend as much on the ability of the fungus to overcome the presence of inhibitory substances as the capacity to produce the enzymes necessary to degrade particular plant polymers. For this reason these materials have become known as **modifiers** (Swift, 1977) and may account for the high degree of selectivity or taxon specificity of some fungi.

In 1947 Gilliver tested water extracts from 1915 plant species and showed that about 23% contained antifungal substances. Antifungal compounds are particularly widespread in woody plant species, e.g. *Betula pendula*, *Populus tremula*, *Acer platanoides*, *Fagus sylvatica* and conifers (Melin, 1946). The antifungal properties of plant tissues may be due to the presence of polyphenols, phenolic acids and aldehydes, quinones and flavonoids (Mansfield, 1983). In the heartwood of trees stilbenes, tropolones and other antifungal substances are often present (section 6.1).

As early as 1911 Cook and Taubenhaus showed that a number of fungi were inhibited when crude extracts of tannin, a polyphenol, were incorporated into culture media. The inhibitory activity of tannins seems to be partly due to the presence of gallic acid, which is a component of hydrolysable and some condensed tannins, both widely distributed in plants (Lindeberg, 1949; Dix, 1974). Condensed tannins may also include in their structure certain flavanonols, e.g. taxifolin (dehydroquercitin), with antifungal activity, although the major component of most condensed tannins, flavan 3,4-diols (leucoanthocyanidins), appear to have no inhibitory properties.

Catechol, benzoic acid, *p*-hydroxybenzoic acid and *p*-hydroxybenzaldehyde are a few of the lignin-related phenolics with antifungal properties. At high concentrations they are inhibitory even to those litter-inhabiting fungi which normally respire them. The chemical structures of some of these compounds are illustrated in Figure 3.1.

Phenolics such as these can inactivate respiratory enzymes and some of the extracellular enzymes of fungi (Lyr, 1961). The inhibition of extracellular cellulases, xylanases and pectinases by phenolics seems to be associated with the presence of enzyme SH (sulfhydryl) groups. Normally SH groups function to activate the enzyme but they tend also to react with oxidized phenols and so become inactivated (Lyr, 1965). The inhibitory action of tannins is probably due also to the readiness with which they tan or form complexes with extracellular enzymes (Goldstein and Swain 1965). The antifungal properties of these compounds can be very variable, depending on the fungus species and physical conditions; their effectiveness is particularly conditioned by pH. They appear to have no common chemical structure which can be correlated with antifungal activity. In flavonoids the possession of 1 or 2 polar groups (but not more) seems to be a necessary feature and a

50 STRUCTURE OF FUNGAL COMMUNITIES

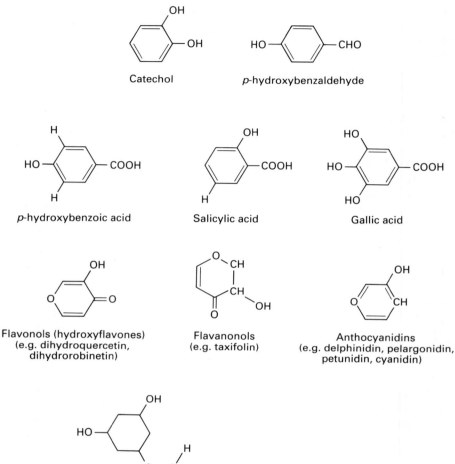

Figure 3.1 Some inhibitory phenolic compounds.

tendency to polymerize readily may also be an important characteristic (O'Neill and Mansfield, 1982).

Fungal polyphenol oxidases oxidize many simple hydroxyphenols, like gallic acid to quinones in general reactions of the type described in Chapter 2. Quinones are respiratory poisons but tend to form polymers by spontaneous reaction and these, because of their large size, cannot penetrate hyphae and are consequently harmless to fungi. These polymers are the brown pigments that form in culture when phenols are oxidized. Polyphenol oxidases therefore tend to detoxify and it is seen as significant that these enzymes are produced abundantly by basidiomycetes decomposing leaf litter or wood resources rich in polyphenols.

Basidiomycete litter decomposers and white-rots of wood produce extracellular laccases (Davidson, Campbell and Blaisdell, 1938; Lindeberg, 1948; Giltrap, 1982), while the tyrosinases of basidiomycete brown-rots of wood oxidize phenols within hyphae (Lyr, 1962). It is worth noting also, in this connection, that in basidiomycete white-rots the enzyme cellobiose oxidoreductase will catalyse the reduction of quinones to possible respirable products (Chapter 2).

Polyphenol oxidases are not inhibited by phenols due to the low dependence of the enzymes on free SH groups for activity (Lyr, 1962, 1965). However, polyphenol oxidases complex with tannins and many believe that the protective properties which tannins impart to certain woods owe more to their tendency to complex with fungal polyphenol oxidases than to any direct effect on fungal respiration.

Plants produce a great variety of defensive chemicals which may give resistance to fungal disease (Mansfield, 1983). They are found in a number of different chemical families and include compounds such as avenacin (a triterpenoid saponin), tomatine (an alkaloid saponin), tuliposides (lactones) and mustard oil (an isothiocyanate). Many of these substances occur as antifungal glycosides which only become active when oxidized or hydrolysed by fungal or plant enzymes.

In addition to antifungal compounds, plant tissue may contain flavonoids and other phenolics which stimulate fungal growth. There is evidence that these can be strongly specific in their action. Aqueous extracts of the leaf litter of several conifers and *Populus tremula* have been found to stimulate the growth of basidiomycetes which decompose their litters (Olsen, Odham and Lindeberg, 1971; Theander, 1978; Lindeberg *et al.*, 1980). In *Pinus sylvestris* the stimulatory compounds have been identified as certain non-glycosidal phenolics, catechin, gallic acid, etc. and taxifolin glucoside (dihydroquercitin 3-glucoside). The latter markedly stimulated the growth of the conifer litter-decomposing agarics *Marasmius androsaceus* and *Micromphale perforans* but interestingly not non-litter-decomposing fungi from the same habitat or fungi which decompose other types of litter (Lindeberg *et al.*, 1980). Ferulic acid, a lignin-related phenolic, which can accumulate in soil under mor-forming conditions, also has stimulatory properties. It is particularly abundant in decomposing conifer litters and associated soils. Significantly, it has been shown to enhance the growth of *Penicillium* species of acid soils and *Dothichiza pythiophila* and *Thysanophora penicilloides*, two fungi of conifer litter, while depressing the growth of some fungi of angiosperm litter (Black and Dix, 1976b). Similarly extractives from elm have been shown to stimulate preferentially basidiomycetes which rot elm wood (Rayner and Hedges, 1982).

Plant tissues also give off volatiles which may affect fungal growth. Alcohols and aldehydes are commonly present in vapours from decomposing plant tissues (e.g. alfalfa, *Medicago sativa*). In low concentrations some of these substances will stimulate fungal growth but at high concentrations they become inhibitory (Gilbert *et al.*, 1969; Owens *et al.*, 1969). Cabbages (*Brassica oleracea*) and other cruciferous plants give off gases as they decay containing many inhibitory sulphur compounds such as *S*-methylcysteine, homocysteine and isothiocyanate (Lewis and Papavizas, 1969, 1970). Antifungal volatiles are also produced by plants in the Umbelliferae (Singh, Singh and Gupta, 1979).

In woody tissues a number of volatiles have been detected which affect fungal growth. *Pinus* species give off fungitoxic volatiles from fresh wood and stimulatory volatiles from dry wood (Flodin and Fries, 1978). The inhibitory effects are thought to be due to terpenoids (Hintikka, 1970); the monoterpenes, limonene, myrcene and α- and β-pinene occur abundantly in volatiles emanating from *Pinus sylvestris* wood and have been shown to inhibit the growth of *Stereum sanguinolentum, Lenzites saepiaria, Schizophyllum commune* and other wood-rotting Hymenomycetes (De Groot, 1972; Flodin and Andersson, 1977; Flodin and Fries, 1978). Toxicity may vary with the fungus and concentration. *Heterobasidion annosum* (*Fomes annosus*), for example, is more tolerant of myrcene than *S. sanguinolentum*, while at low concentrations α- and β-pinene may stimulate growth (Flodin and Fries, 1978).

Dried wood of *Pinus* gives off fewer monoterpenes and the detectable stimulating effects produced by volatiles from the dry wood have been traced to the presence of unsaturated fatty acids, e.g. linoleic and oleic acid, and their corresponding aldehydes and alcohols. The aldehydes hexanal and nonanol (pelargonaldehyde) have been shown to be strongly stimulatory to hyphal growth. These are among the chief auto-oxidation products of the unsaturated fatty acids of *P. sylvestris* wood and increase considerably when pine wood is dried by heating (Fries, 1961; Glasare, 1970; Rice, 1970; Flodin and Andersson, 1977; Flodin and Fries, 1978).

Fungi show negative and positive tropic responses to volatiles given off by *Tilia vulgaris* and *P. sylvestris* woods and in some cases these responses may be specific to the particular wood tissue. This may be ecologically significant in that it may enable a fungus to locate selectively and grow towards suitable resources while avoiding unprofitable encounters, thus enhancing both the competitiveness and the survival of the fungus (Mowe, King and Senn, 1983).

3.2.3 Influence of physical factors

a) Temperature

Temperature is one of the cardinal factors affecting fungal growth. The majority of fungi are **mesophiles** and grow in the temperature range 5–35°C, with optima between 20 and 25°C according to species. Extraordinarily little is known about temperature changes in the mesophilic range on the structure of mesophilic communities, since most observations have been made on monocultures at uniform temperatures near to the optima for growth. Fungi are normally subjected to a range of temperature and through the day temperature may rise and fall by 20–30°C. Wood, for example, can quickly warm up to over 40°C on a summer's day and fall to below 10°C at night, fluctuations which apparently have no adverse effect on the growth of wood-inhabiting species (Morton and Eggins, 1977). However, studies indicate that for some species small changes in temperature can markedly stimulate growth and have emphasized the probable importance of temperature fluctuations in shaping community structure (Jensen, 1969; Weidensaul and Wood, 1974). Further, it should not be assumed that the distribution of species necessarily closely follows optimum temperatures for growth. It has been shown, for example, that because of competition certain *Penicillium* species are commoner on barley grain at the extremes of their temperature range than they are at their optima (Hill and Lacey, 1984). There is also evidence that interspecific interactions can affect temperature response and cause shifts in optima for physiological activity (Webster, Moran and Davey, 1976).

When temperatures are persistently above the mesophilic range, the growth of mesophiles becomes restricted and may be prevented altogether. The uneven distribution of *Armillaria* species in Africa, for example, has been connected with the failure of the fungus to produce rhizomorphs under experimental conditions when the temperature rises above 26°C (Rishbeth, 1978).

At high temperatures **thermotolerant** and **thermophilic** fungi with much higher optimum temperatures for growth and which can grow at temperatures up to 45°C and even up to 60°C become dominant. In temperate climates these fungi become the dominants in fungal communities in well-insulated composts and they are among the most important microbial agents causing serious deterioration of badly stored agricultural products. In the tropics they can colonize a wide range of materials with moisture contents above 18% of dry weight, where, if insulation is good, insolation or self-heating can raise the temperature.

Heat-tolerant fungi are also quite common in temperate soils and even thermophiles, with higher minimum temperature requirements for growth than thermotolerants, may colonize once the sun warms the soil above 20°C (Apinis, 1963; Tansey and Jack, 1976). Rather surprisingly, perhaps, heat-tolerant fungi are not more numerous than mesophiles in tropical soils even where the soil is highly insolated (Gochenaur, 1975). Here, lack of moisture at certain seasons may be the limiting factor.

Cold-tolerant fungi with lower optimum temperatures for growth dominate communities where temperatures are persistently low. Plant litter of the Arctic and Antarctic tundras supports a reduced but varied mycoflora consisting of **psychrotolerant** strains of common mesophiles, many sterile forms, some with clamp connections, and a few **psychrophilic** species (Latter and Heal, 1971; Flanagan and Scarborough, 1974). The unusually high incidence of sterile forms may be due to the absence of antagonistic *Aspergillus* and *Trichoderma* species which favour temperate and tropical soils.

Psychrotolerants grow down to about −3°C but are most active at temperatures above 5°C. Consequently, in the tundra the main period of fungal growth is confined to a few months in the summer. For details of the physiology of thermophiles and psychrophiles see Chapter 12.

b) Aeration

Except under waterlogged conditions, many of the spaces and pores of the substratum of the terrestrial fungus will be gas-filled, but with mixtures that will be quite different from those of air. Carbon dioxide concentrations will be higher and oxygen levels lower, water vapour levels may be high and organic volatiles from microbial metabolism and decomposing organic matter will also be present, some of which are liable to inhibit fungal growth.

Most terrestrial fungi can grow well at lower oxygen and higher carbon dioxide levels than those of air (Durbin, 1959; Klotz, Stolzy and Dewolf, 1963; Brown and Kennedy, 1966; Covey, 1970) and studies in culture and in soil have found that low partial pressures of oxygen have little effect on growth or germination down to about 4% oxygen (Griffin, 1966; Griffin and Nair, 1968; Macauley and Griffin, 1969a). Some species are more tolerant of low oxygen concentrations than others and a number of terrestrial fungi are undoubtedly facultative anaerobes. These include *Fusarium oxysporum*, *F. solani*, *Trichoderma viride*, *Penicillium* and *Mucor* species (Tabak and Cooke, 1968; P.J. Curtis, 1969).

In soil, fungal growth is seldom likely to be impaired through lack of oxygen since the proportion of oxygen in the gas-filled pores of soil near the surface seldom falls below 10% (Macauley and Griffin, 1969a). Even at a depth of 15 cm in beech litter the oxygen concentration in the soil atmosphere has been registered at over 19.5% (Brierley, 1955). Shortage of oxygen is not likely, therefore, to be a major factor limiting growth on the surfaces of substrata in soil.

Within the substratum, oxygen concentration will depend upon the porosity of the substratum and the ease with which oxygen can diffuse to the interior. If spaces and pores become filled with water, the flow of oxygen to the interior will slow down due to the low solubility and slow rate of diffusion of oxygen in water compared with air. This may lead to shortage of oxygen in waterlogged substrata, and in saturated soils when respiring microbes are present it has been calculated that there would be no oxygen in the centre of a soil crumb no more than 3 mm in diameter (Greenwood, 1961; Greenwood and Berry, 1962). In bulky substrata, oxygen shortages may arise due to the length of the diffusion path to the air. The atmosphere within decaying wood, for instance, may contain as little as 1% oxygen, with carbon dioxide levels above 10% (Thacker and Good, 1952). Some wood-rotting fungi have become adapted to these conditions and can tolerate low oxygen concentrations at very high carbon dioxide levels (Gundersen, 1961; Hintikka and Korhonen, 1970; Boddy, 1984).

Aquatic fungi range from strict aerobes to facultative anaerobes (Emerson and Natvig, 1981). The latter are widespread in stagnant waters at the bottom of ponds, in marshes, wet ditches, etc. Known facultative anaerobes occur in the Blastocladiales (*Blastocladia*) and in the Leptomitales (*Rhipidium*, *Sapromyces*, *Mindeniella* and *Aqualinderella*). Aero-aquatics in the Deuteromycotina that can survive prolonged anaerobic conditions must also be presumed to be facultative anaerobes (Field and Webster, 1983). The sewage fungus *Leptomitus lacteus*, which can grow in waters heavily polluted with organic matter, is not, however, a facultative anaerobe.

Some facultative anaerobes (e.g. *Aqualinderella fermentans* and *Blastocladia ramosa*) are obligate fermenters and have no capacity for aerobic respiration. They have only rudimentary mitochondria and reduced cytochrome systems (Emerson and Held, 1969; Held et al., 1969; Held, 1970). They fix carbon dioxide and are dependent on very high concentrations of carbon dioxide (5–20%) for growth.

The only known strictly anaerobic fungi are the remarkable rumen 'chytrids' found in the digestive tracts of ruminants (Theodorou, Lowe and Trinci, 1992).

Unless the carbon dioxide concentration exceeds 10% in air–gas mixtures containing 20% oxygen, it usually does not reduce fungal growth (Macaulay and Griffin, 1969a), and in many cases increases in carbon dioxide up to about 7%, at atmospheric concentrations of oxygen, may stimulate growth (Mitchell and Zentmyer, 1971; Gardner and Hendrix, 1973) since most fungi have a capacity for heterotrophic carbon dioxide fixation. In the gas-filled spaces of the soil in continuity with the atmosphere, the concentration of carbon dioxide is thought to lie usually between 0.2 and 2% (Russell, 1961), but higher figures have been recorded in waterlogged soils and it is possible that the carbon dioxide concentration may reach 10% under these conditions. At 10% carbon dioxide concentration, the growth rate of most fungi is likely to fall to about half. Rishbeth (1978) recorded a 52% drop in the growth rate of rhizomorphs of one isolate of *Armillaria* and a 22% drop in another isolate at this concentration of carbon dioxide, and similar results have been recorded for other fungal species (Griffin and Nair, 1968; Macauley and Griffin, 1969a).

Higher concentrations of carbon dioxide occur with increasing depth in the soil and there appears to be some correlation between the distribution of fungi in soil and their tolerance of high carbon dioxide concentrations. Durbin (1959) isolated three strains of *Rhizoctonia solani* infecting plant roots from different levels in soil and measured their growth at 20% carbon dioxide concentration. The growth of two strains whose distribution was in the upper layers of the soil was reduced by 52% and 80%, while the growth of the strain from deeper in the soil was reduced by only 3%. Burges and Fenton (1953) measured linear spread and mycelial production in cultures of three soil fungi isolated from different depths in the soil at different

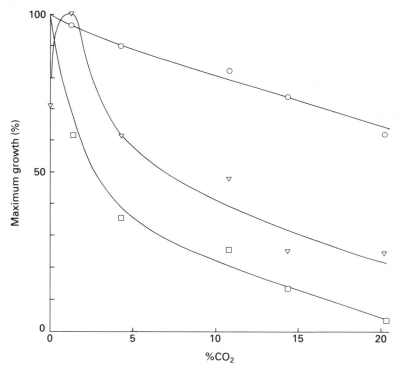

Figure 3.2 Growth of fungi at different concentrations of carbon dioxide. ○ = *Zygorrhyncus*; ▽ = *Gliomastix*; □ = *Penicillium*. (Reproduced with permission from Burges and Fenton, 1953.)

carbon dioxide concentrations and obtained similar results. *Zygorhynchus vuillemini*, which is comparatively more abundant in the lower horizons of the soil, grew best at the higher concentrations of carbon dioxide, producing more than 50% of its maximum growth at carbon dioxide concentrations up to 20%. *Penicillium nigricans*, from the upper soil horizons, grew poorly at carbon dioxide concentrations above 5%, while *Gliomastix convoluta*, from the middle soil horizons, showed an intermediate response (Figure 3.2).

Griffin (1966) investigated the fungal colonization of plant materials in contact with soil when held at gas mixtures with reduced oxygen levels combined with high concentrations of carbon dioxide. Little change in the pattern of colonization compared with that in air was noticed until the $CO_2:O_2$ ratio became greater than one. Under these conditions, colonization by *Curvularia* and *Rhizoctonia* species declined while the frequency of *Zygorhynchus vuilleminii* increased. Since $CO_2:O_2$ ratios greater than one are normally found only in soil water, Griffin concluded that there would only be a significant shift in the activity of certain fungal populations in soil if the soil pores became filled with water. This must be the major factor which determines the effect of gases on fungal communities in soil and plant substrata.

The inhibiting effect of high concentrations of carbon dioxide on fungal growth is probably due to the formation of bicarbonate ions in solution which affect fungal metabolism (Cantino, 1966). Fungi have been found to respond more directly to the bicarbonate ion concentration in solution than to the partial pressure of carbon dioxide in the atmosphere above (Griffin, 1965). Bicarbonate ion concentration can increase in solution as the pH rises, therefore a high concentration of carbon dioxide in the substratum is likely to suppress fungal growth more under alkaline than in acid conditions (Macaulay and Griffin, 1969b).

c) Hydrogen ion concentration

Most fungi prefer acidic conditions but will tolerate a wide range of pH (Table 3.3). Most basidiomycetes are exceptions to this, with few growing well above pH 7.5 (*Coprinus* species, which favour alkaline conditions appear to be rather exceptional; Fries, 1956). *Pythium* species also have a rather narrower range and are seldom found in great numbers in soils at pH below 5.5 or above 8, preferring soils where the pH is between 6 and 7. Consequently they are more frequent in agricultural and managed soils than in undisturbed soils, which tend to be of lower pH (Adams, 1971; Lumsden, Byer and Dow, 1975; Lumsden *et al.*, 1976; Dick and Ali-Shtayeh, 1986).

Optimal pH for the growth of the majority of fungi is usually between pH 5 and 6, but often there is no very conspicuous optimum and growth

Table 3.3 Limits of pH for the growth of some common fungi

Species	Limits of pH
Amanita muscaria	2.5 – 6.8
Amanita pantherina	2.9 – 7.5
Aspergillus candidus	2.0 – 7.7
A. flavipes	2.5 – 9.0
A. fumigatus	3.0 – 8.0
A. niger	1.5 – 9.8
A. repens	1.8 – 8.5
Boletus edulis	2.5 – 6.4
Boletus subtomentosus	3.5 – 5.8
Botrytis cinerea	2.0 – 8.0
Cladosporium herbarum	3.1 – 7.7
Fusarium oxysporum	2.0 – 9.0
Lactarius deliciosus	3.0 – 6.8
Mucor racemosus	2.0 – 8.4
Paxillus involutus	2.2 – 6.0
Penicillium cyclopium	2.0 – 10.0
P. italicum	1.6 – 9.8
P. luteum	1.4 – 8.3
Sclerotinia sclerotiorum	2.0 – 10.0
Stemphylium solani	2.0 – 10.0
Suillus grevillei	3.6 – 6.4
Trichoderma koningii	2.5 – 9.5

Sources: Panasenko (1967) and Hung and Trappe (1983).

occurs uniformly well over several pH units (Hung and Trappe, 1983).

Trichoderma species and *Penicillium* species tolerate lower pH better than many other fungi and predominate in acid soils (Warcup, 1951a). Optima for mycelial growth as low as pH 2.5 have been reported for some isolates of *Trichoderma viride* (Brown and Halsted, 1975).

It is interesting to note that in some genera there is a division between species which tolerate low pH and those tolerating high pH. *Absidia glauca* is a species of alkaline soils while *A. orchidis* is a fungus of acid soils, similarly *Mortierella ramanniana* (*Mucor ramannianus*) and *M. isabellina* are characteristic of acid soils, whereas *M. alpina* and *M. minutissima* are fungi of alkaline soils (Warcup, 1951a; Sewell and Brown, 1959).

Because of the effective buffering system which exists in hyphae, external hydrogen ion concentration has little direct effect on fungal metabolism but may have several indirect effects on growth. Hydrogen ion concentration in the environment controls ionization of salts in bathing solutions and hence the availability of ions to the fungus. Ion uptake may also be influenced by the effect of hydrogen ion concentrations on the permeability of the plasmalemma. Enzyme activity is pH-dependent and at non-optimal pH the efficiency of extracellular enzyme catalysis will be reduced. The effect of pH on fungal growth is therefore likely to be rather complex and any changes in the growth responses of fungi to shifts in pH are usually difficult to ascribe to any single factor.

Plant remains usually have a net negative charge and attract cations. The local pH of litter will depend upon the base saturation or the proportion of the negatively charged sites which are occupied by basic cations (Mg^{2+}, K^+ or Ca^{2+}) compared with H^+. Plant remains generally have a base saturation of about 50%, giving a pH in the range 5.5–6.5, but conifer litter is acidic with a pH in the range 3.5–4. The basic cation content of litter depends upon the base saturation of the soil in which the plant grew. Plants growing in soils with a low base content tend to give rise to litters with a lower pH than the same species growing in alkaline soils (Coulson, Davies and Lewis, 1960; Davies, Coulson and Lewis, 1964).

As decomposition of litter proceeds, H^+ ions are produced as a result of the formation of organic and mineral acids. These H^+ ions are exchanged for basic cations in the soil solution bathing the litter where they will in turn be exchanged for cations held on the soil particles. An adequate buffering effect by the soil solution as decay proceeds will therefore depend upon the base exchange capacity of the soil particles. In soils of low base status, such as highly leached soils, the buffering capacity of the soil will be too low to prevent soil pH falling and this will lead eventually to a drop in the pH of the litter as H^+ ions gradually replace cations held at negatively charged sites on cell walls. This will reduce the rate of decay and lead to the accumulation of humus with a low pH.

d) Water relations and water availability

Water activity (a_w)

The **concentration of water** is one measure of its availability to organisms. Concentration is best expressed as a mole fraction and is directly related to the depression of the vapour pressure:

$$C = \frac{N_w}{N_w + N_1} = \frac{P}{P_O}$$

where C is concentration; P is partial pressure of the solution; P_o is partial pressure of pure water; N_w is number of moles of water; and N_1 is number of solute molecules.

This equation only applies for solutions of non-electrolytes and to a limited extent to very dilute solutions of electrolytes. For concentrated solutions of electrolytes there is a difference between the theoretical concentration of water and the effective concentration due to solute interference. The effective water concentration of a solution is known as the **water activity**. Water activity is the same as percentage equilibrium relative humidity (ERH) but on a scale of 0–1. Thus a_w = ERH ÷ 100.

Water potential (ψ)

For the fungal ecologist this is the most useful term that describes water availability. Whereas water activity only describes water status or concentration, **water potential** is an expression of energy status and takes into account all the forces that affect energy and increase or decrease flow to lower energy levels (lower concentrations) such as hydrostatic pressure and matric potential (see below). It is therefore the only expression directly relevant to the movement of water into the fungus from an environment where these forces exist. In those circumstances it cannot be calculated but must be determined by measurement (see Hamblin, 1981, for example). The unit of measurement is the megapascal, formerly bar (10 bar = 1 MPa).

Water activity and water potential are related by the equation

$$\psi = RT/V \log_e a_w$$

where R is the universal gas constant; T is temperature in degrees Kelvin; and V is partial molal volume of water. See Table 3.4 for some conversions.

Even where matric forces etc. need not be considered, water potential may be preferred to a_w as a measure of water availability. Because of differences in scale, water potential is the more sensitive, which is advantageous when determining the behaviour of very sensitive fungi, e.g. aquatic species. For a fuller account of the definitions of water activity and water potential the reader is recommended to turn to Papendick and Mulla (1986) or to Brown (1990).

At atmospheric pressure, pure water is defined as having zero water potential. Water potential becomes increasingly more negative as the amount of solute present increases (**solute** or **osmotic potential**). Water potential is also reduced by interactions at the water–solid, water–gas interfaces in the matrix of the substratum (**matric potential**). In nature, water may be unavailable to a fungus due to low osmotic or matric potential or sometimes both, but for terrestrial species the latter is usually the more important since at low absolute water contents matric potential becomes the major component of water potential. As water content falls, only the finer part of the pore system of the substratum will be filled; adhesion forces will therefore increase and the matric potential will

Table 3.4 Equivalent water potentials of some water activities at 25°C

Water activity	Water potential (MPa)
0.999	−0.138
0.995	−0.690
0.99	−1.38
0.98	−2.78
0.97	−4.19
0.96	−5.63
0.95	−7.06
0.90	−14.50
0.85	−22.40
0.80	−30.7
0.75	−39.6
0.70	−40.1

become very negative. This can be illustrated with reference to soils of differing particle size (Figure 3.3). In sand, with larger pores, the water potential is between 10^3 and 10^4 greater than that in clay at the same water content.

Water potential is temperature-dependent and becomes more negative as the temperature rises.

Individual fungi are capable of growth over a range of water potentials, but for some the range is wider than for others, optima are generally high and as the water potential falls, the rate of growth decreases. Usually, the minimum requirements for growth and sporulation are higher than for germination and tend to increase in all cases as the pH falls below optimum (Magan and Lacey, 1984a).

The effect of water potential on growth is strongly influenced by temperature and vice versa. The greatest tolerance to low water potentials is usually close to the optimum temperature for growth (Ayerst, 1969; Magan and Lacey, 1984a) and maximum tolerance to extremes of temperature is best at optimum water potentials (Ayerst, 1969).

Water uptake depends upon the ability of the fungus to adjust the internal osmotic potential of hyphae. Over the range at which they grow, fungi have the capacity to osmoregulate and can maintain internal turgor by the uptake of ions and non-electrolytes from the external environment as the external water potential falls (Adebayo, Harris and Gardner, 1971; Luard and Griffin, 1981). These adjust the internal osmotic potential and keep it at a level below that of the external environment, thus allowing water to pass into the hypha (Luard, 1982a; Wethered, Metcalf and

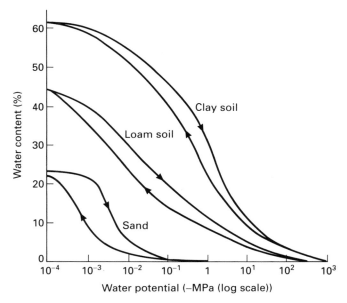

Figure 3.3 Water sorption isotherms for three soils, showing hysteresis between drying and wetting boundary curves. (Reproduced with permission from Griffin, 1981b.)

Jennings, 1985; Wethered and Jennings, 1985).

Hyphae under water stress synthesize osmoregulatory polyols, mainly glycerol, mannitol and arabitol. Another more important function of these substances may be to protect enzymes from damage at low water potentials. Certain aquatic lower fungi in the Oomycetes and *Mucor hiemalis* (Zygomycotina) synthesize proline in addition to or instead of polyols (Luard, 1982b; Wethered and Jennings, 1985). This is thought to have a similar function.

Growth rate reduces as the water potential of the environment falls and there is a direct relationship between reduced growth rate and falling hyphal turgor (Eamus and Jennings, 1986). Poor growth at low water potentials could also be due to a reduction in protein synthesis or poor enzyme functioning at low internal osmotic potentials, and it seems likely that the need to divert resources from growth to the synthesis of polyols also has an effect on growth. It has been speculated that a major difference between those fungi which grow at very low water potentials and those which cannot may lie in the more efficient utilization of resources in the former together with major differences in the allocation and budgeting of energy. At present there is little experimental evidence to support this, except to report that it seems significant that Wilson and Griffin (1975) found higher respiration rates per unit of growth at decreased water potentials for *Phytophthora cinnamomi*, a fungus relatively intolerant to water stress, compared with the rates recorded for more tolerant fungi.

For a fuller account of the physiology of adaptation to growth at low water potentials see section 12.2.3.

As might be anticipated, water moulds are the fungi least tolerant of water stress. Thus the Saprolegniaceae generally have a lower limit for growth at about -1.4 MPa while growth begins seriously to decline at about -0.5 MPa. Only slightly more tolerant are the Pythiaceae, most of which are capable of some growth down to about -2.6 to -4.0 MPa (Kouyeas, 1964; Sommers *et al.*, 1970; Harrison and Jones, 1971, 1974; Sterne and McCarver, 1978). Optimal values for the growth of water moulds are about -0.1 MPa (Stern, Zentmyer and Bingham, 1976; Sterne and McCarver, 1978).

These measurements fit very closely with observations on the distribution of these fungi. The Saprolegniaceae occur predominantly in fresh water and soil, there are very few marine species, and they are rarely found in brackish water. By contrast representatives of the Pythiaceae occur both in freshwater and marine environments (Te Strake, 1959; Johnson and Sparrow; 1961; Fell and Master, 1975).

Saprolegnia and *Pythium* species require high water potentials for reproduction, typically an osmotic potential of at least -0.5 MPa for the formation of zoosporangia (Johnson and Sparrow, 1961; Harrison and Jones, 1971, 1975; Bremer, 1976). *Phytophthora* species require even higher matric potentials for reproduction, about -0.03 MPa, at least to form zoosporangia and near zero matric potential before zoospores are released (Bernhardt and Grogan, 1981). It is well known that the incidence of plant disease caused by pathogenic fungi in the Pythiaceae increases at high matric potentials and this is obviously linked with an increase in inoculum potential as zoosporangia form and release zoospores (Cook and Papendick, 1972; Kuan and Erwin, 1981).

The water relations of water moulds have been reviewed comprehensively by Duniway (1979).

The majority of terrestrial fungi grow well at water potentials down to about -6 MPa and many germinate and make slow growth down to about -14.5 MPa (Griffin, 1963, 1972). Optimum for growth of these fungi is about -1 MPa (Griffin, 1981b). Most of those, from several different taxonomic groups, studied by Tresner and Hayes (1971) grew in the presence of 10% NaCl giving rise to an osmotic potential of about -11.5 MPa in the growth medium. In other terms this would be the same as growth at 92% equilibrium relative humidity at 25°C.

Basidiomycetes inhabiting wood, litter or soil in temperate regions have generally been found to be less tolerant of water stress than other terrestrial fungi, although there are some exceptions (Wilson and Griffin,

1979). Of the 104 examined by Tresner and Hayes (1971), 90% could not tolerate an osmotic potential of −4.0 MPa (5% NaCl), equivalent to an equilibrium relative humidity of 97% at 25°C. These figures place these fungi near to aquatic species as among the least tolerant of fungi to water stress. Optimum water potentials for growth often exceed −1 MPa (Boddy, 1983; Koske and Tessier, 1986), values which would not be attained in undecayed wood or leaves until the fresh weight water content was greater than about 40% (Dix, 1985) or 60–70% in the case of straw (Papendick and Campbell, 1981; Magan and Lynch, 1986).

However, a very incomplete range of basidiomycetes has so far been tested and it would be surprising, for example, if *Galleropsis desertorum*, *Gyrophragmium dunaii* and *Montagnea arenaria* and other xeromorphically adapted agarics of gasteroid form that inhabit arid places described by Watling (1982) did not turn out to be xerotolerant.

Exceptionally, terrestrial fungi grow at water potentials as low as −40 MPa (Griffin, 1963, 1972). These are the **xerotolerant** and **xerophilic** species which dominate fungal communities on very dry substrata. A similar group are the **osmotolerant** fungi which grow on jams, syrups, nectars, sugary fruits and salty foods with low osmotic potentials. These fungi have lower optima for growth, around −4 MPa, and xerophilic species fail to grow if the water potential is above −3 MPa. Maintaining turgor at very low water potentials must require great powers of osmoregulation. The not insignificant amounts of water produced during respiration and the capacity which fungi have to imbibe water from the atmosphere may also be important in the survival of fungi under these conditions, making them perhaps independent to some extent of the need to take up water from the substratum. Many of these fungi are *Aspergillus* and *Penicillium* species, the troublesome biodeteriorants of dried agricultural products in stores.

For further information on xerotolerant and osmotolerant fungi refer to section 12.2.

Adebayo and Harris (1971) found that both *Alternaria tenuis* (*A. alternata*) and *Phytophthora cinnamomi* ceased growth at a higher matric potential compared with osmotic potential; differences were about −7 MPa for the former fungus and about −1.0 MPa for the latter. Similar results were obtained by Magan and Lynch (1986) studying the water relations of fungi decomposing cereal straw, and smaller but significant differences have also been measured for other fungi (Cook, Papendick and Griffin 1972; Cook and Papendick, 1972). This seems to be generally true and should be kept in mind when absolute values for tolerance to low water potentials are being discussed, because most experiments quoted in the literature relate to measurements of growth at low osmotic potentials. The reason for this difference in sensitivity can be traced to several indirect effects of low matric potentials on fungal growth. At low matric potentials many of the pores of the substratum on which the fungus is growing will be empty of water. The overall effect will be a reduction in diffusion of solutes to the fungus mycelium and a reduction in extracellular enzyme activity, both of which will in turn reduce growth. These effects are likely to begin to be felt at surprisingly high water potentials for it has been calculated that at matric potentials as high as -0.015 MPa no water would be present in soil pores with radii greater than 100 μm (Griffin, 1981a).

Terrestrial fungi may grow over a range of water potentials in culture but most are probably uncompetitive at the lower limits for growth, and at lower water potentials xerotolerant species become dominant. Since xerotolerant species are less numerous, numbers of species will tend to fall as the water potential is reduced. This is well illustrated by the experiments of Chen and Griffin (1966), who recorded a drop from 51 species (22 genera) to 38 species (7 genera) colonizing hair from soil when the water potential was reduced from near optimum to near to the limits for mesophilic growth. However, it does not follow that the highest frequency of occurrence of a species will always coincide with the optimum water potential for growth. For uncompetitive species that can tolerate low water potentials, a reduction in the number of competitors may be the most important factor determining frequency of distribution. *Aspergillus repens* for example, competes unsuccessfully with other storage fungi in culture and predominates on grain at an a_w of 0.85 although its optimum a_w for growth is about 0.95 (Magan and Lacey, 1984b) and some *Penicillium* species are known to be commoner on grain at water potentials closer to their limits of tolerance than at water potentials close to their optima for growth (Hill, 1979).

The water relations of the substrata on which fungi grow are undoubtedly complex. Water is received mainly by rainfall and is lost principally by evaporation. Substrata in contact with soil can take up water by capillarity and lose it by gravitational force. Too little water stops growth but waterlogged conditions which restrict aeration can be equally detrimental. Substrata in contact with soil can be sheltered from rain by tree canopies but generally speaking they are likely to provide more favourable water regimes than exposed standing stems or leaves attached to plants.

Dew formation is also an important source of water and there are few nights when dew does not form on cooling surfaces. Continuous wetting by dew and subsequent rapid evaporation as temperature rises subjects those fungi colonizing exposed standing stems and their leaves to wet–dry cycles which have a very profound effect on the distribution of fungi on these organs (Chapter 5). Rates of evaporation and uptake of water will vary according to the physical and chemical characteristics of the substratum and are likely to vary considerably for different components of the litter. The water content of different parts of the same component can also differ greatly according to the degree of exposure, contact with the soil, etc.

Boddy (1984) refers to details of some examples.

Different components of the litter are also likely to have different water potentials at the same water content due to varying chemical and physical characteristics (Dix, 1984a). In general, for substrata with comparable water loadings, i.e. those at the same water content per unit volume, those with lower densities will have higher matric potentials than those of greater density and those with larger voids will have higher matric potentials than those with smaller voids. We must also expect variation in this respect in different parts of the same litter component due to the uneven distribution of voids by size, and this may have a profound local effect on the growth of hyphae. It follows that as decay proceeds and cell walls are degraded, voids will increase in size and number and water potential values will increase correspondingly at the same water content (Dix, 1985). However, the theoretical improvement in the water potential of the substratum as decay proceeds may be offset by a tendency for water to evaporate more readily from decayed tissues.

The water potential status of the substratum is subjected to changes induced by temperature so that temperature has both an indirect and direct effect on growth; these effects are difficult to separate. As temperature increases, the water potential of the substratum will fall even if the water content stays the same (i.e. there is no evaporation). Reverse changes occur as the temperature falls. The magnitude of these changes increases with decreasing water potential. A rise in temperature from 10 to 15°C would depress the osmotic potential of a solution at -5.69 MPa to -5.84 MPa, but at water potentials at about -10 MPa a similar shift in temperature would depress the water potential by about 0.15 MPa. A 20°C change at this water potential would increase or decrease the water potential by about 0.7 MPa. Thus water potential changes caused by fluctuating temperatures are small in the growth range of most species but could be ecologically significant for sensitive species or for those growing at the extreme of their range.

3.2.4 Interspecific interaction

Critical to the structure of fungal communities are the outcomes of interactions that develop between mycelia occupying or attempting to establish themselves on the same resource.

Studies in culture indicate there are a number of possible outcomes to interactions between species, ranging from stimulation to one or both of the mycelia, through mutual tolerance and degrees of intolerance, to the development of highly aggressive forms of competitive behaviour.

Examples of stimulation are rarely reported but they are probably commoner than is presently realized. Some examples of stimulation, some of them possibly mutually beneficial, have been described in section 3.2.2.

Others have been described by Morton and Eggins (1976). Often hyphae of different species intermingle in culture with no perceived benefit or disadvantage to either mycelium, but if some degree of mutual intolerance exists, opposing mycelia will eventually stop growing, even though this may be after contact between the two mycelia has been established. Such mycelia remain visibly distinct, forming discrete units. They may respond on contact by constructing defensive hyphal barriers at the mycelial fronts to limit the advance of the aerial mycelium (Figure. 3.4). These restrictions on growth arise as a result of mycelial responses to the release of metabolic inhibitors, like antibiotics, or to limitations imposed on growth by competition for nutrients, enzyme substrates, oxygen perhaps or even space. These are referred to as **indirect antagonisms** since they do not require hyphal contact to take effect, but act in advance of the extending mycelium. In the more aggressive forms of behaviour the hyphae of one competitor will advance into the mycelium of the other and destroy it by overgrowth and through hypha-to-hypha interactions. These are known as **direct antagonisms** since they rely on contact, and cover such phenomena as mycoparasitism and hyphal interference.

a) Indirect antagonism

Competition for substrates and space

If one fungus is more successful in competing for available substrates this will prevent supplies reaching others and thus this will create an antago-

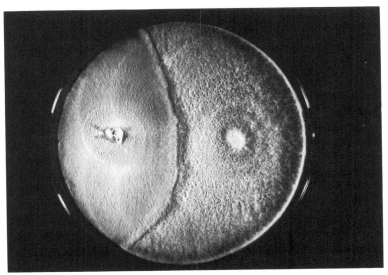

Figure 3.4 A pigmented hyphal barrage formed between *Stereum hirsutum* (left) and *Coriolus versicolor* (right).

nism or a harmful effect. When substrata containing adequate supplies of available nutrients and respirable substrates become available for colonization, rampant growth will ensue. This has been called the **flare-up phase** of colonization. During this phase early establishment and rapid growth by some species will competitively exclude others. At the end of this phase, when soluble carbohydrates and nutrients have become seriously depleted, intense competition will arise for any remaining supplies. This will cause a shift in favour of populations that can successfully scavenge for nutrients and utilize cellulose or lignin for energy supplies. These fungi will flourish and by competitive antagonisms they will eliminate those that cannot compete due to nutritional stress. However, some fungi without the necessary enzymes to release energy locked up in complex polymers may survive in these circumstances if they can enter into mutualistic associations with those that can, as discussed earlier in this chapter.

Nitrogen concentrations in particular may become critical. Although fungi can grow over a range of C:N ratios they have optima for maximum growth and metabolic activity. As nitrogen supplies become depleted, the competitiveness of certain species will change (Park, 1976a,b,c).

When nitrogen levels fall, spore germination will be inhibited by competition for nitrogen uptake (Blakeman and Brodie, 1976). This is one of the principal causes of soil fungistasis as discussed in section 7.3 If nitrogen supplies are initially very high, this can increase demand and competition for other nutrients such as phosphorus (Dowding, 1981).

Space represents captured resources and as such is crucial to establishment and survival. During the initial phase of the colonization of vacant substrata competition for space is usually low and may remain so for many different types of resource. Calculations show that fungal hyphae, together with the cells of other microorganisms, may occupy <0.5% of the available area or volume of the resource (Lockwood, 1981). The area occupied by yeast cells on *Acer platanoides* leaves calculated from the data of Breeze and Dix (1981) are fairly typical. Yeasts reached a peak of 12 700 cells cm^{-2} on the adaxial surfaces of senescing leaves on 21 October in 700 colonies averaging 1–2 cells in size. They would have covered only approximately 0.28% of the adaxial surface area. Data produced by Lockwood (1981) indicate that hyphal species occupy a similar amount of space. Rather surprisingly, these proportions do not appear to change markedly when fallen leaves enter the litter layer (Lockwood, 1981). Even on roots, where microbial numbers are higher due to the rhizosphere effect, fungal hyphae have been estimated to occupy no more than 3% of the root surface of grasses and herbs and 2–3.7% of the root surface of *Pinus nigra* seedlings (Rovira et al., 1974).

On roots and leaves a relatively high number of individuals are in little competition for space due predominantly to nutritional shortages and a high level of competition for substrates (see Chapters 4 and 7 dealing with the colonization of leaves and roots). By contrast, the ever-expanding mycelia of a limited number of macrofungi growing in wood, unhampered by lack of resources, soon reduces available space to a premium. Zone lines form marking out the boundaries of adjacent mycelia, and antagonistic interactions develop to define captured space (Chapter 13). Ultimately in this situation the strongest competitors replace other species, capturing their resources and space.

Antibiosis

Many fungi produce **fungistatic** or **fungicidal metabolites** (antibiotics) which diffuse from hyphae and slow or stop the growth of competitors from some distance away. If two opposing species produce inhibitory metabolites, mutual inhibition may result (Figure 3.5). Inhibition by antibiosis is often species-specific and a response only occurs when appropriate species meet. Antibiosis phenomena are associated with the capture of resources and with the defence of already secured resources from takeover by would-be competitors.

Berdy's review of 1974 lists 768 fungi producing antibiotics, but this is far from a definitive list since there are tens of thousands of species

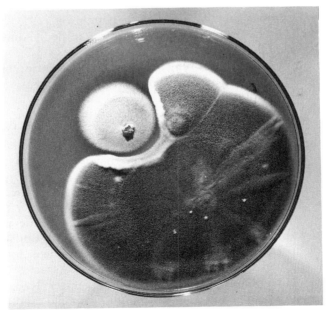

Figure 3.5 Mutual inhibition between *Collybia peronata* (left) and *Penicillium janthinellum*. Note the thickened mycelial margin of the latter. (Reproduced with permission from Mohamed, 1986.)

which have yet to be tested or for which there are no published records. Berdy's list shows that by far the largest number of producers, over 500, are in the Deuteromycotina, 140 species are in the Basidiomycotina and only 14 in the aseptate groups. This distribution reflects the alternative lifestyles that have been adopted by fungi included in different taxonomic groups. Aggressive competitors in the Basidiomycotina typically capture resources and replace other species by overgrowth and hyphal contact and rarely employ antibiotics to antagonize other species (see below).

Aseptate species are mainly ruderals ('weed' species) for which successful establishment on virgin substrata ahead of competitors is all important. They are adapted to achieve wide dispersal, rapid spore germination and high growth rates.

Antibiotics consist of both volatile and non-volatile substances. They may be complex secondary metabolites or some of the simple products of primary metabolism such as alcohols and aldehydes. The latter are autotoxic and their accumulation in culture media can lead to the arrest of growth, metabolic disorganization, breakdown of organelles and lysis. This may happen when plentiful supplies of nutrients and carbon sources are still present, a fact recognized by early mycologists, who described the phenomenon as **staling**.

Volatile antibiotics of many kinds are commonly produced by fungi in all taxonomic groups and the reviews of Hutchison (1971) and Fries (1973) contain lists of some that have been identified. Among these, acetaldehyde, ammonia, ethylene, ethanol and hexa 1,3,5-triyne are known to be inhibitory to fungi at certain concentrations (Glen, Hutchinson and McCorkindale, 1966; Smith, 1973; Ko, Hora and Herlicska, 1974). Robinson and Park (1966) and Dennis and Webster (1971a) have shown that acetaldehyde can be inhibitory at concentrations as low as 15 ppm. *Trichoderma* species are particularly active producers of several antifungal volatiles. These were investigated by Dennis and Webster (1971a) and some have now been identified (Claydon *et al.*, 1987).

Non-volatile antibiotics are produced by many fungi and cover a whole range of chemically diverse secondary metabolites. Soil fungi in the genera *Aspergillus*, *Penicillium* and *Trichoderma* are very prolific producers. Often several kinds are produced by the same species, resulting in a very broad action spectrum. Citrinin, patulin, gliotoxin and penicillic acid are some of these produced by *Penicillium* and *Aspergillus* species. The mode of antifungal action of most of these is unknown (Ciegler, Kadis and Ajh, 1971). Some, such as aflatoxin from *Aspergillus flavus* and rubratoxin from *Penicillium rubrum*, are also well known for their toxicity to mammalian cells. *Trichoderma* species produce peptide antibiotics and trichodermin (Godtfredson and Vangedal 1965; Dennis and Webster, 1971b) that significantly augment their mycoparasitic activity (see below). Hadacin from *Penicillium stoloniferum* and fusidic acid from *Fusidium* inhibit gene function and enzyme synthesis in a wide range of organisms. Hadacin prevents the elaboration of adenine nucleotides and their polymerization into nucleic acids, while fusidic acid binds to ribosomes causing translation malfunction and errors in polypeptide manufacture.

b) Direct antagonisms

These are contact phenomena that lead to the destruction of mycelia and the elimination of competitors in successions (section 3.3). They are associated with the combative strategies of secondary resource capture outlined in section 1.1.4 (see also Rayner and Webber, 1984).

Overgrowth

Many aggressive litter-decomposing and wood-decay basidiomycetes destroy the mycelia of other fungi simply by growing over them and by doing so capture their resources by replacement. In most cases the mechanism appears to be purely physical, but some, such as *Collybia confluens* and *C. dryophila*, can inhibit other species by antibiosis (Mohamed and Dix, 1988) and may antagonize sensitive species by a combination of physical challenge and chemical poisoning. In typical examples of physical challenge the mycelium of the aggressor advances on a broad front over the mycelium of the victim. Frequently, a band of thickening is produced at the colony margin of the aggressor by the repeated branching of unmodified aerial hyphae in order, it seems, to overcome any resistance by sheer mycelial bulk. Overgrowth of the victim is accompanied sooner or later by hyphal lysis. Lysis may quickly follow contact at the mycelial fronts, suggesting that in some cases lysis may be a result of enzyme action by the aggressor as opposed to autolysis following metabolic disturbance (Figures 3.6 and 3.7).

In some species, for example *Collybia peronata*, overgrowth is sometimes initiated by the production of runners which traverse the mycelium of the victim in advance of the general mycelium. These are clearly functionally analogous to the more elaborate strands produced by wood-decay basidiomycetes for the purpose of capturing resources and spreading over inhospitable terrain. Stimulation of runner formation in *C. peronata* is species-specific. (section 2.2.3).

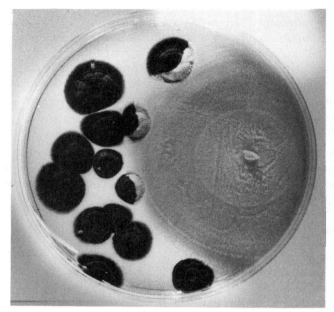

Figure 3.6 *Collybia peronata* (right) overgrowing *Cladosporium cladosporioides*.

Mycoparasitism

Mycoparasites fall into two physiological groups, *biotrophs* and *necrotrophs* (Barnett and Binder, 1973), and are represented in every taxonomic group (Madelin, 1968; Hawksworth, 1981; Lumsden, 1981).

Biotrophs are highly specialized species that do not destroy host hyphae and cannot grow in the absence of a host. They are therefore of little relevance to the present discussion. Necrotrophs are widespread saprophytes that overgrow and destroy other competitors, producing toxins and wall-degrading enzymes for this purpose. As is implied by the term necrotroph, replacement by this mechanism is believed to involve nutritional exploitation of the mycelium of the dead victim.

The soil fungi *Gliocladium roseum* and *Trichoderma* species are typical necrotrophs. *G. roseum* has been studied by a number of workers, notably Barnett and Lilly (1962), Walker and Maude (1975) and Pachenari and Dix (1980). It parasitizes a large number of other fungi, with spore-bearing structures and the spores of the host being especially vulnerable. When attacking *Botrytis* species, the mycelium of the parasite completely overgrows that of the host, hyphal organization becomes disrupted, vacuolation and leakage occur and lysis of hyphal walls and organelles soon follows. Wall-degrading glucanases and chitinases and low molecular weight short-range toxins are secreted by the mycoparasite. Penetration of the host from

Figure 3.7 Lysis of hyphae of *Lacrymaria velutina* by *Collybia dryophila* in the interaction zone between two opposed mycelia. The wider hyphae of *C. dryophila* (Cd) are darkly stained, those of *L. velutina* (Lv) are narrower, lightly stained, vacuolated with granulated cytoplasm. (Reproduced with permission from Mohamed, 1986.)

appressoria, and coiling of the hyphae of the mycoparasite around those of the host are occasionally seen.

The mycoparasitic behaviour of *Trichoderma* species is somewhat similar to that of *G. roseum* except that *Trichoderma* species produce mixtures of several powerful volatile and non-volatile antibiotics that cause hyphal bursting, vacuolation and coagulation of protoplasm, inhibiting growth some distance away (Dennis and Webster, 1971a,b,c). Production of antibiotics mixtures significantly increases the range of inhibitory activity and *Trichoderma* species antagonize a great number of other fungi.

The majority of *Trichoderma* isolates exhibit coiling around host hyphae (Dennis and Webster, 1971c). Penetration is sometimes seen (Figure 3.8), but is not common and may be restricted by the diameter of host hyphae. Coiling and similar contact phenomena have no obvious significance in necrotrophic attack. In *Trichoderma* species, a direct relationship has been established between necrotrophic vigour and production of antibiotics; coiling, on the other hand, appears unconnected and was found to occur just as frequently among isolates with low antagonistic vigour as among the most aggressive examples (Dennis and Webster, 1971c). It has been suggested that coiling may merely be a response to host resistance

74 STRUCTURE OF FUNGAL COMMUNITIES

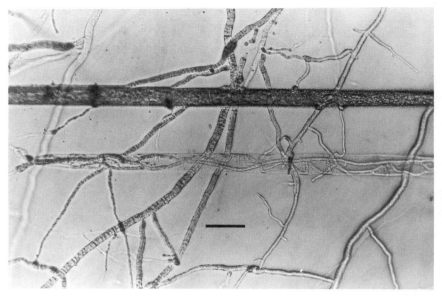

Figure 3.8 Parasitism of *Pythium ultimum* hyphae by *Trichoderma viride*. Two parallel hyphae of *Pythium* are shown. *Trichoderma* hyphae pass over or under the uppermost *Pythium* hypha and branches occasionally wrap around it. The lower *Pythium* hypha has lost its contents. It contains *Trichoderma* hyphae and two of them can be seen passing through the hyphal tip to the left. Scale bar = 10 μm.

(Deacon, 1976). In *T. hamatum* lateral branches are produced which grow directionally towards the host (Hubbard, Harman and Hadar, 1983)

Dennis and Webster (1971c) found that penetration was most frequently observed with isolates of *T. viride* and *T. harzianum*. In *T. harzianum* penetration is effected by appressoria produced from short hook-shaped contact branches (Chet, Harman and Baker, 1981; Elad *et al.*, 1983b). *T. harzianum* produces β1–3 glucanases and chitinases when grown in cultures with cell preparations of hosts (Elad, Chet and Henis, 1982) and vesicles which are seen to accumulate at the tips of penetrating hyphal branches are thought to carry these enzymes to assist in the mechanical penetration of host walls. Hosts may respond by producing a sheathing matrix which encapsulates and temporarily restricts hyphal penetration (Elad *et al.*, 1983a,b).

The antagonistic properties of *Gliocladium* and *Trichoderma* have potential for exploitation in biocontrol measures against fungal crop and tree pathogens and wood-decay fungi (Lundborg and Unestam, 1980; Bruce and King, 1983; Bruce, Austin and King, 1984; Papavizas, 1985; Whipps, Lewis and Cooke, 1988).

A number of **fungicolous species**, i.e. those that grow on the fruit-bodies

DEVELOPMENT OF FUNGAL COMMUNITIES   75

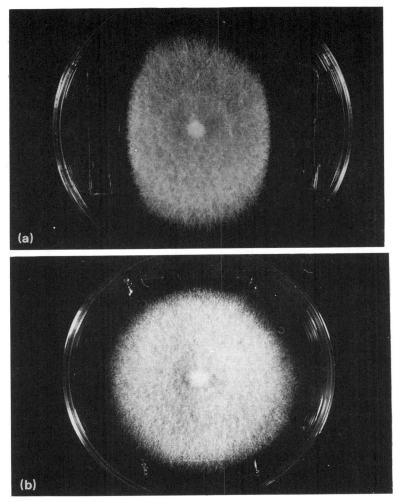

Figure 3.9 Growth of *Stereum hirsutum* 68 h after inoculation on malt extract agar plates. (a) Growth towards and on radii at right angles to wells containing diffusates from a *Hypomyces aurantius* culture. (b) Growth towards and on radii at right angles to wells containing uninoculated control medium. (Reproduced with permission from Kellock and Dix, 1984a.)

of higher fungi, are necrotrophs. *Hypomyces aurantius* (*Cladobotryum varium*) found growing on *Coriolus versicolor* and other members of the Aphyllophorales antagonizes hosts by the production of a powerful fungicidal toxin (Kellock and Dix, 1984a,b). This disrupts the internal organization of host hyphae causing coagulation of protoplasm and

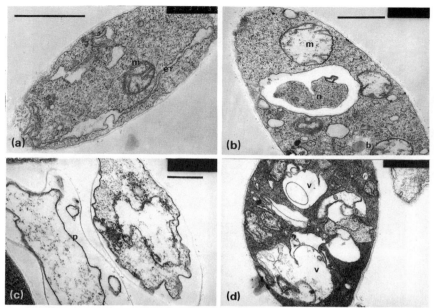

Figure 3.10 Appearance of *Coriolus versicolor* hyphae after 2 h exposure to concentrated culture filtrates containing diffusates from *Hypomyces aurantius*. (a) Swollen mitochondria (m) and cisternae of endoplasmic reticulum (er). (b) Nucleus (n) with enlarged perinuclear space, disintegrating mitochondrion (m) and lipid body (lb). (c) and (d) Later stages showing shrinkage of plasmalemma (p) from hyphal wall and many large vacuoles (v). Scale bar = 1 μm. (Reproduced with permission from Kellock and Dix, 1984a.)

destruction of organelles (Figure 3.9). Initially lipid bodies accumulate, followed by the swelling of mitochondria, cisternae of the endoplasmic reticulum, and enlargement of perinuclear spaces (Figure 3.10a,b). Later, many membrane-bound vacuoles appear, derived from mitochondria that have lost their contents, and the plasmalemma shrinks from the hyphal walls (Figure 3.10c,d). These changes resemble those produced by hyphal interference.

Hyphal interference

Hyphal interference was first described in detail by Ikediugwu and Webster (1970a,b), who observed it with *Coprinus heptemerus*. Since then it has been shown to be a characteristic form of antagonistic behaviour in basidiomycetes. The essential feature of this type of antagonism is that it depends on contact or near contact between hyphae; the antagonistic effect seems incapable of being transmitted over distances greater than a few micrometres. Hyphal growth of the victim slows dramatically, membrane function is impaired leading to changes in permeability and loss of turgor. Swelling and loss of contents from organelles and invagination of the plasmalemma is eventually followed by localized death of segments of the hypha in contact with those of the antagonist (Ikediugwu, 1976). Multiple contact between hyphae gradually causes the death of the whole mycelium. No penetration of the victim occurs and no toxins or destructive enzymes have been isolated. For a more detailed account of the behaviour of *C. heptemerus* see section 8.1.1(c).

c) *In vivo–in vitro* correlation

Strong antagonistic behaviour in culture is not necessarily a reliable guide to the outcome of interspecific interaction in nature. This has important implications for the use of antagonistic fungi as biocontrol agencies against fungal pathogens.

The competitive saprophytic ability of a fungus, *sensu* Garrett (1956), that determines the outcome of interactions with other fungi will depend not only on an ability to translate antagonistic potential into antagonistic action but also upon an ability to resist the antagonisms of other species. The latter is obviously of overriding strategic importance, for a high susceptibility to the antagonistic behaviour of others negates the ability to behave antagonistically oneself. A spore whose germination is completely inhibited by an antibiotic will never grow into a mycelium producing antibiotics. It is, in fact, the genetic potential of the fungus to perform well under antagonistic conditions which will play the major role in determining the outcome of colonization. This has been borne out by several observers. Park (1955) found that the colonization of substrata by soil fungi when introduced into sterile soil (no antagonism) bore no relation to their success in the colonization of the substrata in unsterilized soil. In other experiments, Dwivedi and Garrett (1968) found that the colonization of an agar medium pre-inoculated with soil could not be positively correlated with the colonization of the same medium without soil inoculum by the same fungi. Successful colonization could, however, be correlated with the ability of the fungi to grow in culture filtrates containing the metabolic products of some of the soil microorganisms.

In the context of outcome, the role of antibiosis is particularly controversial due to frequent poor correlation between behaviour in culture and competitiveness in nature. While Bruehl, Miller and Cunfer (1969) found that strains of *Cephalosporium gramineum* producing antibiotics in culture competed and survived better on buried straw than non-antibiotic-producing strains, others have found no strong connection between antibiotic production and competitiveness outside of culture. Webber and Hedger (1986), for example, reported that several saprotrophs from elm bark that successfully antagonized *Ceratocystis ulmi* by antibiotic production in culture were ineffective as competitors in the natural substratum except in those cases where antibiotic production was allied with mycoparasitic behaviour. Similarly, Magan and Lacey (1985), studying the colonization of cereal grain, found that while *Penicillium brevicompactum* and *Epicoccum nigrum* were active competitors through antibiotic production in culture, competitiveness was lost when inoculated with other fungi on to wheat grains.

There may be several reasons for these anomalies. First, while there is ample evidence to show that fungi can produce antibiotics *in vitro* (Brian, 1957), conditions may not always be optimal for production *in vivo*. Rich carbohydrate supplies must be available because synthesis requires heavy expenditure of energy. Antibiotic production also tends to be limited by the fact that antibiotics are secondary metabolites and as such are only produced abundantly on substrata rich in carbohydrates when growth slows because nitrogen or other essential nutrients are limiting (Bu'lock, 1961).

Other ecological problems associated with antibiotic production centre round the practical questions of their probable effectiveness given the right conditions for their production. Can they be produced in sufficient quantity? Are they rendered ineffective by immobilization or inactivated by adsorption to the substratum or colloidal particles in soil?

3.2.5 Effects of animals

Despite the increasing interest in the effect of animals on the growth of fungi, this is an area where much still remains to be discovered.

Of the many animals that come into contact with fungi only a comparative few have any direct effect on the structure of fungal communities. Foremost among these are the detritivorous arthropods which feed on fungal mycelium. Experimental studies on the effects of hyphal grazing by these animals on fungal growth have given somewhat conflicting results, but it is generally accepted that light grazing can, under certain circumstances, markedly stimulate mycelial growth, but that over-grazing reduces growth and leads to an increase in bacterial populations (van der Drift and Jansen, 1977; Hanlon and Anderson, 1979). There is some uncertainty as to the cause of the stimulation of mycelial growth after grazing. It may be due to the beneficial effects of recycling nutrients in faeces and other ejecta. However, Hanlon (1981) found that stimulation only occurred when ample nutrients were still available in the substratum and it may be that the ejecta contain too little that can be utilized for growth after the mycelium has been digested by the animals. On the other hand, the stimulatory effects of grazing could be direct. Growth could be stimulated by physiological mechanisms analogous to those that are switched on when plants are stimulated into growth in order to replace biomass lost to grazing herbivores. It has been speculated that stimulation of growth may also occur because removal of hyphae by grazing stops the production of staling substances by ageing hyphae and in time this may allow accumulated levels to fall.

The composition of fungal communities may be affected by selective grazing, and some examples are known where, when arthropods show a preference for the mycelium of a particular fungus, this seriously impairs the ability of that species to colonize litter in competition with other fungi. Parkinson, Visser and Whittaker (1979) demonstrated that when in competition with a basidiomycete, the capacity of a sterile fungus to colonize leaf discs was considerably reduced by preferential grazing by the collembolan *Onychiurus subtenuis*. Visser (1985) has speculated that some selection by grazing animals may be due to the production of toxic metabolites by some fungi. It is perhaps significant in this connection that *Trichoderma* and *Aspergillus* species, known toxin producers, are consistently avoided in choice experiments by most of the fungivores that have been investigated (for details see Shaw, 1992).

Studies such as these can sometimes be correlated with field observations. Newell (1984a,b) found that *Marasmius androsaceus*, a small toadstool that colonizes conifer needles, was confined to needles in the top litter layers. In the F1 it is replaced by a competitor, *Mycena galopus*. Newell suggested that the reason for this is that the collembolan, *Onychiurus latus*, which grazes *M. androsaceus*, preferentially ranges more abundantly in the F1 (for details of this experiment see section 13.3.1).

Theoretical considerations suggest that one of the effects of random low to intermediate level grazing by animals would be to increase species diversity in the community. More species can exist when the biomass of individuals is kept low by harvesting (Yodzis, 1978). However, work with fungi has not tended to support this hypothesis, mainly because the activities of the grazing animals also have a direct effect on the substratum on which the fungus is growing. Leaf-eating aquatic invertebrates preferentially ingest leaf tissues made palatable by the growth of aquatic hyphomycetes in the tissues, but in doing so they not only eat the fungus but also the resource on which the fungal community depends. The net result is that there is direct competition between the fungi and the animals for resources and under these circumstances fungal species numbers decline and have been found to correlate negatively with the feeding activity of the animals (Bärlocher, 1980). Exceptions may be during the early stages of colonization where wounding of leaf tissues may promote fungal colonization (Bärlocher, Kendrick and Michaelides, 1978). Similar results to the above were obtained by Wicklow and Yocom (1982) when investigating the effect of grazing by the sciarid fly (*Lycoriella mali*) on coprophilous fungal communities on rabbit dung. They discovered that as the numbers of eggs of the fly added to the dung increased, the numbers of species of fungi that could be isolated from the dung decreased. They suggested that in this case disturbance caused by the burrowing activities of the larvae could have been responsible. For further details, see Chapter 8.

Other consequences may follow from the activities of animals. The role of small invertebrates in the dispersal of spores is well known (Visser, 1985). Transport could increase species diversity which in turn might lead to changes in community structure through interspecific competition. By burrowing and feeding in the substratum, small animals bring about changes in the moisture-holding capacity, in aeration and chemical composition by the added nitrogen and other nutrients from exuviae, faeces and other rejecta, changes that will produce new microhabitats for fungal growth. Special cases are also known where arthropods stop fungal growth by producing fungistatic and fungicidal metabolites (see Visser, 1985).

3.3 SUCCESSIONS

Studies on the structure of fungal communities on various substrata carried out over a period of time reveal that in general they tend to change both quantitatively and qualitatively. Species diversity tends to be richest and numbers of individuals usually greatest during the earliest stages of colonization; thereafter, following a period of relative stability, species and total numbers begin to decline. Pioneer communities are typically composed of a large number of different species each occurring in low frequency with no obvious dominant. Mature communities contain fewer species with one or two obvious dominants common to nearly all samples of the same substratum at a similar stage in the development of the fungal community.

Time-related changes in community structure are the so-called **fungal successions**. Some of these trends are illustrated in the data in Table 3.1 taken from an analysis of fungal populations growing on French bean roots (Dix, 1964). As the bean roots aged, the number of fungal species colonizing the roots increased from 15 on the apices to 19 on the older parts of the root. On decaying roots only 6 species were present, 2 of which were dominants and were present on more than 60% of all roots sampled. This contrasts with the apices of young roots, where the majority of species were present in low frequency.

Many of the factors which influence these changes have been identified and the sequence of events involved is now fairly well understood. When dead organisms become available for colonization a struggle immediately arises among potential saprophytic colonists for establishment. This is a very critical time, for it has been shown that once a small number of fungi become established it becomes very difficult for others to gain a foothold (Barton, 1960, 1961; Bruehl and Lai, 1966). In some cases these effects may follow as little as 24 h after exposure to colonization (Dix, 1969). This is known as the **prior colonization effect**, the underlying mechanisms of which are not fully understood, principally because they are not easy to investigate. The speed with which the phenomenon develops suggests that antibiosis is not involved. More likely the rapid occupation of the substratum creates a shortage of unoccupied space on the surface, giving rise to severe localized competition for nutrients and supplies of soluble carbon.

The rapidity with which the struggle for the early occupation of the substratum is settled produces a selection pressure for species with fast germination and rapid growth. It is not surprising to find, therefore, that a feature of pioneer communities is that they contain many 'weed' species associated with a ruderal or 'R' type behaviour embodying these adaptations, enabling them to capture new resources ahead of competitors and to compete efficiently for ephemeral energy supplies.

Pioneer 'weed' species are not necessarily exclusively ruderal. Phylloplane fungi that exist in the harsh physical conditions on the leaf surface, where nutrients are usually also in short supply, and soil fungi that colonize dry raw leaf litter are good examples of some that are also adapted to stressed conditions. Pioneer species can also be highly antagonistic to other 'weed' species.

Weak parasites are usually prominent in pioneer communities. In soil, the pioneer species of *Pythium* and *Fusarium* are capable of invading living roots. These species have the advantage that they have the possibility of colonizing senescing plant tissues ahead of competitors as soon as resistance to infection drops low enough. Leaves, shoots and roots of plants may also harbour harmless or mutualistic non-obligate endophytes that are in place and ready to take over when the plant dies (Chapter 4).

The reasons for the loss of many **pioneer colonizers** as the community matures during successions is a subject which has been studied by many mycologists (Frankland, 1981). The discovery that some fungi produce no enzymes capable of degrading lignin or cellulose led to an ecological classification of fungi based on substrate utilization and to the idea that pioneer colonizers, dependent upon simple organic sources, disappeared from communities when supplies of these became exhausted (Garrett, 1951). The survivors, or later dominants, were thought to be those fungi that could utilize the more enduring cellulose and perhaps lignin of plant substrata. This view became widely accepted and, at one time, the term pioneer colonizer became practically synonymous with the term 'sugar fungus', a fungus incapable of degrading plant cell wall polymers.

The nutritional hypothesis has since, however, been found to be an unsatisfactory explanation of fungal successions because it has not always been possible to correlate the observed behaviour of fungi in successions with their nutritional physiology or with any dramatic change in levels of particular substrates in the resource. Not all pioneer colonizers are, in fact 'sugar fungi' in the strict sense and many are now recognized as being able to hydrolyse hemicellulose and even, in some cases, cellulose. Among soil fungi, prominent pioneer colonizers such as *Mucor*, *Rhizopus* and *Mortierella* species have been shown to hydrolyse xylan (Flanagan, 1981). The genus *Pythium* contains several cellulolytic species (Deacon, 1979), while the majority of *Fusarium* species are highly active in the hydrolysis of cellulose and carboxymethylcellulose (Venkata-Ram, 1956; Domsch and Gams, 1969; Flanagan, 1981). The pioneer leaf colonizing saprophyte *Cladosporium herbarum* hydrolyses xylan (Flanagan, 1981) and some isolates show moderate cellulolysis (Siu, 1951). The disappearance of fungi such as these from resources when plant cell wall polysaccharides are still available cannot, therefore, be explained by the disappearance of sugars. Furthermore, some 'sugar fungi' can enter into commensal relationships with cellulolytic fungi and persist in successions long after any sugars that were present have disappeared (for examples refer to section 3.2.1). The disappearance of many pioneer colonizers from successions clearly requires another explanation.

The development of antagonistic phenomena seems to be a logical alternative. The accumulation of staling and antibiotic toxins in the substratum may stop growth. As a result, hyphae may decline and then undergo autolysis or become parasitized by the hyphae of other species. Pioneer species that are predominantly ruderal have a low tolerance of antagonism; the survivors and later dominants on the other hand are typically more tolerant of antagonisms and are themselves the more active producers of metabolic inhibitors (Park, 1963). These later dominants are also notably very active as mycoparasites and in hyphal interference and related phenomena, destroying the hyphae of other species. They are characteristically species that have evolved a competitive or 'C' type strategy for survival. With these attributes they are able to oust non-antagonistic pioneer species and defend captured resources and space successfully from most competitors. Their superior enzyme equipment enables them to survive successfully on substrata with declining resources, utilizing cellulose and lignin as prime energy sources. By the conservation of resources and successful scavenging they can grow on substrata where the concentration of nitrogen and other minerals would be limiting for others.

Good examples of fungi adopting this lifestyle are *Coprinus* species on dung and composts (Harper and Webster, 1964; Ikediugwu and Webster, 1970b; Hedger and Hudson, 1974) and *Phanerochaete velutina* and *Phlebia merismoides* on wood (Rayner, 1977b, 1978).

A vexed question which has occupied the attention of mycologists for several decades is the subject of the order of establishment of species in successions. That is to say, whether the sequence of events observed in successions represents differences in the timing of the arrival of species or whether these events merely reflect changes in the development of the community as slower growing fungi become more dominant. In the past, the fungi of pioneer communities have been referred to as **primary colonizers**, those of the mature community as **secondary colonizers**, implying that the changes observed during the development of communities are due to successive waves of invasion by fungi.

Many difficulties in interpreting successions have been caused by the reliance in the past on the appearance of sporulating structures as evidence of the presence of a species. We now understand, however, that the appearance of sporulating structures is an event which may bear little relationship in time to the appearance or disappearance of the mycelium. Actively growing mycelia may be present but never sporulate or sporulation may be delayed, sometimes for a long while. Clearly descriptions of successional changes can only be accepted as accurate if based on the presence or absence of actively growing mycelia. The succession on dung is a clear example where reliance on the timing of the appearance of sporulating structures once gave a misleading impression (section 8.1.1). Here it has now been shown that the sequence of sporulation of individual species is a direct result of the speed at which they are capable of forming their fruit-bodies and does not reflect a sequence in colonization (Harper and Webster, 1964). Furthermore, the appearance of fruit-bodies on dung has been shown to be strongly influenced by water availability. At reduced humidities basidiomycetes and many ascomycetes that are present do not fruit and their fruit-bodies are replaced by those of species not normally associated with fruiting successions on dung (Kuthubutheen and Webster, 1986a).

Studies of successions on wood by Rayner (1977b) have provided similar evidence for the early establishment of species which later come to dominate the mature community. On the other hand, older views must prevail for substrata such as leaves, which are clearly subjected to successive waves of invasion as they pass from the tree to the litter.

The tendency for the climax of successions to become dominated by one or two highly antagonistic species may also result in changes in the rate of decomposition as the succession develops. When species diversity is richest at the beginning of the succession this is likely to correspond with the highest rates of decomposition, since the greater the genetic diversity, the greater the enzyme diversity. There is also a greater likelihood that mutualistic associations and synergistic interactions can develop that can promote decay (section 3.2.1).

Eventually, the rate of decomposition at the climax of the succession will become that of the most vigorous competitor. These often have slow growth rates demanding lower metabolic activity and, if this is the case, the rate of decomposition will slow down.

The course of successions is undoubtedly influenced by a variety of other factors. Selective grazing by invertebrates may deflect successions in favour of certain fungi. The activities of such animals can also change the chemical and physical nature of the substratum, so favouring the development of certain species (section 3.2.5). The colonization of plant tissues by some fungal species at a certain time may be linked to enablement phenomena, such as decreases in the levels of plant-formed antifungal inhibitors, perhaps brought about by the metabolic activities of some of the earlier fungal colonizers or to the establishment of commensal relationships (section 3.2.1). Nitrogen availability is also important. Time-course studies of the colonization of agar plates containing cellulose as a carbon source and different quantities of nitrogen by fungi from soil and litter showed distinct differences in the pattern of fungal development (Park, 1976a). High nitrogen levels in the plates favoured the early development of populations of *Penicillium* species, *Monodictys levis*, *Mortierella vinacea*, *Humicola* species, *Cylindrocarpon magnusianum* and *Chrysosporium pannorum*. After a short while *Humicola* species and *C. magnusianum* declined, *Penicillium* species and *M. vinacea* became dominant, and later *Trichoderma harzianum* and *Scytalidium lignicola* appeared. At low nitrogen levels, *Humicola* species and *C. magnusianum* persisted longer. *Penicillium* species and *M. vinacea* appeared later and some different species, such as *Pyrenochaeta acicola*, were recorded.

The importance of water availability has already been mentioned and there is a possibility that some successional changes could be directly related to improvements in water availability which arise in the substratum as decay proceeds. The partial decay of leaf litter by pioneer colonizing microfungi may, for example, facilitate the subsequent invasion by basidiomycetes which require higher water potentials for growth. Such changes may also regulate basidiomycete successions on wood. In bulky substrata improved aeration, due to the production and enlargement of voids, as decay proceeds is probably another important change that facilitates colonization.

4
COLONIZATION AND DECOMPOSITION OF LEAVES

4.1 DEPOSITION

The decay of the aerial parts of plants, in which fungi play an important role, is completed in the soil but begins on the senescing organs before they are shed.

Throughout the life of plants, fungal spores are being continuously deposited on their surfaces from the atmosphere. The number of spores present in the atmosphere at any time is subject to seasonal weather patterns and follows a cycle which is mainly dictated by temperature and rainfall. Fungal spores build up in the atmosphere in spring as overwintering mycelia begin to sporulate, and reach their highest levels during the late summer and autumn. Figure 4.1 shows the build-up of *Alternaria* spores to a summer maximum near Cardiff, UK (Hyde and Williams, 1946). Records such as these may be obtained by exposing to the wind microscope slides coated with glycerine or Petri dishes containing an agar medium, although methods employing instruments which draw in air for sampling are now more often employed. In the **Hirst spore trap** (Hirst, 1952), air is drawn in by a suction pump and any spores present are impacted onto greased slides, enabling them to be counted and identified. These instruments, which operate at about 90% efficiency, can give more reliable and informative results than other methods, although identification is much more difficult than by the Petri dish method.

In conjunction with time control mechanisms, instruments can be used to monitor the numbers and types of spore present in the atmosphere over short periods and the results related to changing atmospheric conditions. On warm days in summer, numbers tend to reach a maximum in the early afternoon but different species reach their maxima at different times.

86 COLONIZATION AND DECOMPOSITION OF LEAVES

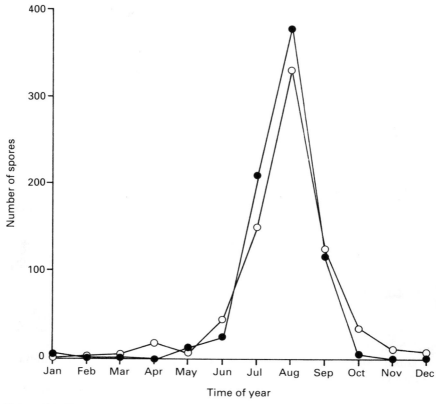

Figure 4.1 Numbers of *Alternaria* spores caught each month on daily exposed glycerine-coated slides. ●–● = 1943; ○–○ = 1942. (Drawn from the data of Hyde and Williams, 1946.)

Ascospores and basidiospores, both of which require moisture for discharge, generally reach a maximum in early morning when dew forms; on the other hand, dry spore forms reach their maxima in the afternoon. If rain falls, spore numbers quickly become reduced due to wash-out.

Fungal propagules are deposited on plants mainly by **wind impaction**, although **sedimentation** and **rain wash-out** from the atmosphere and **splash dispersal** are undoubtedly also important (Chapter 1).

Spores being carried by the wind may be deflected by turbulence created by objects in their path, and whether spores are deposited or not depends very much on the amount of turbulence, the position of the spore relative to the object in its path and the speed at which the spore is travelling. Spores which are travelling fast and in a direct line with the object in their path may not be swept aside by turbulence but continue on their way, penetrating the boundary layer surrounding the object. The greater the wind speed the faster the momentum of the spore and therefore the greater the chances of deposition. On the other hand, the chances of deposition are reduced by large irregular objects which create more turbulence. Wind tunnel experiments have shown that the most efficient impaction of spores occurs when large spores are travelling in high wind speeds towards small cylinders. At any given wind speed large spores have a greater momentum because of the more favourable surface area-to-mass ratio.

Generally, spores are less efficiently impacted on leaf laminae than on petioles and stems because of the greater turbulence which surrounds the former due to their shape and orientation in space. Nevertheless, a great many spores do manage to find their way on to leaf surfaces.

The various sorts of hairs present on the surfaces of some leaves may protect leaves from fungal colonization by holding any spores which get on to the leaf surfaces up in less favourable environmental conditions, preventing germination and increasing the chances of 'blow off' by turbulent currents which from time to time dip down into the boundary layer. Hydrophobic hairs may trap spores in water films where, if they germinate, they may perish before they can reach the leaf, if the water films dry out. Paradoxically, these structures may sometimes actually increase the chances of spore deposition by increasing the boundary layer thickness over the leaf surface which, once penetrated, allows spores to fall onto the leaf by gravity (Forster, 1977).

Further protection is provided by the waxy surfaces of many leaves, making them highly hydrophobic. Wettable spores are particularly susceptible to run-off when waxy leaves are wetted. However, spores which are sticky may not be washed off.

A high impaction efficiency of spores is not compatible with wide dispersal and may be a biological disadvantage. Gregory (1952) speculated that spores of about 10 μm, which is a common spore size, represents a compromise between efficient dispersal and deposition. Fungal spores of this size would be near to the size limits for impaction on grass in meadows where the fastest wind speed is not likely to often exceed 6.5 km h^{-1} close to the ground. The spores of *Agaricus campestris* (7–8 x 4–5 μm) and many common wood-rotting members of the Basidiomycotina, e.g. *Ganoderma adspersum* (10–11 x 6–7 μm), *Piptoporus betulinus* (5–6 x 1–2 μm) and *Polyporus squamosus* (10–15 x 4–5 μm) are close to this size. The deposition of the spores of wood-rotting species on the trunks and branches of trees is possible because of the greater wind speeds that occur above ground level. Fungi with smaller spores, e.g. *Penicillium* and *Mucor* species, would have difficulty in impacting on even the smallest objects in the highest wind speeds and appear primarily to be adapted to wide dispersal and deposition by other methods. This is in keeping with the non-specific resource requirements of many of these fungi.

For further general reading on the behaviour of spores in the atmosphere and deposition on leaves see Gregory (1973). For details of deposition, retention and germination in relation to leaf morphology see Allen *et al.* (1991) and Juniper (1991).

4.2 COLONIZATION OF LIVING LEAVES

4.2.1 Techniques for study

Eventually, some of the spores that land on the plant surface begin to grow in the unique habitat created by the plant and its interaction with the environment. The special nature of aerial plant surfaces as habitats for fungal growth was first recognized by Last (1955), who more or less simultaneously with Ruinen (1956) used the word **phyllosphere** to describe it. Later, a more precise word, the **phylloplane**, meaning the actual leaf surface became more commonly used.

The colonization of the aerial organs of plants has been investigated using a variety of techniques (Dickinson, 1971). A **cultural method** which is often employed involves the washing of samples of plant tissue in several changes of sterile water to remove extraneous spores and then counting and identifying the fungi which arise from the tissue after plating out in a suitable nutrient agar. Maceration of the tissues prior to plating out improves the isolation of the fungi. A useful adjunct to these techniques is to plate out the washing water containing the washed-off propagules, which gives an idea of the numbers and species of fungi being deposited from the atmosphere. Cultural methods are, in most respects, less satisfactory than direct observational techniques. The major disadvantage is that the former do not allow a very accurate distinction to be made between fungi which were actually growing on the plant at the time of plating out and those which were present only as dormant propagules but which subsequently germinated in the agar medium. Nevertheless, the assumption is often made that what has not been washed off tissues using washing techniques must be growing on or in them. When combined with a suitable surface sterilization technique, such as a quick immersion for about 1 min in 0.1% $HgCl_2$ followed by a wash in sterile water, cultural techniques are useful in following the internal colonization of the tissues.

Among **direct observational techniques**, the incubation of tissues in a damp atmosphere for a few days and then examining for sporulating structures has proved valuable for those fungi which sporulate readily. It is important that the incubation time is kept short to ensure that only those species which are actually growing at the time of collection have time to sporulate.

Thin organs like leaves may be mounted and studied directly under the microscope, if they are first cleared using chlorine gas or chloral hydrate solution (Shipton and Brown, 1962). Latterly this method has been replaced by leaf surface impressions. Nail varnish, Bexacryl or similar materials are coated over the surface of the leaf and, when dry, the hardened film is stripped off, bringing with it many propagules from the leaf surface. These films may then be stained, and examined microscopically. A more rapid technique, which is even less troublesome in preparation, involves the use of double-sided tape (Langvad, 1980). More recently, the scanning electron microscope has proved useful for more detailed *in situ* studies of the distribution of propagules on plant surfaces.

These direct observational methods have the distinct advantage that a fairly accurate picture of the distribution and condition of the fungal propagules on the surface of the plant may be obtained. Their one obvious disadvantage is that the identification of the fungi must be made from the propagules alone. However, many are quickly identifiable, to the generic level at least, by the experienced mycologist. Another disadvantage is that direct observational methods cannot separate spores which are merely dormant from those which are dead. It is becoming clear, moreover, from the disparity which can arise between the number of spores on a leaf surface calculated by the culture of plated out washing water and the number arrived at from the examination of leaf surface impressions, that many spores on plant surfaces are, in fact, dead. One thing at least is obvious: for comprehensive studies of the colonization of plant tissues by fungi a variety of different investigative techniques must be employed.

4.2.2 The development of the fungal communities of attached leaves

In spite of the technical difficulties outlined above, a remarkably detailed picture of the growth of fungi on leaves and the aerial organs of living plants has been built up from the observations of many workers. From these studies Hudson (1968) constructed some readily recognizable general patterns of behaviour.

The earliest invaders of living leaves are **symbiotic endophytes**, some of which are host-specific. Many are parasites that cause disease, but symptomless infections of leaf tissues by fungi are also very common. These have been described as **symptomless parasites** or better **neutral symbionts**.

Neutral symbionts have now been isolated from over 300 species of phanerogam, having been most extensively studied in ericas, conifers and grasses (Petrini, 1986). Samples of leaves from five evergreen shrubs revealed that as many as 75% of the leaves of some were infected, with many leaves containing more than one fungal species (Petrini, Stone and Carroll, 1982).

A great variety of endophytic fungal taxa have been listed in more than 30 different genera. The commonest in woody plants include *Phyllosticta, Phomopsis, Leptostroma, Phoma, Cryptosporiopsis* and the anamorphs of

Xylaria and *Hypoxylon* species (Carroll, Müller and Sutton, 1977; Petrini Müller and Luginbühl, 1979; Petrini, 1984). Numbers of isolates tend to be large but only a few taxa are present in significant numbers and these represent the characteristic endophytic mycoflora of the plant species irrespective of plant distribution. For more details see Petrini (1992) and Chapter 6. In grasses *Balansia, Epichoe* (*Acremonium*) and *Atkinsonella* in the tribe Balansiae of the Clavicipitaceae are the commonly isolated genera (Clay, 1988).

A remarkable feature of some of the endophytes of grasses and of fir and spruce needles (*Rhabdocline parkeri* and *Leptostroma* species) is that they produce alkaloids and other metabolites toxic to herbivorous insect larvae. These substances kill larvae or slow their rate of development to such an extent that fewer become feeding adults, thus limiting damage to the plant (Clay, Hardy and Hammond, 1985a,b; Miller, 1986; Clark, Miller and Whitney, 1989).

The presence of endophytes in grasses has been shown to increase plant biomass, improve resistance to drought and increase competitiveness. The defensive role of endophytes may even extend to the antagonism of microbial pathogens. An apparent example is the exclusion of *Lophodermium seditiosum* from *Pinus sylvestris* needles by *Lophodermium conigenum* (Minter, 1981). Other possible examples are quoted by Clay (1991). For these reasons some plant endophyte associations are considered to be mutualistic. However, a defensive role for endophytic associations applies to only a few of the many hundreds of examples and this has led to speculation that other types of mutualism may be involved. Carroll (1991a) discusses the possibilities for nutrient recycling and the removal of competitors where fungal endophytes of the host are potential pathogens of other plants.

For a fuller account of the effects of endophytes on host physiology and details of the production and action of toxic metabolites see Clay (1991, 1992) and Siegel and Schardl (1992). Unfortunately, space does not allow discussion of the possibilities for the deployment of endophytes in biological control except to mention that to date the artificial infection of plants with potentially protective endophytes has only achieved limited success (Leuchtmann and Clay, 1988).

Other early colonizers are certain **saprophytic** or **weakly parasitic fungi** which are mainly restricted to the phylloplane (or leaf surface) until the leaf becomes senescent. One of the remarkable features of this group is that although the spores of many fungal species abound in the atmosphere only a selected few colonize the phylloplane. These species are very widely distributed, occurring on a wide range of plants under different climatic conditions; the only major distinction being between those which are found on the leaves of angiosperms and those which occur on the leaves of gymnosperms. As soon as the leaf buds unfold the first saprophytic fungi to

COLONIZATION OF LIVING LEAVES 91

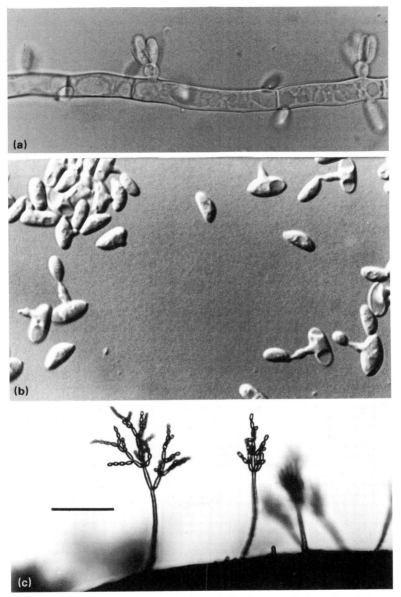

Figure 4.2 Some common fungi of the phylloplane and decaying leaves. (a) *Aureobasidium pullulans*: mycelium-producing blastoconidia. (b) *Sporobolomyces roseus*: yeast-like cells, some with attached ballistospores. (c) *Cladosporium cladosporioides*: conidiophores with chains of conidia. Scale bar = 20 μm (a,b); 80 μm (c).

appear on the young leaves of angiosperms are certain yeasts. These rapidly build up to large numbers persisting as dominants in the community throughout the summer and autumn. Yeasts also colonize conifer needles but their occurrence here is not so well documented. The red yeasts *Sporobolomyces* and *Rhodotorula*, and *Cryptococcus* and other white yeasts are the common genera (Figure 4.2)

Sharing the very early stages of the succession with the yeasts on angiosperms is *Aureobasidium pullulans*, a hyphal fungus which becomes very active on leaf surfaces during the early summer and autumn (Figure 4.2). Its counterpart on conifers is *Dothichiza pythiophila*. This is soon followed by other hyphal forms, whose spores become deposited in increasing numbers on leaf surfaces as their spores concentrate in the atmosphere from June onwards. On angiosperm leaves the commonest of these are *Cladosporium* species, mainly *Cladosporium herbarum* and *C. cladosporioides* (Figure 4.2). Also occurring fairly frequently are *Botrytis cinerea*, *Epicoccum purpurascens* (*E. nigrum*) and several *Alternaria* and *Stemphylium* species. On *Pinus* species the corresponding fungi are *Lophodermium pinastri*, *Ceuthospora pinastri* and *Naemacyclus niveus*.

In the early summer the germination of most of these species is poor on leaf surfaces and mycelial development is rather spasmodic until conditions improve in late summer and early autumn. At this time improved climatic conditions coincide with an increase in nutrients supplied from leaking senescing leaf cells.

Those phylloplane species capable of some growth under the more fluctuating nutritional and climatic conditions prevailing earlier in the year appear to be ecologically separate from those that require more favourable conditions in order to grow on leaves. Thus some authors have separated the earliest colonizers, the true phylloplane inhabitants (yeasts, *Aureobasidium pullulans* and *Cladosporium* species) from later colonists, the phylloplane invaders or primary saprophytes (*Botrytis* and *Alternaria* species, etc.)

Although *Alternaria alternata* (*A. tenuis*), *Aureobasidium pullulans*, *Cladosporium* species and *Epicoccum purpurascens* (*E. nigrum*) are essentially epiphytic while the leaf is alive, they can become endophytic (Pugh and Buckley, 1971a; Carroll and Carroll, 1978; Petrini, Müller and Luginbühl, 1979) and under certain circumstances it is possible that some may even behave as weak parasites (Dickinson, 1981). The main route of entry is via stomata (O'Donnell and Dickinson, 1980) and attempts to penetrate via the epidermis are usually easily repulsed by plant defence mechanisms. However, if energy has to be diverted to this purpose this will affect plant growth and it is significant that clearing leaves of saprophytic epiphytes with fungicides has been shown to improve grain yields in cereals (Smedegaard-Petersen and Tolstrup, 1986). Increased yields in these treatments has also been associated with delayed flag leaf senescence after the removal of saprophytes (Dickinson and Wallace, 1976).

The colonization of pea (*Pisum sativum*) leaves by fungi, which has been described by Dickinson (1967), is generally typical of the leaves of herbaceous and woody angiosperms in temperate regions. Using cultural and direct observation techniques Dickinson regularly sampled leaves throughout the season. Fungal propagules were detected on the first sampling date (28 April) but in the early stages of the succession, which were dominated by yeasts and a parasitic hyphal species (*Ascochyta pinodes*), the numbers of fungal propagules remained low. On 12 May, for example, the number of propagules present in leaf impressions was 170 cm^{-2} for yeast cells, and 413 cm^{-2} for *A. pinodes* spores. No other species were detected. During July, August and September the number of fungal propagules increased dramatically and by 4 August *Cladosporium* species had become the dominant hyphal forms. By 29 September 170 153 yeast cells and 3274 *Cladosporium* spores cm^{-2} of leaf were recorded. Germinating spores of saprophytic hyphal forms were first observed in June but these did not reach a significant proportion of the actively growing community until September.

Numerous studies have been made of phylloplane populations but detailed analyses of the dynamics of phylloplane community development are surprisingly sparse. Studies on Norway maple (*Acer platanoides*) made by Breeze and Dix (1981) show that during June, July and August the phylloplane community tends to be rather unstable; losses of large numbers of germinated spores occur from time to time resulting in considerable fluctuations in the mycelial biomass. During this early period leaves seem to be subjected to several overlapping waves of invasion by spores which make a series of transitory attempts to establish stable, actively growing fungal populations.

From September to November the numbers of spores deposited on Norway maple leaves increased several-fold over the summer values and this was accompanied by large increases in the mycelial biomass as more spores germinated, growth increased and fewer germinated spores were lost. As a result a more stable community gradually became established which maintained the actively growing biomass at high levels throughout the autumn.

On leaves which remain attached to the plant and physiologically active for several years, such as holly (*Ilex aquifolium*), these events become cyclic with phylloplane populations declining in winter and spring and increasing again in summer and autumn. As such leaves age there is no tendency for phylloplane populations to build up progressively as the years pass, although there is some tendency for older leaves to support a larger and more varied autumn fungal community than younger leaves on the same shoot (Mishra and Dickinson, 1981).

The phylloplane population studies of Breeze and Dix (1981) reveal a number of interesting relationships which are probably a common feature of phylloplane communities on angiosperms. While *Cladosporium* species were the most numerous of the hyphal fungi present, other genera were found to contribute similar or larger amounts to the total mycelial biomass of the phylloplane of Norway maple on several sample dates. This applied particularly to *Epicoccum* and *Stemphylium* species on adaxial leaf surfaces during August and September. In addition, it was found that on all sample dates the total biomass of the yeast-like cells present exceeded the mycelial biomass of the common genera *Cladosporium*, *Epicoccum* and *Alternaria* combined, and during June and July the ratio of the biomass of the yeast-like cells to the combined mycelial biomass was often more than 50 to 1.

Present information suggests that the way in which fungal communities develop on leaves is in response to several seasonal events and can best be explained as follows. Most spores have the potential to germinate on most living green leaves but in early summer germination is generally depressed by dry atmospheric conditions during the day and nutritional shortages. Phylloplane fungi can make only limited use of substrates, such as cutin, in the cuticle (Dickinson and Macnamara, 1983) and the competition for external supplies of suitable respiratory substrates is high.

Some spores, such as those of *Alternaria alternata*, may make limited growth during favourable climatic periods during the summer using endogenous food reserves and the contents of older hyphae which are autolysed and recycled to the actively growing hyphal tips. Alternatively, spores may germinate to produce secondary spores or undergo microcyclic conidiation. This type of response is commoner among those fungi with smaller types of spores such as *Cladosporium cladosporioides*, and may have some survival value for these species (Oso, 1972; Skidmore, 1976; Dickinson and Bottomley, 1980). If antifungal substances from leaves are present these will also restrict growth. In time these populations die, or form resting structures, as endogenous nutrients become exhausted or unfavourable climatic conditions return. Dead spores and mycelium are eventually washed off by rain to be replaced by other spore populations which may repeat the cycle. If a large amount of pollen accumulates on leaves, as can happen for example in close standing plants in cereal crops, this can act as an exogenous source of nutrients and sustain germ tube growth.

In late summer and autumn favourable climatic conditions become more prevalent and many leaf cells become progressively 'leaky' and leaves eventually senesce, releasing substrates on to the leaf surface. The production of antifungal substances by leaves will also decline. These factors combine to produce more favourable conditions for germination and growth and lead to the establishment of a more stable and actively growing fungal community on the leaf surface.

Some of the factors influencing the colonization of leaf surfaces by fungi are discussed in further detail in section 4.3.

If leaves have been invaded by fungal parasites this may also bring about significant quantitative changes in the fungal populations which can be supported on the leaf surface. Increases in the numbers of *Sporobolomyces* species and other yeasts on leaves infected by mildews and rusts have been reported by Brady (1960), Pady (1973, 1974), Hayes (1982) and Last (1970). On *Antirrhinum* leaves increases in yeast populations coincided with the rupture of the leaf epidermis by rust uredosori (Collins, 1982) and are presumably due to the release of nutrients into the phylloplane from damaged tissues and the leakage of metabolites from rust hyphae (Shaw, 1963).

Some epiphytes are more prevalent on infected material; others on non-infected. For example, *Cladosporium* species and *Alternaria alternata* were recorded more frequently on the rust-infected leaves of wheat (*Triticum aestivum*), plum (*Prunus domestica*) and poplar (*Populus generosa*), whereas *Botrytis cinerea* and *Phomopsis perniciosa* occurred more commonly on uninfected leaves (McKenzie and Hudson, 1976). The reasons for these differences are probably quite complex but Hudson (1978) suggested, supported by some experimental evidence, that species more commonly recorded on infected leaves may increase due to their ability to utilize certain fungal metabolites, such as trehalose and erythritol synthesized by the rust from the host photosynthates. Fungal species that do not increase on infected leaves may lack this ability and so become less competitive. Quantitative differences in the fungal populations on non-infected plants and plants infected by fungal parasites may also be related to the mycoparasitic activity of certain fungi. Thus McKenzie and Hudson (1976) detected increases in the populations of *Gonatobotrys simplex* on poplar and plum, *Ramularia* sp. on plum and *Verticillium lecanii* on wheat associated with the infection of the plants by rusts, which in turn act as hosts for these mycoparasites. It is interesting to note in this connection that Omar and Heather (1979) have shown that *Alternaria* and *Cladosporium* species can parasitize some rusts, including the poplar rust, *Melampsora larici-populina*.

Later in the season on more enduring substrata, such as the leaves of woody plants, a **secondary saprophytic phase** may develop in which fungi of the Deuteromycotina and Ascomycotina are again of major importance. The distribution of these fungi tends to be more restricted than that of the primary colonizers although certain patterns involving the same secondary colonizing species are discernible on similar substrata. Often this phase does not reach full development until the senescent plant remains reach the litter layer of the soil where the secondary colonizers complete their development and continue to flourish for some time (Hudson, 1968) (section 4.4).

The most detailed studies of fungal successions on gymnosperm leaves have been carried out on *Pinus* species. Other studies have been made of *Larix europaea* (McBride and Hayes, 1977) and *Pseudotsuga menziesii* (Sherwood and Carroll, 1974).

The fungal populations of conifer needles tend to be dominated by parasites which are characteristic for each tree species, while few of the numerically important saprophytes are shared with angiosperms. Young needles of *Pinus* species may first become infected by *Lophodermella sulcigena* and *L. conjuncta*, which cause premature die back and casting of the needles. Needles infected by these parasites generally also become colonized by *Lophodermium pinastri* (imperfect state *Leptostroma pinastri*) which invades moribund tissues. *L. pinastri* also colonizes many of the uninfected needles at senescence so that at the end of the third year of life, when the needles are naturally shed, most have become colonized by this fungus.

Aureobasidium pullulans may also be present in low numbers on conifer needles at a very early stage in the succession but on pine needles, *Sclerophoma pythiophila*, whose anamorph *Dothichiza pythiophila* resembles *Aureobasidium pullulans*, and has often been confused with it, is numerically more important.

Ceuthospora pinastri is another early colonizer of pine needles. The low number of records for this fungus in the past may be accounted for by the fact that it seems to have been frequently misidentified as *Fusicoccum bacillare*.

Early studies of the colonization of pine needles by fungi are notable for the lack of records for yeasts. More recently, however, yeasts have been found in the phylloplane of larch (*Larix europaea*) (McBride and Hayes, 1977) and Corsican pine (*Pinus maritima*) (Mitchell and Millar, 1978). Populations of white, pink and yellow yeasts were all recorded abundantly, with the pink yeast, *Sporobolomyces roseus*, being the most numerous. These results suggest that yeasts are numerically important on all conifers but that previously their presence has been overlooked, possibly due to the use of inappropriate isolation techniques.

Cladosporium herbarum has been isolated from conifer needles but it appears to be relatively rare except on European larch (McBride and Hayes, 1977). On pine needles it has been found most frequently growing on the moribund asocarps of *Lophodermium pinastri* and *Lophodermella sulcigena* (Lehmann and Hudson, 1977; Mitchell and Millar, 1978).

As in the case of angiosperm leaves the pattern of succession on conifer needles may be modified to some extent when needles are colonized by parasitic fungi. On Corsican pine needles Mitchell and Millar (1978) noted several important changes occurring in the behaviour of saprophytes under these circumstances. *Ceuthospora pinastri* occurred more frequently on needles invaded by *Lophodermella sulcigena*, and *Hendersonia acicola*, which was otherwise more commonly associated with needles in the litter, was found in higher numbers associated with lesions on attached needles produced by *Lophodermella conjuncta* and *L. sulcigena*.

Phylloplane populations are frequently affected by pollutants, in particular by lead and sulphur dioxide emissions, but probably also to some extent by ozone and the oxides of nitrogen, both directly and as contributors to acidic precipitation. A significant body of information has been accumulated on this subject (see Magan and McLeod, 1991; Magan, 1993) but some of it is somewhat difficult to interpret due in part to the different methods of investigation that have been used but also to weaknesses in experimental design, particularly with regard to control of pH. As a general statement it appears safe to conclude that exposure to elevated concentrations of sulphur dioxide reduces total fungal biomass. The growth of both white and red yeasts is suppressed and the germination of the spores of some hyaline filamentous species is reduced. By contrast, some pigmented species are more tolerant and this can lead indirectly to increases in their populations. There is some evidence also that some species may be stimulated at low concentrations, e.g. *Alternaria* and *Cladosporium* species.

Lead pollution allows *Aureobasidium pullulans* to become more dominant and greatly reduces the numbers of *Sporobolomyces roseus* cells and bacteria (Mowll and Gadd, 1985).

Reports on the effects of ozone are more conflicting, partly due to flaws in the design of some of the experiments. It seems to have little direct effect on fungal growth but it may have an indirect effect through its effect on the plant. To date there have been no studies on the effects of the oxides of nitrogen on phylloplane communities.

The action of pollutants in disturbing the balance between fungal populations on leaf surfaces may affect interactions between leaf pathogens and phylloplane saprophytes and have consequences for the part believed to be played by saprophytes in the natural regulation of plant disease (section 4.3.3). This disturbance may even affect the subsequent rates of decay of leaf litter (Prescott and Parkinson, 1985) (section 4.5).

Fungi also colonize the buds of plants, which bear a similar mycoflora to leaves. The surfaces of the outer bud scales of all angiosperms that have been examined have been found to be extensively colonized by phylloplane fungi, especially by yeasts and *Aureobasidium pullulans*. *Cladosporium herbarum*, *Epicoccum purpurascens* and *A. pullulans* have also been isolated from the inner bud scales of some trees, and yeasts have even been isolated from the shoot primordia of ash (*Fraxinus excelsior*) and lime (*Tilia platyphylla*) (Hislop and Cox, 1969; Pugh and Buckley, 1971a; Warren, 1976). *Lophodermium pinastri* colonizes the surfaces of the bud scales of pines and other conifers and *Sclerophoma pythiophila* has been isolated from the developing strobili of *Pinus sylvestris* (Whittle, 1977).

These persistent bud populations indicate that there must be some continuity in the fungal populations on the leaves, for some fungi at least, from one season to the next. This behaviour could underlie the predominance of yeasts and *A. pullulans* as early colonizers of young leaves. In fact, it appears to be essential for *A. pullulans* to be able to exploit leaf surfaces as soon as the young leaves unfold, for if it comes into contact with the hyphae of *C. herbarum* and other phylloplane species in culture, it is antagonized, stops growing and forms microsclerotia.

The active growth phase of *A. pullulans* on leaf surfaces may therefore be relatively short and its continuous isolation from leaves throughout the season may owe more to the persistence of microsclerotia and endophytic populations than to the existence of a large continuously growing population on the leaf surface.

The flowers of angiosperms also eventually become available for fungal colonization but despite their diversity and their short life fungal colonization is not very unlike that already described for leaves. However, like many fruits, they may develop a characteristic yeast flora. Among hyphal species *Botrytis cinerea*, most races of which are non-specific parasites, tends to dominate.

Although a great deal of information has now accumulated on the subject of the colonization of leaves by fungi, most of this information is concerned with the identification of species and successions on various plants and very little with the details of distribution either on whole plants or individual leaves.

Such facts as are known of the distribution on the whole plant relate mainly to trees, where fungal populations tend to be more numerous on leaves which are lower and more peripheral in the canopy (Carroll, 1979; Andrews, Kenerley and Nordheim, 1980).

On individual leaves, fungal colonization may follow slightly different patterns on leaves where one leaf surface is positioned above the other. With this orientation spores are usually deposited in greater numbers on the upper (**adaxial**) leaf surface, although in some plants, growth is more extensive and the proportion of the spore populations which germinate greater on the lower (**abaxial**) surfaces (Ruscoe, 1971; Breeze and Dix, 1981). Better growth on the abaxial surface is probably due to the more sheltered conditions which prevail there and the effects of reduced competition. Other factors which may also be important are the retention of substrates for growth and the higher numbers of stomata present on abaxial surfaces, which when open help to keep the relative humidity high.

In some plants differences in morphology influence distribution on the two leaf surfaces. Foster (1977) found that more spores were deposited on abaxial than adaxial surfaces of Sitka spruce (*Picea sitchensis*) and sycamore (*Acer pseudoplatanus*) because of the presence of prominent, spore-trapping, wax structures on abaxial surfaces.

On the smaller scale a notable feature of some leaves, for example *Acer pseudoplatamus*, is that spores, resting structures and yeast colonies tend to be distributed along veins (Pugh and Buckley, 1971b). This is probably because veins are channels into which spores become washed and find more favourable conditions for growth. Veins may also be regions of higher levels of leaf exudation or exudates may accumulate there; veins may also provide more shelter in a fluctuating physical environment. In the sand dune plant *Hippophaë rhamnoides* fungi become distributed in a deep midrib groove overhung by hairs (Lindsey and Pugh, 1976).

Leaf surfaces are uniquely complex and bear, distributed in specific patterns, hairs, glands and wax extrusions from the cuticle. They are traversed by the fluid-conducting veins of the vascular bundles and the epidermis is perforated by stomata. Some leaves produce special glands called **hydathodes** which secrete water, and other glands secrete watery fluids containing sugars and nutrients. On reflection it is obvious that the environment on leaf surfaces will not be uniform and will be made up of a variety of quite distinct microhabitats, varying from species to species, the physical characteristics of which will depend upon the nature and distribution of structures, such as those listed above, on the surface. Some of the effects of these structures on the microclimate on the leaf surfaces are discussed in section 4.3, but much remains to be discovered about the conditions in these microhabitats in relation to the distribution and growth of fungi on leaf surfaces.

4.3 FACTORS AFFECTING THE DEVELOPMENT OF FUNGAL COMMUNITIES ON LEAF SURFACES

4.3.1 Cell leakage

One of the most important factors influencing the growth of microorganisms on plant surfaces is leakage from cells. Moisture which falls on the plant from precipitation or condensation tends to adhere to the leaf surfaces as a film held by surface tension, especially on hairy non-waxy surfaces. This film of water probably already contains a diverse range of materials, including pollutants, from the atmosphere, and into it diffuse organic materials and nutrients from the cells of the plant. In this solution can be found all of the essentials for the growth of the majority of microorganisms: sugars, amino acids, minerals and growth factors (Tukey, 1971). Any variation between plant species in the composition of the solution is more likely to be quantitative than qualitative. In an analysis of dew collected from the leaves of several plants the quantity of amino acids varied, for example, from 16 μg cm^{-3} in lettuce (*Lactuca sativa*) to

52 μg cm^{-3} in water melon (*Cucumis melo*); and the carbohydrate content ranged from 19 μg cm^{-3} to 74 μg cm^{-3} respectively (Chet, Zieberstein and Henis, 1973).

Solutions on leaf surfaces take up molecules, anions and cations from the plant by exchange between the surface solution and the fluid of the cell free space, which bathes the plant cells outside the plasmalemma. The water films on the leaf surface compete with the metabolic activities of the plant cells for solutes in the free space, and any factor which reduces this competitive effect tends to increase leakage into the surface solution. Thus the addition of fertilizers to the soil, for example, may lead to increased leakage on to plant surfaces.

Leakage from leaves tends to increase as the tissues age, primarily due to the changing physiological condition of the leaf as it matures. In young tissues, the metabolic demand for nutrients and the metabolites required for growth is high, which in turn reduces their availability in the free space and cuts down leakage to the surface solution. When the leaf is mature it ceases to grow and becomes less competitive and consequently leakage into the surface solution increases. Leakage resulting from insect damage with the deposition of honeydew by aphids also contributes significantly to the nutrition of fungi on living leaves.

Sooner or later leaves and other organs senesce, that is deterioration in metabolic efficiency sets in, the defects of which become accumulative and ultimately lead to death. This has important consequences for microbial development on leaf surfaces because it leads progressively to the leakage of greater and greater quantities of solutes into water films on the leaf surface. Finally, the disorganization of organelles and the loss of the integrity of the cell through the disruption of membrane function leads to the free passage of soluble materials from cells on to the leaf surface. This increasing tendency for solutes to leak from leaves as they age seems to explain why there is a great burst of fungal growth on leaf surfaces in early autumn at the time of the onset of leaf senescence.

Since leakage can only occur into water films, leaves which are water repellant, because of their shape and/or the presence of a waxy cuticle, e.g. conifer needles, will be poorer environments for microbial growth. The wettability of leaf surfaces increases with age and non-wettable leaves may become wettable in time due to the deterioration of the cuticle. Wax is removed by weathering (Forster, 1977) and is also degraded by lipase-producing epiphytic yeasts and bacteria (Ruinen, 1966; McBride, 1972)

4.3.2 Competition and the pollen effect

Since it is well known that most spores will germinate in distilled water without added nutrients or metabolic substrates, it is not immediately obvious why spores should germinate so poorly on leaf surfaces in early summer (as described in section 4.2.2).

The first clues to the solution of this puzzle came from studies made by Blakeman and Fraser (1971) and Blakeman (1972) on the germination of *Botrytis cinerea* spores on chrysanthemum (*C. morifolium*) and beet (*Beta vulgaris*) leaves. In the course of these experiments it was observed that the failure of spores to germinate on young leaves was associated with the multiplication of epiphytic bacteria within the water droplets in which the spores were suspended. In several critical experiments Blakeman and co-workers readily established that the inhibition of spore germination depended upon the presence of bacteria, and that spores could be successfully germinated on young leaves pretreated with anti-bacterial antibiotics.

Brodie and Blakeman (1975) subsequently discovered that when ^{14}C-labelled spores were placed in water on glass slides with bacteria, some of the radioactive content leaked into the water and was taken up by the bacteria. Leakage from dehydrated structures like spores and seeds on contact with water is a common phenomenon (Simon, 1974). Spores have been shown to lose as much as 10% of their dry weight following a brief immersion in water (Lingappa and Lockwood, 1964). This evidence suggested that the mechanism of inhibition is nutritional competition. Nutritionally independent spores lose nutrients and so come into competition with bacteria for external metabolic substrates.

In a series of experiments Brodie and Blakeman (1975) discovered that the key factor in this story is that spores take up certain amino acids that are essential for germination much less rapidly than competing bacteria. Whereas leaf-inhabiting bacteria removed over 80% of labelled amino acid

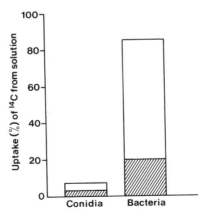

Figure 4.3 Percentage uptake of ^{14}C-labelled glutamine (0.1 µmol cm^{-3}) over 24 h by spores of *Botrytis cinerea* and bacteria. ▨ = Proportion of ^{14}C in cells; □ = proportion of ^{14}C evolved as CO$_2$. (Reproduced with permission from Brodie and Blakeman, 1975.)

from solution over 24 h, *Botrytis cinerea* spores took up only 10% in the same time (Figure 4.3). This was reflected in competition experiments where a decrease in the germination of *B. cinerea* spores on glass slides occurred in solutions containing labelled amino acids and glucose in proportion to the removal of amino acids and the number of leaf-inhabiting bacteria present.

Similar results were obtained using other leaf-inhabiting bacteria and a *Sporobolomyces* yeast as the competing organism. This may explain why yeasts dominate the mycoflora of young leaves.

B. cinerea spores were found to remove glucose from solution more efficiently than bacteria (70% compared with 30% in a 24 h period), but in mixtures of amino acids and glucose, germination did not appear to be related to the removal of glucose, although the rate of uptake of amino acids by spores was found to increase slightly in the presence of glucose.

Pollen can stimulate spore germination and growth on leaf surfaces by reducing nutritional competition. The abundance of pollen in the atmosphere coincides with the increase in the numbers of spores during the summer months and pollen and spores are deposited on plant surfaces together in great numbers.

Pollen grains behave like fungal spores and other dehydrated structures, losing some of their contents when they first come into contact with water; carbohydrates, most kinds of amino acid and even proteins have been detected in leachates and these become available to microorganisms in the environment.

In the field the development of the mycoflora on plant surfaces can be very closely related to the deposition of pollen. Diem (1974) observed that on barley leaves (*Hordeum vulgare*) the frequency of occurrence of fungal

Table 4.1 Fungal colonization of barley leaves

Date (1971)	Frequency of colonies cm^{-2} leaf		Area of colonized leaf surface (mm^2) cm^{-2} leaf		Mean colony area (mm^2)		Pollen grains cm^{-2} leaf	
	A	B	A	B	A	B	A	B
16 June	0.3	17.2	0.006	0.774	0.019	0.045	50	300–600
20 June	1.2	11.7	0.027	0.594	0.022	0.051	50–100	100–300
26 June	0.4	0.5	0.003	0.014	0.009	0.028	50	50
30 June	0.2	0.1	0.003	0.002	0.015	0.018	50	50
2 July	0.3	0.2	0.005	0.002	0.017	0.011	50	50

A: Station in the centre of a barley field; B: station surrounded by Gramineae.
Source: Diem (1974).

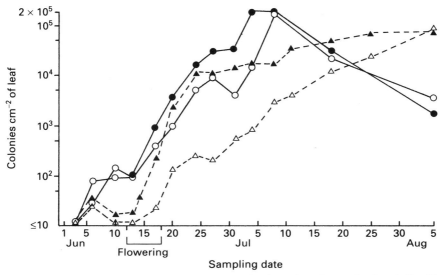

Figure 4.4 Effects of deflowering on the colonization of rye leaves by fungi. ●- -● = Pink yeasts (pollen present); ○- -○ = pink yeasts (no pollen); ▲- -▲ = *Cladosporium* (pollen present); △- -△ = *Cladosporium* (no pollen). (Redrawn from Fokkema, 1971.)

colonies and their size could be correlated with the amounts of pollen deposited. Heavy pollen deposition at one site coincided with good fungal growth, whereas low pollen counts were associated with poor fungal development at another (Table 4.1).

Fokkema (1971) studied the effects of pollen on the development of the mycoflora of rye (*Secale cereale*) leaves with special emphasis on the development of pathogens (Figure 4.4). Rye plants were grown in two well-separated plots under similar conditions except that in one plot the ears were removed before flowering to prevent pollen from the plants falling on to the leaves. The mycoflora developing on the leaves was studied by a leaf-washing technique and pollen deposition by direct microscopic examination. The numbers of colonies of *Cladosporium herbarum*, *Aureobasidium pullulans*, white and pink yeasts increased on the leaves of all plants throughout the summer, but by June there were about three times as many isolations of these fungi made from the leaves with flowers compared with those without, a difference that corresponded to a difference in the amounts of pollen distributed on the leaves of the two sets of plants. As the season advanced, the difference in the number of isolates tended to diminish until by the end of August it had disappeared, probably due to the accumulation of pollen from other plant species on the deflowered plants. Also leaves may have become leaky as they aged and released increasing amounts of nutrients on to the leaf surface.

Fokkema also found that pollen can stimulate pathogen development. Again using deflowered and untreated rye plants it was found that when spores of *Helminthosporium sativum* (the *Drechslera* state of *Cochliobolus sativus*) were inoculated on to the plants after flowering on 12 June, infection was stimulated on the untreated plants compared with deflowered no-pollen plants (Figure 4.5). Later inoculations on 30 June showed less difference between the plants, possibly due to the development of an antagonistic saprophytic mycoflora on the leaves (section 4.3.3).

The effect of adding extra pollen to similar plants at the time of inoculation was also investigated. Before flowering (2 June) extra pollen greatly stimulated pathogen development on both deflowered and untreated plants compared with no extra pollen on similar plants. Immediately after flowering, this effect was considerably reduced, somewhat surprisingly in the case of deflowered plants. Later in the season, however, (30 June) extra pollen added at inoculation time stimulated infection on deflowered plants compared with deflowered plants and no extra pollen. There was no comparable stimulation at this time in the case of the untreated plants. A possible explanation may be that there was a greater development of an antagonistic saprophytic mycoflora on the leaves of the untreated plants compared with deflowered plants due to the absence of natural pollen in the latter.

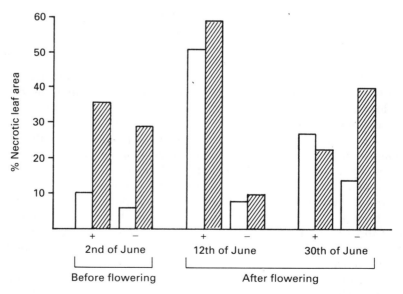

Figure. 4.5 Comparison of the development of *Cochliobolus sativus* lesions on rye (untreated +) with (deflowered −); with ⁄⁄ and without ☐ the addition of extra pollen. (Data from Fokkema, 1971.)

4.3.3 Interspecific competition

The antagonistic potential of several of the common phylloplane fungi towards other fungi has been demonstrated in a number of separate experiments. Antagonistic fungi include *Aureobasidium pullulans*, *Cladosporium* species, *Epicoccum nigrum* and yeasts in the genera *Sporobolomyces*, *Cryptococcus* and *Tilletiopsis* (McBride, 1971; Fokkema, 1973; Skidmore and Dickinson, 1976; Collins, 1976; Hoch and Provvidenti, 1979; Brown, Finlay and Ward, 1987).

In vitro experiments such as those conducted on *Antirrhinum* leaves indicate that interspecific antagonisms are likely to be important factors regulating fungal growth in the phylloplane (Table 4.2).

Not unnaturally, the antagonistic behaviour of phylloplane fungi has attracted the attention of plant pathologists as a possible means of controlling plant disease, and several interesting examples of experiments which seek to explore this potential can be found in reviews of the literature (Blakeman and Fokkeman, 1982; Blakeman, 1988). For instance, van den Heuvel (1971) found that the number of lesions produced on *Phaseolus vulgaris* leaves by *Alternaria zinniae* could be considerably reduced if the spores of the pathogen were simultaneously inoculated with those of *Alternaria tenuissima*.

The germination of *A. zinniae* spores on leaves was delayed by the presence of *A. tenuissima* and after 5 h, following simultaneous inoculation of both fungi, the percentage germination of *A. zinniae* spores was 18%,

Table 4.2 Reduction in numbers of *Sporobolomyces roseus* cells by interspecific antagonism with *Cladosporium cladosporioides* on *Antirrhinum* cultivars

Cultivar	Site of inoculation	Inoculation treatment	*Sporobolomyces* (no. cm^{-2} leaf)	
			at 14 days	at 21 days
A. majus (Nanum)	Older leaves	*S. roseus*	2.5	14.5
		$S + C$	0.0	4.0
	Young leaves	*S. roseus*	0.0	4.9
		$S + C$	0.0	0.5
A. majus (F_1 hybrid)	Older leaves	*S. roseus*	2.7	15.6
		$S + C$	0.0	3.0
	Young leaves	*S. roseus*	0.0	4.8
		$S + C$	0.0	0.8

$S + C$ = *Cladosporium cladosporioides* + *Sporobolomyces roseus*.
Source: Collins (1976).

compared with 85% in controls in the absence of *A. tenuissima*. At 72 h the respective figures were 87% and 99%. This short delay in germination seems to be critical in preventing the establishment of the infection. Parasitic *Alternaria* species have also been shown to be antagonized by other phylloplane saprophytes, notably by *Aureobasidium pullulans* (van den Heuvel, 1969; Pace and Campbell, 1974; Rai and Singh, 1980; Brame and Flood, 1983).

In other experiments Dickinson and Skidmore (1976) found that several common saprophytic phylloplane fungi have the potential to antagonize *Septoria nodorum*, a pathogen of barley (*Hordeum vulgare*), inhibiting spore germination and germ tube growth when inoculated on to leaves with the pathogen.

In most of these experiments it has been found that the antagonistic effects can be considerably enhanced if the antagonist is inoculated on to the leaves at least 24 h in advance of the pathogens.

Experiments have shown that artificially increasing the saprophytic fungal populations on plants has potential for controlling fungal disease on a large scale in the field or greenhouse. *Cladosporium herbarum* sprays have been used effectively to reduce *Botrytis* rots of soft fruits (Newhook, 1957; Bhatt and Vaughan, 1962) and infection levels of *Cochliobolus sativus* and *Septoria nodorum*, leaf diseases of wheat (*Triticum aestivum*), have been reduced using yeast cells suspended in nutrient solutions (Bashi and Fokkema, 1977; Fokkema *et al.*, 1979).

The effects of the antagonistic activities of phylloplane fungi are not ameliorated by the presence of pollen. Thus when pollen was added to sugar beet leaves (*Beta vulgaris* var. *altissima*) the number of lesions produced by *Phoma betae* rose dramatically from 22% of the total inoculations in controls without pollen to 100% with pollen, but in the presence of *Aureobasidium pullulans*, *Torulopsis candida*, *Sporobolomyces pararoseus* and *Cladosporium cladosporioides*, the number of lesions produced by *P. betae* in the presence of pollen dropped to 10% (Warren, 1972).

Fokkema (1973) reproduced similar effects on rye (*Secale cereale*), where *Aureobasidium pullulans* and the yeasts *Sporobolomyces* and *Cryptococcus* species were especially effective in reducing the number of *Drechslera sorokiniana* (*Cochliobolus sativus*) lesions on leaves in the presence of pollen.

These observations explain why, in trials, the effects of pollen in promoting lesion development by pathogens on field-grown plants has always been found to be less than on plants reared in the greenhouse, which of course have much smaller natural saprophytic fungal populations.

The nature of the antagonisms produced by phylloplane fungi is subject to continuing investigation and much remains to be explored and explained. The failure of pollen to ameliorate antagonistic effects, as described above, may indicate that non-nutritional antagonisms can be involved.

Several yeasts, *Aureobasidium pullulans*, *Cladosporium herbarum*, *C. macrocarpum*, *Alternaria alternata* and *Stemphylium botryosum* are some of the species that have been shown to produce antibiosis effects against other phylloplane fungi in paired cultures (Fokkema, 1973; Skidmore and Dickinson, 1976). Fokkema (1973) found that many isolates of *A. pullulans* and *Sporobolomyces* species and several *Cryptococcus* species were very active against *Drechslera sorokiniana* (*Cochliobolus sativus*).

These results confirm earlier reports that isolates of *Sporobolomyces ruberrimus* and *A. pullulans (Pullularia pullulans)* were active producers of antifungal antibiotics in culture (Yamaski, Satomura and Yamamoto, 1951; Baigent and Ogawa, 1960). The majority of *Alternaria* species are also known to be producers of antifungal antibiotics (Lindenfelser and Ciegler, 1969).

Inhibitions which developed after hyphal contact were studied by Skidmore and Dickinson (1976) using the simple device of growing paired cultures on Cellophane spread over an agar medium. After contact had been established between the mycelia the Cellophane was lifted and portions in the region of hyphal contact were prepared for microscopic examination. This revealed that some contact inhibitions were due to the development of hyphal interference-like effects (section 3.2.4(b)). *Septoria nodorum*, for example, caused bursting of hyphae in *Botrytis cinerea*, and a loss of selective permeability in the hyphae of *Drechslera graminea* and *Alternaria alternata*.

Relevant to antagonistic interactions on leaf surfaces and their special importance in limiting the development of plant disease is the increasing use of fungicides as the most effective method of controlling potential pathogens. Many fungicides such as Zineb and Dithane are non-selective and drastically reduce the populations of saprophytic fungi in the phylloplane (Dickinson, 1973; Dickinson and Wallace, 1976; Mehan and Chohan, 1981). When these are applied the incidence of disease may actually increase (Fokkema *et al.*, 1979). There is evidence that the use of more selective fungicides such as Benlate and Maneb may be more beneficial than those that indiscriminately eliminate the entire fungal population. Mehan and Chohan (1981) found that when Benlate was used to control *Mycosphaerella* leaf spots of groundnut (*Arachis hypogaea*), more effective control of the disease was obtained than when Dithane was used. An examination of the saprophytic fungal populations revealed that on Benlate-treated leaves, after an initial setback the leaves eventually supported a larger saprophytic mycoflora than Dithane-treated leaves or untreated controls.

Some of the SBI fungicides, e.g. triadimefon, have less effect on saprophytes than Benlate or Maneb and may be even more appropriate to use in this context (Dik, 1991).

The interaction between saprophytes and fungicides as it affects plant pathogen control is likely to prove fairly complex (Dik, 1991). Reduction in saprophytic populations by fungicides can lead to an increase in pathogen activity not only because of a general reduction in direct interspecific antagonism but especially because smaller populations will have an indirect effect, reducing nutritional competition and greatly benefiting pathogen growth. Increased nutritional benefits to pathogens may in turn reduce the efficacy of fungicides applied against them.

4.3.4 Plant inhibitors

Antifungal inhibitors that are constitutive in plant tissues as a result of normal metabolic activity are referred to as **preformed inhibitors**, as opposed to those called **phytoalexins** that are synthesized in response to the presence of a fungus. Some of the kinds of preformed antifungal substances that exist in plant tissues have been described in section 3.2.2.

In the past, mycologists have been largely concerned with the effects of preformed inhibitors on the growth of fungi in litter (Melin, 1946; Lindeberg, 1949). The fact that many of the simpler forms of these substances are water-soluble and could possibly play a role in the development of fungi on the surfaces of plants has only relatively recently been seriously investigated, although previously Melin (1946) had detected water-soluble antifungal substances in the leaves of many trees including

Table 4.3 Seasonal growth of fungi on leaf surfaces of Norway maple

Sampling date	Mean no. yeast cells per colony	Cladosporium spp.					
		No. conidia cm^{-2}		% germination		Mean germ tube length (μm)	
		Adaxial	Abaxial	Adaxial	Abaxial	Adaxial	Abaxial
10 June	1.0	32	21	–	–	–	–
3 July	1.7	85	43	–	–	–	–
17 July	3.0	469	117	4.5	–	15	–
6 August	4.0	139	320	23.0	–	26	–
4 September	4.3	501	224	6.4	4.7	36	27
16 September	7.0	85	106	25.0	50.0	75	90
13 October	7.0	11	32	–	66.0	–	100

Source: Irvine, Dix and Warren (1978).

Table 4.4 Seasonal change in the relative concentrations of gallic acid in leaf washings of Norway maple

Date of extract preparation	No. of leaves washed	Surface area of leaves (cm^2)	Total GLC units of acid $(\mu l^{-1})^a$	GLC units of acid cm^{-2} leaf
5 August	30	6 418	22 854	3.56
22 August	30	7 275	8 009	1.10
10 September	138	13 744	183	0.01

[a] 1 μl of 10^{-3}M gallic acid standard = 166 520 integrator units.
Source: Irvine, Dix and Warren (1978).

Norway maple (*Acer platanoides*), and in a comprehensive survey of woody plants Topps and Wain (1957) demonstrated that the leaves of many contained antifungal substances which readily passed from the leaves into water. Extracts of ash (*Fraxinus excelsior*), beech (*Fagus sylvatica*), chestnut (*Castanea sativa*), hazel (*Corylus avellana*) and oak (*Quercus* species) were all active, with privet (*Ligustrum vulgare*) and elder (*Sambucus nigra*) extracts the most potent. Later, Dix (1974) identified one of the antifungal substances present in dew on the leaves of Norway maple as gallic acid. This was tentatively linked with the poor growth of *Cladosporium* species in the phylloplane of Norway maple.

The effect of the antifungal inhibitors present in Norway maple leaves on the development of the phylloplane mycoflora was subsequently more thoroughly investigated by Irvine, Dix and Warren (1978), who found that the number of fungal propagules and yeast colonies counted by direct observation on leaf surfaces was low compared with other plants. Numbers reached a maximum of only 3.4×10^3 cm^{-2}, while on rye (*Secale cereale*) white yeasts alone may reach 3×10^5 cm^{-2}, and pink yeasts, *Cladosporium* species and *Aureobasidium pullulans* may each obtain a level of 2×10^5 cm^{-2} (Fokkema, 1971). On Norway maple the number of cells in each yeast colony was small, and the germination of hyphal species was generally poor and hyphal growth limited until late in the year (Table 4.3). Leaf washings were shown to contain several substances active against the fungal phylloplane populations of lime (*Tilia platyphylla*) in laboratory experiments and in the field. The inhibitory activity of the washings declined as the summer advanced and this appeared to be correlated with the recorded increase in the growth of fungal species in the phylloplane during September. Table 4.4 shows the seasonal decline of gallic acid in the leaf washings of Norway maple as determined by gas–liquid chromatography analysis.

A similar situation appears to exist in birch (*Betula* species), where antifungal compounds have also been detected in leaves (Warren, 1976). *Cladosporium* species germinated poorly on birch leaves in spite of the high sugar levels which could be collected in leachates, while the leaves of *Tilia platyphylla*, whose leachates contain lower sugar levels but no antifungal substances, supported a much richer mycoflora.

Doubtless many other antifungal substances occur on leaf surfaces and will be identified eventually. Evidence to date suggests that these include several terpenoid and methylated flavonoid compounds (Bailey, Vincent and Burden, 1974; Harborne *et al.*, 1976).

In some instances the fungistatic properties of leaves are due to substances in cuticular waxes. Martin, Bant and Burchill (1957) separated an ether-soluble acid fraction from the leaf wax of a number of apple varieties which proved to be strongly inhibitory to the spores of the apple mildew fungus (*Podosphaera leucotricha*). Chloroform extracts of waxes collected from the leaves of broad bean (*Vicia faba*), sugar beet, red beet (*Beta* species), birch (*Betula* species), lettuce (*Lactuca sativa*), tomato (*Lycopersicum esculentum*), blackcurrant (*Ribes nigrum*), and chrysanthemum (*C. morifolium*) have all been shown to be inhibitory to the germination of spores of *Botrytis cinerea* (Blakeman and Atkinson, 1976; Blakeman and Sztenjnberg, 1973). In chrysanthemum leaves, the inhibitory substances appear to be polar compounds associated with the leaf waxes (Blakeman and Atkinson, 1976).

In broad bean there is evidence which indicates that inhibitory substances may only be released from cuticular waxes as a result of interaction between the leaves and the invading fungus (Rossall and Mansfield, 1980).

Under field conditions the inhibitory effects of waxes probably diminishes as the leaves age due to erosion by weathering, removal by microbiological activity, and the amelioration of their effects by the increasing tendency for ageing leaves to leak nutrients and organic substances on to the leaf surface. This point is brought out by the observations of Rossall and Mansfield (1980), who found that the inhibitory properties shown by broad bean leaves from plants reared in the protective environment of a greenhouse could not be reproduced when field-grown plants were tested.

In a number of plants antimicrobial or potential antimicrobial substances are located in the epidermal glands and hairs of leaves. The glands of *Chrysanthemum parthenium* contain parthenolide, a sesquiterpene lactone, with antifungal properties (Blakeman and Atkinson, 1979), while several plants have been shown to possess specialized epidermal hairs containing what are thought to be phenols (Blakeman and Atkinson, 1981).

It is not clear, however, what effect, if any, substances stored in glands and hairs have on fungal growth on leaf surfaces. It is not certain that they are released on to the leaf surface, other than at leaf senescence, or whether they could ever reach significant concentrations. For a review of the subject see Blakeman and Atkinson (1981).

Phytoalexins are only synthesized by plants in response to the invasion of tissues and as such do not appear to have any influence on the growth of the epiphytic saprophytic mycoflora (Mansfield, Dix and Perkins, 1975; van den Heuvel, Verheus and Kruyswijk, 1978).

4.3.5 Climatic factors

The stems and leaves of terrestrial plants, projecting as they do some way up into the atmosphere, present a rather inhospitable environment for fungal growth. Due to the vagaries of the climate and the cycle of light and darkness and its effects on plants, rapid changes occur in the physical conditions prevailing on the leaf surface, with temperature and humidity fluctuating, often to extremes. Water from rain and dew wets plant surfaces periodically but may rapidly evaporate so that fungi can be continually subjected to wetting and drying cycles.

While the importance of environmental factors have long been recognized, relatively little factual information is available on the effects of environmental changes on the colonization of the surfaces of attached leaves, due to the difficulties of undertaking meaningful experimental work in this area. This is particularly unfortunate since evidence is beginning to accumulate which suggests that the physical environment at the leaf surface may be of paramount importance in its effect on fungal colonization, outweighing even the effects of leaf age and the influence of competition (Dickinson and O'Donnell, 1977; Breeze and Dix, 1981).

While conditions on the leaf surface will generally tend to follow the physical changes taking place in the atmosphere, the physical environment on the leaf surface will be greatly influenced by complex interactions between the plant and the atmosphere surrounding it. As solar energy falls on the leaf the temperature of the tissues will rise and act as a heat store. The temperature at the leaf surface will, therefore, depend upon the amount of solar energy in the form of heat that falls on the leaf, the amount of heat that is absorbed and the rate of heat loss from the tissues. Heat is lost by the evaporation of water, which is controlled by the humidity gradients over the leaf surface. At low atmospheric humidities the humidity gradients over the leaf steepen and the rate of evaporation of water rises, with the result that the rate of cooling of the leaf increases. High wind speeds also steepen humidity gradients over the leaf surface and favour heat loss, but even at low humidities, on windy days, the temperatures at the leaf surface during daylight hours are likely to exceed air temperatures due to the absorption of solar energy. At night, when the absorption of heat is low, heat loss will exceed heat gain and the temperature of the leaf surface will fall near to that of the atmosphere.

However, in summer leaf surface temperatures at night are seldom likely seriously to limit fungal growth, and during daylight temperatures must often be close to the optimum for most phylloplane fungi.

The humidity at the leaf surface depends upon the rate of water loss from the plant, wind speed and the humidity of the atmosphere. In low atmospheric humidities or high wind speeds, as a result of high rates of evaporation of water from the plant surfaces, water stress may develop causing the plant stomata to close. Under these conditions, when the leaf cuticle can give up no more water from the interstitial spaces, low humidity levels will develop above the leaf and very dry air can be brought extremely close to the leaf surface by a pulsating mechanism of exchange of air at the leaf surface with that of the atmosphere above (Sherriff, 1973). During the summer, low humidity conditions probably occur not infrequently on leaf surfaces and seem to be one of the important factors restricting fungal growth at this time.

When the temperature falls at night, dew may collect on the cooling aerial parts of exposed plants. In summer when daytime temperatures are usually high, dew probably forms on more nights than it does not, and this respresents an important source of water for phylloplane microorganisms, especially at a time of the year when rainfall is normally lower. The turgid spores of some fungi will only germinate if liquid water is present.

Temperature and humidity at the plant surface may also be affected by certain variable features of the anatomy and morphology of plant organs. The higher numbers of stomata that occur on the abaxial (lower) leaf surfaces in most dicotyledonous plants could lead to higher humidities or lower temperatures there compared with the adaxial (upper) leaf surface, where stomatal numbers are usually lower.

The hairy and waxy protuberances present on the leaves of some plants reduce wind speed and increase the thickness of the boundary layer above the leaf. The rate of evaporation of water will then be slowed, reducing heat loss and causing the temperature and humidity above the leaf to rise. Leaves of this type might seem to favour fungal colonization, but at the same time an important function of these structures is to restrict colonization by preventing spores from actually landing on the leaf epidermis.

Leaf shape is also important: the greater the leaf perimeter in relation to its surface area the cooler the leaf surface will tend to be. Thus leaves which are greatly divided and have large perimeters tend to be cooler than undivided leaves of the same area. This is due to an edge effect that reduces the thickness of the boundary layer at the leaf margin, speeding up water evaporation and heat loss, and as this is transmitted over the leaf, the whole leaf surface tends to become somewhat cooler.

Due to the non-random distribution of hairs and waxy protuberances and their distribution in distinct patterns on the leaf surface, temperature and humidity levels are unlikely to be uniform over the whole leaf surface, giving rise to a variety of microclimates. Temperatures over the surface of a hop (*Humulus lupulus*) leaf have been found to vary in the range 18.8–21.5°C in an air temperature of 15.6°C (Burrage, 1971). The veinal distribution of *Aureobasidium pullulans* and *Cladosporium herbarum* on the leaves of some plants could, for example, be related to the more sheltered conditions and the consequential higher temperatures and humidities which may prevail there (Pugh and Buckley, 1971b).

Bulkier organs like stems have a higher heat storage capacity and a lower rate of heat loss than leaves due to their lower surface area-to-volume ratio. The darker surfaces of woody stems will absorb more and radiate less heat than leaves and the uneven surface and the fissures which occur in the bark of older woody stems will tend to maintain higher humidities and temperatures by reducing evaporation. The generally less favourable climates on leaves as compared with stems may be important in preventing the colonization of leaves by fungi that colonize stems.

Variation in plant habit must also give rise to a range of temperatures and humidities at the plant surfaces. Gradients of temperature and humidity can form along stems from the ground upwards or among leaves at different heights above the ground. In the case of senescent tissues this is reflected in moisture content which can vary according to the height above the ground (Chapter 5). All these factors can have an important bearing on the patterns of fungal colonization on organs at different levels of the plant, or between leaves at different heights above the ground. Again, the pattern of fungal growth on the leaf surfaces of plants which have a rosette or tussock form could be quite different from that on the leaves of plants with a taller or more open form of growth.

In spite of the fluctuating physical environment on the leaf surface and rather harsh conditions which can exist there, leaf-inhabiting fungi successfully colonize the phylloplane. It is not surprising to find, therefore, that these fungi show several adaptations to growth and survival on leaves.

Primary colonizers of the phylloplane grow well in the temperature range 10–25°C and germinate successfully at relative humidities below 90% at 20°C (Table 5.2). Even in the temperature range 10–15°C they will germinate well at relative humidities between 90 and 95% (Dickinson and Bottomley, 1980). Calculations from the data of Magan and Lacey (1984a) indicate minimum water potentials for germination to be between -17.4 and -22.3 MPa and between -14.5 and -17.4 MPa for growth. Once germinated, these fungi show remarkable powers of survival for short periods at extremely low relative humidities. The germ tubes of *Cladosporium herbarum*, *C. cladosporioides*, *Alternaria tenuis* (*A. alternata*), *Helminthosporium sativum* (the *Drechslera* state of *Cochliobolus sativus*) and *Stemphylium botryosum* were shown by Diem (1971) to survive exposure for up to 8 h at relative humidities as low as 40%. However, the survival rate decreased as the germ tube length increased. A feature of some phylloplane fungi is that their hyphal apices are very tolerant of desiccation, and when other parts die they often survive to resume growth on rewetting (Park, 1982a). These unique adaptations enable them to survive during the frequent dry spells which occur on leaves.

Another important adaptation is that many phylloplane fungi can form thick-walled resting structures. The chlamydospores and microsclerotia of *Epicoccum purpurascens* (*E. nigrum*), *Cladosporium herbarum* and *Aureobasidium pullulans* are found commonly on leaves; particularly in veins (Pugh and Buckley, 1971b). Although little is known about conditions stimulating the production of such structures, there seems little doubt that their biological value lies in their ability to carry the population through unfavourable conditions.

The hyphal and spore walls of many phylloplane species are pigmented and it is thought that this must be a protective device and another adaptation to life on plant surfaces where fungi are exposed to near-ultraviolet radiation which may be harmful. In experiments carried out by Pugh and Buckley (1971b) this was found to be generally true, with the pigmented spores of *Alternaria*, *Epicoccum* and *Cladosporium* surviving exposure to ultraviolet light better than the hyaline cells and spores of *Aureobasidium pullulans* and the phylloplane yeasts. Hyaline forms appear to be protected by the formation of colonies and microsclerotia which have a higher survival rate when exposed to ultraviolet light than their individual cells or spores.

Detachment is also an obvious problem facing fungi in this environment. Yeasts, *A. pullulans* and other species produce copious quantities of extracellular sticky polysaccharide slimes which prevent their spores and cells being washed off. Slimes may also catch and hold the spores of other species and so may be instrumental in the development of communities. Dickinson (1986) suggested that slimes may possibly also protect cells from near-ultraviolet radiation, reduce water loss and assist rehydration. Adaptation to physical conditions may be the most important factor which restricts the list of fungal colonizers of the surfaces of attached leaves and similar habitats to a few widely distributed species.

4.4 FUNGAL SUCCESSIONS IN LEAF LITTER

Senescent leaves eventually fall on to the soil surface where they decay, releasing nutrients as their remains become incorporated slowly into the soil structure. Fungi undoubtedly play an important role in this, although relatively little is known of the role of individual species. The general features of fungal successions on leaf litter are, however, well described.

Angiosperm phylloplane fungi persist on fallen leaves and *Cladosporium herbarum*, *Aureobasidium pullulans* and others may be isolated from leaf litter several months after leaf fall and several may go on to produce their sexual stages there (Hogg and Hudson, 1966; De-Boois, 1976). The role of these fungi in litter decomposition has hardly been investigated and is somewhat obscure. Their ability to hydrolyse native cellulose is rather variable, most strains of *A. pullulans* are non-cellulolytic, *Cladosporium* species generally show only rather weak activity while *Epicoccum purpurascens* and *Alternaria alternata* are rather more active (Siu, 1951; Hogg, 1966; Park, 1982b; Godfrey, 1983). In culture, *A. alternata*, *A. pullulans* and *Cladosporium* species decompose tree leaf litter very poorly (Hering, 1967; Godfrey, 1983), and Frankland (1969) found that *A. pullulans* caused only a very small dry weight loss from bracken petioles after 6 months' incubation. Many are inhibited by gallic acid (Dix, 1979), and phenols present in the leaf litter of trees may restrict their growth, while in culture they can be antagonized and overgrown by litter-decomposing agarics (see Figure 3.6). All in all, most appear to be poorly adapted for sustained growth in litter and the ability to decompose certain sorts of leaf litter may be rather slight. It has been suggested that the high frequency of isolation of some phylloplane species from the more enduring sorts of leaf litter months after leaf fall may owe more to the survival of mycelial resting structures and to the remains of a heavy spore load than to the existence of actively growing mycelia.

More enduring leaf litters typically develop a secondary flora of litter microfungi, the sporulating structures of which usually appear about a year after leaf fall (Hogg and Hudson, 1966). These are undoubtedly responsible for a certain amount of the physical and chemical change that is seen to occur in the first year after leaf fall.

Once in the litter, leaves also quickly become colonized by typical soil-inhabiting fungi and as time passes these become dominant as leaves are buried and pass into the deeper layers of the litter. Soil-inhabiting microfungi include several genera that are ubiquitous in distribution such as *Penicillium*, *Humicola*, *Trichoderma*, *Fusarium*, *Gliocladium*, *Doratomyces*, etc. However, species in these genera are known to be variously distributed in different soils (section 7.2). These all have continuous actively growing resident populations in soil and are the so-called **autochthonous species** of the soil. Many metabolize litter phenolics, and especially important among these in this respect are certain *Penicillium* species. The latter have been shown to be generally tolerant of phenols and in culture they can utilize tannin phenolics as a sole source of carbon (Cowley and Whittingham, 1961; Lewis and Starkey, 1969; Grant, 1976; Dix, 1979). Most authochthonous soil fungi can hydrolyse polysaccharides; nevertheless, *Trichoderma*, *Penicillium* species and others cause little weight loss of tree leaf litter in monocultures after long incubation periods, and some of the commonest autochthonous soil fungi associated with tree leaf litter appear to play only a minor part in its direct decomposition (Hering, 1967; De-Boois, 1976). However, it has been shown that these fungal populations have an important indirect role to play in tree leaf litter decomposition. General residential microbial populations, including *Penicillium* and *Fusarium* species, can synergistically increase rates of decay of leaf litter when co-cultured with leaf litter-decomposing agarics (Dix and Simpson, 1984; Dix, unpublished). The reason for this is that interaction between species can increase the efficiency of the production and action of polysaccharidases as described in section 3.2.1. These interactions are frequently mutually beneficial.

Associated with autochthonous fungi on all types of angiosperm litter are the widely distributed **zygmogenous** soil-inhabiting species. Typical of this group are species in the genera *Mucor* and *Mortierella* which are non-cellulolytic and exist in soil as opportunists, remaining dormant in the form of resting structures for long periods but capable of rapid growth and colonization when favourable conditions allow. *Mucor hiemalis* has been shown to be inhibited by gallic acid (Dix, 1979) and the growth of many zygmogenous forms in litter is probably further restricted when phenolics are present. The part played by zygmogenous species in the direct decomposition of litter is probably very minimal. This is borne out by the experiments of De-Boois (1976), but like autochthonous species, they may have some indirect part to play in litter decomposition.

While the general pattern of the microfungal succession on angiosperm leaves in litter as described above is widely accepted, to date, very few successions have been completely worked out, so that for most plants many details of secondary leaf-inhabiting mycofloras and the invasion of leaves by litter-inhabiting fungi, in particular, remain to be discovered.

One microfungal succession which has been thoroughly worked out is that occurring on beech (*Fagus sylvatica*) leaves studied by Hogg and Hudson (1966). These workers found that some of the common fungi associated with the living leaves persisted in the litter, and eventually completed their life cycles there. The teleomorphs of *Cladosporium herbarum* (*Mycosphaerella tassiana*), *Aureobasidium pullulans* (*Guignardia fagi*) and those of the beech leaf parasite *Discula quercina* (*Apiognomonia errabunda*) became common in the litter and reached maximum numbers in the spring, following leaf fall. In addition, some fungi not encountered on leaves sampled directly from the trees, or present only in low numbers, first appeared or increased in numbers relatively soon after leaf fall. *Mycosphaerella punctiformis* belongs to this group. It was only detected in low frequency on living leaves, but after leaf fall numbers increased and by the following summer its ascomata were present on the majority of leaves in the litter. *Mollisina acerina* and *Discosia artocreas* also belong to this group. They were not found on living leaves but fruit-bodies appeared on leaves in the litter in the following spring, reaching maximum numbers during the first summer after leaf fall. These appear to be fungi that invade living leaves but whose rate of growth is so slow that they can only be detected after a long period of development in the litter. *M. acerina* and *D. artocreas* must invade the leaves well before leaf fall, for if leaves are detached from trees in the early summer and placed in the litter the fruit-bodies of these fungi still appear in the following spring (Hogg, 1966).

Later, Hogg and Hudson (1966) found that leaves in the litter were invaded by a variety of litter-inhabiting fungi which first appeared in the late summer of the year following leaf fall. This phase reached full development in the late autumn of the same year, with many species persisting in the litter through the second winter to the following spring. Most of these fungi have been recorded on several different sorts of angiosperm litter and few if any appear to be very specific in distribution. *Polyscytalum fecundissimum*, *Endophragmia* species in the Deuteromycotina and the ascomycete *Microthyrium microscopicum* were the most frequently observed species. The discomycete *Helotium caudatum* (= *Hymenoscyphus caudatus*) also became common at this stage. The pattern of fungal development on beech leaves is illustrated in Figure 4.6.

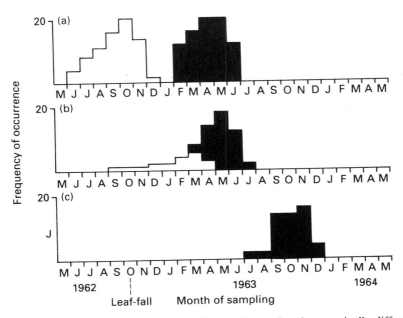

Figure 4.6 Occurrence of fungi on beech leaves illustrating three typically different types of behaviour during the fungal succession. Frequency of occurrence was measured on 20 leaves as recorded by the appearance of sporulating structures. □ = Conidia; ■ = ascocarps. (a) *Apiognomonia errabunda*; (b) *Mycosphaerella punctiformis*; (c) *Microthyrium microscopicum*. (Reproduced with permission from Hogg and Hudson, 1966.)

There is evidence that suggests that the patterns of development recorded on beech leaves are representative of similar patterns occurring on other angiosperm tree leaf litters. Hering (1965) reported, for example, a somewhat less detailed but similar account of the succession of fungi on the leaf litter of a number of other deciduous trees. *Mycosphaerella punctiformis* appeared early in the leaf litter of *Fraxinus excelsior* (ash), *Quercus robur* (oak) and *Corylus avellana* (hazel). The leaf parasites *Gnomonia setacea* and *G. gnomon*, corresponding to *Apiognomonia errabunda* on beech, produced teleomorphs in the litter on birch (*Betula* sp.) and hazel leaves respectively.

Associated with the final stages of decay where leaf litter accumulates in bulk are the litter-inhabiting basidiomycetes, mainly agarics (toadstools) in the genera *Mycena*, *Marasmius*, *Collybia* and *Clitocybe*. Characteristically, these fungi strongly hydrolyse plant cell wall polysaccharides and most also degrade lignin (Lindeberg, 1944, 1947). Many are active producers of laccases (Lindeberg, 1948; Giltrap, 1982) or extracellular polyphenol oxidases capable of detoxifying litter phenolics (Lyr, 1962, 1963). For these reasons, litter-inhabiting agarics are considered to be among the most active decay agents of leaf litter, especially tree leaf litter (Table 4.5), and

Table 4.5 Losses (%) caused by litter-decomposing agarics (incubated 165–171 days) at 25°C

Species	Lignin	Cellulose	Total dry matter
Fagus sylvatica leaves			
Clitocybe alexandri	32.3	6.4	14.2
Lepiota procera	55.1	28.4	22.3
Lepiota amianthina (*Cystoderma amianthinum*)	53.3	43.2	32.1
Collybia butyracea	76.5	15.6	28.5
Collybia dryophila	75.8	70.1	37.1
Mycena epipterygia	56.9	73.9	39.0
Mycena rosella	19.4	33.8	16.5
Mycena vulgaris	55.3	49.6	31.0
Glyceria maxima straw			
Clitocybe clavipes	41	52.3	25.9
Clitocybe geotropa	66.4	84.0	46.7
Clitocybe infundibuliformis	77	82.5	50.3
Clitocybe nebularis	66	35.4	33.3
Clitocybe odora	65.2	71.9	30.8

Source: Lindeberg (1947).

they form a significant proportion of the fungal biomass present in woodland litter. Frankland (1975) estimated that *Mycena galopus* accounted for about 10% of the total biomass of the mycelial standing crop present in the leaf litter of an oak wood. She estimated that the total annual production of mycelium by this fungus amounted to about 1.5 kg ha^{-1} after taking into account loss of mycelium due to lysis and consumption by soil animals. From culture experiments it was ascertained that this represented a conversion efficiency of oak leaf litter into fungal mycelium of about 20%.

In temperate woodlands, agaric mycelium is most active in the deeper layers of the litter. This is most probably because few can tolerate the dry conditions at the surface where uncompacted litter dries out too quickly and is very often at water potentials that are too low to permit growth (Dix, 1984a,b). Because agarics are generally absent, this may be the main reason why decay of the surface litter in such woodlands is very slow.

As leaf litter decays, its physical properties change. As a consequence the water-holding capacity increases and water potential values will improve because partially decayed leaves have higher water potentials at the same moisture content compared with undecayed leaves (Dix, 1985). A feature of many of the microfungi that are found on leaves in litter is that they are able to grow when the water potential is low. Non-xerotolerant species will grow fairly actively down to about -6.0 MPa and make some growth down to about -14.5 MPa. Xerotolerant *Penicillium* species can continue to grow at water potentials as low as about -30 MPa. Many of these moulds would be able to flourish on dry litter at the soil surface, since at a fresh weight moisture content of about 7% the water potential of undecayed deciduous tree leaves lies somewhere between -15.0 and -17.5 MPa (Dix, 1985), and tree leaf litter at the soil surface would not fall below these values except perhaps for short spells in the driest months of the year. Microfungi that are able to grow in dry uncompacted litter will probably make some improvement in the water relations of the litter in preparation for agaric colonization.

The majority of leaf litter-decomposing agarics are unselective and grow on a range of plant litter within different habitats, i.e. they are **resource-specific** but not **taxon-specific**. The majority of deciduous woodland inhabitants that fruit in mixed deciduous woodlands can occur in conifer woods as well, but some may show a preference for one or the other. *Collybia peronata* and *Mycena pura*, for example, are not unknown in conifer woods but they occur predominantly in deciduous woods, the latter preferring beech. Only a few species are considered to be taxon-specific. *Clitocybe odora* and *Collybia dryophila* are two examples that are associated almost exclusively with beech leaf litter. When considering distribution, it is of course advisable to bear in mind that distribution patterns are records of fruiting and may not necessarily reflect the distribution of mycelia. For further information on the distribution of basidiomycetes in litter see section 13.3.

For conifers, the pattern of development of the fungus flora on needles in litter has some general features in common with the mycoflora developing on the leaf litter of angiosperm trees. One similarity is that the leaf-inhabiting fungi of the phylloplane persist on the needles in the litter for several months after needle fall and some may go on to produce their sexual stages there. Again, there are among these a small group of needle-inhabiting fungi which first appear on living needles in very low numbers but become more abundant once the needles reach the litter. In the litter, the decaying needles are invaded by litter-inhabiting fungi which complete the decomposition of the needles and correspond to the fungi fulfilling a similar ecological role in angiosperm leaf litter.

The microfungal succession on Corsican pine needles (*Pinus maritima*) described by Mitchell and Millar (1978) serves well to illustrate these points. These authors continued to isolate some of the common fungi of the phylloplane and living needles such as *Sclerophoma pythiophila*, *Ceuthospora pinastri* and *Leptostroma pinastri* in abundance from needles for about 6 months following the passage of needles into the litter in September. By December, when the first direct observational records were made, *Lophodermium pinastri*, the teleomorph of *Leptostroma pinastri*, was already abundant on the needles with ascocarps persisting through the winter and surviving until the winter of the following year. *Hendersonia acicola*, which had first been recorded in low numbers on living needles prior to needle fall, grew plentifully on some of the needles in the litter during the autumn and persisted in high numbers until the following summer. *Verticicladium trifidum*, the anamorph of *Desmazierella acicola*, is one of the important agents of decay. This fungus first appeared on needles in the litter during the first winter and continued to be observed fruiting on needles in abundance through to the following spring and summer. This fungus belongs to a group of needle-decomposing species which appear first to colonize needles in the litter. Other litter-inhabiting fungi which appear at this stage are *Sympodiella acicola*, *Troposorella monospora* (*Helicoma monospora*) and *Kriegeriella mirabilis* (*Lophostroma pinastri*). These fungi form networks of hyphae on the needle surface but do not invade the interior and their part in the actual decay of needles may be small.

Few agarics decay conifer litter exclusively and several of the common species of conifer woods, such as *Mycena galopus*, *Clitocybe nebularis* and *Collybia maculata*, also decay deciduous litter. However, *Micromphale perforans* and *Marasmius androsaceus* are two small species that are exceptions. The latter spreads by conspicuous black rhizomorphs colonizing the interior of needles utilizing both lignin and needle polysaccharides (Figure 2.4). It is also common on ericaceous litter in conifer woods. Other agarics that seem particularly to favour conifer needle litter for fruiting include *Cystoderma amianthinum*, *Agaricus langei* and *Hygrophoropsis aurantiaca*.

The succession on *Pinus sylvestris* needles described by Kendrick and Burges (1962) is generally similar to that on Corsican pine, with the needles supporting many of the same fungal species.

As is the case with angiosperm litter, soil-inhabiting microfungi seem to play a relatively minor role in the early stages of needle decomposition. *Penicillium* and *Trichoderma* species may be isolated from decaying needles, especially from the F2 layers of the soil, but colonization appears to be largely superficial. Their ability to colonize the interior of needles appears to be restricted by the prior occupation of the interior by some of the early colonizing litter-inhabiting fungi such as *Desmazierella acicola* (Black and Dix, 1977). Kendrick and Burges (1962) suggested that soil microfungi are present in conifer litter mainly as a heavy spore load which is unrepresentative of their capacity effectively to decompose the needles.

While the general pattern of the succession of fungi on conifer needles in litter may bear some resemblance to the general features of the fungal succession on the leaf litter of angiosperm trees, there are great differences in detail. The most striking contrast is in the composition of the two communities, with only a few of the fungal species of angiosperm leaves occurring on conifer needles. The unique nature of the mycoflora of conifer needles has naturally aroused special interest and has been linked with the high phenolic content of the needles and the tendency for phenols to inhibit fungal growth. However, some of the simple phenols widely distributed in conifer needles are readily utilized by many of the common fungi found in angiosperm litter (Henderson, 1960, 1961). Further, when pine needles are buried in garden loam and soil from deciduous woods they become colonized by the resident soil fungi, and species such as *Gliocladium roseum*, which are not normally associated with conifer litter, may be isolated frequently from needles buried in these soils (Black and Dix, 1977). Probably a more important ecological effect of the presence of phenolic materials in conifer litter is the tendency for phenolics to accumulate in a free state in soils with a low base status, causing the pH to drop and limiting fungal growth. It is significant that Lehmann (1976) and Lehmann and Hudson (1977) were able to increase the incidence of *Cladosporium* species on *Pinus sylvestris* needles and effect colonization by *Epicoccum purpurascens* by adding urea (in urine) to the litter, which in addition to adding nitrogen causes the pH to rise, while the addition of ammonium sulphate, which causes the pH to fall, had no effect on the populations of these species.

When these workers raised the pH of the litter by adding sodium carbonate or potassium hydroxide, some completely unexpected fungi appeared, such as *Ascobolus denudatus*, a fungus found on angiosperm litter only occasionally.

There have been few detailed studies of the fungal successions on lower plants other than those of Kilbertus (1968), who studied the moss *Pseudoscleropodium purum*, and Frankland (1966, 1969) and Godfrey (1974), who studied bracken (*Pteridium aquilinum*). Little of fresh significance has emerged since Frankland reviewed the subject in 1974. The early stages of the colonization of these plants show few special features and *Cladosporium* species and most of the fungi associated with the phylloplane of angiosperms appear on living leaves from May onwards. Kilbertus (1968) failed to record *Aureobasidium pullulans* on *P. purum* and it would be interesting to learn if this exception is general for all mosses.

However, some notable differences occur in the mycoflora of the litter of lower plants, with bracken fronds, in particular, showing some peculiar features. Fronds decay noticeably more slowly than angiosperm leaves and by contrast fronds are invaded early in the fungal succession in the litter by cellulose- and lignin-decomposing basidiomycetes and members of the Deuteromycotina, which become the dominants by the end of the second year. Prominent members of this community include *Phialophora* species and several other well-known wood soft-rots and the litter-decomposing basidiomycete *Mycena galopus* (Frankland, 1966, 1969). Soil-inhabiting moulds increase late in the succession and *Mucor hiemalis* and other members of the Zygomycotina finally come to dominate the well-rotted tissues by the end of the fifth year. The interesting feature of this succession is that in many respects it resembles the decay of wood; the early invasion by lignin decomposers, the appearance of soft-rots, and the slow rate of decay are all characteristics of wood decay. This is presumably due to the chemical and anatomical similarities which the fronds share with wood, notably a low nitrogen level and a high lignin content due to the presence of many lignified fibre cells.

4.5 DECOMPOSITION OF LEAF LITTER

Plant litter may remain on the soil surface for several months and even years before it completely decays and the residues have been finally incorporated into the mineral soil. Under certain circumstances, in plant communities where litter production is high and rates of decay are slow, litter may accumulate on the soil surface in layers several centimetres deep. In Britain this frequently occurs in forests, where several tonnes of litter from trees falls on the forest floor annually. Figures for litter production by oak (*Quercus petraea*) at 3.8 t ha^{-1}year^{-1} and Norway spruce (*Picea abies*) at 5.7 t ha^{-1}year^{-1} are fairly typical for mature trees of angiosperms and conifers in temperate zones, although these figures cannot compare with those for litter production by trees in equatorial forests, where mean annual litter production is about twice this amount (Bray and Gorham, 1964). About 65–75% of the litter that falls on to the forest floor is leaf material.

Leaves accumulating in the upper layers of the mass of litter on the soil surface in the Aoo horizon are uncompacted and very much affected by the fluctuating physical conditions in the atmosphere above. These leaves dry out quickly after wetting and conditions are generally unfavourable for fungal growth in the long term. Growth is slow and spasmodic, and little sign of decomposition of the litter is apparent. As more litter falls, the leaves become buried and pass into the Ao horizon. During this process, the moisture content increases, as some protection from desiccation is afforded by the layers above, and the rate of decay consequently increases. In the upper F1 fermentation layer of the Ao there is little alteration in the physical appearance of the leaves but a considerable drop in the tensile strength of the material is noticeable, indicating that decay is proceeding and that the structural polymers of the cell walls are being degraded. In the deeper F2 fermentation layer decay symptoms become more recognizable and the leaves become fragile and progressively more fragmented.

The final product of microbial decay of the leaves is **humus**, which becomes slowly mixed with the mineral soil in the A horizon below the fermentation layers. Humus is a complex mixture of polymeric phenolic residues, mainly derived from plants, combined with carbohydrates and nitrogenous materials of plant, animal and microbial origin. The nitrogen content is fairly high, about 5% on average, and about 30% of the total carbohydrate content can be extracted as simple C_6 and C_5 sugars. In spite of the relatively high nitrogen and carbohydrate levels, the rate of further decomposition is slow, and in undisturbed soils, under certain physical conditions, humus may accumulate as an H layer, beneath the F2, in the Ao horizon as the rate of formation exceeds the rate of removal. Humus decay can be so slow that it has been estimated that the older humic fractions of some soils may be several hundred years old. The stability of humus is due to the high phenolic content, which inhibits microbial growth, to the presence of a number of recalcitrant molecules, which cannot be respired by most microbes, and to a tendency for the phenolic substances in the humus to complex with the more readily degradable components, so protecting them from microbial attack. Quinones, for example, arising from the oxidation of plant phenols in the soil by microbial polyphenol oxidases, stabilize polypeptides, amino acids and amino sugars in humus by the formation of complexes in which the amino group of these compounds becomes attached to the aromatic nucleus of the quinone (Haider and Martin, 1970; Bondietti, Martin and Haider, 1972). The effectiveness of phenolic complexes of this type in resisting degradation is well known, and artificially formed complexes between tannins and substances which are normally readily hydrolysed, such as cellulose, hemicellulose or gelatin, become equally resistant to decomposition when added to soil (Benoit and Starkey, 1967; Benoit, Starkey and Basaraba, 1968).

It is probable that some of the phenolic material in humus is also of microbiological origin. In culture, several soil fungi synthesize simple phenols such as resorcinol from sugars and oxidize these to quinones from which polymeric brown melanic substances form (Martin and Haider, 1969; Martin, Haider and Wolf, 1972). Some of these reactions have already been described in Chapters 2 and 3 in relation to the oxidation of aromatic compounds by polyphenol oxidases. The humic acid fraction of humus contains resorcinol and compounds whose chemical properties and infrared spectra resemble fungal melanins (Haider and Martin, 1967; Filip et al., 1974). According to Schitzer and Negroud (1975), however, humic acids produced by fungi are relatively complex, containing aliphatic and aromatic compounds, and have a low phenolic component. As such, they are not very comparable either with the humic or fulvic acid fraction of soil humus or with the fungal compounds of Martin, Haider and Wolf (1972).

The rate of decomposition of leaf litter and its incorporation into the mineral soil are processes whose rates vary enormously depending on physical conditions and plant species. The leaf litter of trees such as *Alnus glutinosa, Fraxinus excelsior, Ulmus* species, *Tilia* species and *Acer pseudoplatanus* generally decay relatively quickly and take less than a year to become reduced to humus. On the other hand, leaf litter from trees such as *Pinus sylvestris, Fagus sylvatica* and *Quercus robur* decay relatively slowly (Mommaerts-Billiet, 1971). Generally, the leaf litter of grasses, herbs and some shrubs decay relatively rapidly, the leaves of deciduous trees decay less rapidly, and the needles of conifers and the shoots of mosses decay the slowest of all (Mikola, 1954). The high durability of certain leaf litters has been correlated with a high polyphenol content and high C:N ratios, which tend to make the litter unpalatable and unacceptable as food for soil animals. In food selection experiments, the preference of earthworms for certain leaves corresponded well with low free polyphenol levels and low C:N ratios in the leaves but less closely with texture or hardness (Satchell and Lowe, 1967). Qualitative differences may also be important; Anderson (1973) found that soil animals preferred *Castanea sativa* litter to that of *Fagus sylvatica*, although the polyphenol content of the former was twice as high. Food selection experiments indicate that tannic and protocatechuic acids if present in high concentration may discourage feeding while the catechol and gallic acid content appear to be less important (Satchell and Lowe, 1967).

The significance of these experiments relates to the fact that the rate of decomposition of many types of leaf litter, especially that of trees, is dependent markedly upon the rate at which litter becomes fragmented. This is largely the work of the myriads of small soil animals that ingest plant litter, comminuting it as they feed. Slugs, millipedes, some mites and collembola, insect larvae, and the larger forms of earthworm, such as *Lumbricus terrestris* and *Allolobophora longa*, are the most important. Their part in litter decomposition and nutrient recycling in soil is well documented in the reviews of Jensen (1974), Wallwork (1976) and Anderson and Ineson (1984). The fragmentation of the leaves and the gradual reduction in size of the components of the litter operates to speed decomposition by promoting microbial growth in two ways. The fragmented litter compacts more easily and consequently a marked improvement in the water-holding capacity of the litter follows. This is a notable feature of the transition of the litter from the Aoo, where it is more or less intact, uncompacted, very dry and with a small animal population, to the F2 where the litter is much wetter, fragmented, more compacted and supports a large animal population. The second important improvement brought about by animal feeding is that fragmentation increases the surface area for microbial attack. This is especially important for bacterial decomposition, for unlike fungi, whose hyphae can penetrate tissues, bacterial growth is largely confined to the surface and as such is dependent upon the area of exposed tissue.

Fungi play little or no part in the fragmentation of leaf litter and when soil animals are excluded from the decomposition process the rate of physical change occurring in the litter can be seriously affected. Edwards and Heath (1963) and Heath, Edwards and Arnold (1964) found that when tree leaf litter was placed in nylon mesh bags on the forest floor little loss or obvious physical change occurred over one season in bags whose mesh size was 0.003 mm, which exposed the leaves to microbiological invasion but was small enough to exclude the soil animals. At larger mesh sizes above 0.5 mm, which were big enough to admit the soil animals, the leaves were broken up and disappeared. However, the action of soil animals in accelerating the rate of decay of leaf litter appears to be more important in the case of the tougher leaves from trees than is the case for the softer leaves of herbs. In further experiments, Heath, Arnold and Edwards (1966) found that increasing the mesh size of bags containing leaves of kale (*Brassica oleracea*), beet (*Beta vulgaris*) or lettuce (*Lactuca sativa*) did not significantly increase the rate of weight loss from leaves, which disappeared equally fast from all bags irrespective of mesh size. Curry (1969) concluded from similar experiments that soil animals may contribute little to the rate of disappearance and decay of grassland herbage.

Nevertheless, flourishing animal communities in the soil are recognized as having an important part to play in promoting decay of many sorts of leaf litter by fungi and other microorganisms and where grazing animal populations are reduced through lack of suitable food, litter decomposition is greatly slowed down.

The growth of fungi in leaf litter is also influenced by the feeding of soil animals in other ways. Flourishing animal communities in soil enhance microbial growth by greatly improving nitrogen supplies. The grazing animals void faecal pellets which have a high nitrogen content and in effect, as the animals feed on the litter, they concentrate nitrogen, and locally increase its abundance in a nitrogen-limited environment.

The growth of fungi on the leaf litter also produces important reciprocal effects for the benefit of soil animals. Fungal respiration lowers the C:N ratio and polyphenols become oxidized and transformed to innocuous substances. Fungal hyphae increase the protein content of the litter, and cellulose and lignin which cannot be digested by most soil animals are transformed into digestible fungal carbohydrates. As a result, the litter becomes more palatable and more nutritious for the soil animals.

The life of animals and fungi in the soil is thus strongly bound together, and where fungal growth is inhibited for any reason (low pH, low moisture content or the presence of inhibitory phenols), the process of rendering the litter palatable to the soil animals is slowed down. Moisture plays a key role in this story and it is possible that one of the main reasons why some leaf litters such as pine decay more slowly than others is because they often lack bulk, hold little water and dry to critically low water potentials for fungal growth faster than other leaf species (Dix, 1984a). If animals cannot feed, the litter remains unfragmented, loose and at low water potentials for much of the time. Under these conditions the agarics, the major decomposers of the litter, which require relatively high water potentials for growth, cannot invade.

Leaf litter with a tendency to dry out quickly, with a high C:N ratio and a high polyphenol content therefore tends to decay slowly and may accumulate over the mineral soil. In these circumstances it may take many years before a leaf falling onto the soil surface completely decays. It has been calculated, for example, that for *Pinus sylvestris* needles, which have a high polyphenol content and an exceedingly high C:N ratio of the order of 66:1, it may take more than 10 years for the residues finally to enter the humus layer (Kendrick, 1959).

The level of simple polyhydroxyphenols and tannins in plants apparently may also vary with the base status of the soil, and beech (*Fagus sylvatica*), birch (*Betula* species), oak (*Quercus* species), sycamore (*Acer pseudoplatanus*), Douglas fir (*Pseudotsuga taxifolia*) and Scots pine (*Pinus sylvestris*) litter has been found to contain more polyhydroxyphenols when growing on soils with a low base status (Coulson, Davies and Lewis, 1960; Davies, Coulson and Lewis, 1964). Plants such as these producing litter with a high phenolic content on soils with a low base status give rise to a **mor** type of soil with a low pH and a high free polyphenol content in which microbiological activity is inhibited and the populations of soil animals feeding on the litter are greatly reduced. In such soils decomposition is exceedingly slow and litter and humus consequently accumulate. Mor conditions in Britain are associated with forests on uplands where the base status of the soil is low due to the high rainfall which leaches the soil. On base-rich or **mull** soils the soil is buffered and litter decomposition tends to be relatively fast since the free polyhydroxyphenols which are released from the litter polymerize at the higher pH into more innocuous substances (Coulson, Davies and Lewis, 1960).

To date there have been few studies of the effects of pollutants on leaf litter populations and the consequences for litter decomposition and nutrient recycling. By contrast, corresponding studies of the effects of pollutants on phylloplane populations have been much more numerous (Magan and McLeod, 1991).

Data for the effects of exposure to sulphur dioxide on the fungal communities of several different sorts of leaf litter indicate that there can be selective effects, as there are when phylloplane populations are exposed to pollutants. The exposure of angiosperm leaf litters to sulphur dioxide in the range 0.010–0.030 $\mu l\ l^{-1}$ in the field by Newsham *et al.* (1992a,b) resulted in an increase in the abundance of some fungal species and a decline in others. Little nett change was noted in the extent of fungal colonization, in spite of which there was a reduction in the rate of litter decomposition, presumably due to a reduction in the diversity of species. Newsham *et al.* (1992a,b) found little evidence that isolates of species from sites more heavily polluted with sulphur dioxide were naturally more resistant to sulphur dioxide exposure than were isolates of the same species from less heavily polluted sites.

5
DEVELOPMENT OF FUNGAL COMMUNITIES ON HERBACEOUS STEMS AND GRASSES

The annually produced leafy stems of small plants offer a practical opportunity for studying the effects of physical variables and differences in age of leaves and shoots on the development of fungal communities on whole plants. Few, however, have attempted any detailed comparative studies of this nature and in this respect the investigations of Webster (1956, 1957) on *Dactylis glomerata* remain exemplary. This grass, commonly known as cocksfoot, increases in size annually by means of tillers. These form during the growing season from buds at the base of the shoots and overwinter as a number of short internodes with a large number of associated leaves that collectively form a large basal tussock. In the following spring new shoots develop by the expansion of some of the basal internodes of the tiller. Each expanded shoot consists of about six internodes, the developing inflorescence and a corresponding number of leaves. The latter are typical of grasses and are joined to a leaf sheath which is attached at a node and partially encircles the stem. At maturity the stem with its inflorescence stands about 100 cm above the ground and is consequently more exposed to climatic changes at the top than at the base in the tussock.

During the development of the shoot the internodes expand unevenly and those at the base remain rather short with internodal distance increasing progressively up the stem. The leaves unfold as the stem slowly elongates and those produced by nodes at the base are formed fully 3 months before those at the apex. There are never more than about five functional leaves present on the stem at any one time and leaves at the base senesce as new leaves form above, so that by the time the inflorescence has opened, usually in May, several dead leaves and sheaths are already present at the base. Once seeds have formed, senescence gathers apace, progressing upwards at a fast rate such that by about late September all the leaves and their sheaths at the upper internodes are in an advanced state of senescence or are dead. As a consequence of these events, the life of the leaves of the upper nodes is rather short compared with those below. Stem senescence progresses from the inflorescence to the base, the lowest internodes within the tussock remaining alive the longest.

5.1 DISTRIBUTION OF FUNGAL POPULATIONS ON *DACTYLIS GLOMERATA*

Webster (1956, 1957) followed the fungal colonization of the plant for almost 2 years after flowering, covering a period of about 18 months while the stems remained erect and continuing past the second winter on stems that had collapsed. Colonization was followed on plants collected from three localities and the fungi present recorded by the recognition of sporulating bodies by microscopic examination after incubation of whole stems in damp chambers for 2 days. While this method of recording is generally satisfactory, it is now recognized that it does have some weaknesses and records may not always be fully representative. If the sporulating structures are small or not easily recognized, they can be overlooked and a further problem is that a fungus may be present as mycelium but may not sporulate for some reason.

For the purposes of recording, internodes were numbered from the base of the stem upwards (Figure 5.1). All internodes and leaves at the base of the tiller formed before stem elongation were labelled internode I. Fungi began to sporulate on leaves and sheaths at lower internodes soon after flowering, spreading rapidly upwards as senescence progressed. Numbers of sporulating colonies reached a peak at all internodes during September and October with the total numbers and numbers of species being greatest at the inflorescence and at internodes V and VI (Figure 5.2). At the lowest internodes, I and II, numbers of sporulating colonies remained very low throughout this period and whereas at upper internodes fungi colonized both leaves and stems, here colonization was largely confined to the leaves. Numbers of sporulating colonies declined to low levels at all internodes during the winter, but this was essentially a temperature effect, as opposed to nutritional shortages, since the pattern repeated itself in the following year. Second peaks were more pronounced at internodes II and IV compared with internodes V and VI and numbers of sporulating colonies reached higher levels at all lower internodes compared with those attained there previously. The species richness increased at all internodes as species not previously recorded sporulated. Most stems collapsed to the ground in

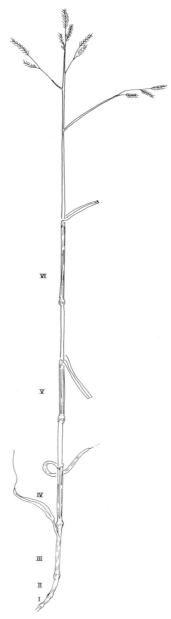

Figure 5.1 *Dactylis glomerata* showing form of tiller and internode numbering system.

FUNGAL POPULATIONS ON D. GLOMERATA 131

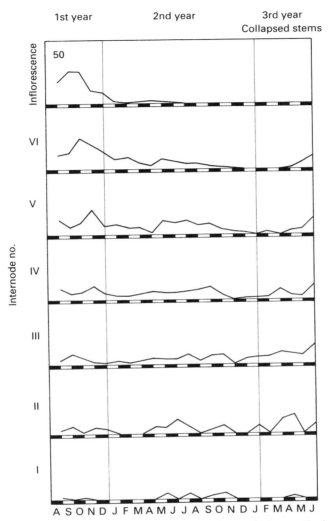

Figure 5.2 Monthly records of total numbers of fungi sporulating on the internodes of ten stems of *Dactylis glomerata* after flowering. For each internode the height of the vertical axis represents 50 species. (Redrawn from Webster, 1956.)

the second winter but supported sporulating colonies again in the spring at all internodes (Figure 5.2).

Detailed analysis of the fungal populations revealed several distinct patterns of behaviour. On upright stems Webster (1956, 1957) recognized several groups. For simplicity these have been reduced to four, separating those that were recorded at upper internodes from those confined to lower

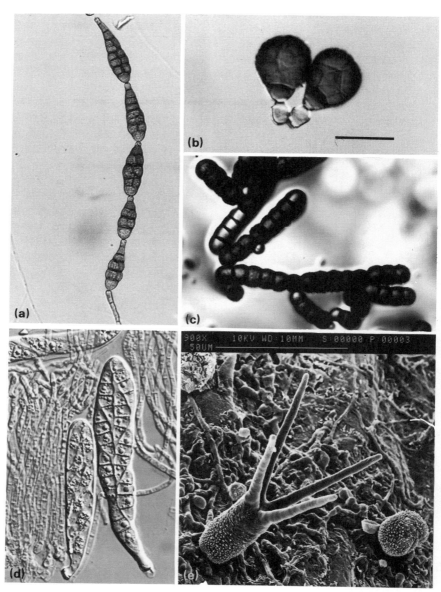

Figure 5.3 Primary and secondary saprophytes of *Dactylis glomerata*. (a) *Alternaria alternata* (= *A. tenuis*): conidiophore with chain of conidia. (b) *Epicoccum nigrum*: two conidiophores with conidia about to separate. (c) *Torula herbarum*: branching chains of conidia. (d) *Phaeosphaeria eustoma* (=*Leptosphaeria microscopica*): asci and pseudoparaphyses. (e) *Tetraploa aristata*: mycelium with one developing and one mature conidium. Scale bar = 30 μm (a); 20 μm (b,c,d); 50 μm (e, separate scale).

internodes and distinguishing between primary and secondary colonizers among both.

The most frequently recorded group were the primary saprophytes of the leaves and stems, the commonest members of which were the anamorphs *Cladosporium herbarum*, *Alternaria tenuis* (*Alternaria alternata*) and *Epicoccum purpurascens* (*Epicoccum nigrum*) (Figure 5.3). *C. herbarum* records almost certainly also included *C. cladosporioides* since these two species are equally common but could not have been separated by the methods employed. Sporulation of primary saprophytes was first recorded in low frequency on leaves at lower internodes in May and progressed upwards as the season advanced. In due course many were recorded at upper internodes at 100% frequency and heavy infections of *C. herbarum* in particular developed on the senescing leaves at the same time as it largely disappeared from lower internodes. After the winter decline, a second flush of these species appeared at upper internodes in the following spring and summer. Altogether they persisted at the upper internodes for about 15 months until the stems collapsed in the second winter, after which they did not reappear. The records for *Alternaria tenuis* are shown in Figure 5.4a.

Leptosphaeria microscopica (*Phaeosphaeria eustoma*) (Figure 5.3) and *Pleospora vagans* (*Phaeosphaeria vagans*), ascomycetes recognized by their pseudoperithecia and pycnidia, are also members of this group. These were recorded less frequently than other members of the group but had a similar distribution pattern. Records of pseudoperithecia reached a peak in August and September after flowering with sporadic records in June and July. Since pseudoperithecia take some time to form, their appearance must have been the result of colonization that had taken place several weeks beforehand. *Aureobasidium pullulans* and *Botrytis cinerea* were only recorded in low frequencies by the methods employed in this study.

The sporulating structures of *Selenophoma donacis* and *Mycosphaerella recucita* were absent or only sporadically recorded in the summer of flowering but these species fruited abundantly at upper internodes the year after. They appear to be secondary saprophytes of upper internodes, never appearing below internode III. They are probably weak competitors and significantly they disappeared quickly from collapsed stems.

Among the primary saprophytes that were recorded immediately after flowering at lower internodes, some were identified as species which did not spread to upper internodes as senescence progressed. Their behaviour was typified by *Acrothecium* sp. and *Leptosphaeria nigrans* (*Phaeosphaeria nigrans*) and is illustrated in Figure 5.4b.

The former was confined to internode IV and below, the latter spreading as far as internode V in very low frequencies. Both reappeared and increased in frequency at lower internodes in the following spring. They declined after stems collapsed, *Acrothecium* sp. became very sparse but *L.*

134 FUNGAL COMMUNITIES ON HERBACEOUS STEMS AND GRASSES

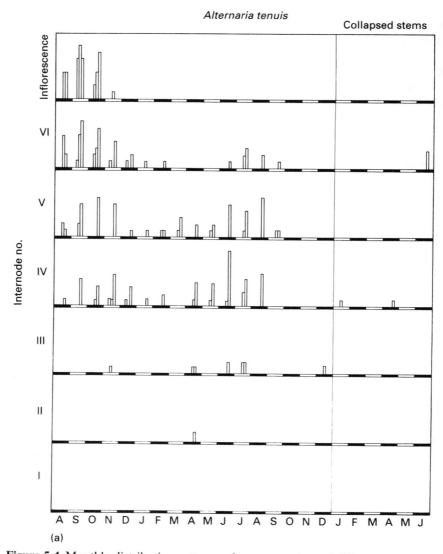

Figure 5.4 Monthly distribution patterns of representatives of different ecological groups of fungi sporulating on the internodes of *Dactylis glomerata* after flowering. Records are for three localities and are for combined collections on stem and leaf sheath (maximum 10). (Redrawn from Webster, 1957.)

nigrans held on and sporulated on all internodes. Both are evidently weak competitors.

These were followed at lower internodes by *Tetraploa aristata* (Figure 5.3) and *Mollisia palustris*. These are typical of the secondary saprophytes

FUNGAL POPULATIONS ON D. GLOMERATA 135

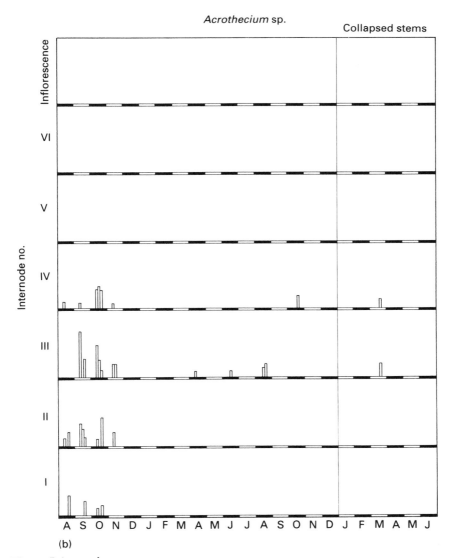

Figure 5.4 *contd.*

of the lower internodes of the grass. Their sporulating structures were not recorded until the spring following flowering, occurring rarely above internode IV or V respectively. They increased in frequency on collapsed stems where they were able to spread to upper internodes (Figure 5.4c). Strong competition from these species may be one reason why primary saprophytes disappeared from upper internodes after the stems collapsed.

136 FUNGAL COMMUNITIES ON HERBACEOUS STEMS AND GRASSES

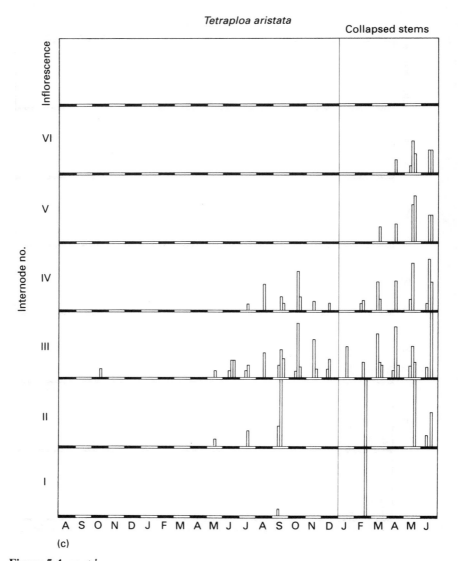

Figure 5.4 *contd.*

5.2 INTERPRETATION OF DISTRIBUTION PATTERNS

Two of the more important observations which stand out from this study are first, that fungal populations sporulating on erect grass stems and their leaves show clear qualitative and quantitative differences in distribution, and secondly, that as time passes the structures of these communities appear to change as the sporulating structures of additional species are detected while others disappear. It is not difficult to suggest why these differences exist. Contrasting physical conditions must occur at different levels of the stem which must have an important selective effect on colonization. Differences in age of the various leaves and parts of the stem, the likely effect of this on nutritional status at senescence and the changing nutritional conditions in the tissues as time passes must also be influential.

The greater numbers of fungi sporulating on leaves and stems at upper internodes immediately after flowering and the greater species richness of the community there compared with lower internodes is undoubtedly due to the fact that the primary colonizers do not sporulate at lower internodes for very long but continue to sporulate until winter at the upper internodes. This may reflect a difference in the nutritional status between the lower and upper internodes at senescence. The fact that at lower internodes stems remain alive longer and so are able to resist colonization may also be important in this respect and may explain why numbers increase substantially at lower internodes in the summer of the second year after the lower internodes die.

A decline in the nutritional status of upper internodes following the first flush of fungal growth is probably the main reason why lower numbers sporulate at upper internodes in the following year.

The division of the fungal populations into those that sporulate on upper internodes and those that do not may be controlled largely by purely physical factors and, of these, moisture is probably the most important. Plant cover, even if dead, reduces water loss from the soil, and near to the ground the atmosphere around plants can often be saturated with water up

Table 5.1 Mean moisture content (% fresh wt) of erect stems of *Dactylis* in the year after flowering

Internode	Date						
	11 Sept	16 Mar	25 Apr	5 Jun	11 Jul	7 Nov	26 Nov
Inflorescence	3.6 (10.4)[a]	17.8	–	–	7.2	16.9	16.8
VI	4.6 (35.4)	14.5	11.0	8.7	9.3	13.0	17.4
V	4.9 (63.4)	14.8	12.1	8.7	11.2	14.3	36.3
IV	7.7 (63.6)	15.8	11.8	9.7	12.5	22.4	61.1
III	51.3 (67.2)	33.2	17.6	10.6	19.6	52.2	72.9
II	67.7 (62.1)	74.2	54.0	18.2	31.1	58.7	77.9
I	–	–	76.8	62.6	43.6	32.0	–	78.0

[a]Numbers in parentheses are values for stems of the current year in the same tussock.
Source: Recalculated from Webster (1956).

to about 10 cm above soil level. Beyond this, however, the humidity gradient becomes very steep, declining rapidly to low levels with increasing height above the soil. In *Dactylis glomerata* a steep declining humidity gradient will occur from the basal internodes sheltered in the tussock to the exposed inflorescence. At certain times of the day in summer water saturation deficits in a tussock 7.5 cm above soil level can be 3–5 times lower than at comparable heights in non-tussock-forming grasses (Luff, 1965). High moisture contents were recorded at the base of dead stems on many occasions during the year when at the same time values below 20% of the fresh weight were recorded at upper internodes and at the inflorescence of *Dactylis* (Table 5.1). The upper internodes of erect stems receive more rain and dew than lower internodes but due to exposure they tend to dry out very quickly and are thus more liable to be subjected to moisture extremes in wetting and drying cycles than lower internodes. Dead and senescing leaves at upper internodes have lower moisture contents than living leaves. Consequently they heat up more rapidly when surface water has evaporated than living leaves that have an internal water supply to cool them (Dowding, 1986). These physical extremes make senescing leaves at upper internodes very inhospitable habitats for fungal growth.

Physiological studies have shown that fungi colonizing the upper internodes of the grass are capable of germination and growth at lower humidities than those that are confined to lower internodes (Webster and Dix, 1960). Inhabitants of upper internodes germinate when the relative humidity is 92%, or lower in some cases, and if nutrients are available, this figure can be reduced considerably (Table 5.2). Generally, higher relative humidities are required for the full development of mycelia and sporulation (Magan and Lacey, 1984a). At 20°C 89% relative humidity is equivalent to a water potential of -15.8 MPa, 92% is -11 MPa, values which arise in undecayed leaves when the fresh weight moisture content lies somewhere between 7 and 10%. The latter assumption is based on measurements taken on deciduous tree leaves by Dix (1985) and estimates made from the data of Myrold *et al.* (1981) indicate that these values would not be expected to differ greatly for grasses. Thus it seems probable from some values obtained for the moisture content of erect stems that the fungal colonists of upper internodes would, at times at least, find sufficient water available there for growth (Table 5.1) especially when it is appreciated that the above assumptions are based on measurements made on undecayed material and that decayed material will usually have a greater water potential than undecayed material at the same moisture content.

The general decline in the moisture content of stems from one season to the next, as illustrated in Table 5.1, will undoubtedly have an impact on fungal growth and may contribute to the lower levels of sporulation recorded for fungi at upper internodes in the second season after flowering.

Table 5.2 Minimum relative humidities (%) for germination and growth of some leaf saprophytes

Species	Germination (with nutrients bold)	Linear growth
Alternaria alternata (*A. tenuis*)	89 **84.5**	(88)
Cladosporium cladosporioides	(86)	(88)
Cladosporium herbarum	89 **84.5**	(90)
Epicoccum purpurascens (*E. nigrum*)	92 **88**	(89)
Leptosphaeria microscopica (pycnospores and ascospores)	92	93
Pleospora vagans		
Ascospores	89	93
Pycnospores	92	
Leptosphaeria nigrans		
Ascospores	95	94
Pycnospores	100	
Tetraploa aristata	98	94
Torula herbarum	100 (H_2O)	95
Acrothecium sp.		94

Sources: Data from Dix (1958) at 20°C and in parentheses from water activity measurements of Magan and Lacey (1984a) at 25°C.

Fungi confined to lower internodes on the other hand demand that water be more readily available and show no adaptation to life in exposed dry conditions. These require water or at least 98% relative humidity for germination (or a water potential of about -2.7 MPa). On the same assumptions as above, this is equivalent to a moisture content of grass tissues of about between 25 and 30% of the fresh weight. The opportunities for these fungi to colonize upper internodes at times when temperature conditions are also favourable would therefore appear to be very limited. By contrast, lower internodes sheltered in the tussock would provide a much more suitable environment for their growth (Table 5.1). It is significant that many of these fungi can spread to upper internodes after stems collapse. While stems remain erect, these physical differences persist and keep the populations of the two communities apart and account for the unusually long length of time that primary colonizers persist on the stems as vigorous saprophytes.

Fungi which inhabit the upper internodes of this grass are well-known primary colonizers of the leaves and stems of plants. Although the number of species is rather small, they are cosmopolitan, occurring commonly in similar situations on many plants. They seem to show several adaptations to an exposed life in the phylloplane and on non-woody plant surfaces. These features are discussed in detail in Chapter 4 but it is worthwhile drawing attention to some of these here. All form some kind of resting structure by means of which they survive shorter or longer periods of unfavourable climatic conditions. In some cases these structures take the form of **sclerotia**, specialized mycelial masses with a resistant outer covering; in others, reproductive **pseudoperithecia** or **pycnidia** serve a similar function. Both structures have the capacity to produce and release spores quickly when favourable conditions return. These structures thus serve two purposes: they carry the population through unfavourable conditions and they enable the fungi to spread rapidly to occupy available substrata. These fungi are opportunists and these structural adaptations give them a special advantage in the lifestyle that they have adopted.

A straightforward interpretation of Webster's records for *Dactylis* would suggest that simple successions are operating such that over a period of time some species disappear, perhaps because competition becomes too intense or the quality or quantity of available nutrients becomes unfavourable, as new species colonize and join the communities. However, when presence or absence of a species is dependent on sporulation it is conceivable that spores may not be produced even if the mycelium is present. This may be due to unsuitable nutritional conditions and may be an explanation of why some species 'disappear' as stems age or to the fact that certain sporulating structures take a long time to appear even when conditions are ideal. This could mean that some species persisted longer than was recognized while others may have colonized earlier than was recorded.

Those that have made similar studies on other plants but who have also used culture techniques, plating out washed fragments of the plant, found that *Alternaria tenuis*, *Cladosporium herbarum* and other primary saprophytes may be present as mycelium at lower internodes some time after they were unable to sporulate at these levels (Hudson and Webster, 1958; Yadav and Madelin, 1968).

5.3 DISTRIBUTION PATTERNS ON OTHER PLANTS

Studies on *Dactylis glomerata* were followed by those on another grass, *Agropyron repens* (Hudson and Webster, 1958), and these revealed a remarkably similar picture of the pattern of colonization, differing only in some qualitative aspects of the mycoflora. The fact that *A. repens* (couch grass or twitch) is non-tussock-forming but still showed up major differences in the distribution of species at different levels on the stems suggests that moisture content or the nutritional status of stems are more important regulators of fungal growth than atmospheric humidity.

On *Agropyron* direct observation and cultural techniques were employed to follow colonization, the latter showing up the presence of *Aureobasidium pullulans* in large numbers. This has since become recognized as another widely distributed dominant primary colonizer of leaves and stems. This fungus can easily be overlooked by direct recording techniques because of the form and size of its sporulating structures and this must explain why it was not recorded in great numbers by Webster (1956, 1957) on *Dactylis glomerata*.

Seasonal studies of the fungal colonization of whole plants were carried further by Yadav (1966) and Yadav and Madelin (1968), who used the methods of Hudson and Webster (1958) to investigate populations present and sporulating on *Heracleum sphondylium* (hogweed) and *Urtica dioica* (nettle) respectively. These dicotyledonous herbs make an interesting comparison with monocotyledonous grasses, with which they show some similarities. Nettle and hogweed produce tall flowering stems standing about 1 m high. Hogweed is generally the taller, often exceeding a metre in height with fewer but longer internodes. The shoots begin to die back soon after flowering and become available for fungal colonization from about August onwards.

In these plants senescence is **basipetalous** and fungal colonization of shoots begins at the top of the plant and progresses towards the base. As with grasses, *Alternaria tenuis* (*Alternaria alternata*), *Epicoccum purpurascens* (*Epicoccum nigrum*) and *Cladosporium herbarum* were among the early invaders in the summer of flowering. *Aureobasidium pullulans* was identified on hogweed when cultural techniques were employed. *Botrytis cinerea* was recorded more frequently at this stage on both nettle and hogweed compared with grasses and this may be a reflection of the higher moisture content of their tissues at senescence.

One essential difference was that *Leptosphaeria microscopica* and *Pleospora vagans* were not recorded on nettle or hogweed. Corresponding species were *Leptosphaeria acuta* on nettle and *Phomopsis asteriscus* on hogweed. These fungi seem to occur in similar niches on different plants and presumably represent a more specialized group of primary colonizers whose distribution is restricted by interaction with hosts (they may be weak parasites) and possibly also by unique growth requirements.

As time passed, the early colonizers of the nettle and hogweed were joined in the same year by other species which became numerically important members of the community. Apart from *Torula herbarum*, these populations were qualitatively very different from species found on grasses, but hogweed and nettle have some of these species in common. Some, such as *Dendryphion comosum* and *Periconia cookei*, became evenly distributed on stems but others, such as *Torula herbarum*, were confined to lower internodes where the community was more varied compared with the corresponding group represented by *Acrothecium* sp. and *Leptosphaeria nigrans* on *Dactylis glomerata*. At this stage of the decomposition of the stems, several populations were also present which were recorded on only one of the plants. For example, *Stachybotrys atra* and *Hormiscium* sp. were found only on hogweed while *Leptosphaeria doliolum* and *Pleurophragmium simplex* were recorded only on nettle.

In the second year examination of the lower internodes of upright stems yielded many more species, most of which were recorded only in low frequencies. This group also contained a greater variety of species than the corresponding group on grasses, most of which were present only on one of the plant species. Only *Ophiobolus erythrosporus*, *Dictyosporium toruloides* and *Trichocladium opacum* were recorded on both. In this group *Dasyscyphus sulphureus* and *Pyrenopeziza urticola* on nettle seem to correspond to *Dasyscyphus grevillei* and *Pyrenopeziza revincta* on hogweed and perhaps occupy similar niches on their respective plants.

Similar patterns of succession have been observed on plants growing in other climates. In warmer climates there are some species differences and there is a general tendency for diversity of species to increase. Hudson (1962) studied sugar cane (*Saccharum officinarum*), a plant which apart from a difference in scale resembles temperate grasses in form. In one season it produces up to 30 very large leaves, about 1 m in length, on very tall stems. In this case the fungal succession on the leaves followed much the same course irrespective of the position of the leaves on the stem. Very early colonizers of green leaves included *Leptosphaeria sacchari*, *Guignardia citricarpa* and other parasites. On senescing leaves these were joined by *Cladosporium herbarum*, *Alternaria tenuis* (*Alternaria alternata*), *Nigrospora sphaerica* and *Curvularia lunata*, all of which first appeared on basal leaves and spread upwards. These saprophytes have also been recorded on other tropical monocotyledons (Meredith, 1962; Lal and Yadar, 1964). Hudson (1962) did not record *Epicoccum nigrum* or *Aureobasidium pullulans*, a major difference compared with temperate grasses. The early colonizers of sugar cane were followed 2–3 months later by *Periconiella echinochloae* and a few other species which were in turn followed by a third, larger invasion, which included *Spegazzinia tessarthra*, *Lophodermium arundinaceum* and *Anthostomella minima* among the commonest species. These fungi appeared first on leaves at lower internodes and in time of appearance resembled the secondary colonizers of the temperate grasses. All of these species eventually appeared on leaves at all internodes. The fact that there is no distinction between the populations at upper and lower internodes may be a sign that all of these fungi are tolerant of dry conditions. On the other hand, it may mean that by contrast with temperate grasses there is always sufficient moisture present in the stems and leaves at upper internodes. The larger leaves and more bulky stems of sugar cane could have something to do with this. Sugar cane and temperate grasses have one secondary colonizing species in common, *Tetraploa aristata*, and it is interesting to compare its behaviour in this connection. On temperate grasses it remains at lower internodes but on sugar cane it behaved as all other fungi and colonized leaves at all levels on the stem.

Several general conclusions can be drawn from the above studies. When the stems and leaves of herbs and grasses become senescent, they are first colonized by a small group of vigorous primary saprophytes which eventually colonize all parts, following senescence either up or down the stem. On plants growing in temperate climates they sporulate longer at upper internodes and probably persist for a longer period of time there. There is some evidence that they are adapted to life in exposed conditions, especially with respect to their ability to grow and survive at low water potentials. Several, notably *Cladosporium herbarum* and *Alternaria alternata*, seem to be cosmopolitan in distribution and are associated with plants growing in temperate and warm climates. Some are essentially specialized parasites which are associated with specific hosts. Primary saprophytes of plants growing in temperate conditions also include species which are confined to lower internodes and are suspected of being intolerant of dry conditions. Significantly, they readily spread to the upper internodes of stems once they collapse. More species are counted as primary saprophytes on dicotyledons than on monocotyledons.

Stems which remain erect until the following year come to support the sporulation of an increasingly diverse community, many species of which occur in low frequencies. In many examples there is a clear separation into those that occur on dicotyledons and those that occur on monocotyledons, while the distribution of some may be even more specific. A distinction can also be made between populations occurring on monocotyledons in temperate and tropical climates. Allowing for the fact that in some cases the late appearance of the sporulating structures of these species may be due to a long maturation period, most of this group can probably be correctly regarded as representing a later wave of invasion by secondary colonizers. On all plants studied in temperate regions, some species were noted in this group which did not spread to upper internodes while stems remained erect.

More work is perhaps needed to help explain some of the differences in the patterns of development of these communities, but we are probably fairly safe in assuming that the general trends can be explained by the uneven distribution of moisture in the stems, adaptations or otherwise of the fungi to low moisture availability, climatic variation and physiological differences between plant species.

A conspicuous feature of all these studies is that the records for soil fungi were very low. However, it is probable that they were present in greater numbers than recording by direct observation would indicate. On hogweed and nettle they were commonly isolated from the mycelium present in plated washed stem fragments.

If fragments of the plant are washed and plated out these frequently show that the mycelia of soil fungi are present and studies by Yadav (1966) and Yadav and Madelin (1968) show that the presence of soil fungi increases as decay proceeds. They possibly play a more important role in the decomposition of upright stems than is suspected. They have been recognized as prominent members of the community on the old leaf bases of *Carex paniculata*. This plant holds its leaves upright in a dense tussock for several years after they die (Pugh, 1958).

6
COLONIZATION AND DECAY OF WOOD

6.1 WOOD AS A RESOURCE

Wood is a substance of considerable anatomical and chemical complexity that can support a rich community of fungal species in a variety of microbial niches.

Wood decays slowly because of the major limitation to microbial growth imposed by the presence of **lignin** in the cell walls of the bulk of the tissues. Lignin coats cell wall polysaccharides and chemically combines with them to form **lignocellulose**, a substance that is very resistant to microbial degradation. This is due to the strong and random bonding between the phenyl-propane units that make up the lignin polymer and the chemical diversity of the lignins of different plants. Lignin therefore protects the cell wall polysaccharides from microbial hydrolysis except in the case of those microorganisms that can first chemically modify or degrade lignin. These are the only microorganisms that can initiate any significant decay of wood. They are mainly highly specialized fungi in the Basidiomycotina, but additionally some fungi in the Ascomycotina, notably in the Xylariaceae, can also initiate the degradation of lignocellulose.

To grow on wood, fungi also need to be able to tolerate certain chemical and physical stresses associated with wood. Typically high levels of tannins, phenols and other antifungal aromatics are present in wood. Concentrations are particularly high in heart wood. The presence of these substances is believed to be a major factor accounting for the selectivity of wood-decay fungi for certain wood species (see below). Significantly, heart-rot basidiomycetes produce laccases and tyrosinases (polyphenol oxidases) and other enzymes in order to detoxify wood (Lyr, 1962, 1963) and, as would be expected, heart-rotting fungi are very tolerant of any specific inhibitor found in the woods that they can decay.

In bulky wood tissues high carbon dioxide concentrations and low oxygen concentrations are usual. Even in the drier parts of wood where gas exchange with air is better, concentrations of carbon dioxide can be as high as 16–20% and oxygen concentrations as low as 1% have been measured (Thacker and Good, 1952).

Wood-decay fungi are adapted to these conditions. The growth of some wood-decay basidiomycetes is actually stimulated when CO_2 concentration is increased up to about 10% by volume and most that have been tested can make at least 50% of their growth in air at CO_2 concentrations between 30 and 40% by volume. By contrast, the growth of non-wood-decay basidiomycetes falls sharply above about 15% CO_2 (Hintikka, 1982).

Similarly, reduction of partial pressures of oxygen down to about 40 mm (5% by volume) has little effect on growth. Oxygen concentrations as low as 1.3% retard growth only moderately and O_2 concentrations have to fall as low as 0.4% to stop growth altogether, a figure which may be significant for growth in the interior of very bulky tree trunks only (Scheffer, 1986).

Little difference has been found between the tolerance of heart-rots, sapwood-rots and wood-stainers to low partial pressures of O_2 (Highley et al., 1983; Scheffer, 1986). This may be because heartwood can be much drier than sapwood and as a consequence can be much better aerated than might be expected. Survival of basidiomycetes decaying prepared timbers (product rotters) is more severely affected at low partial pressures of O_2 than heart-rots (Scheffer, 1986). This must be related to the better aeration of drier prepared timbers compared with heartwood.

Another adverse feature of wood which limits its utilization as a resource by fungi is its low nitrogen content. Undecayed wood, excluding bark, contains a maximum of 0.20–0.30% nitrogen by weight. The average value is about 0.09% and values as low as 0.03% have been measured (Allison, Murphy and Klein, 1963; Woodwell, Whittaker and Houghton, 1975; Swift, 1977). There is little difference in nitrogen concentrations between deciduous or conifer wood or between heartwood or sapwood in the same tree species, although slightly higher levels may exist in the bark. These figures compare with a range of about 0.6–6% nitrogen found in non-woody stems and foliage (Woodwell, Whittaker and Houghton, 1975; Fitter and Hay, 1981; Crook and Holden, 1948). Further, these comparisons do not take into account the non-random distribution of nitrogen in wood or its availability. Higher levels of nitrogen are likely to be present in the parenchyma ray cells, for example, than in vessels or tracheids, and much of the nitrogen of wood is insoluble or combined in aromatic complexes and is therefore not available.

A feature of the adaptation of wood-decay basidiomycetes is that they can grow at very low nitrogen concentrations, producing mycelia with a very low nitrogen content (Levi and Cowling, 1969). Cellulase output can be maintained at high levels at low nitrogen availability and cellulolysis may continue in culture even when the C:N ratio stands at 2000:1, well above the C:N ratio of wood (normally in the range 300–1000:1). This is in marked contrast to non-wood-decay fungi where present evidence shows that cellulase production ceases when the C:N ratio exceeds 200:1 (Levi and Cowling, 1969). In some fungi causing soft-rot of wood it has actually been found that the highest decay activity occurred at the lowest nitrogen levels tested (Butcher, 1975).

Such results indicate a very efficient recycling of nitrogen within the mycelium with remarkably high rates of intracellular protein turnover. Furthermore, when older parts of mycelium fall into disuse, and the contents are autolysed, it is believed that nitrogen is not lost, but that it is translocated to actively growing hyphae (Levi et al., 1968). Thus as growth proceeds, and nitrogen is collected from the substratum, it tends to accumulate in the actively growing hyphae at the mycelial front.

Some wood-decay basidiomycetes are highly antagonistic to other fungi and for these the heterolytic breakdown of the mycelium of other fungi during interspecific interactions in wood-decay communities may be important in releasing supplies of nitrogen with which they can supplement their own (Griffith and Barnett, 1967). Such antagonistic interactions are described in section 3.2.4.

Hohenbuehelia and *Pleurotus* species are wood-rotting basidiomycetes that destroy and digest nematodes (Thorn and Barron, 1984, 1986). The latter are usually abundant in rotting wood feeding on bacteria and they probably represent a significant supplementary nitrogen supply for some wood-decay fungi. *Hohenbuehelia* species have been linked with *Nematoctonus* clamp-bearing anamorphs in the Deuteromycotina that capture nematodes with sticky appendages (section 10.3). Active lysis of bacterial cells may be another potential source of nitrogen for some. *Pleurotus ostreatus* will attack and destroy bacterial colonies in culture which then serve as a nutrient source for the fungus (Barron and Thorn, 1987). For a review of the possibilities for the acquisition of nitrogen from other organisms see Barron (1992).

Nitrogen fixation by bacteria in wood (Sharp and Millbank, 1973) and in sporocarps (Larsen et al., 1978; Spano et al., 1982) probably also constitutes a further important source of nitrogen for wood-decay fungi. It has been speculated that but for the presence of these bacteria, decay would proceed exceedingly slowly. However, attempts by Sharp (1974) to show that nitrogen fixed by bacteria in wood could be transferred to fungi were unsuccessful.

It has also been suggested that bacterial hydrolysis of urea, a fungal waste product, may make ammonia ions available to wood-decay fungi that do not possess ureases (Stevens, 1987).

A notable feature of wood-decay fungi, especially heart-rot basidiomycetes, is the relatively high degree of resource specificity or selectivity which many show, fruiting being associated, with rare exceptions, with a particular host or fairly narrow range of hosts. For example *Phaeolus schweinitzii* and *Heterobasidion annosum* are selective for conifer wood, fruiting on a range of species. They have been recorded only occasionally on hardwoods. *Laetiporus sulphureus* is most commonly recorded only on hardwoods, notably *Quercus* and *Salix*, but it may sometimes be found on *Taxus* and other gymnosperms. Other heart-rot basidiomycetes are even more selective. *Piptoporus betulinus* is confined to *Betula*, *Phellinus pini* to *Pinus*, and *Fistulina hepatica* usually to *Quercus* and *Castanea*. *Oudemansiella mucida* is found mainly on *Fagus* but may be encountered on a very small range of other hardwoods. These distribution patterns seem to be related to a number of interacting factors, of which any affecting spore germination seem to be the least important, with spores often germinating freely on both 'host' and 'non-host' woods (Paine, 1968; Tsuneda and Kennedy, 1980).

For parasitic species, the ability to overcome host resistance must be of paramount importance. For others distribution may reflect the diverse nature of wood and the enormous variation in anatomical, chemical and fine structure differences which exist between woods of different trees. Although lignin is common to all woody tissues, it is chemically and structurally diverse and probably highly characteristic for individual plant families. Typical conifer lignins or 'guaiacyl' lignins are formed almost entirely from units derived from coniferyl alcohol, while those of angiosperms are mainly 'syringyl' lignins with a high proportion of sinapyl alcohol-derived units.

The quality of hemicellulose present in the cell wall matrix is also variable. In angiosperm wood the major hemicellulose is glucuronoxylan and only very small amounts of glucomannan are present, but in gymnosperm woods, galactoglucomannan is the major hemicellulose. About 20% of the cell wall polymers in conifers are glucomannans. Gymnosperms also contain significant amounts of xylan but as arabinoglucuronoxylan.

How far these chemical differences between wood species, and in particular the specific types of lignin present, influence the distribution of wood-decay fungi is uncertain. All the evidence suggests that the lignolytic systems of white-rot fungi are rather non-specific (Kirk, 1983) and brown-rot fungi, which have no enzymes capable of oxidizing lignin and which may depend upon a chemical attack to remove lignin, would arguably be unaffected by differences in the chemical composition of lignins.

Other chemical differences between woods include a variety of non-structural phenolic compounds and other aromatics, collectively called **extractives**, which can inhibit fungal growth. These may be the more important with regard to distribution, since wood-decay fungi differ in their sensitivity to specific compounds and there is some evidence that tolerance to certain extractives can be correlated with distribution (Hintikka, 1971b, 1982). These substances are believed to be responsible for the resistance of woods to general decay. Extractives of a general distribution include phenolic acids, e.g. vanillic, gallic, caffeic and ferulic acid; flavan-3-ols and flavonol constituents and residues of flavonoid compounds, such as catechin, quercetin, pyrogallol and catechol. These are all in some part chemically related to either tannin or lignin. The resistance of chestnut (*Castanea sativa*) to fungal pathogens, for example, and the small number of fungi which can grow on this wood is said to be due to high levels of tannin which are present in the tissues compared with the wood of other deciduous trees. Oaks seem to be similarly protected (Hart and Hillis, 1972). In other woods resistance to decay and high durability is due to the presence of fungicidal compounds and powerful antioxidants more specific in their effects and in distribution. Substances in gymnosperm woods with antifungal properties include terpenes (e.g. ferruginol in *Podocarpus ferrugineus*), phenolic resins (e.g. podocarpic acid in *Podocarpus* species), flavanonols (e.g. taxifolin in *Pseudotsuga* species), stilbenes (e.g. various forms of pinosylvin in *Pinus* species) and tropolones (e.g. thujaplicin and nootkatin from *Cupressus*, *Chamaecyparis* and *Thuja* species). Obtusaquinone and obtusastyrene are two highly potent antifungal aromatics associated with the high durability and natural bioresistance of certain tropical woods (Eslyn, Kirk and Effland, 1981).

For a fuller list of extractives found in wood and further information see Wise and Jahn (1952), Scheffer and Cowling (1966), Hart, (1981), Rayner and Boddy (1988) and section 3.2.2.

Sometimes the decay resistance of a particular wood seems to be due to several antifungal compounds acting together to give a synergistic effect (Scheffer and Cowling, 1966).

Antifungal substances in wood are thought to be produced as a result of secondary metabolic activity in senescing parenchyma cells in the sapwood. From here they diffuse out and accumulate in the walls of the heartwood cells. Consequently, as a general rule, heartwood is much more resistant to decay than sapwood, and heart-rot fungi are predominantly S-selected. Outer heartwood is usually more resistant to decay than inner heartwood (probably because antifungal substances deteriorate with age) and resistance increases from the apex of the tree to the base. There is evidence in the case of stilbenes that they are produced by tissues following wounding and other stresses (Hart, 1981).

Alternatively, the selectivity of some wood-decay fungi may relate to the presence of extractives which actually stimulate growth (Rayner and Hedges, 1982).

6.2 COLONIZATION OF WOODY TISSUES

6.2.1 Endophytes

It has been increasingly discovered that the living stems of woody plants harbour **neutral fungal symbionts** as endophytes. Neutral symbionts are latent infections which are, in effect, the primary fungal colonizers of the stem. The ecology of neutral endophytes has been extensively reviewed by Petrini (1992) and reviews by Petrini (1986) and Carroll (1986, 1988) contain many references to the original investigations. It is suspected that most, if not all woody plant species harbour neutral stem endophytes but because they are symptomless they are difficult to detect and can only be successfully surveyed by plating out carefully prepared surface sterilized tissues. Histological studies in conifer needles have shown that endophytes of this type consist of only a few hyphae confined to single cells until senescence and this makes them very difficult to find by microscopic examination (Stone, 1987).

Extensive sampling has shown that endophyte communities are potentially very large for each plant species. Isolations from the xylem and bark tissues of the roots and stems of *Alnus*, for example, yielded 85 different fungal taxa, a fairly typical number (Fisher and Petrini, 1990). Only a few of these species are dominants. So far the list of woody plant species with stem endophytes cover the genera *Ilex*, *Hedera*, *Ruscus*, *Ulex*, *Pinus*, *Fagus*, *Juniperus*, *Fraxinus*, *Quercus* and *Alnus* (Table 6.1). Other genera (e.g. those of the Ericaceae, Petrini, 1984) have many endophytes in their leaves but their stems have yet to be investigated. Many of these will undoubtedly also eventually be found in the stems since few endophytes are known to be organ-specific.

Some neutral endophytes may be host-specific but the majority have been isolated from a range of plant species and are at most restricted to species within the same plant family. Individual endophytes are usually present both in the bark and wood but the frequency of isolation is usually much higher from the living cells of the bark (Boddy and Griffith, 1989).

Lists of endophytes isolated from stems frequently include soil fungi such as *Penicillium* and *Trichoderma* species. These records should perhaps be treated with some caution in view of the technical difficulties of preparing clean tissue for plating out.

Of direct relevance to the subsequent development of the fungal community is the presence of many recognized wood-decay fungi among endophytes. Prominent among these are members of the Xylariaceae and their anamorphs, *Daldinia* and *Hypoxylon* species (*Nodulisporium*) and *Xylaria* and *Hypoxylon* species (*Geniculosporium*). These are mostly non-specific in distribution (Petrini and Petrini, 1985). Their

Table 6.1 Records of isolates of common endophytes from woody stems

Species	Buxus	Ilex	Hedera	Ruscus	Ulex	Pinus	Fagus	Juniperus	Fraxinus	Quercus	Alnus[a]
Aureobasidium pullulans	+	+			+	+	+				+
Coniothyrium		+	+	+	+		+		+	+	
Cryptocline		+				+					+
Cryptosporiopsis (Pezicula)					+	+	+			+	+
Cladosporium		+					+	+	+	+	+
Daldinia (anamorph)				+	+	+					
Epicoccum nigrum			+			+	+	+	+	+	+
Fusarium lateritum			+		+	+	+				
Geniculosporium					+	+		+			+
Hormonema						+	+	+			+
Hypoxylon (anamorphs)					+	+	+				+
Nodulisporium			+	+		+	+	+			
Phoma	+	+	+	+		+	+	+		+	
Phomopsis (Diaporthe)	+	+	+	+	+	+	+	+	+	+	
Xylaria (anamorphs)							+		+	+	

[a] Additional information from Fisher and Petrini (1990).
Sources: Data from Boddy and Griffith (1989).

presence as latent infections in the bark, from where they can invade the wood, undoubtedly accounts for the rapidity with which such species appear as colonists of dead attached twigs and recently felled logs. The basidiomycetes *Peniophora* species and *Vuilleminia comedens* are more restricted in distribution but exhibit a similar type of behaviour (section 13.2.1. The leaf-inhabiting fungi *Epicoccum nigrum* and *Aureobasidium pullulans* are also frequently isolated from woody stems but soon disappear as the decay communities develop.

An interesting aspect of the lists of endophytes recorded from woody plants is the frequent appearance of decay species in unexpected hosts. Outstanding examples include *Daldinia*, which has been isolated from *Pinus*, and several *Xylaria* species, which are commonly isolated from ericaceous hosts and have also been found in *Quercus*. The explanation of this may lie in the discovery by Chapela, Petrini and Hagman (1991) that in the Xylariaceae spore germination is triggered by monolignol glucosides produced by the plant. These chemical messengers selectively stimulate spore germination but the correct chemical signal for an individual fungal species may be produced by a range of plants, including some non-hosts. One could speculate and suggest that the reason why wood-decay endophytes are present in a wider range of stems than those on which they are known to fruit is because whilst they can exist in the living cells of the bark of 'non-hosts' they cannot go on to invade the wood when the stem dies. Perhaps the presence of toxic phenols which are outside the normal range that can be detoxified or the lack of suitable enzymes to degrade specific lignins may prohibit colonization.

Some woody endophytes are probably instrumental in protecting plants from bark-boring and foliage-feeding insects. Insect feeding and reproduction are adversely affected associated with the production of toxic metabolites by the fungus. Studies indicate that *Phomposis oblonga* in elm stems and *Rhabdocline parkeri* in conifer needles may behave in this way. Other possibilities include *Hypoxylon* and *Leptostroma* species. For further details see Carroll (1991b) and section 4.2.2.

6.2.2 Wood-decay macrofungi

a) Establishment

Entry into living trees is usually effected via wounds. Other probable routes of entry include lenticels, leaf scars and via tissues weakened by drought or microbial damage. Entry into felled or fallen timbers is facilitated by the exposure of massive amounts of unprotected tissues. Colonization may be by germinating spores or alternatively in the case of roots and timbers in contact with the ground by migratory mycelia or cords (section 2.2.3). Decay of standing trees originating from roots causes **butt rot**; that beginning in above ground parts **top rot**.

Once entry has been gained axial colonization can be very rapid. Hyphae can grow unhindered by cell wall barriers in vessels which can be several millimetres long. Radial spread takes place more slowly through disrupted pit membranes in lignified cell walls and through the cell walls of non-lignified wood parenchyma, especially that of the medullary rays. Wood parenchyma and the medullary rays contain starch and the distribution of rays and the amount of wood parenchyma present markedly influence the rate of decay of different woods. The spatial patterns that develop in the initial stages of decay also vary and often reflect the numbers, size and distribution of the vessels, the main channels of invasion. The decay patterns of ring-porous wood are very characteristic compared with those of diffuse-porous wood, where the larger vessels are more randomly distributed.

b) Colonization of standing trees

The pioneer decay macrofungi of living standing trees are parasites or

Table 6.2 Some common basidiomycete colonizers of wood

Species	Type of rot
I. Fruit bodies appearing on standing trees	
Heterobasidion annosum	White-heart rot of conifers
Phaeolus schweinitzii	Brown-heart rot of conifers
Armillaria mellea	White-rot of heartwood and sapwood of hardwoods (corresponding species on conifers is *A. ostoyae*)
Oudemansiella mucida	White-rot of sapwood of beech (*Fagus sylvatica*)
Fistulina hepatica	Brown-rot of heartwood of *Quercus* spp. (rare on other hardwoods)
Fomes fomentarius	White-rot of sapwood and heartwood of *Betula* spp. and *Fagus sylvatica*
Ganoderma applanatum	White-heart rot of *Fagus sylvatica* (occasionally on other hardwoods)
Pleurotus ostreatus	White-heart rot, common on *Fagus sylvatica*. Attacks a range of hardwoods. Rare on conifers
Polyporus squamosus	White-heart rot of *Ulmus* spp., *Acer* spp. and occasionally other hardwoods
Piptoporus betulinus	Brown-rot of sapwood of *Betula* spp.
Laetiporus sulphureus	Brown-heart rot of hardwoods. Usually on *Quercus* spp., one of the few fungi decaying *Castanea sativa*
Hypholoma fasciculare	Heart-rot of a range of hardwoods, occasionally conifers
II. Fruit bodies appearing on exposed stumps and felled timbers	
Bjerkandera adusta	White-rot of most hardwoods
Coriolus versicolor	White-rot of most hardwoods
Stereum hirsutum	White-rot of most hardwoods, common on *Quercus* spp.
Stereum sanguinolentum	White-rot of conifers
Chondrostereum purpureum	White-rot of hardwoods especially *Fagus sylvatica* and *Betula* spp. Silver leaf disease of plums
Daedalea quercina	Brown-rot of *Quercus* spp. (occasionally other hardwoods)
Pseudotrametes gibbosa	White-rot of *Fagus sylvatica*
Coniophora puteana	Brown-rot of all types of wood, common on timber in buildings

aggressive saprophytes. The most economically important of these are certain basidiomycetes (mainly Aphyllophorales) (Table 6.2). These cause brown- or white-rot in the heartwood of the trunk or main branches of the tree. (For an explanation of these terms see section 6.3). Major wounds caused by the loss of wind-thrown branches and cutting or trimming operations expose heartwood and provide a commonly used entry route. Heart-rot fungi are uniquely tolerant of the chemical and other stresses existing in heartwood and many are highly selective and only grow on the wood of a few tree species. One reason for this may be the variety and distribution of antifungal phenols in heartwoods as explained previously in this chapter. Some heartwood rots such as the common root-rot of hardwoods, *Armillaria* species, grow in sapwood or heartwood and can invade heartwood via exposed sapwood. The wetter conditions encountered in sapwood and the relatively poor aeration in sapwood usually confines heart-rots to heartwood. Heart-rots seldom attack fallen timbers. This is probably because microenvironmental changes which follow the exposure of the tissues to air favour the establishment of more aggressive non-specialized decay competitors.

Other basidiomycete wood-decay fungi of living trees are opportunistic invaders of sapwood. Establishment of these species depends upon major wounds that cause extensive disruption of water columns in xylem vessels and tracheids, leading to the drying of exposed tissues and most importantly the entry of air, since water-saturated functional sapwood would otherwise be too poorly aerated to allow fungal growth (Rayner, 1986).

Decay columns of sapwood rots spreading from wounds are limited by the extent to which air has been able to move into tissues and by the defensive barriers of gums, resins, tyloses, etc. that the tree may produce to seal off the wound.

Other wood-decay species establish themselves first as latent infections and do not require a major wound to gain entry. Certain species causing die-back of twigs and branches in the canopy are believed to gain entry via minor wounds, caused by insects etc., and circulate as propagules in the transport system of the tree. Decay is then thought to develop in any organ which is physiologically weakened, perhaps through shading or due to water stress, as opportunity arises. A feature of this behaviour is the rapidity with which these rots develop in senescing tissues without obvious infection loci (Boddy and Rayner, 1981, 1983a,b). *Exidia glandulosa, Stereum gausapatum* and *Vuilleminia comedens* are examples of some white-rot fungi of heartwood that behave in this way. These can cause considerable weight loss from branches and twigs while still attached to the tree (Swift et al., 1976).

Common ascomycetes in the pioneer phase on hardwood timbers are members of the Xylariaceae, some of which persist for several years (Coates and Rayner, 1985a,b; Chapela, Boddy and Rayner, 1988). In

COLONIZATION OF WOODY TISSUES 155

Figure 6.1 Fruit-bodies (perithecial stromata) of Xylariaceae. (a) *Daldinia concentrica* on a trunk of ash (*Fraxinus excelsior*). (b) *Hypoxylon multiforme* on fallen branch of birch (*Betula pendula*). (c) *Xylaria hypoxylon* on dead branch of gorse (*Ulex gallii*). Scale bar = 4 cm (a); 2 cm (b,c).

advanced attacks these may be detected by their conspicuous perithecial stromata on the surface of the wood (Figure 6.1). Some of these fungi are important tree pathogens, e.g. *Ustulina deusta*, a parasite of beech (*Fagus sylvatica*) and other hardwoods and *Hypoxylon mammatum*, a parasite of aspen (*Populus tremula*) (Rogers, 1979). *Hypoxylon rubiginosum* is an aggressive saprophyte that invades beech and other hardwoods when bark had been damaged by *Nectria* species: *Hypoxylon punctulatum* behaves similarly, invading oak trees attacked by the wilt fungus (*Ceratocystis fagacearum*) (Shigo, 1958, 1964). Many saprophytic species in this family are highly selective and fruit on the wood of a single tree species or family only (Whalley, 1985).

The pioneer basidiomycetes and ascomycetes persist as dominants in the wood for several years, with many of the parasites behaving saprophytically in the later stages of the death of the tree.

c) Colonization of felled and fallen timbers and prepared wood

Exposed stumps of felled trees, cut timbers and wind-thrown branches become rapidly colonized by relatively non-selective saprophytic polypores, resupinates and toadstools that establish themselves from spores. *Chondrostereum purpureum*, *Coriolus versicolor*, *Corticium* species, *Pseudotrametes gibbosa*, *Phlebia radiata* and *Stereum hirsutum* are very common on hardwoods in this condition. Non-basidiomycete species occurring on hardwoods include the discomycete *Ascocoryne sarcoides*, whose purple fruit-bodies are particularly common on freshly exposed stumps.

Other early arrivals on stumps and timber in contact with the ground are the **strand-** and **rhizomorph-formers** which invade from exploratory mycelial networks of strands present on the forest floor. Strands grow out from food bases and have a high inoculum potential. They locate the scattered resource without waste of biomass and are particularly advantageous in the establishment of mycelia on resources low in readily available energy supplies (section 2.2.3). Strand-formers include familiar toadstools like *Armillaria* species, *Hypholoma fasciculare*, *Tricholomopsis platyphylla*, the resupinate *Phanerochaete velutina*, and the gasteromycete *Phallus impudicus* (stinkhorn). Stinkhorns cause white-rots of hardwoods and conifers (Dix and Cairney, 1985). They are seldom found actually fruiting on wood but their strands can usually be traced to decaying wood beneath the ground.

Saprophytes whose fruit-bodies are associated with the later stages of the succession include several toadstools such as *Mycena galericulata*, *Pluteus cervinus* and *Psathyrella hydrophila* on hardwoods and *Paxillus atrotomentosus* and *Tricholomopsis rutilans* on conifers. Gasteromycetes may also be represented by the appearance of the puffball,

Lycoperdon pyriforme, and occasionally *Sphaerobolus stellatus*.

On wetter, well-rotted woods, the fruit-bodies of the Dacrymycetales, e.g. *Calocera cornea* and *Dacrymyces stillatus*, appear. Also common are certain discomycetes such as *Dasyscyphus niveus, Mollisia cinerea, Mollisia melaleuca* and *Catinella olivacea*, whose small apothecia appear in swarms on the wood.

As wood decays, water potential values increase, becoming higher than those in undecayed wood at the same moisture content and the late appearance of the fruit-bodies of these species may in some cases be a sign that decayed wood provides a better physical environment for their colonization, growth or development.

On worked timber, colonization by basidiomycetes depends very much on the prevailing moisture conditions. If the wood is either too damp or too dry none will usually develop. If conditions are satisfactory for growth those that do appear include species which are quite different from those normally colonizing standing trees or felled timber. The wood of buildings may be colonized by *Fibroporia vaillantii* and other resupinates in the genus *Poria* (*sensu lato*). Wet-rot (*Coniophora puteana*) and dry-rot (*Serpula lacrimans*) may also develop. These both produce brown-rot in structural timbers and dry-rot in particular can be very serious. The principal basidiomycete decaying untreated poles and posts in contact with the soil is *Lentinus lepideus*. However, this fungus can be effectively controlled by wood preservatives and in practice microfungi causing soft-rots are of greater economic importance in the decay of preservative-treated posts in the soil (Dickinson, 1982).

For a more detailed treatment of the colonization of woody tissues by macrofungi see section 13.2.

6.2.3 Microfungi

Associated with all stages of the succession of the larger fungi on wood, especially that in contact with the ground, is an ancillary succession of non-decay microfungi (Table 6.3). Moulds commonly found in litter and soil in the genera *Cylindrocarpon, Fusarium, Cylindrocladium, Penicillium, Cladosporium, Trichoderma, Aureobasidium, Mucor* and *Mortierella* adopt a pioneer role, colonizing non-lignified cells in the wood in advance of wood-decay species within a few days of exposure. These are essentially fungi of a ruderal habit that exploit disturbance and the opening up of new opportunities for colonization.

Colonization by the majority is superficial, even in wetter wood in contact with the soil (Banerjee and Levy, 1971). They cannot degrade lignin or, except possibly to a very limited extent in the case of *Trichoderma* species, attack polysaccharides in lignified cell walls. They therefore produce no great structural change in wood and are not responsible

Table 6.3 Colonization (%) of birch and oak stumps by microfungi

Microfungal species	After 6 months		After 12 months		> 24 months	
	Birch	Oak	Birch	Oak	Birch	Oak
Agyriella sp.	0	0	0	33	0	28
Acremonium sp.	0	0	0	0	9	0
Botrytis cinerea	0	0	0	20	0	0
Calcarisporium sp.	0	17	7	20	9	6.8
Ceratocystis sp.	0	0	0	0	2	0
Chaetomium sp.	0	0	0	0	0	1.7
Fusarium sp.	0	0	29	0	2	0
Gliocladium sp.	0	0	14	0	0	0
Graphium calicioides	50	17	14	0	0	3.4
Humicola spp.	0	0	29	0	49	0
Mucor sp.	0	0	0	0	13	3.4
Trichocladium opacum	0	0	29	0	6	0
Cladosporium sp.	0	0	0	20	9	44.1
Paecilomyces sp.	0	17	0	0	0	3.4
Scytalidium sp.	0	0	0	0	2	3.4
Cylindrocarpon/ Cylindrocladium sp.	62.5	8.0	64.2	7.0	27.6	0
Arthrobotrys sp.	0	0	21	0	11	0
Doratomyces nanus	13.0	0	42.8	7.0	0	
Dothiorella sp.	0	58.3	21.4	0	6.4	5.1
Rhinocladiella spp.		0		7.0	0	42.4
Penicillium spp.	50.0	66.7	21.4	53.3	12.8	37.3
Phialophora fastigiata	25.0	0	28.6	0	53.1	
Trichoderma spp.	62.5	41.7	28.6	20.0	40.4	20.3
Total no. of stumps examined	8	12	14	15	47	59

Source: Data from Rayner (1977a).

directly for any great amount of decay, but their role in wood decay may not be inconsequential (see below).

Leptodontium, Rhinocladiella and others have an interesting ecology and occur deep in the decaying wood in the narrow bands of undecayed tissues between the zone lines which mark out the boundaries of the basidiomycete mycelia (Rayner, 1976).

The appearance of non-decay moulds is closely followed some weeks later by the wood-decay microfungi that cause soft-rots. In this group *Phialophora* species are common on standing trees and stumps while on worked timbers fungi in the genera *Chaetomium, Humicola* and *Doratomyces* are more common. Several pioneer non-decay moulds such as *Aureobasidum pullulans* and *Ceratocystis* species are known as wood sapstainers, and the development of discoloration is a characteristic feature of the early decay of the sapwood, especially that of conifers. Following colonization by soft-rot fungi the number of non-decay moulds generally declines somewhat, possibly due to diminishing supplies of simple carbohydrates, but later there is often an upsurge in numbers associated with the establishment of basidiomycete rots in the wood. The more rapid type of decay produced by the basidiomycetes probably releases surplus sugars, some of which become available to the moulds, and this allows their numbers to increase again. This is an example of the commensal type relationship which develops between fungi in many similar situations where plant polymers are being degraded. Among the later appearing secondary moulds are many cellulolytic species due to the improved access to cell wall cellulose. These features of the microfungal colonization of wood are illustrated in Figure 6.2, a spatial and time-course study of the colonization of buried birch wood stakes (Clubbe, 1980).

In general, the pattern of succession of microfungi is similar on all woody substrata where they are present in any numbers and the minor variations which exist can usually be attributed to some important environmental difference. The wetter the timber, for example, the greater the number of microfungi which may be found growing on it. On drier exposed wood the

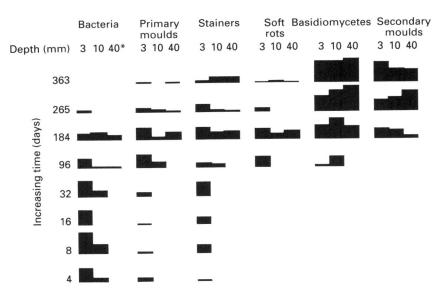

Figure 6.2 Colonization of buried birch wood by microfungi. * = Sample depth in wood. Frequency of colonization, 1 mm = 5%. (Redrawn from Clubbe, 1980.)

succession may not progress beyond the primary mould–sapstain stage. Rotting posts in the ground generally carry a much heavier load of microfungi than the drier timbers of dead standing trees. As time passes, the number of microfungi tends to increase on all types of woody substrata with the same species often persisting for several years.

Non-decay moulds have a potential role in wood decay as modifiers of wood, and together with soft-rot fungi may help to prepare wood for invasion by basidiomycetes. They disrupt pit membranes and thus assist the passage of hyphae through the tissues. The action of soft-rot enzymes on cell walls will raise the matric potential of the tissues and increase water availability at lower moisture contents.

Some microfungi of wood can also be involved in the removal of toxic phenols (Shortle and Cowling, 1978; Carey and Savory, 1979). Soft-rot fungi seem to be more effective in this respect than some of the non-decay moulds, and the basis of the effective protection of wounds in trees from sapwood-decaying Hymenomycetes using *Trichoderma harzianum* seems to be that the latter can out-compete and prevent the establishment of soft-rots like *Phialophora* species but has a low capacity to remove inhibitory phenols (Smith, Blanchard and Shortle, 1981). However, in spite of this it should be mentioned that the preconditioning of wood is not essential for the establishment of all basidiomycete decay fungi and Rayner and Boddy (1988) draw attention to many such examples.

Mixed fungal communities decay wood faster and the microfungi of wood are undoubtedly important in this respect (Hulme and Shields, 1975). Non-decay moulds in particular have an important role to play. They can enter into mutualistic associations with wood-decay basidiomycetes and accelerate the utilization of lignin and cellulose, as explained in section 3.2.1.

Under certain circumstances, non-decay fungi may antagonize wood-decay species and may restrict decay, especially if they colonize the wood first (Hulme and Shields, 1975). *Scytalidium album* is an example of a possible antagonist of wood-decay macrofungi. Rayner (1978) found that it could replace a number of wood-decay basidiomycetes in culture.

6.2.4 Fungus–invertebrate interactions

Dead wood is the natural habitat of the larvae of several species of beetle and fly that feed and burrow in the wood. As they do so they introduce changes which promote the growth and spread of decay fungi. Wood is transformed to particulate matter, increasing the surface area on which fungal extracellular enzymes can act. Their faeces locally increase nitrogen concentration in a nitrogen-deficient environment and channels are produced which greatly improve aeration and facilitate the spread of the mycelia into the wood. Fungal growth in turn improves conditions for the animals. The C:N ratio of the wood falls as cell wall polymers are respired and fungal extracellular enzymes oxidize toxic phenols and soften cell walls. These changes improve the palatability of the wood and its nutritional value weight for weight. The mutual benefits derived from these interactions are similar to those arising from the interactions of invertebrates and fungi in communities in decaying leaf litter, and here again they can significantly increase the rate of decay.

Inocula can be transported deep into wood by acarid species and springtails as they move about in the boreholes left by the larvae of beetles. The increased activities of the former as time passes probably account for the numbers of soil-inhabiting fungi increasing on well-decayed wood (Swift and Boddy, 1984). Arthropods may further facilitate the establishment of soil fungi by reducing competition from wood-decay fungi by selective grazing.

6.3 TYPES OF WOOD ROT

Wood-rotting fungi are, by definition, those which can bring about significant weight loss and structural change in woody tissues. Three basic types of rot are recognized, each conspicuously different and taking its name from the general appearance of the decayed wood. Oxidation of lignin in **white-rot** causes the wood to take on the diagnostic white or bleached appearance. In **brown-rot** only cellulose and other wood carbohydrates are utilized and in this case the decaying wood remains predominantly brown. These types of rot are caused mostly by Basidiomycotina but some of the Xylariaceae (Ascomycotina) cause white-rots. The Ascomycotina and Deuteromycotina (fungi imperfecti) produce a third type of rot known as **soft-rot** in which the wood loses mechanical strength and becomes wet and spongy.

At the microscopic level several major differences arise between the three basic types of rot because of the differences in the range of enzymes produced, their behaviour, and the variation that exists in the chemical composition of wood cell walls. The distribution of lignin, because it restricts access to cellulose, is particularly important and determines which parts of the wall are attacked preferentially by brown- and soft-rot fungi.

A mature plant cell wall has three parts: the **middle lamella** and the **primary** and **secondary walls** (passing from outside to the cell lumen). The secondary wall is the thickest and is divided into the S1 and S2 layers and an S3, nearest the cell lumen. The latter is usually very thin and difficult to distinguish from the inner part of the S2 (Figure 6.3).

In lignified cells lignin is present in the matrices of all cell wall layers. In hardwoods it is generally present in its biggest proportion of the whole in the middle lamella and primary wall. About 90% of the total lignin content of the cell is located here in hardwoods (Meier, 1962). In softwoods only

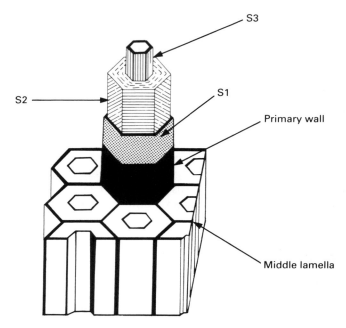

Figure 6.3 The cell wall layers of a mature tracheid. (Redrawn from Timell, 1967.)

about 60% of the total lignin content is found here and other cell wall layers, notably the S3, are more heavily lignified than in hardwoods.

Cellulose is the principal carbohydrate constituent of all cell wall layers of lignified cells in both hardwoods and softwoods (Table 6.4). However, the cellulose content of some lignified cell wall layers can be quite low. A typical *Pinus* tracheid contains only about 2% of the total cellulose content of the cell in its S3 layer, compared with about 47% in the S2 (Côte et al., 1968). In softwoods the major hemicelluloses that combine with cellulose in the cell wall are glucomannan and glucuronoarabinoxylan; in hardwoods the major hemicellulose is glucoronoxylan. Glucomannan is virtually absent from hardwoods. The proportions of the different carbohydrates present vary slightly from wall layer to layer. In tracheids, cellulose tends to be a bigger proportion of the carbohydrate content in the S2 and S3, the major hemicelluloses tend to be in lower proportions in the middle lamella and primary wall (Table 6.4).

Differences in the distribution of carbohydrates and lignin in the cell wall layers of hardwoods and softwoods may lead to differences in the way in which rots produced by the same fungus develop in different woods.

Table 6.4 Distribution (%) of polysaccharides in tracheid wall layers

	Middle lamella + primary wall	S_1	$S_{2\ outer}$	$S_{2\ inner} + S_3$
Birch				
Galactan	16.9	1.2	0.7	0.0
Cellulose	41.4	49.8	48.0	60.0
Glucomannan	3.1	2.8	2.1	5.1
Arabinan	13.4	1.9	1.5	0.0
Glucuronoxylan	25.2	44.1	47.7	35.1
Pine				
Galactan	20.1	5.2	1.6	3.2
Cellulose	35.5	61.5	66.5	47.5
Glucomannan	7.7	16.9	24.6	27.2
Arabinan	29.4	0.6	0.0	2.4
Glucuronoarabinoxylan	7.3	15.7	7.4	19.4

Source: Data from Meier (1962).

6.3.1 Soft-rots

Ascomycotina and fungi imperfecti causing soft-rots are among the first fungi to invade exposed wood. They can tolerate very wet conditions and become the principal decay organisms in very wet wood where wood-decay basidiomycetes cannot flourish. They are of major economic importance and decay posts, timbers and joinery subject to constant wetting, submerged in water or in contact with soil, often in spite of the wood having been treated with a preservative.

Over 300 species of fungi have now been recognized as causing soft-rots (Duncan and Eslyn, 1966; Seehann et al., 1975) and the literature on this subject is now voluminous. Species in the genera *Acremonium*, *Sporocybe*, *Phialophora*, *Cephalosporium* and *Doratomyces* are commonly isolated from soft-rotted wood and have been shown in trials to cause significant weight losses in timbers under suitable conditions within a few weeks. To date there has been only one recorded instance of a basidiomycete fungus causing a type of soft-rot in wood, *Oudemansiella mucida* (Daniel, Volc and Nilsson, 1992).

The early descriptive work and some of the more thorough investigations have been carried out on soft-rots caused by *Chaetomium globosum* (Savory, 1954; Savory and Pinion, 1958; Levi and Preston, 1965). When *Chaetomium globosum* attacks the softwoods of conifers like *Pinus sylvestris*, infection is established first in the parenchyma ray cells and spreads into the lumen of the adjacent tracheids through the simple pits which connect the two cells through the common wall. From cell lumina the spread of the fungus continues through the xylem tissue by penetration of tracheid walls, the hyphae reducing in size as they pass through to the lumen of the neighbouring cell. Alternatively, hyphae may stop growing after passing through the S3 layer of the tracheid wall and branch in the S2 layer by growth at right angles in two directions from the penetrating hypha (Corbett, 1965). These are the characteristic hyphal T branches that form elongated cavities around themselves by enzyme action. Frequently these cavities are seen to occur in chains, due, it is believed, to start–stop oscillatory growth as hyphae progress in their longitudinal growth (Hale and Eaton, 1985). Figure 6.4 shows soft-rot cavities in wood caused by the aquatic hyphomycete *Tricladium splendens*.

Fungi causing soft-rots degrade cellulose and hemicellulose by producing cellulase and a variety of hemicellulases (Nilsson, 1975). Some soft-rotting fungi also partially degrade lignin, considerably modifying it in increasing amounts in the later stages of decay (Levi and Preston, 1965; Eslyn, Kirk and Effland, 1975). Most, however, may have only a limited effect on lignin, releasing CO_2 only from methoxyl groups and side-chains (Haider and Trojanowski, 1975).

Branching in the S2 wall layers of conifer tracheids is thought to relate to the higher deposits of cellulose here compared to the lower cellulose and higher lignin content of the S3 layers. The shape of the hyphal cavity is probably due to the fact that the hyphal branches follow the orientation of

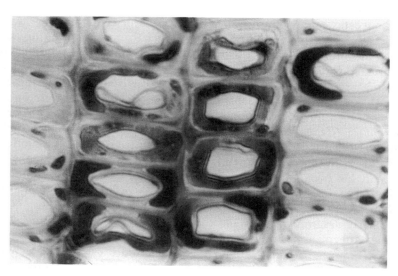

Figure 6.4 *Tricladium splendens* causing soft-rot cavities in wood. (Reproduced with permission from Gunasekera, Webster and Legg, 1983.)

the cellulose microfibrils which are parallel to the long axis of the cell in the S2. The wall-degrading enzymes of the fungus seem able to diffuse more readily in the same direction rather than across and between the microfibrils. Consequently an elongated type of cavity arises as the wall is degraded.

While this description of soft-rot attack is typical for conifer woods, a number of morphological variations exist for other types of wood. Courtois (1963) has listed no less than 14 of these, although some seem to be very minor and may be stages in the development of others. Notable differences occur between hard- and softwoods and when *Chaetomium globosum* attacks a hardwood like birch (*Betula*), fewer T branch cavities form in the S2 layers of the lignified cells and the typical decay symptom of the S2 is a V-shaped notch. By contrast with softwoods, principal decay is by erosion of the S3 layers which are attacked from the cell lumina (Corbett, 1965). This is called type 2 attack to distinguish it from cavity formation in type 1 attack. Often the type 2 erosion of the cell wall in soft-rot is indistinguishable from white-rot attack (see below). These differences between soft-rots developing in angiosperm and conifer woods seem to be fairly general (Liese, 1964) and appear to be due to the fact that the S3 layers of the cell walls of hardwoods contain a higher proportion of cellulose and possibly more readily oxidizable forms of lignin than those of softwoods (Table 6.4).

Some soft-rot fungi attack only angiosperm wood but while many others also attack conifer wood the latter is generally more resistant. Duncan (1960) tested 99 isolates of soft-rot fungi, all of which produced significant weight losses in the angiosperm wood of sweetgum (*Liquidambar styraciflua*), but fewer than half of which attacked *Pinus* species and none the wood of *Sequoia sempervirens*. The number of soft-rot fungi attacking the wood of *Pinus* species increased following leaching, indicating that water-soluble inhibitory substances are present which protect the wood. In the case of *S. sempervirens* leaching had no effect and soft-rot attack only developed if the wood was kept continuously wet and in contact with the soil, suggesting that perhaps more persistent substances inhibit growth in this wood and that these may only be removed through the metabolic activity of soil microorganisms.

In angiosperm wood soft-rot decay proceeds at a faster rate than in conifers, a fact which seems to be related to important anatomical differences in the structure of the two types of wood. Corbett (1965) observed that the spread of *Chaetomium globosum* through the tissues of *Pinus sylvestris* occurred more readily through the simple pits connecting the parenchyma ray cells with the tracheids than through the bordered type of pit which connects tracheid with tracheid. In gymnosperm wood parenchyma is confined to the medullary rays, but in angiosperms it is more plentiful and in addition to that in the medullary rays it is also scattered generally in the xylem. Simple pits associated with parenchyma cells are therefore more numerous in angiosperm wood and facilitate the rapid spread of the hyphae through the tissues and thus speed up the rate of decay.

6.3.2 White-rots and brown-rots

Brown-rot and white-rot fungi attack living trees, decay dead wood and cause damage to unprotected timbers in buildings.

Hyphae first invade parenchyma cells of the wood exposed by cuts and wounds and spread through the tissues in the lumina of vessels and tracheids and by passing transversely through disrupted pit membranes or by boring through cell walls.

Fungi causing brown-rots are relatively few in number, comprising less than 10% of all wood-decay basidiomycetes. They are mainly distributed in the northern hemisphere and are mostly members of Polyporaceae, growing mainly on conifer woods, where they degrade cellulose and hemicellulose primarily, and have only a slight modifying effect on lignin, demethylating aromatic methoxyl groups by an unknown mechanism (Kirk and Farrell, 1987). The polysaccharides of the S2 wall layers are the first to be attacked and as these are hydrolysed the cell walls, as seen by electron microscopy, take on the appearance of expanded polystyrene. Typically the more lignified S1 and S3 layers and the middle lamella are initially less affected and the structure of the wood tissue is maintained by these until in the final stages of decay they collapse. It has been suggested that brown-rot fungi overcome the problem of the presence of lignin, which would be expected to protect the wall polysaccharides, by the non-enzymatic solubilization of lignin by hydrogen peroxide, with free iron from the wood acting as a catalyst (Koenigs, 1974a,b). Oxidation by hydrogen peroxide with peroxidases acting as catalysts has now been accepted as the mechanism used by white-rot fungi to oxidize lignin (section 2.3.5(a)).

Poria placenta cannot hydrolyse crystalline cellulose unless hemicellulose or sugars are present (Highley, 1977) and when brown-rot fungi attack wood, hemicellulose is usually the first carbohydrate to be lost in any measurable quantity. It may be that hemicellulose hydrolysis provides enzyme substrates for hydrogen peroxide production and is the necessary first step in the attack on lignocellulose by fungi causing brown-rots (Highley and Kirk, 1979).

White-rot fungi metabolize carbohydrates and lignin in wood and lignin oxidation cannot proceed without the metabolism of some cellulose or other carbohydrate (Kirk and Farrell, 1987). In white-rot caused by *Coriolus versicolor* a higher proportion of the total lignin content of the wood is oxidized during decay than the proportion of the total carbohydrate respired (Cowling, 1961). Amounts of carbohydrate lost are still high, however, because the carbohydrate content of wood is the greater. Utilization of individual carbohydrates by this fungus is approximately proportional to the amounts present in the sound wood (Cowling, 1961).

It is now generally recognized that there are two different types of white-rot fungus. Some, such as *Coriolus versicolor*, bring about the simultaneous decay of both lignin and cellulose, while others (e.g. *Ganoderma tsugae*) selectively oxidize lignin, respiring hemicellulose, prior to the removal of significant amounts of cellulose (Blanchette, 1984; Otjen and Blanchette, 1985).

Unlike the wall-degrading enzymes of brown-rotting fungi the extracellular enzymes of the fungi of white-rot act close to hyphae. They may be bound to hyphal walls or held close in a mucilage sheath (Palmer, Murmanis and Highley, 1983). As a result, the type of rot produced is strikingly different from that of brown-rot. Hyphae lying in the cell lumina degrade the S3 layer initially and as the attack progresses lignin is removed from the lignocellulose fibre complex of the wall, exposing the cellulose fibrils. At first the effects are very localized and grooves or lysis furrows arise around the hyphae, but as enzyme action progresses, the furrows of adjacent hyphae merge, and in time a general thinning of the whole of the S3 occurs (Figure 6.5). Eventually all cell wall layers are progressively thinned and finally tissues distintegrate as cells separate.

Degradation of lignin by white-rot fungi is a strongly oxidative process

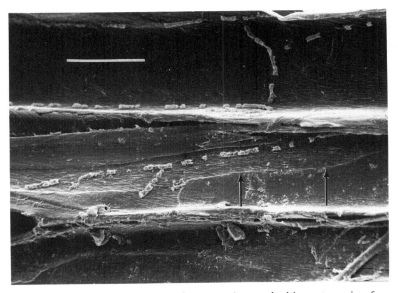

Figure 6.5 Hyphae of *Phanerochaete chrysosporium* and white-rot erosion furrows (arrowed) in Scots pine wood. Scale bar = 20 μm. (Photograph by M.D. Hale.)

that is believed to take place through secondary metabolic pathways. These switch on when growth ceases due to the exhaustion of some essential nutrient like nitrogen. Thus most of the oxidation of lignin occurs at the end of the primary growth phase (Keyser, Kirk and Zeikus, 1978).

Of the few hundred species that have been examined, the majority of fungi causing white-rot occur only on angiosperms (hardwoods), a few occur only on conifers (softwoods), with very few species occurring on both (Cowling, 1961; Seifert, 1968).

Attempts have been made to relate the preference shown by brown- and white-rot fungi for certain woods to hemicellulase production. *Ganoderma applanatum*, which causes white-rot of hardwoods, readily hydrolysed hardwood xylans but caused little hydrolysis of glucomannan, the major hemicellulose of softwood. By contrast, two fungi destroying softwood, the brown-rot *Serpula lacrimans* (dry-rot) and the white-rot *Stereum sanguinolentum*, were shown to have a high activity against softwood glucomannans and xylans (Lewis, 1975). A similar correlation was produced by Keilich, Bailey and Liese (1970).

There is evidence too which suggests that these enzyme systems may be highly specific. Lewis (1975) found that whereas *Serpula lacrimans* could not hydrolyse xylans from hardwoods, *Stereum sanguinolentum* could. The most likely explanation of this is that different xylanases are responsible for the hydrolysis of xylans from different woods and that *S. sanguinolentum* produces at least two of these. However, so little work has been done on this subject that it would be unwise to generalize at this stage. Indeed exceptions have come to light which show that there is very little qualitative difference between the hemicellulases produced by brown- and white-rots (Highley, 1976).

More extensive reviews of the literature covering wood rots in general and the mechanisms of wood decay may be found in Liese (1970), Levy (1975), Highley and Kirk (1979), Montgomery (1982) and Kirk and Fenn (1982). The decay of commercial timbers and of wood in contact with the ground has been reviewed by Dickinson (1982) and Levy (1982) respectively.

6.3.3 Xylariaceous rots

Relatively little research has been carried out on wood decay caused by the Xylariaceae, compared with that for basidiomycetes, but representatives of a number of genera have been shown to decay hardwoods slowly (Blaisdell, 1939; Merrill *et al.*, 1964). *Ustulina deusta, Xylaria hypoxylon, Daldinia concentrica* and *Hypoxylon* species, among several others, produce a type of decay similar in appearance to that of white-rot (Cartwright and Findlay, 1942; Merrill *et al.*, 1964; Coates and Rayner, 1985c). Nilsson and Daniel (1989) observed that in addition, *Xylaria* species will also form cavities in secondary cell walls, typical of the type 1 soft-rot produced by other wood-decay ascomycetes in hardwoods, whereas *Daldinia* and most *Hypoxylon* species did not. Thus the type of decay produced by xylariaceous rots will depend upon the species of fungus. Lignin oxidation has been demonstrated for some species using C^{14}-labelled lignin (Sutherland and Crawford, 1981). Nilsson and Daniel (1989) noted that carbohydrates were preferentially metabolized and that, of the components of lignins, syringyl propane units were selectively oxidized by the species they investigated. Consequently, little loss of weight was recorded for pine wood compared with birch which probably explains why the natural substrata for xylariaceous species are hardwoods and not gymnosperm woods, rich in guaiacyl lignin.

6.4 WATER RELATIONS

The development of wood decay communities is greatly influenced by water content and relatively small changes can have dramatic effects on the progress of disease in living trees and rates of decay in general (Cartwright and Findlay, 1950; Henriksen and Jørgensen, 1952; Liese and Ammer, 1964; Etheridge, 1958). It may be too high for fungal growth in wood saturated with water and in the functional wood of living trees (Rayner, 1986). The major adverse effect of too much water is poor aeration. On the other hand the water content may be too low and this is very often the limiting factor restricting the establishment of decay communities and slowing rates of decay. This is because wood-decay basidiomycetes, responsible for brown-rot and much of the white-rot of wood, are mostly very sensitive to water stress. Apart from soft-rots, these are the two principal causes of the destruction of wood. Of the wood-decay basidiomycetes that have been tested, the lowest water potential allowing any measurable growth lies between about -5.0 and -7.0 MPa, depending on fungal species (Boddy, 1983; Dix, 1984b).

These values lie well below the fibre saturation point of wood which corresponds to a water potential of about -0.1 MPa (Griffin, 1977). In this condition all free water is held in the micropores of the cell wall and none exists in cell lumina or any void above about 5 µm in diameter, making it difficult for decay basidiomycetes to grow, since their cell wall degrading enzymes are too large to enter cell wall micropores (Griffin, 1977).

For undecayed wood -5.0 to -6.0 MPa corresponds to about 30% dry weight moisture content. Between -7.0 and -8.0 MPa undecayed wood has about 25% dry weight moisture content (Dix, 1985). These figures will vary slightly according to wood species. Wood held at about 20% moisture content can therefore be considered to be protected from basidiomycete decay. However, it is important to distinguish between establishment of decay and continued growth on decayed wood. Once decay has begun growth may continue at low moisture contents because there can be a considerable difference between the water potential of undecayed and decayed wood at the same moisture content depending on the state of decay. For example at 22% moisture content the water potential of undecayed oak has been measured at −9.5 MPa while that of decayed oak stood at −3.8 MPa. At 11% moisture content the respective figures were −18.2 MPa and −6.1 MPa (Dix, 1985).

Dead wood can be subjected to great fluctuations in water content and can often be below critical water potential values for growth. Standing wood, although in contact with soil moisture, may suffer from extreme exposure and wood on the forest floor will become very dry if no rain falls for some time or if sheltered by tree canopies that intercept showers. It is therefore not surprising that a strategy for mycelial survival has evolved among wood decay basidiomycetes. Table 6.5 gives some idea of the capacity of a random selection of wood-decay basidiomycetes to survive for considerable periods of time at low water potentials. These experiments were conducted by growing the fungi on Cellophane over malt agar and

Table 6.5 Basidiomycete survival time (weeks) at low water potentials

Species	Water potential			
	−17.5 MPa	−20 MPa	−25 MPa	−30 MPa
Heterobasidion annosum		32[b]		
Pleurotus ostreatus		32[b]		
Flammulina velutipes (−2.7 to −5.3)[a]		32[b]		
Schizophyllum commune (−5.4 to −6.0)		27[b]		
Coprinus atramentarius		24[b]		
Ditiola radicata (−3.0)			7[b]	6[b]
Calocera cornea (−4.5)			11[b]	6[b]
Dacrymyces stillatus (−6.0)			11[b]	6[b]
Stereum hirsutum (−5.4 to −6.0)	7[b]			
Phanerochaete velutina (−4.4 to −7.1)		20[b]		
Auricularia auricula-judae (−4.0)	2		1	1

[a] Minimum ψ (MPa) for growth, where known, shown in brackets (Boddy, 1983; Dix, 1984b; Dix, unpublished).
[b] Still alive at the end of period. *S. hirsutum* was dead after 30 weeks at −20 MPa, *P. velutina* after 23 weeks at −20 MPa.

then transferring the mycelium attached to the Cellophane to open Petri dishes supported over KCl solutions in sealed vessels held at 25°C (Dix, unpublished). In most cases the results could not be brought to an end-point and this is quite remarkable considering that the rapid rates of drying experienced in the experiments are unlikely to occur naturally and would be less conducive to survival than slow adjustments to water loss (Savory, 1964).

Adaptation to survival at low water potentials is especially advantageous ecologically to species growing in dead twigs and branches exposed to drying to very low moisture contents in tree canopies. *Schizopora paradoxa*, a colonist of decorticated attached oak branches (Boddy and Rayner, 1981), has been reported as surviving for 6 years in wood held at a relative humidity of 35% (Theden, 1961).

There are reports in the literature of wood-decay fungi resuming growth after being held in wood kept at very low moisture contents for years (Findlay, 1950; Findlay and Badcock, 1954; Theden, 1961). Findlay (1950) recorded survival times exceeding a year for several wood-decay basidiomycetes of importance in the decay of timbers in buildings when grown on wood and placed in relative humidities of 60 and 45%. A calculation of the theoretical water potential of the wood in equilibrium with 60% relative humidity would be about -70 MPa, at 45% relative humidity it would be about -110 MPa.

7
FUNGI OF SOIL AND RHIZOSPHERE

7.1 TECHNIQUES FOR STUDYING FUNGI OF SOIL AND ROOTS

Devising quantitative and qualitative methods for studying soil fungi is fraught with difficulties. Traditional isolation and enumeration techniques, like the **soil dilution plate**, and its modifications, which in essence involve adding soil, or a diluted suspension, to Petri dishes and covering with a suitable agar medium, are of very limited value in many respects. Usually one cannot distinguish between colonies that arise on the plates from spores and those that come from hyphae, and many inactive propagules will germinate once they are placed in suitable conditions. This makes it difficult to say much about the activity in the soil samples of the species recorded on the plates. Such records can only give very general information about the species present and their distribution in the soil. Warcup (1955) tried to resolve this problem by the microscopic examination of the young colonies arising from the soil particles on the plate. This is very time-consuming and not too successful.

Further disadvantages of soil plates are that they tend to exaggerate the numerical importance of heavily sporulating species while tending to be selective and exclude others. The overwhelming numbers of spores in soil can be reduced by washing samples before plating out (Watson, 1960), either by hand or in a mechanically operated apparatus such as that constructed by Williams, Parkinson and Burges (1965). This apparatus both washed the soil and sieved it in series separating the various soil components with their fungal populations. Unfortunately, however thorough the washing, some spores will still persist due to the difficulty of removing spores trapped in minute crevices in the soil debris.

Some idea of hyphal activity in soil can be gauged by measuring **hyphal productivity**. This can be determined by measuring hyphal biomass. Several direct and indirect methods have been developed to measure biomass but none has proved to be entirely satisfactory.

A direct method that has been generally favoured is to make crushed soil suspensions in molten agar, which, when solidified, can be sectioned for microscopic examination. This method was introduced by Jones and Mollison (1948). Hyphal length is measured and the biomass (volume x density) calculated. For this purpose hyphae are assumed to be cylindrical and their density to be about 1.5 g cm^{-3}. It is essential to use vital stains, such as fluorescein diacetate (Søderstrom, 1977) or autoradiographical techniques (Bääth, 1988) in order to distinguish between living and dead hyphae. Other workers have estimated biomass by microscopically examining stained soil smears.

The major disadvantage of these techniques is that in most cases they do not allow a positive identification of the hyphal fragment to be made, which means that biomass estimates cannot be made qualitative. Not even the most experienced mycologist can identify and assign to different species all the many different hyphae seen in soil preparations.

In order to overcome this problem, attempts have been made to develop more specific methods using **immunofluorescence** (Frankland *et al.*, 1981). This method depends on the complexing of antibodies prepared in rabbit serum with their corresponding fungal antigens when the two are brought together. If the antibodies are coupled with fluorescein isothiocyanate, when the antibody–antigen reaction occurs living hyphae become stained and are readily recognized by ultraviolet microscopy. Alternatively, the antibody label can be an enzyme such as a peroxidase, as in the **ELISA method** of analysis. This method has great potential as an investigative tool but its use at present is somewhat limited because of 'cross-reactions' that are liable to develop even when specially selected and purified antigens are used to prepare the antisera. This lack of specificity is due to the many antigens that fungi share in taxonomically unrelated groups. Extra refinement of the technique by absorption of antisera with cross-reactive fungi could bring about the necessary improvements (Chard, Gray and Frankland, 1985a,b). A better approach might be to try to produce antibodies against specific antigens using monoclonal antibody techniques.

All direct techniques are time-consuming and for rapid surveys of biomass a number of indirect techniques have been developed, each depending for accuracy on the assumption that a constant relationship exists between the measured parameter and fungal biomass. Of these the **measurement of ATP** is probably the most reliable. Sound chemical methods have been worked out for the extraction and measurement of ATP in soil samples (Jenkinson and Oades, 1979; Tate and Jenkinson, 1982) and reproducible results have been obtained for ATP-to-biomass conversion factors for various groups of soil microorganisms (Ausmus, 1973). Since ATP decays quickly in soil, measurements relate only to living populations.

Another indirect method measures the release of **fluorescein dye** from fluorescein diacetate following hydrolysis catalysed by fungal esterases (Swisher and Carroll, 1980; Schnürer and Rosswall, 1982). **Chitin** and **ergosterol measurements** have also been used to estimate fungal biomass (Ride and Drysdale, 1972; Seitz et al., 1979). Both chitin and ergosterol estimations suffer from the disadvantage that their hyphal content varies according to the conditions of growth.

Measurements of respiration offer other possibilities for the indirect calculation of biomass. These are based on the soil fumigation studies of Jenkinson and Powlson (1976) and the substrate-induced respiration studies of Anderson and Domsch (1978). The former depends on the assumption that the increase in CO_2 output that arises when soils fumigated with chloroform are re-inoculated with soil microorganisms comes from the respiration of the previously killed biomass by the recolonizing microorganisms. The results can be related to biomass by treating soils containing known amounts of introduced biomass. The method of Anderson and Domsch (1978) measures increases in CO_2 output when glucose is added to soil and is calibrated using the soil fumigation method of Jenkinson and Powlson (1976). By using selective inhibitors this method can distinguish between bacterial and fungal respiration and total biomass can be roughly apportioned between them. Comparisons with direct microscopic measurements of biomass indicate that there is a reasonable correlation with results obtained by respiration measurements.

Fungi colonizing roots can be studied by **cultural techniques** which are essentially the same as those used to study the colonists of leaves, briefly described in section 4.2.1. Washed roots are plated out, using some suitable agar medium, and the fungi that develop are counted and identified. This method suffers from some of the disadvantages of the soil plating techniques outlined above. Washed roots can also be studied by direct microscopy using stains and fluorescent brighteners (Rovira et al., 1974; Johnen, 1978). This approach allows a more accurate estimate of the extent to which roots are colonized than any cultural method.

Direct observational studies on roots and the surrounding soil have been greatly advanced by the use of the **electron microscope**. Low-temperature scanning electron microscopy allows fully hydrated material to be used with the advantage that spatial relationships can be observed on relatively undisturbed material in a near natural condition. Combined with histochemical techniques, scanning electron microscopy opens up the possibility of studying hyphal activity *in situ* (Foster, 1986, 1988). Scanning electron microscope techniques for studying the microbiology of roots have been critically reviewed by Campbell and Greaves (1990).

Those interested in a fuller account of the methods available for the isolation and study of soil microorganisms should consult Parkinson, Gray and Williams (1971) *IBP Handbook No. 19* and several chapters in

Methods in Microbiology (Grigorova and Norris, 1990). These contain practical details and a critical evaluation of each method. Several papers by Ross (1987, 1988, 1989) give further details of the use of soil fumigation procedures to measure microbial biomass.

7.2 FUNGAL DISTRIBUTION IN SOIL

Generally, both numbers of propagules and diversity of fungal species decrease with increasing depth in soil, reflecting changes in the physical and chemical conditions of the soil profile. Fungi follow the distribution of organic matter and are more numerous in the litter of the Aoo and the Ao soil horizons. Surviving populations of leaf-inhabiting fungi dominate fungal communities in the raw litter of the Aoo but in the deeper layers of the Ao and mineral soil in the A horizon below they are replaced by soil-inhabiting species in the genera *Trichoderma*, *Penicillium*, *Mortierella*, *Mucor*, *Fusarium* and others.

In the mineral layers of undisturbed soils much of the active fungal mycelium is associated with the presence of plant roots, with the germination of fungal spores being elsewhere generally inhibited by the fungistatic property of soil (section 7.3).

At the bottom of deep A horizons and in the horizons below, numbers of fungi decline markedly. Species lists are much reduced and frequently new dominants arise. These changes may reflect changes in soil aeration and the composition of gases in the soil atmosphere. In a podsol under pine, for example, *Oidiodendron fuscum*, *Paecilomyces* and *Phialophora* species replaced *Penicillium* species as the dominants in the B and C horizons (Burges, 1963). Very few species increase in number with depth, the main exception being some sterile fungi (Bissett and Parkinson, 1979a). In a heathland soil, *Mycelia sterilia* accounted for more than half the total number of isolates from the B and C horizon (Sewell, 1959). Some of these trends are illustrated in Table 7.1.

A large quantity of mycelium present in the lower soil horizons is dead, a trend which is particularly noticeable in the transition from the Aoo horizon and the fermentation (F layers) of the Ao horizon to the mineral soil in the A horizon. Nagel-de Boois and Jansen (1971) measured fungal activity in the soil of a mixed oak wood (Meerdink Wood) in Holland where growth in the different soil horizons was compared by estimating percentage colonization of nylon mesh squares 3 weeks after they had been introduced into the soil. The highest levels of colonization were recorded throughout the year in the raw litter of the Aoo horizon and fermentation layers of the Ao horizon, with comparatively low colonization levels in the humus layer of the Ao horizon and in the A horizon. Seasonal peaks of activity occurred in spring, early summer and autumn, with low levels occurring from February to April, and again from August to September.

Table 7.1 Vertical distribution of microfungi in the soil horizons of a spruce forest (calculated as percentage of the total number of isolations from each horizon)

	$A00$	$A0$		A	B
		F	H		
Acremonium griseoviride	3.3	6.8	1.1	0.3	0
Candida sp.	2.8	3.4	1.1	0	0
Chrysosporium pannorum	0	1.0	1.9	3.9	1.2
Humicola fuscoatra	0	0.5	3.6	0	0
Mortierella echinula	3.8	5.1	3.3	0.8	0.3
M. macrocystis	0.9	1.5	4.2	7.0	6.8
M. minutissima	0.2	0.7	1.4	4.5	0.6
M. nana	0.2	0.2	1.7	6.1	12.6
M. parvispora	4.0	5.6	10.0	7.0	1.8
M. ramanniana	17.7	13.4	2.8	1.4	0.6
M. vinacea	0.5	0.5	2.2	8.9	8.6
Mucor hiemalis	1.9	9.7	0.3	0	0.3
Mycelium radicis atrovirens	0	0.2	1.9	1.1	2.5
Oidiodendron sp.	0	0	1.1	2.5	3.1
Penicillium brevi-compactum	4.2	3.4	1.4	0.6	0.6
P. daleae	1.2	0.5	13.9	19.6	15.7
Tolypocladium geodes	0	1.0	3.3	5.6	6.5
Trichoderma polysporum	12.5	7.3	5.8	3.6	1.5
T. viride	5.0	6.1	3.3	2.2	0.6
Verticillium bulbillosum	1.9	3.9	2.8	0.3	0.3
Sterile mycelia species	2.8	3.6	3.3	3.6	8.6
Total number of isolates	424	411	360	358	325

Source: Söderstrom (1975).

The total fungus biomass per gram of litter or soil, including both living and dead mycelium, was greatest in the humus (H) layer of the Ao, but about 95% of this mycelium was dead. The highest amount of living mycelium was present in the Aoo horizon and this was about 50 times higher than that in the A horizon, which had the lowest value recorded (Table 7.2). The amount of mycelium present in each horizon fluctuated with time, but the variations recorded were not correlated with the seasonal growth patterns, and presumably reflected the changing rates of decomposition and loss of hyphae occurring through the year.

Frankland (1975) obtained somewhat similar results in a deciduous wood (Meathop Wood) in Lancashire, England, during July. Biomass values for

Table 7.2 Distribution of mycelium in the soil horizons of deciduous woods

Soil horizon	Length of mycelium (m g^{-1} dry soil)				Biomass (μg dry wt g^{-1} dry soil)[a]	
	Meathop Wood		Meerdink Wood		Meathop Wood	Meerdink Wood
	Living	Total	Living	Total		
Aoo	1482	4118	1591	2014	2908	1422
Ao(H)	885	2530	256	4279	1786	3023
A	145	393	25	514	277	363
B	26	96	–	–	67	–

[a]Assuming moisture content = 85%; hyphal radius = 1 μm, specific gravity = 1.5.
Sources: Meathop Wood (UK): Frankland (1975).
Meerdink Wood (Netherlands): Calculated from data of Nagel-de Boois and Jansen (1971).

mycelial standing crops showed the highest values per gram of litter or soil in the Aoo horizon and were lowest in the subsoil of the B horizon. The highest values for living mycelium were also recorded in the Aoo, which were almost twice that recorded in the H layer of the Ao horizon and more than 50 times that recorded in the B horizon. The values for living mycelium remained fairly constant at about 30% of the total in each horizon (Table 7.2).

Soil fungal communities are among the most diverse but it is recognized that characteristic populations are associated with the plants of different ecosystems. As Christensen (1989) so aptly put it, 'a beech forest can be recognised as readily by its mycoflora as by its vegetation cover'. An analysis of over 30 different surveys of soil populations covering deserts, grasslands, forests, heaths, tundras, etc. by Christensen (1981) provided clear evidence for this. In temperate regions, *Fusarium*, *Papulaspora* and *Humicola* species are typical of grasslands, *Paecilomyces carneus* and several *Oidiodendron* species are characteristic of forest soils and there are differences between the *Mortierella* species of grasslands and forest soils. Some of the latter which favour forest soils with a lower pH also occur in heath soils. Forest soils also support their own characteristic *Penicillium* populations and carry a greater diversity of these species compared with grasslands. Differences with respect to these species also exist between the populations of coniferous and deciduous forest soils (Widden, 1979). These differences in distribution patterns are strikingly illustrated in Table 7.3 for three genera taken from the data of Christensen (1981).

In Arctic tundra soils mycelial forms make up a high proportion of the isolates (Flanagan, 1981) and *Chrysosporium* and *Tolypocladium* species

Table 7.3 Distribution of three genera occurring as principal species in five ecosystems

Species	Deserts and desert grassland	Grasslands	Forest	Heath	Tundra
Fusarium semitectum	+				
F. acuminatum	+	+			
F. solani	+	+			
F. nivale	+	+			
F. oxysporum	+	+	+		
F. sambucinum		+			
F. avenaceum		+			
Mortierella macrocystis			+		
M. isabellina			+	+	
M. parvispora			+	+	+
M. alpina		+	+		+
M. gracilis		+			
M. minutissima		+	+		+
M. vinacea			+	+	
Oidiodendron citrinum		+			
O. echinulatum			+		
O. cereale			+		
O. chamydosporum			+		
O. maius			+		
O. tenuissimum			+		
O. griseum			+		
O. flavum			+		
O. periconioides			+		

Source: Christensen (1981).

are marker populations. These species also appear in soils in temperate regions where the climatic conditions are similar to tundra – the cold wet uplands of Britain for example (Widden, 1987). In deserts and desert grasslands *Aspergillus* species predominate (Christensen, 1981).

Many soil fungi are pandemic in distribution but the principal species in tropical and subtropical communities differ from those found in temperate zones. Diversity tends to be much reduced in northern ecosystems, with the elimination of genera such as *Aspergillus* and *Fusarium* (Christensen, 1981).

Fungal communities in soil usually show seasonal variation (Bissett and Parkinson, 1979b). Some fungi are summer species whereas others are more abundant in the winter. For example, Widden and Abitol (1980) found that in a Canadian spruce forest soil *Trichoderma polysporum* occurred most frequently in autumn and winter, *T. viride* was most abundant in spring, while *T. koningii* increased in the summer. In general, the communities of winter and spring tend to differ from those of summer and autumn, which tend to be rather similar. Widden (1986b) suggested that those species that increase in the more unfavourable times of the year may be poor competitors but rather good survivors.

As population changes are seldom unidirectional, total numbers in the community rarely show more than minor fluctuations with changing season.

7.3 FUNGAL ACTIVITY IN SOIL

Mineral soil is a poor medium for fungal growth due to the generally low concentration of suitable metabolic substrates and the hostile environment created by antagonistic microbial activity.

In mineral soil spore germination tends to be restricted to plant rhizospheres (section 7.4) and to the sparsely distributed microsites of organic debris. Elsewhere, spores lie dormant and slowly become moribund if organic substrates do not reach them. The failure of spores to germinate in mineral soil is known as **soil fungistasis**.

From the sites of activity, hyphae extend out into the mineral soil to scavenge for substrates. There is no doubt that many fungi can function as **oligotrophs**, that is they can grow at very low organic C concentrations in low nutrient flux environments, producing very fine mycelial networks to facilitate substrate collection (Wainwright, 1988, 1993). Much of the mycelial network of soils is probably supported in this fashion, utilizing the low concentration of C substrates diffusing into the soil solution and atmosphere from roots and the decomposition of plant debris. A wide variety of substrates are available and can be utilized, including volatile hydrocarbons, aromatic aldehydes and acids, and many aliphatics. The heterotrophic fixation of CO_2 is also a distinct possibility.

The mycelial networks of soil are extensive; several hundred metres per gram of soil is not unusual (Elmholt and Kjøller, 1987), although the total biomass represents only about 0.02–0.1% of the soil mass. The mycelial networks of soil have an important function in binding the soil particles together, helping to maintain a good soil structure for plant growth (Lynch and Bragg, 1985). The role of fungi in mineral cycling in soil is well known. They are primarily concerned with the mineralization of organic C and N but the importance of fungal activity in soils in mineral transformations such as the oxidation of reduced sulphur compounds should not be overlooked (Cromack and Caldwell, 1992; Wainwright, 1992).

The antagonistic property of mineral soil to spore germination was discovered more or less simultaneously by Dobbs and Hinson (1953) and Chinn (1953), but it was the former who coined the term **soil mycostasis** (**fungistasis**) to describe it and who investigated it in more detail, showing it to be a general property of soils. Dobbs and Hinson (1953) examined soils for fungistatic properties by spreading spores on to the surface of permeable Cellophane which was then placed, spore surface uppermost, in close contact with the soil. Chinn (1953) used a slide technique to similar effect, first coating the slide before burial in soil with spores in a suspension of 0.5% cooled molten peptone agar, both techniques permitting the easy recovery of the spores for examination at the end of the experiments.

Since these original studies, the phenomenon of soil fungistasis has been demonstrated using different techniques and has been shown to operate against the spores of many different fungi (Ledingham and Chinn, 1955; Lingappa and Lockwood, 1963; Dix, 1972).

Early on, it was realized that there was a link between fungistasis and microbiological activity in soil, and Dobbs and Hinson (1953) in their original studies demonstrated that the phenomenon disappeared if soils were sterilized by heating. Subsequently, Lingappa and Lockwood (1963) and others restored the inhibitory activity of sterile soils by adding isolates of soil fungi, bacteria and actinomycetes.

Initially, research centred on the search for a specific fungistatic substance, such as an antibiotic, which might be involved in an antibiosis effect in soil, but all efforts in this direction failed. An alternative nutritional hypothesis based on the observations by Dobbs and Hinson (1953) and Chinn and Ledingham (1957) that fungistasis could be overcome by the addition of sugar or various degradable organic substances to soil was advanced by Lockwood (1964) following the discovery by Lingappa and Lockwood (1964) that dry spores are initially leaky when placed in water and that a single brief washing could extract as much as 10% of their dry weight. In soil, such spores would be brought into direct competition with other soil microorganisms for scarce resources in the environment in order to replace lost soluble food reserves. Further experiments by Lockwood and co-workers strengthened the hypothesis (Ko and Lockwood, 1967; Steiner and Lockwood, 1969) by linking low sensitivity to fungistasis indirectly to large food reserves (large spore size). Low sensitivity was also negatively correlated with an ability to germinate in distilled water without the addition of metabolic substrates.

A nutritional hypothesis has also been put forward to explain the failure of spores to germinate under similar conditions on young leaves (Blakeman and Fraser, 1971; Blakeman and Brodie, 1976). The experimental evidence which supports this is described in detail in section 4.3.2 and adds considerable force to the now generally accepted argument that the nutritional status of spores and competition for substrates between microorganisms are directly involved in both these phenomena.

While the loss of metabolic substrates from spores and the chronic shortages of replacements in soil are important reasons why spore germination is inhibited in soils, other factors probably nearly always contribute. Experimental evidence suggests that inhibitory volatiles are also involved. Balis and Kouyeas (1968) observed that the germination of *Arthrobotrys oligospora* spores in contact with soil improved considerably when suspended over silver nitrate or mercuric perchlorate solutions, substances which form complexes with unsaturated hydrocarbons. Subsequently Hora and Baker (1970) inhibited spore germination in several fungi by placing spores on agar discs which had been previously 'activated' by incubation for 24 h over soil when resting on sterile slides or Cellophane. Inhibition on the activated discs, however, was never as great as when spores were buried in the same soil on Millipore filters (Figure 7.1).

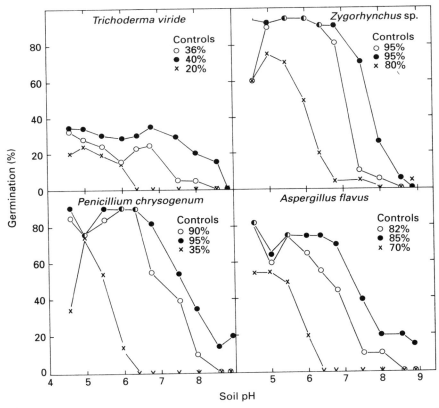

Figure 7.1 Germination of spores in various fungistasis assay techniques in soils of different pH. ●–● = Agar discs preincubated on soil over sterile slides; ○–○ = agar discs preincubated on soil over Cellophane; x–x = spores incubated in soil on Millipore filters. Note: inhibition of germination increases with increasing pH. (Reproduced with permission from Hora and Baker, 1970, *Nature*, 225, 1071–2. Copyright 1970 Macmillan Magazines Limited.)

Actinomycetes and certain fungi such as *Trichoderma* species, known to produce a range of antifungal volatiles, were shown to be able to reproduce this effect when allowed to recolonize sterile soils, but mixed cultures of bacteria or fungi not actively producing volatiles had no effect (Hora and Baker, 1972).

In untreated soils the fungistatic effect was found to be slight under acid conditions, but stronger in alkaline soils which favour the development of actinomycetes. Mixtures of substances which have specific effects appear to be involved, for while Gupta and Tandon (1977) found that the volatiles produced by *Streptomyces halstedii* and *S. griseus* both inhibited the growth of *Curvularia lunata*, only the volatiles produced by *S. griseus* inhibited the mycelium of *Rhizoctonia solani*.

Experiments by Bristow and Lockwood (1975) distinguished between a volatile component in an alkaline clay and a non-volatile factor in a loam. Spores held over soils on agar discs were inhibited in the former but not in the latter. Glucose amendments added to the loam restored germination to high levels in spores on discs placed on the soil but not in similar experiments with the clay.

A large number of potentially inhibiting volatiles are produced by soil microorganisms (section 3.2.4). It has been suggested that microbiologically produced ethylene may be an important regulator of microbial growth in soil (Smith, 1976). However, as yet, there is no direct evidence which links ethylene with soil fungistasis. On the contrary, it has been found that ethylene does not inhibit spore germination at concentrations found in soil nor does pre-treating soils with ethylene increase the fungistatic effect (Pavlica *et al.*, 1978; Schippers, Boewinkel and Konings, 1978). It may be that ethylene acts indirectly through the induction of allyl alcohol and other inhibitors as suggested by Balis (1976).

To date, the only volatile substance which has been convincingly linked with soil fungistasis is ammonia (Ko, Hora and Herlicska, 1974; Pavlica *et al.*, 1978) and even here the evidence is equivocal. Some workers have found that ammonia does not inhibit the germination of all soil fungi and have produced evidence that other, as yet unknown, mycostatic volatiles must exist in the soil atmosphere (Schippers, Meijer and Liem, 1982).

In some soils fungistasis has been found to persist after sterilization (Dobbs and Gash, 1965; Ko and Hora, 1971, 1972). **Residual fungistasis**, as it is called, appears to be a feature of only a minority of soils and, perhaps, is of ecological significance in only a very few (Dobbs, 1971). Residual fungistasis is associated with certain mineral properties of soil: in alkaline soils the effect has been attributed to calcium carbonate and in acid soils to iron oxide or to the salts of aluminium.

It is apparent that soil fungistasis cannot be attributed to a single factor but it should be regarded rather as a manifestation of the sum of all the antagonistic forces of biological and chemical origin which may exist at any one time in a particular soil.

The soil environment is also antagonistic to hyphal growth, although hyphae are usually less sensitive to soil fungistasis than spores of the same species (Steiner and Lockwood, 1969; Mitchell and Dix, 1975).

Hsu and Lockwood (1971) found that hyphal sensitivity was inversely correlated with growth rates and with hyphal diameter and in this respect hyphae show similarities with spores. Hyphae subjected to continuous nutrient stress will ultimately undergo autolysis.

For further reading on the topic of soil fungistasis see Lockwood and Filonow (1981) and Lockwood (1992).

7.4 THE RHIZOSPHERE AND ROOT COLONIZATION

7.4.1 The rhizosphere

Plant roots stimulate microbial growth in their vicinity, with the result that the numbers of microbes in soil around roots increase 50–100 times above those in soil away from the influence of roots. This phenomenon is called the **rhizosphere effect** and the root soil complex where it occurs is known as the **rhizosphere**. Rhizospheres contain about 10^6 fungi, 10^7 actinomycetes, 10^9 bacteria and 10^3 protozoa g^{-1} soil.

The rhizosphere effect is primarily brought about by the release from roots of a variety of organic substances which serve as substrates for microbial growth. Fungal spores which elsewhere in the soil, in the absence of suitable substrates for growth, are lying dormant, under the influence of soil fungistasis (section 7.3), are stimulated to germinate with the result that a flourishing fungal community develops on and around the root. The effectiveness of roots in overcoming soil fungistasis has been demonstrated many times. Spores need only to be suspended in agar films on slides or stuck onto sticky tapes and placed under germinating seeds to show a marked stimulation in germination compared with controls in soil without roots. The more vigorous the seedling, the greater the measurable response (Jackson, 1957, 1966; Barton, 1957; Jackson, 1960).

When considering the rhizosphere effect, it is ecologically important to distinguish between fungal communities on the root surface, or **rhizoplane** as it is called, and those in the rhizosphere. While most fungi which grow in the rhizosphere soil and rhizoplane are widely distributed soil-inhabiting fungi, a small number which grow in the rhizoplane seem to be more or less confined to this habitat and form a separate ecological group of saprophytic or weakly parasitic root-inhabiting species.

Soil-inhabiting fungi include many common examples in the genera

Fusarium, Penicillium, Gliocladium, Mucor and *Mortierella* which are found on all kinds of decaying plant debris in soil. Yeasts also occur in the rhizosphere and some, like *Cryptococcus terreus* and a few others, seem to be true soil inhabitants, although little is known of their ecology (Last and Price, 1969).

Common root-inhabiting fungi include various forms of sterile mycelia, some but by no means all of which are mycorrhizal. Sterile mycelial forms are especially frequent on the roots of grasses and cereals and on the roots of woody plants (Harley and Waid, 1955a; Gadgil, 1965).

Not all of the soil-inhabiting fungi which flourish in the rhizosphere are able to colonize the rhizoplane, and of the commoner fungi occurring in the rhizosphere of bean roots (*Phaseolus vulgaris*), Dix (1964) found that *Trichothecium roseum, Trichocladium asperum, Papulaspora* species, *Stysanus stemonitis* (*Doratomyces stemonitis*), some *Penicillium* species and several mucoraceous fungi were rare or absent from root surfaces. Some of these trends are illustrated in Figure 7.2. The fungi of the rhizosphere soil and the rhizoplane are therefore best thought of as being two distinct but overlapping communities with many species in common.

Rhizosphere fungi seem to have no special nutritional requirements that are satisfied particularly by organic substances released by roots, which may explain why usually only minor qualitative differences exist between the composition of fungal communities in rhizosphere soils and fungal communities occurring elsewhere in the same soil. The order of change is a matter of degree rather than absolute differences. *Penicillium lilacinum*, *Mucor hiemalis* and several *Fusarium* species are characteristic of rhizosphere soils but are also present in significant numbers in non-rhizosphere soil when washed soils are compared (Parkinson and Thomas, 1965). This situation is in contrast to bacterial communities, where it has been shown that the rhizosphere may contain greater numbers of bacteria that are heterotrophic for certain amino acids (West and Lochhead, 1940; Lochhead and Thexton, 1947). Qualitative differences are, of course, more striking when the rhizoplane is compared with the non-rhizosphere soil due to the presence in the rhizoplane of strict root-inhabiting fungi.

The stimulating effect of roots on fungal growth in the rhizosphere is usually expressed quantitatively as the **R:S ratio** per gram of soil (the ratio of fungi in the rhizosphere to fungi in the non-rhizosphere soil). The R:S ratio is very variable and dependent very much on environmental factors and on plant species. In general, the R:S ratio is likely to be higher in soils of low organic content with a pH below neutral, such as dunes and sandy heaths, and lowest in woodland and garden soils. Figures calculated from the data of Peterson (1958) show a fungal R:S ratio of 33:1 for a sandy acid loam and about 9:1 for a neutral clay. Low R:S values are frequently recorded in tropical soils due to the rapid decay of organic matter in these soils which maintains fungal activity at a high level in the non-rhizosphere

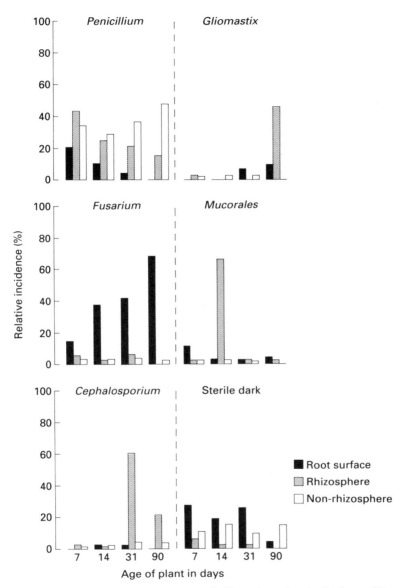

Figure 7.2 Comparison of root surface and rhizosphere fungi of wheat. (Reproduced with permission from Peterson, 1958.)

soil (Odunfa and Oso, 1979). In agricultural soils at near neutral pH in temperate regions R:S values for fungi generally lie between 1 and 3:1 (Timonin, 1939) but higher values of the order 10–20:1 have been recorded

(Katznelson, 1960), although it is probable that the latter reflects some deficiency in the isolation technique used. When dilution plates have been employed, high values usually result because spores from heavily sporulating species increase the counts. R:S values for fungi are usually about 10–100 times lower than those for bacteria (Katznelson, 1960).

No clear boundary exists between the rhizosphere and the non-rhizosphere soil, which makes attempts to measure the extent of the rhizosphere accurately extremely difficult. The maximum effect has been shown to lie between 0 and 6 mm from the root surface in the case of lupin (*Lupinus nootkatensis*) and peanut (*Arachis hypogaea*), with measurable increases in the numbers of fungi over control soils being detected up to 18 mm from the root surface (Papavizas and Davey, 1961; Griffin, 1969). However, these figures are heavily dependent upon plant species, soil conditions, etc.

The rhizosphere effect for individual roots tends to decrease with increasing depth in the soil. With tobacco (*Nicotiana tabacum*) for example, Timonin and Lochhead (1948) recorded only 4000 fungi g^{-1} of rhizosphere soil at depths between 25 and 28 cm compared with about 550 000 g^{-1} at depths between 0 and 12.7 cm; a decrease of more than 99%. This, together with the tendency for roots to be more numerous nearer the surface of the soil, means that the rhizosphere effect is essentially a phenomenon of the upper layers of the soil. Often the density of roots is such that in the upper levels of the soil the greater part must be considered to be in a rhizosphere. Calculations of values for root length per unit volume of soil, or beneath each square centimetre of soil at the surface for depths from 0 to 15 cm are particularly high for grasses and some cereal crops (Newman, 1969; Barley, 1970). Calculations have been made which show that at a depth of 2 cm beneath a sward of *Lolium perenne*, the mean horizontal distance between roots was only 3 mm (Barley and Sedgley, 1961).

There is also some evidence that the rhizosphere and rhizoplane fungal communities change in composition horizontally from the root and vertically following the root downward through the soil. Papavizas and Davey (1961) found that most fungal species declined in numbers horizontally as the distance from lupin (*Lupinus nootkatensis*) roots increased, but *Aspergillus ustus* was still twice as abundant in the rhizosphere soil compared with non-rhizosphere soil at a distance of 18 mm from the roots. On *Calluna vulgaris* roots in the upper layers of the A soil horizon, members of the Mucorales dominated the fungal community (Sewell, 1959). *Penicillium* species were also common on roots from this horizon, but while *Penicillium* species persisted on roots deeper in the A soil horizon, with few exceptions the *Mucor*, *Mortierella* and *Absidia* species of the upper layers disappeared. Below, in the B soil horizon, the numbers of fungal species present on the root surfaces fell dramatically and at this level the fungal community was made up almost exclusively of one or two different forms of sterile mycelium.

The organic matter which passes into the soil from roots, giving rise to the rhizosphere effect, is composed of insoluble cell wall debris, polysaccharide mucilages secreted by cells, mucilages formed from polysaccharide hydrolysates of cell walls, and soluble exudates that pass through the functional membranes of healthy cells and leak from damaged and senescing tissues. It has been calculated that about 12–18% of all photosynthetically fixed carbon passes into soil from healthy cereal roots (Barber and Martin, 1976). The total quantity of all organic matter, cell debris, exudates and mucilages released into the soil has been estimated to be about 50–100 mg g^{-1} root day^{-1} for some plants (Foster et al., 1983). A review by Newman (1985) indicates that the figure may be nearer 200 mg g^{-1} day^{-1}.

A large number of different water-soluble organic substances have been detected leaking from the roots of plants growing under sterile conditions in sand or in culture solutions and identified by chromatographic methods. A table compiled by Rovira and McDougall (1967) for wheat (*Triticum aestivum*) from the results of 10 different workers lists over 40 substances for this single plant species, including 10 different sugars, 21 amino acids and 10 organic acids. Several vitamins, plant enzymes and growth hormones, and a large number of other miscellaneous items have also been collected in exudates from plants. It seems virtually certain that any low molecular weight water-soluble substance present in plant roots can leak out into the soil.

Organic substances released by plant roots into the soil may include compounds with antifungal activity. *Calluna vulgaris* roots produce antifungal substances which inhibit basidiomycetes and this is thought to be the reason why sheathing mycorrhizae develop poorly in *Calluna* heath soils (Robinson, 1972).

The genus *Allium* contains some other well-known examples (Parkinson and Clarke, 1964; Clarke, 1966). Here, organic sulphur compounds have generally been implicated: diallyl-disulphide in garlic (*A. sativum*) and methyl- and propyl-thiosulphinate and thiopropanal-S-oxide in onion (*A. cepa*) (Tariq and Magee, 1990; Brondnitz and Pascale, 1971) respectively.

In other plants, volatile organic acids, aldehydes and unsaturated fatty acids are produced by roots, inhibiting and sometimes stimulating fungal growth (Fries, 1973). Fungi can use volatiles as sole sources of organic carbon and will grow, utilizing minute quantities in the atmosphere (Tribe and Mabadeje, 1972). This may enable fungal growth to take place in the soil some distance away from roots.

Roots in soil produce ethylene, especially under anaerobic conditions (Burg, 1962). Ethylene has been linked with the regulation of microbial growth in soils (section 7.3). It may have significance for the ecology of rhizosphere microorganisms that ethylene production increases when roots are wounded.

The main site of release of organic matter to the soil in young plants is the root tip. Schroth and Snyder (1961) germinated French bean seeds (*Phaseolus vulgaris*) between filter papers and, using specific chemical reagents as sprays, observed that the main sites where diffusible exudates containing sugars and amino acids collected on the papers corresponded with the tip regions of the roots. Pearson and Parkinson (1961) found that the centre of maximum exudation of amino acids from broad bean (*Vicia faba*) root tips was at a point about 28 mm behind the apex below the origin of lateral roots, corresponding with known zones of high proteolytic enzyme activity. It seems likely that this is in connection with xylem differentiation.

The bulk of the organic matter released from young root tips is, however, non-diffusible (MacDougall and Rovira, 1970) and experiments by Rovira (1973) using ^{14}C-labelled plants have indicated that the site of maximum accumulation of insoluble material shed from root tips is the zone of cell elongation. The origin of this material is the mucilaginous sheath surrounding the root apex and the discarded cells of the root cap and the root hairs. These structures are expendable and provide a continuous and varied supply of substrates for microbial growth.

The root cap cells that protect the root tip in its passage through the soil are shed as the root grows forward and are continuously renewed by cell division at the meristem behind. These cells contain a lot of starch. Root hairs are produced as extensions of epidermal cells about 3–10 mm behind the root cap. Typically, their life is very short and after a few days they begin to senesce. They become increasingly leaky during senescence and ultimately they burst and are sloughed off to be replaced by others developing from new epidermal cells arising from the meristem in front.

In many plants, the root tip has been shown to be covered by an external coat of mucilage, forming the so-called **mucigel sheath** (Jenny and Grossenbacher, 1963; Greaves and Darbyshire, 1972). The mucigel sheath is widest at the root apex where it sometimes forms a globule several times wider than the diameter of the root itself. It extends back in a narrowing zone, usually for several millimetres behind the apex, sometimes covering the root hairs, reaching about 1 μm in thickness at its extremity. Mucigel is produced by active secretion from the cells of the root cap, and to some extent by the cells of the epidermis and root hairs (Figure 7.3). Chemically, mucigel is composed of hydrated polysaccharides dispersed in a watery fluid. The mucigel polysaccharides yield a variety of sugar monomers, galactose, arabinose, xylose and fructose on hydrolysis and are potentially a rich source of substrates for microbial growth (Greaves and Webley, 1965; Floyd and Ohlrogge, 1970).

Electron micrographs show progressive penetration and erosion of the

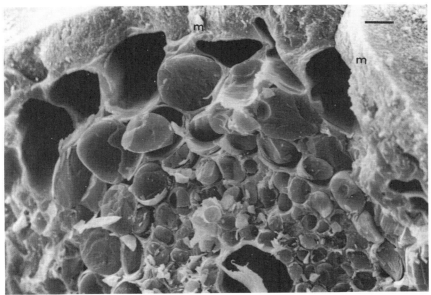

Figure 7.3 Hydrated low-temperature scanning electron micrograph of a fractured wheat root showing the surface mucigel (m). Scale bar = 10 μm. (Reprinted from *Soil Biology and Biochemistry*, 14, R. Campbell and R. Porter, Low temperature scanning electron microscopy of microorganisms in soil, 241–5, copyright 1982, with kind permission from Elsevier Science Ltd, The Boulevard, Langford Lane, Kidlington OX5 1GB, UK.)

mucigel by bacteria. Occasionally, fungal hypha are also seen (Dart and Mercer, 1964; Greaves and Darbyshire, 1972; Foster and Rovira, 1973) (Figure 7.4).

A feature of the development of the rhizosphere mycoflora is that populations increase markedly as plants age. Peterson (1958) recorded an increase in the incidence of *Fusarium* species on the root surfaces of wheat (*Triticum aestivum*) with ageing, while *Cephalosporium* and *Gliomastix* species increased in the rhizosphere soil. Similar changes in ageing plants have been reported for French bean (*Phaseolus vulgaris*), barley (*Hordeum vulgare*) and cabbage (*Brassica oleracea*) (Parkinson, Taylor and Pearson, 1963; Dix, 1964). The increase in fungal numbers around ageing roots is due to the tendency for greater quantities of organic matter to become available for fungal growth from changes taking place in the root tissues with time. As roots age, they suffer increasing amounts of damage by soil microorganisms that invade epidermal and cortical cells from the mucigel sheath and also to some extent through contact with abrasive soil particles, with the result that cells become increasingly more leaky. In

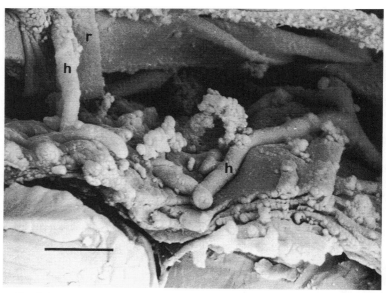

Figure 7.4 Hydrated low-temperature scanning electron micrograph of a fractured wheat root showing fungal hyphae (h) among root hairs (r). Scale bar = 10 μm. (Reproduced with permission from Campbell and Greaves, 1990.)

plants growing under natural conditions, it is common to find that practically all of the cells of the epidermis and cortex are damaged and invaded by fungi and bacteria (Old and Nicholson, 1975). The extent of the invasion is such that it has become obvious that what has been called the rhizosphere is in older roots a continuum extending from the cortical cells of the root out into the soil (Foster, Rovira and Cook, 1983). The incorporation of the root cells into the continuum of the rhizosphere is, in fact, a process which begins remarkably early in the life of the root. The electron micrographs of *Ammophila arenaria* root tissues produced by Marchant (1970) clearly show damage to outer epidermal cell walls and invasion by fungal hyphae in tissues only 4–5 mm behind the root tip. It is usual to find, therefore, that when roots are growing naturally in unsterilized soil, as opposed to root cultures, the parts of the root system which show heaviest exudation correspond to the oldest parts of the root (Dix, 1964).

In many monocotyledons and in some herbaceous dicotyledons, the damaged root cortex has only a limited life and after some time the whole of the primary cortex collapses and is not replaced. In woody plants some or all of the primary cortex is sloughed off during secondary growth. Some idea of the quantity of this material and its importance in promoting fungal growth in the rhizosphere of ageing plants can be gained from the following reports: Bowen and Rovira (1973) published calculations which estimated that of the total amount of photosynthesized carbon translocated to the roots of wheat (*Triticum aestivum*) about 1–2% is released into the soil, 0.2–0.4% as soluble material and 0.8–1.6% as insoluble material; Rogers (1968) calculated that approximately half of the dry matter of the new root growth of apple (*Malus* species) is lost each season by the shedding of cortical tissues. Peanuts (*Arachis hypogaea*) growing in culture solution sloughed off approximately 0.15% of the root tissue each week (Griffin, Hale and Shay 1976).

In addition to all of this there is a tendency for increasing amounts of organic debris to become available for microbial growth in rhizosphere soils due to the natural death of roots. The fine root system of plants has only a limited lifespan and much of it is continuously being replaced. Reviews indicate that as much as 25–92% by weight of the fine root system may be lost annually, contributing several hundred kilograms of biomass per hectare of soil (Head, 1973; Fogel, 1985), an input which has considerably more significance for microbial growth, especially in deeper layers of the soil, than the input from leaves or branch litter.

While the rhizosphere effect tends to increase with time, at any one time the extent of the rhizosphere effect will depend upon a number of other variable factors. Of these the physical environment, soil characteristics, plant species and the physiological condition of the plant can also have a profound qualitative influence on the composition of the rhizosphere mycoflora.

An early investigator of the differences between the rhizosphere mycofloras of different plants was Timonin (1939), who examined the microbial rhizosphere populations of a number of crop plants. In his investigations, considerable variation was found in the numbers of fungi growing in the rhizospheres of different plants with the highest numbers per gram of soil and the highest R:S ratios being recorded for oat (*Avena sativa*). Qualitative differences may also arise. Facultative fungal pathogens may multiply more freely in the rhizosphere of host plants than that of non-host plants and even in some cases between varieties of the same plant. Of the many examples quoted in the literature, the experiments of Reyes and Mitchell (1962) will serve to illustrate this point. These workers discovered that when a number of *Fusarium* species were inoculated into the soil and their multiplication in the rhizosphere of host and non-host plants followed under controlled conditions, a remarkable stimulation in the rhizospheres of their respective host plants was seen compared with numbers associated with non-host plants (Table 7.4).

Good illustrations of the complex changes that can occur in the rhizosphere mycoflora when plants are grown under different environmental conditions in different soils are to be found in the work of Peterson

Table 7.4 Populations of pathogenic *Fusarium* species in rhizosphere soil of host and non-host plants

	No. of propagules ($\times 10^3$ g^{-1} soil)		
	F. solani f. *pisi*	*F. oxysporum* f. *lycopersici*	*F. solani* f. *trifolii*
Expt 1 various crop plants			
Unplanted soil	10b	35b	26b
Pisum sativum	**38a**	37b	38a
Beta vulgaris	6b	42b	
Raphanus sativus	7b	12c	
Lycopersicon esculentum	5b	**60a**	22b
Trifolium incarnatum	5b		**47a**
Expt 2 Leguminosae			
Unplanted soil	10c		
Pisum sativum	**41a**		
Phaseolus vulgaris	12c		
Glycine max	18b		
Vigna sinensis	5c		
Melilotus officinalis	5c		
Medicago sativum	11c		

Numbers followed by same letter are not significantly different. Host–pathogen combinations are in bold.
Source: Data from Reyes and Mitchell (1962).

(1958) and Taylor and Parkinson (1964). The latter compared the independent effects of varying environmental conditions on the development of the root surface mycoflora of French bean (*Phaseolus vulgaris*) and found that *Cylindrocarpon didymum*, *Fusarium culmorum* and *F. solani* were the characteristic fungi of roots in alkaline soil, while *F. oxysporum* and *F. sambucinum* were the counterparts in acid soil. *Cylindrocarpon destructans* (*C. radicicola*) occurred in high numbers on roots in both acid and alkaline soils, but in higher frequencies in acid soil (Table 7.5). *Gliocladium* species and *F. oxysporum* were isolated in higher numbers at 20°C and 50% moisture-holding capacity than at 15°C and 30% moisture-holding capacity. *Penicillium* species, on the other hand, were considerably more numerous at 30% moisture-holding capacity. Peterson (1958) also noted an increase in *Trichoderma* and *Cephalosporium* species on wheat roots in acid soils, with *Gliocladium* species increasing under alkaline conditions.

Physiological changes in the plant are also thought to be very influential in the development of the rhizosphere mycoflora. Light, for example, has

Table 7.5 Frequency of occurrence (%) of fungi on 10-day-old *Phaseolus vulgaris* roots grown in soils of different pH

Fungi	Soil pH	
	4.2	7.8
Fusarium oxysporum	30	2
F. sambucinum	25	–
F. culmorum	1	13
F. solani	–	11
Cylindrocarpon destructans	21	30
C. didymum	–	46
Gliocladium spp.	50	11
Penicillium lilacinum	17	–
Trichoderma viride	4	–
Mortierella vinacea	3	–
Mortierella spp.	2	–
Sterile dark forms	4	6
Sterile hyaline forms	–	8
Total isolates	169	114

Source: Taylor and Parkinson (1964).

been shown to produce some striking changes in the root surface mycoflora of beech (*Fagus sylvatica*) (Harley and Waid, 1955b). When grown at suboptimal light intensities, higher light intensities favoured root colonization by *Trichoderma* and *Gliomastix* species, while lower light intensities encouraged the growth of *Cylindrocarpon* species and several non-sporing forms including *Rhizoctonia* species (Table 7.6).

While little is known of how physical and biotic factors selectively change the rhizosphere mycoflora, circumstantial evidence indicates that they operate by changing the quantity or quality of exudates released by the roots; at least changes which have been shown to influence the mycoflora have also been shown to affect exudation. Qualitative and quantitative differences in the amino acid and organic acid content of exudates occur between different plant species (Rovira, 1956; Vancura, 1964; Vancura and Hanzlikova, 1972) (Table 7.7) and in the same plant species with variation in light intensity and temperature (Rovira, 1959; Smith, 1972).

Increase in exudation may also follow cold shock or water stress

Table 7.6 Frequency of occurrence (%) of fungi on roots of *Fagus sylvatica* grown in different light intensities

Fungi	% Daylight radiation			
	25.1	*14.2*	*10.6*	*6.1*
Trichoderma	72	43	45	22
Gliomastix	21	17	12	7
Sterile hyaline	8	2	1	6
Brown sterile	4	15	14	6
Hyalopus	0	6	1	1
Oospora	0	5	3	2
Calcarisporium	26	11	33	17
Gliocladium	7	6	10	5
Ophiostoma	0	0	8	2
Penicillium	1	4	8	8
Paecilomyces	4	6	10	13
Dark sterile	0	0	0	8
Monilia	3	2	6	6
Rhizoctonia	1	19	22	32
Cylindrocarpon	25	31	51	59

Source: Harley and Waid (1955b).

treatment of plants (Katznelson, Rouatt and Payne, 1954; Vancura, 1967; Vancura and Garcia, 1969).

While difficult to prove, it is easy to see how qualitative changes in the rhizosphere mycoflora may be directly connected with changes in the pattern of exudation, although there is obvious difficulty in the case of changes associated with temperature, pH and the moisture content of the soil, in separating the effects of these on plants from their direct effects on the fungi.

Exudation from plant roots may also increase or decrease according to the availability of certain nutrients to the plants. Instances have been recorded where extra phosphorus supplied to plants leads to an increase in the exudation of soluble matter whereas it was reduced by the addition of extra potassium (Rovira and Ridge, 1973; Trolldernier, 1972). Amino acid exudation may be increased by extra nitrogen but reduced by added potassium or phosphorus (Bowen, 1969; Trolldernier, 1972). However, mycelial abundance on root surfaces has been shown to decrease if extra nitrogen is added to soil and increases when nitrogen is deficient (Turner and Newman, 1984). This seemingly paradoxical result may be due to the

Table 7.7 Relative amounts of amino acids and sugars (as %) in root exudates of barley and wheat

	Barley	Wheat
Amino acids		
Cysteic acid	1	0
Cystine	0	1
Cystathionine	0	2
Glutamine	0	1
Asparagine	3	2
Aspartic acid	3	3
Serine	2	3
Glycine	2	2
α-Aminoadipic acid	1	0
Glutamic acid	3	3
Threonine	3	3
α-Alanine	3	3
β-Alanine	0	1
Proline	1	1
γ-Aminobutyric acid	0	3
Tyrosine	3	2
Methionine (valine)	3	3
Phenylalanine	2	3
Isoleucine	2	3
Leucine	2	2
Sugars		
Oligosaccharides	27.8	26.7
Maltose	5.4	3.1
Galactose	13.6	4.0
Glucose	9.5	16.8
Arabinose + fructose	19.0	17.7
Xylose	15.0	15.9
Ribose	1.3	0.9
Rhamnose	6.8	14.9
Desoxyribose	0.8	–
Desoxysugar	0.8	–

Source: Vancura (1964).

more important effects nitrogen has on carbohydrate exudation. Logically this could increase when nitrogen is in short supply.

Experiments such as these are interesting since they suggest that competition for nutrients between plants and fungi could influence exudation and the abundance and composition of the rhizosphere mycoflora. This possibility has been explored experimentally by Newman and co-workers and reviewed by Newman (1978). In a series of competition experiments in which the grass *Lolium perenne* was grown in the same pot with either of two natural competitors, *Plantago lanceolata* or *Trifolium repens*, the incidence of mycelium on the roots of the grass increased compared with the levels found when the grass was grown separately. In the same experiments, the incidence of mycelium on *Plantago lanceolata* roots also increased but that on the roots of *Trifolium repens* decreased compared with when the plants were grown alone. In such combinations there is usually a tendency for the abundance of root surface mycelium of paired plants to assume levels near to the mean of the monoculture values (Newman, 1985). Changes in the bacterial population on roots in these experiments were found to be positively correlated with percentage changes in shoot nitrogen under conditions of competition. However, no similar correlation existed for the fungal populations, leaving the cause of the changes in the fungal population open to speculation and further experimental work.

The reciprocal effects of fungal growth in the rhizosphere on plant growth are less well researched and are largely outside the scope of this book. Several aspects deserve brief mention but for a more comprehensive treatment see Brown (1975) and Curl and Truelove (1986). The growth of fungi and other microorganisms in the rhizosphere may stimulate plant growth through several mechanisms. Decomposition of organic matter will release nutrients to roots, insoluble minerals like phosphates may be solubilized, saprophytic fungi may reduce pathogenic attack by antagonizing potential pathogens and may break down phytotoxic substances in the soil. Conversely, in addition to the more obvious point that disease is caused by pathogens, fungi may harm plants by competing with roots for nutrients, distort and stunt root growth by the release of growth regulatory substances and secondary metabolites and produce phytoxins by the decomposition of plant debris. There is plenty of evidence to indicate, in fact, that plants grow better in sterile soils.

The effects of microbial rhizosphere populations may thus have far-reaching consequences for plants and Newman (1978) suggested that this might even lead to alterations in the balance of plant communities if under certain conditions fungal populations favoured the rhizosphere of one plant rather than another.

All aspects of the rhizosphere phenomenon have been thoroughly reviewed many times. Some recommended articles are those by Mosse (1975), Hale, Moore and Griffin (1978) and Curl and Truelove (1986) in addition to those mentioned elsewhere in this account.

7.4.2 Root colonization, successions and decay

Radicles emerging from germinating seeds are virtually devoid of microorganisms and present near virgin substrata for colonization by fungi. Early root colonization has been studied on beech (*Fagus sylvatica*), rye grass (*Lolium perenne*), pea (*Pisum sativum*) and broad bean (*Vicia faba*) by Harley and Waid (1955a), Waid (1957), Stenton (1958) and Taylor and Parkinson (1961) respectively. The general picture that emerges is that fungal colonization begins on the older parts of the emerging radicle stimulated by root exudates. It gradually extends towards the apex, reaching to within 1–2 cm of the root tip about 6–7 days after germination. Colonization is at first sparse and numbers of species few, but in time numbers and the total area occupied by hyphae increases and the species list becomes more diverse (Table 7.8). Colonization never completely extends to the root tip and an ascending gradient of density of occupation extends back from a near sterile zone of about 3.0 mm at the root tip (Figure 7.5).

Radicles emerging from seed are in the main colonized by soil fungi and the populations of *Cladosporium herbarum*, *Alternaria alternata* and other leaf-inhabiting fungi present on seed coats normally play little part. Usually the occurrence of these fungi on roots is only occasional and very transitory, presumably because they are poor competitors in soil. They may, however, become important root colonizers in conditions where the soil mycoflora is rather sparse and competition is low, such as in saltmarshes and sand dunes (Pugh, 1967). In these circumstances, the root surface mycoflora may reflect some of the unusual features of the phylloplane populations of plants in these habitats. *Dendryphiella salina* and *Aschocytula obiones* are two phylloplane fungi characteristic of saltmarsh habitats, which are also found on the roots of the saltmarsh plant

Table 7.8 Increase in the colonization of pea radicles with time measured as numbers of root sections from the distal 2 cm of radicles yielding 0, 1, 2 or 3 colonies (%)

No. of colonies	Age of roots (days)							
	7	14	21	28	37	43	50	87
0	29	15	8	4	0	0	0	0
1	60	68	66	76	56	74	82	42
2	10	12	22	18	42	20	18	54
3	4	4	4	2	2	6	0	4

Source: Stenton (1958).

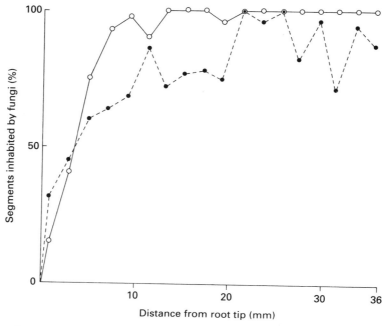

Figure 7.5 Distribution of fungal populations on the apices of rye-grass roots. ●–● = Surface populations estimated by cultural methods; ○–○ = surface populations estimated by direct observation. (Reproduced with permission from Waid, 1957.)

Halimione portulacoides (Dickinson and Pugh, 1965a,b,c).

Colonization of seedling roots seems to take place mostly by progressive invasion by fungi from the surrounding soil and less importantly by the downward growth of hyphae from any already colonized older parts, as shown by the experiments of Taylor and Parkinson (1961). However, in mature root systems, colonization of the new tissues produced at the meristem seems to occur both from the soil and by the downward growth of hyphae from established populations on older parts (Parkinson and Crouch, 1969).

Colonization of the roots of French bean (*Phaseolus vulgaris*) follows a fairly typical pattern and will serve to illustrate some of the general features of the development of the fungal community on roots. Parkinson, Taylor and Pearson (1963) and Dix (1964) followed the succession on the roots of this plant by comparing the fungal species which were present on the apical regions of the root with those on the older parts of the root system of 10-day-old plants, using root washing techniques to isolate species. Their results showed that typical early colonists were *Penicillium* species (notably species in the *Penicillium janthinellum* and *P. lilacinum* series), which tended to dominate communities near apices, *Mortierella* and *Mucor* species, *Trichoderma viride* and *T. koningii*. On the older parts of the root system these species declined while *Gliocladium roseum* and *Fusarium* species (especially *Fusarium oxysporum*), which were present in low numbers on young tissues, increased and became the dominants, often appearing on more than 50% of plated-out root segments. Overall, the numbers of fungal colonists increased on the root surfaces as the root aged but the species diversity tended to decrease (Table 7.9).

Investigations of the mycoflora of the primary root system of more mature plants of *Phaseolus vulgaris* (up to 190 days old) showed that the co-dominance of *F. oxysporum* with *G. roseum* persisted up to about 100 days (with maxima at 70 and 100 days respectively). Thereafter, both species declined as the incidence of *Cylindrocarpon destructans* (*C. radicicola*) and the hyphae of certain sterile dark forms increased. *C. destructans* became the dominant species on the roots after 160 days, reaching a frequency of isolation of 84% at 190 days (Taylor and Parkinson, 1965).

In the same study, Taylor and Parkinson (1965) followed the internal

Table 7.9 Frequency occurrence (%) and distribution of the more common fungi in the rhizosphere and on root surfaces of *Phaseolus vulgaris*

Fungi	*Rhizosphere*	*Root apices*	*Older root*
Fusarium spp.	97	25	97
Pythium ultimum	90	0	7
Trichothecium roseum	87	0	0
Humicola sp.	83	2	12
Gliocladium roseum	80	20	67
Chaetomium globosum	70	7	15
Mucor plumbeus	66	0	2
Trichocladium asperum	60	0	0
Penicillium janthinellum series	56	52	42
P. citrinum series	52	10	7
Trichoderma koningii	52	39	20
Penicillium nigricans	50	0	0
P. expansum series	44	7	2
Mucor hiemalis	44	5	5
Botrytis cinerea	43	22	7
Absidia spinosa	42	2	2
Papulaspora sp.	40	0	0
Trichoderma viride	40	27	17

Source: Dix (1964).

200 FUNGI OF SOIL AND RHIZOSPHERE

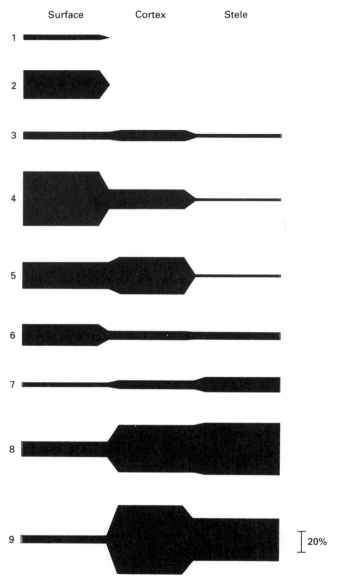

Figure 7.6 Distribution of the commonly occurring fungi on the surface and in the tissues of the roots of French bean (frequency of occurrence %). 1. *Mucor* spp.; 2. *Trichoderma viride*; 3. *Mortierella* sp.; 4. *Penicillium janthinellum* series; 5. *Fusarium oxysporum*; 6. *Gliocladium roseum*; 7. hyaline sterile hyphae; 8. *Cylindrocarpon destructans*, (*C. radicicola*); 9. dark sterile hyphae. (Drawn from the data of Parkinson, Taylor and Pearson, 1963; Dix, 1964; Taylor and Parkinson, 1965.)

invasion of the tissues of *Phaseolus vulgaris* roots using surface sterilization and dissection techniques. Many of those fungi which appeared early in the succession on the root apices and which persisted in low numbers on the older parts of the root achieved at best only limited penetration of the outer cortical tissues of the older parts of the root. Their replacements on the older parts of the root, *Fusarium oxysporum* and to a lesser extent, *Gliocladium roseum*, penetrated the inner cortical tissues, while *C. destructans* and the sterile forms with dark hyphae penetrated the inner tissues of the stele (Figure 7.6).

Penicillium, *Trichoderma*, *Mortierella* and *Mucor* species seem to be part of an ecological group of casual invaders of the root which are more or less confined to the root surface. These are quickly replaced and form stable communities for only a limited time; they appear to play only a minor role in the decay of the root. Some *Penicillium* and *Trichoderma* species from roots are relatively insensitive to soil fungistasis and Dix (1964) tentatively suggested that this could enable them to respond quickly to exudates from roots and account for their early appearance in a pioneer colonizer role on root apices.

Those fungi that replace the early colonizers are the true residential populations of the root, forming semi-stable communities on the root surfaces, penetrating and initiating the decay of the cortex rendering it more or less functionless in older roots. These may be regarded as true root-inhabiting forms, some of which, such as some of the sterile forms with dark hyphae, probably only exist as actively growing mycelia on roots.

Decapitation of the shoots of *Phaseolus vulgaris* at a relatively young age (10 days old) resulted in an increase in the incidence of *Fusarium* species and *Gliocladium roseum* on roots, which caused decay and upset the normal succession such that decapitated roots became colonized by *Volutella ciliata*, a species commonly found in the rhizosphere soil but which otherwise did not invade the root (Dix, 1964).

The large number of other fungal successions that have been described for the roots of other plant species fit very closely into the same general pattern outlined above for *P. vulgaris* and differ only in minor detail. A summary of the general pattern is as follows:

1. Pioneer colonizers of roots always include a large number of casual fungi of which *Penicillium*, *Trichoderma*, *Mucor* and *Mortierella* species are the most common genera isolated.
2. True root-inhabiting fungi quickly follow and eventually replace the casual species. Typically, many of these are pathogens or 'minor pathogens' (Salt, 1979). Early colonizers in this group nearly always include *Fusarium* species and *Gliocladium roseum* (the roots of woody plants seem to be an important exception; Harley and Waid, 1954a; Parkinson and Crouch, 1969). Strains of *F. oxysporum*, *F. solani*, *F.*

culmorum and *F. moniliforme* are pathogenic, causing root-rot and wilt disease in many hosts. Others commonly encountered in this group, depending somewhat on the plant species, are *Microdochium bolleyi* (*Aureobasidium bolleyi*), *Rhizoctonia* species, and certain zoosporic fungi, e.g. *Lagenocystis radicicola*, *Olpidium*, *Pythium* and *Ligniera* species.
3. Somewhat later, sterile mycelial forms, hyaline or darkly pigmented types appear, accompanied on the roots of most plants by *Cylindrocarpon destructans*. Frequently these fungi eventually become the dominants on the older parts of the root system as *Fusarium* species and *G. roseum* decline. On rice (*Oryza sativa*) *Thielavia angulata* is the dominant fungus on mature roots (Das, 1963).

On the roots of trees, grasses, cereals and some other crop plants (e.g. cabbage, *Brassica oleracea*) sterile forms may appear earlier in the succession (Harley and Waid, 1955a; Waid, 1957; Parkinson, Taylor and Pearson, 1963; Gadgil, 1965; Parkinson and Pearson, 1967). On grasses and cereals, dark sterile forms are particularly prominent, invading tissues following early programmed senescence of the root cortex (Deacon, 1987). Some of the darkly pigmented sterile forms on cereals and grasses are now recognized as varieties of *Phialophora radicicola* and the morphologically similar *Gaeumannomyces graminis* (Hall, 1987). The former are mostly non-pathogenic, causing disease in only a very few susceptible hosts, while varieties of the latter invade the stele and cause disease in a number of hosts (Deacon, 1973, 1974).

8
COPROPHILOUS FUNGI

Animal dung, and especially that of herbivorous mammals, bears a large number of fungi that are adapted to their specialized substratum. Various adaptations are commonly found:

- **Mechanisms of violent spore discharge**. These are often phototropically orientated and effective over relatively large distances. They ensure that the spores are carried away from the dung (Ingold, 1971).
- **Adhesive projectiles**. These become attached to herbage and may survive for long periods without being washed off or losing viability.
- **Spores which can survive digestion**, and which may be triggered to germinate following exposure to the chemical and physical environment of the animal gut (Sussman and Halvorson, 1966). In most cases, germination probably does not occur in the gut, but in the voided faeces (Johnson and Preece, 1979; Kuthubutheen and Webster, 1986b). Fungi which have survived digestion and appear on dung have been termed **endocoprophilous** (Larsen, 1971).
- **Specialized nutritional requirements** for substances found in dung which may stimulate growth and sporulation. Some coprophilous fungi also appear to be adapted to grow under the conditions of relatively high pH found in dung.

Dung is a rich substratum for fungal growth. The undigested remains of the herbage ingested by the herbivore initially contain relatively high quantities of readily available carbohydrates in the form of water-soluble organic matter and hemicellulose, in addition to cellulose and lignin. It has a very high nitrogen content, sometimes over 4%, much of which is probably derived from the residues of the large populations of bacteria and protozoa involved in the microbial breakdown of the ingested herbage. It is also rich in vitamins, growth factors and minerals. It has the capacity to retain water, but this capacity declines with age. The actual water content can vary over wide limits, but may be over 700% of the dry weight. The pH is above 6.5. Its physical structure, consisting of short fragments of straw embedded in a mucilaginous matrix, forms a very suitable medium for fungal growth. For references see Lambourne and Reardon (1962), Waksman, Cordon and Hulpoi (1939), Lodha (1974), Webster (1970).

Coprophilous fungi have been the subject of a number of ecological investigations, including succession, interspecific antagonism, interspecific synergism, substrate specificity in relation to different dung types and autecology (Wicklow, 1981, 1992).

Some genera of fungi frequently found on herbivore dung are:

Zygomycotina
Mucorales: *Mucor, Phycomyces, Pilaira, Pilobolus, Utharomyces, Chaetocladium, Piptocephalis, Syncephalis, Kickxella*

Ascomycotina
Pezizales: *Ascobolus, Iodophanus, Cheilymenia, Coprobia, Lasiobolus, Ryparobius, Saccobolus, Thelebolus*
Sphaeriales: *Chaetomium, Coniochaeta, Hypocopra, Lasiosordaria, Podospora, Poronia, Sordaria*
Pleosporales: *Delitschia, Trichodelitschia, Sporormiella.*
Microascales: *Viennotidia (Sphaeronaemella)*

Basidiomycotina
Agaricales: *Bolbitius, Conocybe, Coprinus, Panaeolus, Psathyrella, Psilocybe, Stropharia*
Nidulariales: *Sphaerobolus, Cyathus*

Deuteromycotina
Moniliales: *Stilbella, Arthrobotrys*

Besides these **filamentous fungi** (Eumycota), **slime moulds** (Myxomycota) may also be encountered, including **cellular slime moulds** such as *Dictyostelium* (Acrasiomycetes) and **plasmodial slime moulds** (Myxomycetes). Over 20 genera of Myxomycetes have been reported on dung, and it is presumed that, in most cases, colonization of the dung has occurred after deposition (Eliasson and Lundqvist, 1979). Both groups of slime moulds feed by ingesting microbial cells.

The usual technique for studying coprophilous fungi is to study material immediately after collection or after a period of incubation in a moist chamber. For small mammals, e.g. rodents, it may be more practicable to trap the animals in a non-destructive trap, e.g. the Longworth trap, and to collect dung samples from within the trap. In this way, some unusual members of the Mucorales have been obtained, e.g. species of *Coemansia, Kickxella, Spirodactylon* and *Tieghemiomyces* (Benjamin, 1959; Ingold, 1971). The relatively small mass of rodent dung may not support the fruiting of some of the larger fungi known from other types of dung.

The taxonomy of coprophilous fungi has been widely studied in many countries, and their identification is, in general, not difficult (see Richardson and Watling, 1968, 1969; Bell, 1983; Ellis and Ellis, 1988). Some common coprophilous fungi are illustrated in Figures 8.1 and 8.2.

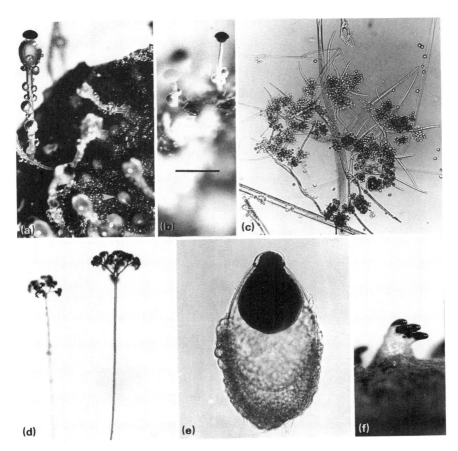

Figure 8.1 Fruit-bodies of some common coprophilous fungi. (a) *Pilobolus* sp.: sporangiophore on rabbit dung. The turnip-shaped structure arrowed is a trophocyst at the surface of the dung. Sporangiophores develop from trophocysts. (b) *Pilaira anomala*: sporangiophores on rabbit dung. (c) *Chaetocladium brefeldii*: branched sporangiophores ending in sterile spines and bearing spherical sporangiola. The fungus is a mycoparasite on other Mucorales such as *Pilaira*. (d) *Piptocephalis cylindrospora*: dichotomous sporangiophore ending in a cluster of cylindrical merosporangia. *Piptocephalis* is also parasitic on other Mucorales. (e) *Thelebolus stercoreus*: apothecium containing a single large multisporous operculate ascus. (f) *Ascobolus immersus*: apothecium with five protruding asci which have bent towards the light. The ascospores are purple in colour. Scale bar = 500 μm (a,b); 120 μm (c); 200 μm (d); 70 μm (e); 1.5 mm (f).

Figure 8.2 Fruit-bodies of some common coprophilous fungi. (a) *Podospora vesticola*: perithecium. The body of the perithecium is transparent and the dark band at its base is a cluster of asci. Asci in the process of elongation can be seen in the upper part. The asci are four-spored. (b) *Chaetomium* sp.: perithecium. The perithecium is surrounded by hairs which are straight around the body and coiled around the neck. (c) *Viennotidia fimicola*: perithecium. The ostiole at the tip of the long tapering neck is surrounded by pointed bristles which support an accumulation of sticky ascospores. (d) *Coprinus heptemerus* basidiocarp on a rabbit pellet. (e) *Stilbella erythrocephala* synnemata bearing sticky conidia. Scale bar = 80 μm (a); 200 μm (b); 40 μm (c); 1 cm (d); 2 mm (e)

8.1 SUCCESSION OF COPROPHILOUS FUNGI

Webster (1970) and Lodha (1974) have discussed the succession of fungal fruit-bodies on dung. If freshly voided dung of a suitable herbivore (e.g. rabbit, hare, sheep, deer, horse) is brought into the laboratory and incubated near to a window in a moist chamber, a sequence of fungus fruit-bodies may be observed over a period of several weeks. Where the dung samples are small and discrete, e.g. in the form of pellets as in rabbits, the successional appearance of fruit-bodies can be expressed in the form of frequency histograms. In Figure 8.3, the frequency of appearance of fruit-bodies on a sample of 10 rabbit pellets is recorded over a period of 60 days (Harper and Webster, 1964). Within about 2 days, the sporangia of *Mucor* appear, followed some 2 days later by those of *Pilaira* and *Pilobolus* (Figure 8.1). *Pilobolus* has phototropic sporangiophores which develop from yellow trophocysts in the dung. It shoots off its sporangia for distances of over a metre towards the light. They adhere firmly to herbage. The end of the phase of activity of Mucorales, which may last for 2–3 weeks, is often associated with the development of parasitic Mucorales such as *Piptocephalis* and *Chaetocladium* (Figure 8.1), which form haustoria on the mycelia and sporangiophores of the other fungi. Ascomycetes next make an appearance in the form of apothecia of such genera as *Ryparobius, Thelebolus, Ascobolus* (Figure 8.1) and *Saccobolus*, and they are followed by pyrenomycetes such as *Sordaria, Podospora* and *Sporormiella*. *Chaetomium* and *Viennotidia* (*Sphaeronaemella*) (both with non-violently discharged asci) may also fruit at this time. The basidiocarps of *Coprinus* (Figure 8.2) appear relatively late, after about 3 weeks' incubation. Deuteromycetes tend to be less conspicuous, but are represented by fungi such as *Stilbella erythrocephala* (Figure 8.2), which form pink synemmata with sticky spores on old rabbit and cow dung, often to the exclusion of other fungi. The *Oedocephalum* conidial state of the minute, flesh-coloured discomycete *Iodophanus carneus* is common on moist dung. Predacious fungi growing parasitically on nematodes, such as species of *Arthrobotrys* and *Monacrosporium* (Chapter 10), appear when dung is incubated under moist conditions which encourage a build-up of nematodes. *Rhopalomyces elegans* is a parasite of nematode eggs, larvae and adults, and may be found on dung which has been in contact for some time with soil (Barron, 1977).

The species which occur and the course of the succession can be controlled by changing a number of environmental variables, especially temperature and water content. The effect of incubation temperature on the appearance of fungal fruit-bodies on rabbit dung has been studied by Wicklow and Moore (1974) by incubating samples of 20 pellets from a laboratory rabbit fed on alfalfa hay, at temperatures of 10, 24 and 37.5°C. Striking differences were found in the fungi fruiting. For example,

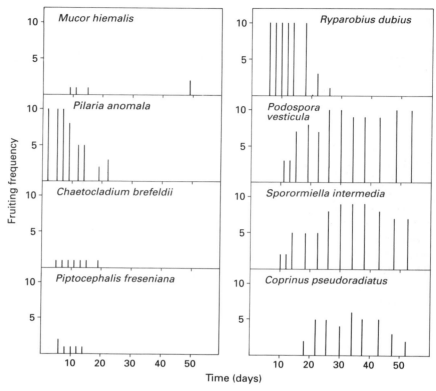

Figure 8.3 Fruiting frequency of eight coprophilous fungi on a sample of 10 freshly voided rabbit pellets incubated in moist chambers in the laboratory (Redrawn from Harper, 1962.)

Thelebolus spp. and *Sporormiella intermedia* were reported only from the pellets incubated at 10°C. This confirms the observations of Wicklow and Malloch (1971) that *Thelebolus* collected in Ontario, Canada, is psychrotolerant. Maximum growth rate for all the species of *Thelebolus* tested fell within the range 15–20°C, and in most cases growth was faster at 0°C than at 25°C. These features may explain why species of *Thelebolus* fruit more readily on dung samples collected in the colder months. However, the findings of Wicklow and Moore (1974) and Wicklow and Malloch (1971) may not be confirmed in material collected in other parts of the world. Rabbit pellets collected during July in Devon, England, produce apothecia of *T. stercoreus* and pseudothecia of *S. intermedia* when incubated at temperatures ranging from 23 to 28°C. Possibly isolates of the same species of fungus from different parts of the world have different temperature optima for growth and fruiting.

At the higher incubation temperatures used by Wicklow and Moore

(1974), a different set of fungi predominated. For example, *Ascodesmis nigricans*, *Coprotus granuliformis* and *Podospora curvicolla* only occurred on material incubated at 24°C. Kuthubutheen and Webster (1986b) have also reported differences between the fungus flora which develops on rabbit pellets incubated at fluctuating lower outdoor temperatures as compared with those incubated at 20°C in the laboratory. Ascomycetes were 2–3 times less frequent on the outdoor-incubated material, and many showed delay in fruiting compared with those on material incubated at 20°C.

The effects of varied water regimes on the fungi fruiting on rabbit pellets was studied by Kuthubutheen and Webster (1986a,b). They incubated field-collected pellets, initially fully imbibed with water, under conditions where water was freely available or where water supply was limited, at a range of relative humidities from 100–81%. They also fed rabbits with sterilized hay liberally inoculated with spores of selected fungi and incubated their pellets under different water regimes. The highest number of species on a sample of 10 pellets was 46 for pellets incubated with free water after 14 days. This was twice the number of species found at 100% relative humidity. Even fewer species fruited at lower humidities (Figure 8.4). Certain species of fungi such as *Pilobolus crystallinus* fruited only when free water was available, or after 7 days at a relative humidity of 100%. This is in marked contrast to the related fungus *Pilaira anomala* which continued to fruit for 45 days even at 81% relative humidity. Other fungi showing dependence on free water or high humidity for fruiting were the ascomycetes *Coprotus granuliformis*, *Saccobolus versicolor* and *Podospora vesticola*. Two basidiomycetes, *Coprinus miser* and *C. stercoreus*, only fruited when water was freely available. Deuteromycetes, on the other hand, fruited over the whole humidity range (Figure 8.4). *Stilbella erythrocephala* and *Penicillium* spp. fruited on a significant proportion of the drier pellets. Competition with other fungi also affected the range of humidities within which some fungi can fruit. *Pilobolus crystallinus*, which needs water in order to project its sporangia, was particularly sensitive. In the absence of competition from other coprophilous fungi, it could fruit at relative humidities down to 86%. When the spores of 10 other fungi were heavily inoculated onto the rabbit food along with the spores of the *Pilobolus*, the numbers of sporangia were greatly reduced, and fruiting did not occur below 100% relative humidity, matching the findings on field-collected pellets. The results of these experiments imply that competition for water is an important factor controlling the fruiting of *Pilobolus*, and probably other fungi.

The availability of water directly affects not only fruiting, but spore germination and mycelial growth rate. Kuthubutheen and Webster (1986b) showed that when spores of coprophilous fungi germinated on dung agar whose water activity (a_w) was lowered by the addition of glycerol,

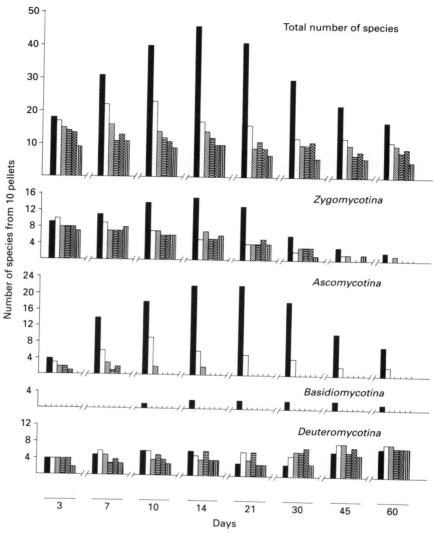

Figure 8.4 Number of species of coprophilous fungi recorded from the presence of fruit-bodies on a sample of 10 freshly voided rabbit pellets moistened and incubated at 20°C at different humidities. ■ Free water; □ 100% r.h.; ▨ 98% r.h.; ▤ 95% r.h.; ▧ 92% r.h.; ▥ 81% r.h. (Redrawn from Kuthubutheen and Webster, 1986, with permission of British Mycological Society.)

percentage germination was depressed with decreasing a_w. Germ tube length and the growth rate of mycelia were also depressed. These effects were less severe for deuteromycetes such as *Stilbella erythrocephala* capable of fruiting on drier pellets.

The effects of variation in microhabitat and microclimate on the communities of coprophilous fungi which develop on rabbit pellets have been studied by Yocom and Wicklow (1980). Pellets from a rabbit fed with alfalfa hay were placed in litter bags at the surface of the ground in a sand dune system which varied from *Ammophila*-built dunes where sites could be exposed or shaded by the *Ammophila*, through relatively exposed drift beach dominated by low-growing annuals and willow seedlings, to a shaded *Myrica* thicket where the soil was saturated, and finally to an oak–maple forest where leaf litter covered the forest floor. The litter bags were set out at different periods so that they were subjected to varying amounts of rainfall. After exposure at the field sites, the samples were incubated in the laboratory for 40 days and the numbers of species of fungi which fruited on them determined. It was found that the degree of exposure affected the richness of the communities fruiting. Twice as many species were found on the oak–maple samples as were found on the dune samples. A significant positive correlation was found between the amount of rainfall which fell during the period of field exposure and the numbers of species which fruited later in the laboratory.

Experimental studies on the effects of water content of cow dung on the growth and fruiting of coprophilous fungi have been carried out by Dickinson and Underhay (1977). The water content of freshly voided cow dung can be very high (400–700% dry wt), and oxygen penetration to the centre of a cow dung pat in the early stages of decomposition may limit mycelial growth. Using approximately 10 g samples of cow dung, Dickinson and Underhay (1977) reduced the water content from 332% to 157%, and followed the appearance of fungal fruit-bodies on the samples when incubated at 18°C at 100% relative humidity over a 10-day period. The effect of lowering the water content was to reduce the proportion of samples bearing fungal fruit-bodies, and also the duration of fruiting. No fruit-bodies were observed after 6 days on the samples at the lowest water content (157%), whilst all the samples at the highest water content (332%) bore fruit-bodies up to 10th day. In other studies in pure culture, a range of coprophilous fungi were subjected to water stress by the addition to the culture medium of mannitol or KCl. One effect of increasing water stress was to reduce the minimum time taken for fruiting by *Pilaira anomala* and *Sordaria humana* (Harrower and Nagy, 1979). Possibly this is an adaptation to ensure that, at times of water stress, rapid reproduction occurs, resulting in dispersal and survival.

It is not surprising that variation in conditions of incubation should affect the range of fungi which fruit on dung. Herbage may contain propagules or mycelia of a great variety of organisms. Bonner and Fergus (1959), in a survey of samples of silages, dried forage and dried grains used for cattle feed, isolated 64 different species of fungi. Not all these fungi survive digestion. The voided dung may also be colonized by fungus spores carried by air, rain, animals such as insects, mites and nematodes, or by contact with the soil, which contains a very wide range of fungi. The different physical conditions imposed by varying the conditions of incubation exert a selective effect on the growth and fruiting of microorganisms present in the dung, and the selection may be reinforced by competitive interactions between them, and other organisms present.

8.1.1 Analysis of the coprophilous fungus succession

There have been a number of attempts to explain the sequence of fruit-body formation on dung. Several workers have pointed out the fact that the succession followed taxonomic groupings, i.e. Zygomycotina → Ascomycotina → Basidiomycotina. They attempted to correlate the succession with what was known about spore germination, mycelial growth rate and nutrition of these major groups of fungi, in relation to published data on the changing chemical content of fungi during decay. A hypothesis to explain the succession, termed the **nutritional hypothesis**, claims that the fungi which fruit early germinate and grow more rapidly than those which fruit later. It is also claimed that the Mucorales are dependent for their nutrition on soluble carbohydrates which are rapidly depleted from the dung, whilst the ascomycetes and basidiomycetes can grow at the expense of cellulose and lignin. However, criticisms of this hypothesis are that it takes no account of the fact that the spores of coprophilous fungi are triggered to germinate by passage through an animal's gut that the depletion of water-soluble organic matter is not as rapid as is claimed, and that the disappearance of the zygomycetes may occur before the levels of water-soluble organic matter have fallen appreciably.

Harper and Webster (1964) made a comparative study of spore germination and germ tube and mycelial growth rate using a number of coprophilous fungi isolated from rabbit dung. In order to simulate the stimulus to spore germination likely to be encountered in a rabbit's gut, spore suspensions were incubated in alkaline pancreatin (pH 9) at 37°C for 5 h, washed, plated out onto dung agar, and incubated at laboratory temperature (about 18°C). The latent period for germination (i.e. the time taken for emergence of germ tubes) and the rate of germ tube extension were measured. Mycelial growth rates were also measured. There were no appreciable differences in latent periods for spore germination. In most cases, spore germination had occurred within 6 h, and in all cases with 12 h. This suggests that spore germination of many of the endocoprophilous fungi occurs in the dung within a few hours of defaecation and that their mycelia may be simultaneously engaged in uptake of nutrients. There is no correlation between mycelial growth rate and time of fruit-body appearance in the succession. This is illustrated by the fact that *Sordaria fimicola*, which has the fastest growth rate of the fungi tested, fruits on dung after about 9 days, as compared with the more slowly growing Mucorales which fruit on dung within 2–4 days. The lack of correlation between mycelial growth rate and time of appearance of fruit-bodies has been confirmed by others, e.g. for fungi from cattle dung by Wicklow and Hirschfield (1979) and by Safar and Cooke (1988).

The simplest explanation of the successional appearance of fruit-bodies of coprophilous fungi is that each fungus has a characteristic minimum time which it requires in order to form a fruit-body. For example, it is likely that the basidiocarp of a coprophilous agaric takes longer to develop than the sporangium of a species of *Pilobolus*. Furthermore, it is possible that the time taken to fruit is little affected by the activities of competing fungi. To test these ideas, Harper and Webster (1964) studied the time taken by fungi isolated from rabbit dung to fruit on pellets incubated on moist filter paper in the laboratory. Their observations were based on pellets obtained from three different sources.

1. Dung freshly collected from the field. These pellets contained the usual range of coprophilous fungi, along with other organisms such as bacteria, protozoa and nematodes normally present in the rabbit's gut.
2. Autoclaved pellets incubated with pre-germinated spores (monoculture). Such dung differs from normal dung in that competitors are eliminated.
3. Pellets collected from caged rabbits in the laboratory fed with sterilized food sprayed with a spore suspension of a single species of fungus. In these pellets, competing fungi which might have been present on normal food were eliminated, but the usual gut microflora and microfauna were included.

The idea underlying the choice of these different treatments was to study the effects of different levels of competition on fruiting time. The results are given in Table 8.1. It is clear that there is a very close degree of correspondence between the time taken to fruit in all three treatments.

Very similar results for minimum fruiting time have been found by Lodha for a range of different animal dung in India (for references, see Lodha, 1974).

Thus, time taken to fruit is a likely explanation of the successional appearance of fruit-bodies. It does not, of course, explain the cessation of fruiting, yet it is clear from the histograms shown in Figure 8.3 that the early fruiters, e.g. the zygomycetes, also cease to fruit early. For an explanation of this phenomenon, **interspecific antagonism** has been suggested as one important factor. Competition by microorganisms inhabiting a nutrient-rich substratum such as dung is likely to be intense. Along with the many species of fungi known to occur on dung, there may also be dense populations of bacteria, protozoa and nematodes. After deposition, the dung may be visited by insects, mites, molluscs and earthworms which feed

Table 8.1 Number of days elapsing before fruiting by coprophilous fungi

Fungi	On fresh dung from time of field collection	On sterile dung monoculture	On fresh dung from feeding experiments
Mucor hiemalis	2	2	2
M. mucedo	2	2	2
Pilaira anomala	2	2	2
Pilobolus crystallinus	4	4	4
Ryparobius dubius	6	6	6
Ascobolus stictoideus	9–10	8–9	8–9
A. crenulatus	10–11	7–8	8
A. glaber	11–12	12–13	13–14
Chaetomium caprinum	9	9–10	9
Sordaria fimicola	9	9	9
Podospora minuta	9–11	10	10
Sporormiella intermedia	10–11	10–11	11
Coprinus heptemerus	9–13	7–8	7–8
C. patouillardii	37	14	11–12
C. radiatus	–	14	11–12

Source: Harper and Webster (1964).

and lay their eggs on the dung, so that an active population of animals, including larval stages, may develop, feeding on the dung and its microbial population. Interspecific competition between fungi on dung may take a number of forms: (a) competition for nutrients; (b) production of antibiotics, which limit the growth or activities of other organisms; (c) hyphal interference, a form of contact inhibition; (d) parasitism; and (e) predation.

a) Competition for nutrients

Studies on the exploitation of nutrient resources in dung by coprophilous fungi have been aided by the use of artificial resource units (**copromes**) prepared by grinding and reconstituting rabbit dung into cylindrical tablets which are then sterilized by irradiation (Wood and Cooke, 1984). The copromes are more uniform in weight and chemical composition than natural substrata, so that changes in composition and calorific value during decomposition can be followed accurately. Safar and Cooke (1988) inoculated germinated ascospores of three coprophilous ascomycetes into copromes in all possible combinations, i.e. singly, as species pairs or all three together. The three ascomycetes chosen were *Ascobolus crenulatus*,

Chaetomium bostrychodes and *Sordaria macrospora*. The numbers of ascocarps were counted after 14 days. The numbers of ascocarps which developed was reduced in the presence of one or two other fungi. The fruiting of *A. crenulatus* was particularly sensitive to the presence of competitors, and it failed to fruit completely in the presence of *C. bostrychodes*. There was no evidence of combative competition between the three fungi, and it was concluded that fruiting in mixtures is reduced because of competition for nutrients. In general, the rate of decomposition, as measured by percentage weight loss, was greater for species mixtures than for single species.

b) Antibiotic production

An example where the suppression of the mycelial growth and fruiting of other coprophilous fungi appears to be mediated by antibiotic production is provided by *Stilbella erythrocephala*. This fungus forms pink synnemata, often to the exclusion of other fungi, especially on dry dung samples of cows and rabbits. Singh and Webster (1973) found that culture filtrates of *S. erythrocephala* could inhibit sporangiospore germination in *Mucor* and inhibit mycelial growth of many coprophilous fungi, including *Pilaira* and *Ascobolus*, often causing disruption of the hyphal tips. The presence of *Stilbella* in rabbit pellets caused significant reduction in the fruiting of *Pilobolus*. An antibiotic from *Stilbella* which causes reduction in mycelial growth rate and also affects bacterial growth has been identified as closely related to the pentaibols antiamoebin or emerimicin (Brückner and Reinecke, 1989).

Unidentified antifungal antibiotics have also been demonstrated in culture filtrates of *Poronia punctata*, a relatively slow-growing ascomycete which fruits late on cattle dung. The antibiotic(s) from *Poronia* inhibited the mycelial growth of seven other ascomycetes which normally fruited earlier in the succession. It has been suggested that antagonism, slow growth rate and late appearance (i.e. fruiting) appear to be correlated in partially defining the niche of *P. punctata* in cattle dung (Wicklow and Hirschfield, 1979a).

c) Hyphal interference

A more subtle type of antagonism between fungi was first discovered in the coprophilous fungus *Coprinus heptemerus* (Figure 8.2). Hyphal interference involves contact between an antagonist (often a basidiomycete) and a sensitive fungus, and results in death and lysis of those cells of the sensitive fungus in contact with the antagonist.

In experiments in which caged rabbits were fed with sterilized food to which spores of coprophilous fungi were added, Harper and Webster (1964) reported that, when mixed spore suspensions including the

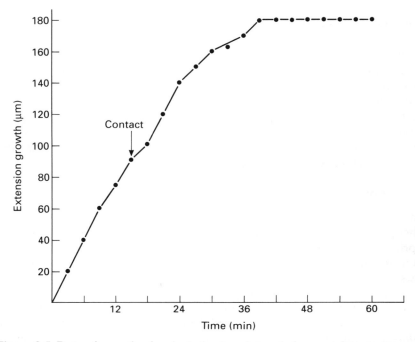

Figure 8.5 Rate of growth of a single hypha of *Ascobolus crenulatus* before and after contact with a hypha of *Coprinus heptemerus*. (Redrawn from Ikediugwu and Webster, 1970a.)

basidiospores of *C. heptemerus* were used, the fruiting of *Pilaira anomala* and *Ascobolus crenulatus* was depressed in the presence of the *Coprinus*. When the same fungi were inoculated into sterilized pellets, it was found that sporangium production in *Pilobolus* was much reduced and, after 8 days, virtually suppressed, when *Coprinus* was present. No diffusible antibiotic could be demonstrated. When colonies of *C. heptemerus* and *A. crenulatus* grew towards each other on agar, there was no reduction in the growth rate of the *Ascobolus* until about 20 min after the hyphae of the two fungi made contact, as shown in Figure 8.5. There was then a dramatic change in the condition of the *Ascobolus* hyphae. They stopped growing, developed refractile walls and vacuolated contents, and lost turgor. The effects were confined to the *Ascobolus* cells in direct contact with *Coprinus*. *Ascobolus* cells in contact with the *Coprinus* absorbed Neutral Red stain, whilst cells not in contact remained impermeable. Plasmolysis tests showed that the affected cells were incapable of plasmolysis, because they had been killed by contact with the *Coprinus* (Figure 8.6) (Ikediugwu and Webster, 1970a).

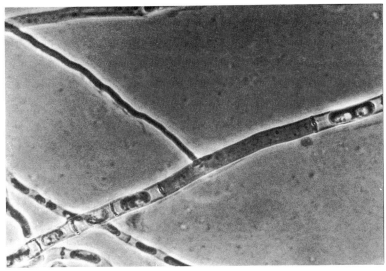

Figure 8.6 The effects of hyphal interference as demonstrated by plasmolysis. The wider hypha of *Ascobolus crenulatus* lying across the field is in contact with a single hyphal tip of *Coprinus heptemerus*. The preparation has been flooded with a concentrated solution of glucose. Cells on either side of the cell in contact with the *Coprinus* hypha have plasmolysed. The cell in contact with the *Coprinus* hypha has not plasmolysed because it has been killed. (Reproduced from Ikediugwu and Webster, 1970a.)

Several other coprophilous basidiomycetes show the hyphal interference phenomenon, some of which are sensitive to the effect themselves. They can be arranged in a 'pecking order' or competitive hierarchy in which *C. heptemerus* proves to be the most effective (Ikediugwu and Webster, 1970b). The effect can be very powerful. A single hypha of *Panaeolus sphinctrinus* has been shown to stop the advance of an entire colony front of *Bolbitius vitellinus*. Investigations of the fine structure of the interface show that, soon after contact with a hypha of *C. heptemerus*, a hypha of *A. crenulatus* shows breakdown of the plasmalemma and cleavage and rounding-off of segments of the cytoplasm (Ikediugwu, 1976).

The physical or chemical bases for the hyphal interference phenomenon are not yet understood. Evidence has been obtained that *Coprinus* can exert an antagonistic influence through a single layer of Cellophane 50 μm thick, but it has not been possible to concentrate and purify any active chemical substance. Whether a single substance or more than one are involved is not known. There is, as yet, no evidence that any substances that may be released from affected cells are taken up by the basidiomycete antagonist. Hyphal interference appears to be a very precisely located form of antagonism. It is most actively produced by growing hyphal tips, and the growing regions of sensitive hyphae are the most susceptible parts. It is clearly a highly effective form of antagonism, limiting the growth and fruiting of competing fungi.

d) Parasitism

The best-known examples of mycoparasitism among coprophilous fungi are the zygomycetes *Piptocephalis* and *Chaetocladium* (Figure 8.1) which are almost exclusively biotrophic parasites of other members of the Mucorales. Many of these parasites have a wide host range. For example, *Piptocephalis fimbriata* is known to parasitize representatives from more than 20 genera of Mucorales, whilst *P. virginiana* is known to parasitize 15 different genera (Berry and Barnett, 1957). There is little evidence to suggest that the parasites are endocoprophilous or even that Mucorales growing on dung are their preferred hosts, although it is known that *Piptocephalis* spores can survive passage through rabbit gut (Wood and Cooke, 1986). *Piptocephalis* species and members of the related genus *Syncephalis* are more frequently isolated from the litter layer of soil than from dung (Richardson and Leadbeater, 1972). It seems likely that the presence of dung on the soil surface stimulates the activity of Mucorales in the soil, and this, in turn, leads to increased abundance of their parasites. The parasites may invade the dung from the soil directly, or possibly by means of spores, which in some cases are air-borne (Ingold and Zoberi, 1963). The conidia of *Piptocephalis* germinate in the absence of a mucorine host, but the germ tubes make only limited growth unless a suitable host mycelium is near enough to permit chemotropically directed growth, penetration and the formation of haustoria within the host cell. Poor growth of the parasite in the absence of a suitable host may possibly be related to a block in the synthesis of polyunsaturated fatty acids such as γ-linolenic acid, which is abundant in the mycelium (Manocha, 1975; Manocha and Deven, 1975). The effects of *Piptocephalis* spp. on the growth rate of their hosts in pure culture are variable. In some cases, the rate of growth of dual cultures differed little from that of uninfected controls; in others it was reduced, whilst in others it was enhanced. These effects are temperature-dependent (Curtis *et al.*, 1978). Wood and Cooke (1986) showed that two species of *Piptocephalis* reduced the growth and sporulation of *Pilaira anomala* over the temperature range 15–30°C, but argued that mycoparasites which develop on herbivore faeces probably have little effect on the development of coprophilous Mucorales. The effect of *Chaetocladium* on its host is often to induce hypertrophy around the point of penetration.

For a general account of mycoparasitism, see section 3.2.4(b).

e) Predation

Dung provides a nutrient-rich substratum on which many animals, especially arthropods, may subsist. Indeed, a number of insects have adapted to living on dung. Within minutes or hours of deposition, eggs of insects such as Muscidae (flies) may be laid on the dung, whilst Sarcophagidae may deposit larvae. In large masses of cow dung, burrowing scarabaeid dung beetles such as *Aphodius* make tunnels. A succession of arthropods has been described (Valiela, 1969, 1974). Arthropod colonization of dung might affect the fungal population in several ways, either directly or indirectly. Insects are known to be vectors of spores and, when they carry propagules such as conidia or oidia which do not require a digestive trigger to germination, they may be important in bringing about colonization, or transference of propagules of differing mating types. Other direct effects may be the ingestion of fungal mycelium, spores or fruit-bodies. Indirect effects may include a hastening of the decay process (Holter, 1979), improved aeration, comminution and mixing of the substratum, stimulation of bacteria or predation of nematodes.

Some of these effects have been studied experimentally in the laboratory and in the field. Breymeyer, Jakubczyk and Olechowicz (1975) added varying numbers of larvae of Scarabaeidae or Anthomyidae (Diptera) to 10 g portions of sheep dung which were then incubated in the laboratory. Counts of fungal colonies on dilution plates after 6 days' incubation with 20 fly larvae were less than one-third of the counts when larvae were excluded. Even lower counts were found after 12 days. In field experiments on cattle dung in the USA, Lussenhop *et al.* (1980) estimated microbial biomass of dung by measuring lengths of fungal hyphae and numbers of bacterial cells. Adult flies or *Aphodius* were added to dung samples which were then incubated in the field at two different sites, an arid short-grass range in Wyoming and a mesic pasture at Michigan. Hyphal density decreased 40% at Wyoming in the presence of *Aphodius*, but the presence of arthropods had no effect at either site on the total number of species of fungi fruiting, and there was no evidence, even at Wyoming, that the activity of *Aphodius* retarded sporulation. However, working with pellets from rabbits fed with alfalfa hay which were inoculated with developing larvae of the coprophilous detritivore dipteran *Lycoriella mali*, Wicklow and Yocom (1982) reported that with increasing numbers of larvae, the number of species of fungi fruiting on the pellets declined.

The effects of colonization of rabbit faeces by the sciarid fly *Lycoriella* on energy transformation have been studied in the laboratory. Helsel and Wicklow (1979) placed samples of about 20 rabbit pellets in flasks with 20 adult flies and incubated them at room temperature. The presence of the flies was accompanied by a loss of about 34% in calorific value during a 30-day period, compared with a 16% loss due to the coprophilous microflora. The fly larvae fed on mycelium and fungal fruit-bodies. It is possible that the growth of certain coprophagous insects is enhanced by the inclusion of fungal cells in their diets.

f) Synergistic phenomena

It would be wrong to conclude that all interspecific relationships between coprophilous fungi are antagonistic. In some cases, evidence has been obtained of synergism, i.e. the activities of one organism or group of organisms stimulates the growth or fruiting of others. Some of these phenomena have been demonstrated in the laboratory on artificial media and, before it can be inferred that similar effects operate in nature, it may be necessary to conduct experiments on natural substrata. A number of different examples are described here.

Viennotidia fimicola and *Eurotium repens*

V. fimicola forms minute, flask-shaped perithecia with long tapering necks ending in an ostiolar fringe of hairs which hold in place a sticky drop of ascospores which have been released from deliquescent asci in the body of the perithecium (Figure 8.2). In nature, it has been found on dung of rabbit, sheep, deer and cow. In pure culture on a variety of natural media, the fungus grows sparsely, forming phialoconidia but rarely perithecia. Cain and Weresub (1957) found that perithecial production was enhanced in cultures contaminated by the hemiascomycete *Eurotium repens*. Culture filtrates of *Eurotium* stimulated growth of *Viennotidia*, but fruiting did not occur in this case. *Eurotium repens* is not a common fungus on dung. When cultures of *Viennotidia* on agar or on sterilized rabbit pellets are grown in the presence of a range of other coprophilous fungi, most of the ascomycetes tested stimulate perithecial production, but not the zygomycetes or basidiomycetes. The stimulus to fruiting is found in cell-free culture filtrates and will also pass through dialysis tubing (J. Follett, 1987, unpublished). Possibly *Viennotidia* needs to be supplied with an exogenous substance before it can complete perithecium development.

Pilobolus and coprogen

Pilobolus spp. can be grown in pure culture on natural media such as dung extract agar, but on synthetic media growth is poor. *Pilobolus* has a number of special requirements, e.g. a pH above 6, and growth on synthetic media is stimulated by the addition of thiamine, haemin and **coprogen**, an organo-iron compound (sideramine) known to be produced by various fungi and bacteria (Hesseltine et al., 1953; Pidacks et al., 1953; Keller-Schierlein and Diekmann, 1970). The role of coprogen in the nutrition of fungi is probably concerned with iron uptake and transport. *Pilobolus* responds to the presence of coprogen by increased growth and by the formation of trophocysts (Figure 8.1) and sporangia. Although the dung of different animals may contain different amounts of substances stimulating sporangium production in *Pilobolus*, there is, nevertheless, a roughly linear response in sporangium production to the addition of differing amounts of dung extract (Singh and Webster, 1976). Harper and Webster (1964) tested the idea that the fruiting of *Pilobolus crystallinus* on rabbit dung might be related to the falling levels of coprogen in the dung with the passage of time. They were, however, unable to demonstrate any diminution in trophocyst or sporangium production in cultures prepared from fresh rabbit pellets as compared with those prepared using pellets up to 30 days old.

Pilobolus and ammonia

Page (1959) made the accidental discovery that, when cultures of *Pilobolus kleinii* were contaminated with *Mucor plumbeus*, a noncoprophilous species, trophocyst and sporangium production were strongly stimulated. It was found that the stimulus to the sporulation of *Pilobolus* was volatile ammonia from the breakdown of asparagine in the culture medium by *Mucor*. Sporulation of *Pilobolus* could be stimulated by the substitution of ammonia for *Mucor*. It is of interest to see whether the '**ammonia effect**' can be demonstrated on the natural substratum. Singh and Webster (1972) had difficulty in confirming the results of the *in vitro* investigations. Although they could confirm that *M. plumbeus* stimulated sporangium production of *P. kleinii* on horse dung, of nine coprophilous species of *Mucor* tested, only one, *M. hiemalis*, caused increased sporulation of *P. kleinii*, and two other species of *Pilobolus* failed to respond. No stimulation could be demonstrated by bacteria known to be capable of liberating ammonia.

The two studies above, i.e. the effects of coprogen and ammonia on asexual production in *Pilobolus*, show how difficult it may be to extend laboratory experiments on synthetic media to the natural substratum.

The growth and fruiting of *P. kleinii* has been studied over a range of concentrations of ammonium (NH_4^+) and acetate, affecting the C:N ratio of the medium (Page, 1960). A simplified version of Page's results is presented in Figure 8.7. The findings are illustrated in the form of 'contours' showing the number of trophocysts per plate. It is clear that both ingredients of the medium affect trophocyst production, with maximal production at $0.02 \text{M } NH_4^+$ and 0.16M acetate.

Whether similar effects occur on dung is not known. If they do, then

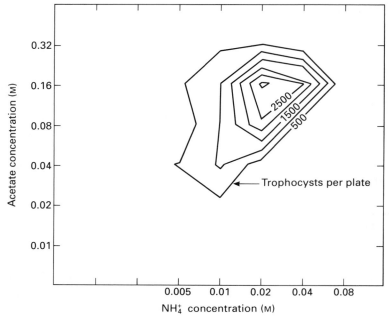

Figure 8.7 The effects of ammonium and acetate concentration on asexual reproduction in *Pilobolus kleinii*. Lines connect points of equal response. Both axes are plotted on a logarithmic scale to base 2. (Redrawn from Page, 1960a.)

they might be taken to support a modified form of the nutritional hypothesis. These results emphasize the need for further study on the physiology of fruiting in order to help to interpret fruit-body succession.

8.2 COMPARISON OF THE FUNGAL FLORA OF DIFFERENT ANIMAL DUNGS

Herbivorous mammals belong to a number of distinct orders with different types of digestion. For example, ruminants – animals which chew the cud, such as the cow and the sheep – have several stomachs, and their food is regurgitated and masticated, which is not the case for horses (perissodactyls) or lagomorph rodents (e.g. rabbit or hare). Another difference which is difficult to evaluate is that rabbits are known to practise **coprophagy**, i.e. they may ingest their own faeces, whereas many grazing animals are known to avoid accumulations of faeces. The length of time for which ingested food remains within the gut may vary from a few hours to several days. It might therefore be expected that different kinds of animal dung should support a different fungus flora. This is true even where animals graze a common territory. For example, *Coprobia granulata* forms its orange apothecia on cow dung, but not on horse dung, even when horses and cows share the same field. Comparisons of the species of fungi on different kinds of dung have been made by Angel and Wicklow (1975), Richardson (1972), Parker (1979), Wicklow, Angel and Lussenhop (1980), Mitchell (1970) and Nagy and Harrower (1979).

Angel and Wicklow (1975) collected dung of several different kinds of herbivores (pronghorned antelope, cattle, rabbit and small mammals) at the Pawnee Site, in north-eastern Colorado, USA, incubated the samples in moist chambers, and recorded the incidence of fungi on them. Criticisms of their technique are that different masses of dung were used for the different mammals, and that the record of fungi is incomplete, Deuteromycotina being excluded. It is also possible that the small mammal faeces may have been derived from more than one species of herbivore. An estimate of the similarity in the species composition among the samples was made using Sorensen's index of similarity which has a maximum value of 100% when two samples support an identical species content. The results of this comparison are given in Table 8.2. The highest index of similarity ($S = 56$) was between antelope and cattle faeces, 15 of the 17 species found on antelope dung being also found on cattle dung. This is not surprising since both animals are ruminants. Low indices were found between dung of small mammals and antelope ($S = 28$) and cattle ($S = 33$), but there was a closer similarity between dung of small mammals and rabbit ($S = 41$).

In a later paper, Wicklow, Angel and Lussenhop (1980) studied the fungal fruit-bodies which developed on the dung of caged rabbit and sheep fed from the same bale of hay. Thirteen species were recorded on sheep dung and 19 on rabbit dung. Their data suggest that differences in the digestive processes of ruminants and lagomorphs may be important selective factors in determining which fungi fruit on different kinds of dung.

Table 8.2 Similarity indices among fungal populations colonizing four types of herbivore faeces

Herbivore	Number of fungal species	Similarity index		
		Cattle	Rabbit	Small mammal
Antelope	17	56 (15)[a]	45 (10)	28 (4)
Cattle	37	–	47 (15)	33 (8)
Rabbit	27		–	41 (8)
Small mammal	11			–

[a]Numbers found on both types of faeces in parentheses.
Source: Based on Angel and Wicklow (1975).

Richardson (1972) used a different approach in his comparisons. He collected a total of 137 dung samples from six common types of animal (hare, rabbit, sheep, cow, horse and roe deer) from a wide range of localities and recorded the fungi which developed on the samples during a period of moist-chamber incubation lasting 2–3 months. The percentage frequency of common fungi on each of the dung samples was calculated. These studies confirm subjective impressions, that some fungi are associated with particular dung types, whilst others show less preference. For example, *Cheilymenia* spp., *Coprobia granulata*, *Ascobolus immersus*, *A. furfuraceus*, *Ascophanus microsporus*, *Lasiobolus ciliatus* and *Podospora curvula* are associated with ruminant dung, whilst *Sporormia bipartis*, *Coniochaeta* spp., *Thelebolus stercoreus*, *Podospora appendiculata* and *P. setosa* are associated with lagomorph dung. Others, such as *Ascobolus albidus*, *Thelebolus nanus* and *Podospora vesticola* (= *P. minuta*), are more general in their occurrence. The reasons for these preferences are not understood. Possibly the different digestive processes exert a selective influence either directly or through interspecific competition. Possibly the chemical and physical properties of the different kinds of dung favour the growth and fruiting of different fungi. Richardson's studies gave evidence of both positive and negative association between species. The underlying causes of both phenomena are likely to be complex. Positive associations might reflect a common preference for physical or chemical features of the substratum. Alternatively, it is possible that one species may be nutritionally dependent on another in respect of growth or fruiting, an example of a synergistic relationship. In the case of negative association, the possibility of antagonistic relationships should also be considered. Examples of both phenomena amongst coprophilous organisms have already been considered.

8.3 AUTECOLOGICAL STUDIES

There have been very few autecological investigations on coprophilous fungi on their natural substrate, although there have been innumerable studies on the physiology and nutrition of coprophilous fungi in pure culture. Wood and Cooke (1987) have studied the nutritional competence of *Pilaira anomala* in culture on liquid and agar media and on reconstituted copromes derived from laboratory-fed rabbits (Wood and Cooke, 1984). The fungus grew well on media containing ammonium or urea, but poorly on media containing nitrate or asparagine. Glucose and fructose were the best carbon sources, and it was unable to utilize starch, cellulose, pectin and protein. On copromes, there was evidence of mycelial activity long after reproduction had ceased. *Pilaira anomala* is regarded as a typical ruderal, being dependent for its nutrition on relatively simple soluble nutrients which are probably quickly exhausted.

9
AQUATIC FUNGI

9.1 INTRODUCTION

A number of different groups of fungi are found in water, including many Mastigomycotina (zoosporic fungi), some Zygomycotina, Ascomycotina, Deuteromycotina, yeasts and a few Basidiomycotina. Some may inhabit water for the whole of their lives, others may be amphibious, with one stage of their life cycle spent in, adapted to and dispersed under water, and another stage dispersed in air. Yet others may have a transient aquatic existence, possibly brought on a substratum by wind or swept by floods into water. The spores of many terrestrial fungi are carried into water by rain, and so may be isolated by conventional mycological techniques. It is therefore necessary to define carefully the term **aquatic fungus**. Park (1972b) has introduced a number of useful terms in distinguishing between the activities of heterotrophic microorganisms in fresh water.

> Microorganisms recorded from an aquatic site by observation or by isolation may or may not have originated there. Those that have, and that maintain themselves in that habitat and spend all their life-history there, may be regarded as **indigenous organisms** or **indwellers**. Other organisms present may normally have an extra-aquatic habitat. Those in which the main habitat is clearly extra-aquatic might be regarded as **immigrants** in relation to water, whilst those that alternate periodically between aquatic and extra-aquatic habitats might be regarded as **migrants**. **Versatiles** might be an appropriate term for organisms in which it could be shown that the movement between aquatic and extra-aquatic habitats is haphazard rather than regular.

Park has summarized these terms in the form of Table 9.1

Dick (1976) has produced a similar classification of the role of fungi in freshwater habitats.

The degree of adaptation of a microorganism to an aquatic environment may vary. **Indwelling organisms** are fully adapted to life in water in the

Table 9.1 Classification of heterotrophic microorganisms resident in water

Category			Presence in water
Resident	Indwellers		Permanent
	Immigrants	Migrants	Periodic ⎫
		Versatiles	Irregular ⎭ Not permanent

Source: Park (1972b).

sense that they are able to maintain their biomass at a more or less constant level from year to year as substrates and nutrients become available, and most are capable of sporulating in the water. **Transients** might arrive at the aquatic habitat and immediately start to decline in activity as a result of changes in conditions, e.g. availability of oxygen, loss of nutrients by leaching, or in the face of competition by organisms better adapted to life in water. Such organisms may arrive already active within their substratum, but unable to sporulate, and so be unable to colonize new substrata within the aquatic habitat. There is obviously a wide range of possible behaviours, and this has been presented by Park in Table 9.2 for decomposer organisms.

As well as decomposer microorganisms, there are parasites attacking hosts such as plankton, algae, macrophytes and various animals.

9.1.1 Aquatic habitats

There is a great diversity of aquatic habitats found on earth, ranging from temporary pools, ponds, lakes, bogs, fens, streams, rivers, estuaries, inland salt lakes, melting ice and thermal springs to seas and oceans. At their margins, these water masses are in contact with rock, soil and mud,

Table 9.2 Possible combinations of categories of presence and activity in decomposers in water

Presence category		Possible activity categories			
		Constant	Periodic	Sporadic	No activity
Residents	Indwellers	+	+	+	−
	Migrants	−	+	+	−
	Versatiles	−	−	+	+
	Transients	−	−	−	+

Source: Park (1972b)

and may be fringed with vegetation. There is also an interface with air. The soil, mud and vegetation contain fungi which may release spores into the water. The water masses may also contain numerous organisms which, in the living or dead state, may provide a source of nutrients for parasitic or saprophytic fungi. It is clear that the ecology of aquatic fungi is an enormous subject, and selection of topics must be made. Valuable review articles on the ecology of different groups of aquatic fungi can be found in Jones (1976a), Sparrow (1968), Dick (1976), Kohlmeyer and Kohlmeyer (1979), Kohlmeyer (1981a), Webster and Descals (1981), Moss (1986a), Bärlocher (1992a) and Shearer (1993).

9.1.2 Techniques in aquatic fungus ecology

As discussed in Chapter 1, there are many different aspects of fungal ecology, and the technique chosen will depend not only on the aspect to be studied, but also on the group of fungi involved. Different techniques applied to the same body of water may yield entirely different results. Park (1972a) used dilution plate, direct observation, baiting and particle plate techniques on the same bodies of water and obtained strikingly different lists of species by the four methods. The baiting technique yielded a preponderance of zoosporic fungi (Oomycetes), and direct observation yielded Oomycetes and aquatic hyphomycetes. Particle plating produced colonies of only one Oomycete (*Pythium*), large numbers of fungi normally associated with moribund, terrestrial plant material (e.g. *Alternaria*, *Cladosporium*, *Aureobasidium* and *Epicoccum*) and no aquatic hyphomycetes. The dilution plates bore many fungi previously recorded from soil, no Oomycetes and only one aquatic hyphomycete (*Heliscus*). This comparative study emphasizes how important it is to use a range of techniques, and to select the most appropriate technique for the investigation. For a discussion on the limitations of techniques for studying the ecology of zoosporic aquatic fungi, see Dick (1976).

In the account which follows, aspects of the ecology of selected groups of aquatic fungi will be considered.

9.2 FRESHWATER FUNGI

9.2.1 Chytridiomycetes

a) Chytridiales

Chytridiales live in the sea and in fresh water, in mud and in soil. Some are biotrophic parasites of higher plants, but these are not considered here. For the most part, their thalli are eucarpic and monocentric, but forms with holocarpic thalli or eucarpic polycentric thalli are also known. They reproduce by posteriorly uniflagellate zoospores. Taxonomic accounts of the group have been given by Sparrow (1960, 1973) and Barr (1989), and figures by Karling (1977). Because of their simple morphology, identification is difficult, and this, in turn, creates difficulties in ecological studies (Miller, 1976; Masters, 1976). Chytrids are of ecological importance as parasites of planktonic algae such as diatoms, desmids and filamentous green algae. They are also ecologically significant as decomposers of the remains of algae, other plants and animals. They can be studied by direct microscopic observations on these substrata, or by baiting water samples with various materials such as pollen grains, Cellophane (cellulose), snake skin, insect wings (chitin) or hair (keratin) (Sparrow, 1960).

Parasitic chytrids

Classical studies on the effects of chytrids parasitizing the phytoplankton diatom *Asterionella formosa* in the English Lake District have been reported by Canter and Lund (1948, 1951, 1953). *A. formosa* forms colonies composed of radially arranged frustules looking like the spokes of a cartwheel (Figure 9.1). The diatom is parasitized by several chytrids. These include *Rhizophydium planktonicum*, *Zygorrhizidium affluens* and *Z. planktonicum*, which are somewhat similar in morphology. Before the distinction between the first two taxa had been pointed out by Canter (1969), they had both been regarded as belonging to *R. planktonicum*. To avoid confusion, the mixed populations are referred to as *R. planktonicum* agg., whilst the 'true' *R. planktonicum* is referred to as *R. planktonicum* Canter emend. (or *R. planktonicum* emend.). The parasite forms an epibiotic sporangium by enlargement of an encysted zoospore on the outside of the host cell, and a thread-like internal rhizoid which may extend throughout the whole length of the diatom frustule. Some 2–3 days after infection, the sporangia mature and release posteriorly uniflagellate zoospores. Infection of the host is enhanced by high light intensity, possibly because it stimulates zoospore motility and also because it causes the host cells to become more susceptible (Canter and Jaworski, 1980, 1981).

One result of infection of an *Asterionella* colony by *R. planktonicum* agg. is to reduce the number of cells per colony (Figure 9.2). This effect is presumably due to the failure of infected diatom cells to divide in the normal way. Epidemics of *Rhizophydium* (i.e. over 25% of *Asterionella* infected) may be associated with drastic reductions in the concentration of cells in the top 5 m of lake water. In Esthwaite Water during an epidemic from 25 October to 6 November 1947, the proportion of infected cells rose to 38%, whilst the number of *Asterionella* cells fell from 121 to 37 cm^{-3} and the mean number of cells per colony fell from over 8 to below 4. In September 1948, more than 90% of *Asterionella* cells in Windermere South Basin were infected (Figure 9.2).

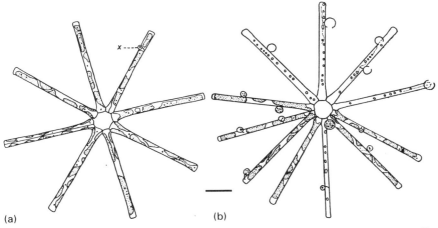

Figure 9.1 *Rhizophydium planktonicum* agg. (a) A healthy colony of *Asterionella* with well-developed chromatophores and an encysted zoospore of *Rhizophydium* (x). (b) Heavily parasitized *Asterionella* colony, with more or less disorganized cell contents and *Rhizophydium* sporangia (live or dehisced). Scale bar = 20 μm. (Reproduced from Canter and Lund, 1948 with permission.)

The reasons for the incidence of epidemics are not clear. They may occur at any time when the numbers of *Asterionella* cells exceed 10 cm^{-3}. Epidemics appear not to be related to the limitations of host nutrients such as available phosphate, nitrate or silica, which are often rising in concentration during epidemics. Clear-cut epidemics occur only in the more eutrophic waters. It is likely that the pathogen may survive in low numbers in *Asterionella* populations at all times in the zoosporangial state.

Canter and Jaworski (1979) showed that certain *Asterionella* clones were highly compatible with the parasite, whilst others appeared to be incompatible, sometimes showing a hypersensitive response in which, soon after penetration, the host cell died rapidly and the encysted zoospore did not enlarge into a sporangium.

Saprophytic chytrids

Pine pollen is a natural substratum for saprophytic chytrids, and samples of pollen scooped from the surface of a lake or pond are invariably colonized by monocentric, epibiotic chytrids, including species of *Rhizophydium* (Figure 9.3). By serial dilutions of lake water with sterile water to which sterilized pine pollen is added, it is possible to make quantitative estimates of the concentration of zoospores of chytrids capable of colonizing pine pollen. This technique, which resembles the **most probable number** (MPN) technique used by bacteriologists to estimate bacterial numbers, has been

230 AQUATIC FUNGI

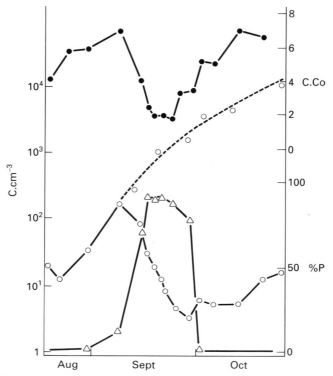

Figure 9.2 Parasitism of *Asterionella formosa* in Windermere South Basin in autumn 1946. ●—● Number of live cells per colony (C.Co); O—O = number of live cells cm^{-3} (C.cm^{-3}, logarithmic scale); O- -O = theoretical numbers of cells which might have been produced in the absence of parasitism. △–△ = Percentage of cells infected by *Rhizophydium planktonicum* agg. (%P). (Redrawn from Canter and Lund, 1951 with permission.)

used by Ulken and Sparrow (1968) in Douglas Lake, Michigan, USA. Douglas Lake is a eutrophic lake surrounded by a conifer–aspen forest. In spring, it receives a large input of *Pinus* pollen, which may remain floating. Estimates of chytrid propagules in the epilimnion of lake water rose from less than 100 to over 900 l^{-1} between 8 and 23 June 1967 (Figure 9.4). This results from the availability of large numbers of pollen grains at that time of the year. The pollen bore numerous sporangia of *Rhizophydium*. A second peak of propagules of about 600 l^{-1} was detected in early July, corresponding to the release of pollen from a different species of *Pinus*. The high numbers declined to about 30 l^{-1} in the epilimnion by mid-August, possibly due to leaching of nutrients from the pollen, sinking of the pollen, and predation by fungi and protozoa. The numbers of propagules estimated in the epilimnion were always greater than those in

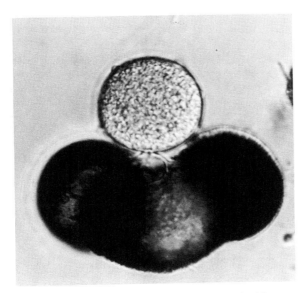

Figure 9.3 *Rhizophydium* sp. Epibiotic sporangium attached by means of rhizoids to the central fertile cell of a pine pollen grain.

the hypolimnion. It is likely that pollen grains sinking through the hypolimnion onto the mud surface develop resting sporangia which give rise to zoospores which provide the inoculum for new infections the following spring at the overturn of stratification of water layers in the lake associated with the melting of ice cover and wind action.

9.2.2 Oomycetes

Oomycetes are not related to 'true' fungi such as Zygomycotina, Ascomycotina and Basidiomycotina. They reproduce asexually by biflagellate heterokont zoospores and are sometimes classified in a separate kingdom (Heterokonta) along with other heterokont organisms (Dick, 1989). We shall consider aspects of the ecology of members of three orders – Leptomitales, Saprolegniales and Peronosporales.

a) Leptomitales

The Leptomitales are freshwater fungi with a characteristic constricted tubular mycelium plugged at the constrictions by spherical plugs of cellulin. Asexual reproduction is by biflagellate zoospores, and sexual reproduction, where present, is oogamous (Dick, 1973b, 1989; Sparrow, 1960). Most members of the group grow on vegetable debris such as twigs or

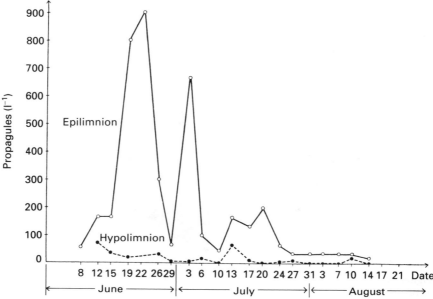

Figure 9.4 Fluctuations of number of propagules of zoosporic fungi in the epilimnion and hypolimnion of Douglas Lake, Michigan, USA during spring and summer 1967. (Reproduced from Ulken and Sparrow, 1968 with permisison.)

fruits. Their ecological requirements are diverse. For example, *Apodachlya* and *Sapromyces* thrive in well-oxygenated water, whilst *Aqualinderella* is obligately fermentative and develops best in CO_2-rich waters on submerged fruits in tropical and subtropical streams. It does not require oxygen for growth, and the presence of free oxygen may indeed inhibit growth. Growth is stimulated by high levels (5–20%) of CO_2. In pure culture the fungus undertakes a homolactic fermentation. It has been considered as 'an oxygen indifferent, facultative anaerobe resembling the lactic acid bacteria' (Emerson and Weston, 1967; Emerson and Held, 1969).

Leptomitus lacteus forms part of a complex of organisms, making up the so-called **'sewage fungus'** associated with rivers polluted with organic matter, especially at acid pH values. Rope-like strands of slimy growth may extend down-river for several kilometres from the source of pollution such as outfalls from distilleries and sewage works. The growths include not only *L. lacteus*, but other fungi such as *Fusarium aqueductuum* and *Geotrichum candidum*, bacteria, ciliates and algae (E.J.C. Curtis, 1969). Although abundant in polluted sites associated with human activities, the fungus has also been recorded in cleaner rivers and around lake margins

(Park, 1972a) and has been reported as an external parasite on freshwater fish (Willoughby, 1978). Propagules have been estimated to occur in low concentrations in Windermere, about 0.7 spores l^{-1} at the lake margin, to less than one in 20 l from the lake centre, implying that the fungus is active in decomposition in the littoral zone (Willoughby and Roberts, 1991).

Studies of the nutrition of *L. lacteus* in pure culture reveal a number of distinctive features. It is unable to utilize sugars as sources of carbon, but can grow on organic and fatty acids. Inorganic nitrogen sources are unable to support growth, and amino-acids are required. By contrast with *Aqualinderella*, it is highly aerobic and, whilst it can survive periods of oxygen deprivation, oxygen is required for active growth (Gleason, 1968). So far, no sexual stage has been discovered.

b) Saprolegniales

The Saprolegniales are the best-known group of water-moulds and are abundant and worldwide in their distribution. An outline of their taxonomy has been given by Dick (1973a, 1989). Their substrata are plant and animal remains in freshwater, mud and soil, and some are parasites of plants, fish, crustacea and other animals. Common genera include *Saprolegnia*, *Achlya*, *Dictyuchus* and *Aphanomyces*.

Varied techniques for studying the ecology of Saprolegniales have been developed, and the effectiveness of some of them has been compared by Dick (1966, 1976).

Baiting

The classical method for detecting the presence of Saprolegniales is by the use of baits (e.g. boiled hemp seeds, dead insects) floated in water. Colonization is dependent on the ability of zoospores to move chemotactically towards the bait. Baits placed in contact with soil or mud may be colonized by mycelium or from zoospores. Dick (1966, 1971) developed the use of hemp seed bait in a quantitative assay which enabled estimates to be made of the minimum propagule number in a water or mud (slurry) sample. The number of species obtained is relatively insensitive to the amount of slurry used. The origin of the inoculum which colonizes the hemp seed bait is uncertain, and could be derived from a zoospore, a vegetative hypha, a germinating gemma or a germinating zoospore.

Diluted slurry can also be incorporated into cornmeal agar plates. The agar is then cut into small blocks, each of which is baited with a hemp seed in water. Using this method, Dick (1971) has estimated minimum propagule numbers between 300 and 1000 g^{-1} dry soil from a slope adjoining Blelham Tarn in the English Lake District. These values are low in relation to those of other groups of soil fungi.

Agar cultures

Willoughby (1962) added samples of lake water to molten dilute oat agar. When cut into sectors, and incubated in sterile water, those sectors of agar which had contained a viable propagule of a member of the Saprolegniaceae developed a coarse mycelium and, in some cases, sporangia. Estimates of propagule number per litre of lake water could be made. Values as high as $5200\ l^{-1}$ (i.e. about 5 propagules cm^{-3}) have been reported for Windermere.

Cold-setting gels

Cold-setting gels, e.g. calcium carboxymethylcellulose (Polycell gel, a wallpaper paste) or the more transparent hydroxyethylcellulose (Natrosol 250) have been used to overcome problems caused by the high temperature of molten agar (c. 45°C) (Willoughby, Pickering and Johnson 1984; Celio and Padgett, 1989). The low concentrations of nutrients minimize the growth of contaminants. Antibiotics such as pimaricin, which selects for Oomycetes and suppresses growth of other fungi, improve the recovery of Oomycete propagules.

Centrifugation

Continuous flow centrifugation of lake water was used by Fuller and Poyton (1964) to concentrate propagules of aquatic fungi prior to plating. This technique has been developed by Hallett and Dick (1981) in a study of Oomycete (mostly Saprolegniales and Peronosporales) propagule number in a freshwater lake.

Direct observation of natural substrata

In some circumstances, it may be possible to assess the frequency of colonization of natural substrata such as the proportion of fish eggs colonized by Saprolegniaceae. Dick (1970) followed the colonization of insect exuviae by Saprolegniaceae in Marion Lake, near Vancouver, Canada.

Using these techniques, information has been obtained on various aspects of the ecology of Saprolegniaceae (see Sparrow, 1968; Dick, 1976).

Distribution

A comparison of the frequency of isolation of Saprolegniaceae from **littoral** (i.e. lake margin) and **benthic** (i.e. bottom) muds was made by Dick (1971) using the hemp seed baiting technique and 10 replicate mud samples of about 10 cm^3, along transects extending from the shore to the open water at Lake Marion near Vancouver (Figure 9.5). The maximum number of species (about 15) was found at the lake margin. This is also the region of maximum abundance as indicated by the fall in the number of negative samples near the lake margin. In the benthic muds, a high proportion (up to 70%) of the samples were negative, indicating a low level of abundance. Although the bottom muds contain viable propagules of Saprolegniaceae, there is little evidence of biological activity. Dick has concluded that most Saprolegniaceae are primarily fungi of the emergent littoral and benthic/littoral interface, and very few are clearly benthic.

Collins and Willoughby (1962) and Willoughby and Collins (1966) have made similar conclusions from a study of the frequency of isolation of Saprolegniaceae in water and mud samples in and around Blelham Tarn in the English Lake District. The high incidence of colony counts following periods of heavy rainfall suggests that spores may be carried into the lake from the surrounding drainage basin. Counts from mud samples near the centre of the lake, taken from the hypolimnion, where the dissolved oxygen content may be zero, were lower than in the water column above, or at the lake margin. Possibly spores do not survive for very long in the anaerobic conditions of the mud beneath the hypolimnion. Comparison of

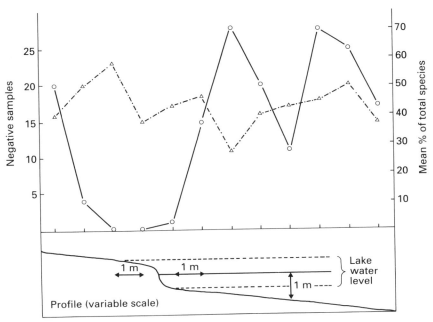

Figure 9.5 Transect from the margin of Lake Marion, Vancouver, Canada, showing the number of negative samples and the mean percentage of the total species of Saprolegniaceae recorded. O–O = Negative samples,: △–△ = mean % of total species recorded. (Redrawn from Dick, 1971 with permission.)

the colony counts taken from inflow streams with lake counts showed the striking effect of a small sewage outfall on the incidence of Saprolegniaceae. Above the sewage outfall, the mean spore concentration was 20 l^{-1} whilst below the outfall the concentration was 312 l^{-1}. A major component (about one-third of all isolates) from the sewage effluent was *Saprolegnia ferax/mixta*, whilst *Achlya* was not represented.

Systematic studies carried out in a series of defined habitat areas permit comparisons of the degree of correlation between the species composition characteristic of the sites chosen. Dick (1971) used an index of similarity to compare the Oomycete populations from different sites. Sites showing a high degree of similarity were *Sphagnum*-dominated or acid humus areas in which *Saprolegnia litoralis* is abundant, associated with *Scoliolegnia asterophora* or *Saprolegnia turfosa*.

The occurrence of Saprolegniaceae may be affected by variations in the pH of the water. However, the pH of a body of water is not constant, and can vary over wide limits due to the photosynthetic activity of algae and macrophytes in the water. The depletion of dissolved carbon dioxide causes a rise in pH, especially in water with low bicarbonate concentration. For this reason, data on frequency of occurrence in relation to pH may be misleading if based only on a single pH reading. Also Saprolegniaceae show seasonal variations in abundance, so that isolated records indicating absence may reflect season rather than adverse effects of pH.

A classical study of the distribution of Saprolegniaceae in relation to pH is that of Lund (1934) who recorded the occurrence of aquatic fungi from a range of aquatic habitats in Denmark. However, his samples and pH records were not repeatedly collected at regular intervals. He classified the aquatic habitats into five groups in relation to pH, and listed the fungi characteristic of each type of habitat:

- **Highly acid** (pH 3.5–4.5), mostly represented by bogs: *Saprolegnia delica, S. diclina, S. litoralis*.
- **Slightly acid** (pH 5.5–6.8); no Oomycete flora was found to be peculiar to these waters; the range of species resembles that of highly acid waters.
- **Neutrally acid** (pH 5.3–7.5): *S. diclina, Dictyuchus sterile*.
- **Neutrally alkaline** (pH 6.5–7.7), e.g. ponds or pools with stagnant water and abundant decaying plant material: *S. ferax, S. monoica, Achlya radiosa*.
- **Constantly alkaline** (pH 7.0–8.4), e.g. certain lakes: *S. ferax, S. hypogyna, Aphanomyces laevis*.

Certain species, e.g. *Achlya racemosa*, occurred over the whole pH range.

Roberts (1963) made regular baitings with hemp seed from water samples with a pH range from 3.6 to 8.0 from 21 sites in Britain over a 19-month period. There was seasonal variation in occurrence, some species being found in summer, others in winter. She classified the fungi found into three groups in relation to pH variation:

- **Acid group**: Species found in soft waters below pH 5.2, e.g. *Achlya americana*, *Saprolegnia litoralis*.
- **Neutral group**: Species found in waters with a wide pH range between 5.2 and 7.4, e.g. *Achlya racemosa*, *Saprolegnia monoica*.
- **Alkaline group**: Species found in waters of pH 7.8 or above, e.g. *Achlya polyandra*, *Saprolegnia ferax*.

In general, Roberts's work confirms the findings of Lund.

It is important to realize that pH, i.e. hydrogen concentration, has complex effects not only on the activity of zoospores (Smith, Armstrong and Rimmer, 1984), mycelial growth, enzyme activity and reproduction, but also affects the availability of salts such as those of calcium, potassium, magnesium, iron and phosphorus, and the form in which nitrogen is present. pH also affects the growth of higher plants, and this in turn affects the substrates available. At low pH, competition with bacteria is also lower than at high pH.

Saprolegniaceae are rare in the sea and are among the most sensitive fungi to decreasing solute potential (i.e. to increased concentration of dissolved salts; Duniway, 1979). Estuaries also represent a hostile environment (Höhnk, 1935, 1939, 1952, 1953). Natural sea water has a salinity of about 35‰, but in estuaries the salinity level can fluctuate from zero at low tide to that of normal sea water at high tide. When suitable baits were added to water samples collected at various points along a salinity gradient provided by an estuary, Saprolegniaceae were found on baits in water samples with up to 2.8‰ salinity values (Te Strake, 1959). Later work, in which baits were left *in situ* at different points along an estuary for periods up to 3 days, gave positive colonization in water where the salinity values varied up to 12‰ (Padgett, 1978a). It is doubtful that the inoculum which colonized baits at the higher salinity levels was released close to them; it had possibly been carried from freshwater sources higher up the river system feeding the estuary. Physiological studies show that zoospore germination, mycelial growth rate, respiration rate and asexual reproduction become progressively reduced as salinity increases (e.g. Te Strake, 1959; Harrison and Jones, 1971, 1974; Padgett, 1978b; Padgett *et al.*, 1988; Smith, Ince and Armstrong, 1990), and that at values around 10‰ (i.e. less than one-third full-strength sea water), plasmolysis of hyphae may occur.

When saprolegniaceous fungi established on floating substrata are swept into the sea, they are not immediately killed. Padgett (1984) showed that Saprolegniaceae in hemp seed and twig baits could survive salinity changes of 0–35‰ for 48 h and, when returned to fresh water, were capable of colonizing fresh baits. The formation of gemmae may aid survival.

An interesting aspect of salinity tolerance is the question of whether

Saprolegnia parasitica, the cause of ulcerative necrosis of salmonid fish, can survive the transition from fresh water to the sea and back as young salmon leave the spawning ground, migrate to the sea and then return to fresh water to reproduce.

Periodicity

Changes in the frequency of isolation using a standard technique or estimates of propagule availability show considerable variations with time, both in the relatively short term (e.g. diurnal periodicity) or in the longer term (e.g. cyclical variation throughout the year). Hallett and Dick (1981), using a centrifugation–plating technique in a small ornamental lake at Reading, England, found two discrete diurnal peaks of Oomycete propagule availability (Saprolegniales and Peronosporales) from 12.00 to 16.00 and 20.00 to 24.00 h. So far, no satisfactory hypothesis to explain these variations has been advanced, and more work on diurnal fluctuation is needed.

Suzuki (1961a,b, in Sparrow, 1968) has claimed that there is a diurnal migration of zoospores of aquatic fungi, closely correlated with the diurnal vertical distribution of dissolved oxygen in spring and autumn. They report that zoospores assemble in the oxygen-rich water layers, a conclusion supported by the finding by Smith, Armstrong and Rimmer (1984) that zoospore production and encystment in *Saprolegnia diclina* are markedly affected by oxygen concentration, being much reduced at low concentrations.

Willoughby (1962) and Clausz (1974) have indicated that there are two peak periods of annual zoospore availability in north temperate latitudes: late spring and autumn, with maximum availability in autumn. However, Hallett and Dick (1981) demonstrated the existence of three, rather than two, periods of peak availability of Oomycete propagules in a Reading lake. Attempts to correlate periods of maximum availability with environmental variables have not met with success.

Wood (1988), working in the River North Tyne, England, using the Polycell gel assay technique, found that the total Saprolegniaceae counts rose sharply in February, and peaked in the summer months June and August. After August, colony counts began to decline towards lower winter levels between December and February.

Using techniques in which standardized baits were added to water or muds at different times of the year, seasonal patterns in frequency of isolation have been reported (e.g. Coker, 1923; Perrott, 1960; Hughes, 1962; Roberts, 1963; Alabi, 1971a,b; Okane, 1978, 1981; Gupta and Mehrotra, 1989). Temperature has an important effect, some species being isolated with equal frequency throughout the year, whilst others are more frequent during periods of low or high temperature. Since colonization of the baits is by zoospores, variations in frequency of isolation may be related to zoospore production and motility. At times when species are not reported, they must be surviving in a different form, e.g. as oospores, and the factors relating to oospore maturation, survival and germination may also be correlated with seasonal periodicity. It has been claimed (e.g. by Ziegler, 1958; Hughes, 1962; Klich and Tiffany, 1985) that seasonal abundance and also latitudinal distribution are correlated with oospore morphology. In some species, there is a large spherical lipid globule placed centrally within the oospore (centric types), whilst in others the globule is subcentric or eccentric. Species with centric and subcentric oospores are isolated more frequently in cooler months, whilst species with eccentric oospores are more frequently isolated in warmer conditions. This is also correlated with the fact that species with eccentric oospores are reported to be more common in warmer countries. The underlying reasons are unknown.

Whilst variations in seasonal periodicity may well be directly controlled by temperature differences, it should be borne in mind that temperature and other correlated features such as daylength also affect surrounding or submerged vegetation in ponds and rivers, and in turn there are seasonal variations in animal populations, all of which may provide substrata for growth of Saprolegniaceae. Seasonal changes in rainfall, water flow, etc., may also exert effects.

c) Peronosporales

The Peronosporales may exist as saprophytes or as parasites of plants in soil or water (Waterhouse, 1973; Dick, 1989). One family, the Pythiaceae, is well represented in freshwater habitats by two genera, *Pythium* and *Phytophthora*. Park (1975) used a selective medium for the isolation of some unusual cellulolytic species of *Pythium* from river water in Northern Ireland. The selective medium makes use of the nutritional preference of these species of *Pythium* for cellulose and the insensitivity of *Pythium* to the polyene antibiotics pimaricin and vancomycin. *Pythium* species isolated using this technique were the zoosporic *P. fluminum* var. *fluminum* and *P. fluminum* var. *flavum*, and a second species in which zoospores were not observed, *P. uladhum* (Park, 1977). *Pythium fluminum* var. *fluminum* may be adapted to growth in rivers where the concentration of dissolved nutrients is often low. Both species, including the two varieties, possess characteristic pigments. The possession of these pigments, and the selective medium for isolation, have enabled Park and McKee (1978) to follow the 'numbers' of these fungi in river water using filter papers, protected in mesh bags, anchored in the river. At intervals, the filter papers were removed and sampled by plating 5 mm discs on selective medium. After 3 days, all discs bore cellulolytic *Pythium*, indicating early and progressive arrival of inoculum. Rates of arrival of propagules can also be calculated and it has been estimated that the arrival rate was about 0.5 propagule $cm^{-2}\ h^{-1}$ per surface exposed. It is also possible to make an estimate of the concentration of *Pythium* propagules in river water by filtering a sample and placing the filter paper on selective medium without cellulose. The pigmented colonies which develop can then be counted and identified. Using this technique, propagule numbers varying from less than 1 to over 4000 l^{-1} have been estimated.

Park (1980b) has extended these studies of the numbers of cellulolytic *Pythium* species over a 2-year period in two rivers in Northern Ireland. The numbers were plotted as $\log_{10}$ against time. Decreases in numbers were found associated with decreased levels of river water, and a general increase was found immediately after a period of heavy rain. No significant correlation was found for the two varieties of *P. fluminum* in either river against pH. They showed different correlations with river temperature. For *P. fluminum* var. *fluminum*, there was no significant relationship with temperature, but for var. *flavum*, high numbers were significantly correlated with low temperature, and low numbers with high temperature. Very significant relationships between numbers of var. *fluminum* and previous rainfall and river flow rate were demonstrated. At the optimum river flow rates, the peak numbers were about 10 times higher than at the lowest flow rates. It is only possible to speculate on the underlying causes of the correlation between numbers and rainfall or flow rate. Possibly substrata containing the fungus could be washed into the river from surrounding habitats, or from soil. Possibly increased flow and turbulence stimulates sporulation, or stirs up sediment so that propagules become suspended.

The colonization of filter paper in a stream has also been studied by Willoughby and Redhead (1973) who observed that, although the filter paper lost more than half its initial dry weight in 50 days, its nitrogen content increased on a percentage basis from about 0.2% to 1.47%, and in absolute amounts from $9 \times 10^{-6}\ g\ cm^{-2}$ of filter paper at 8 days to $190 \times 10^{-6}\ g\ cm^{-2}$ at 50 days. The capacity of microorganisms to concentrate nitrogen in a substratum decomposing in a stream is an important phenomenon which will be discussed more fully in relation to leaf decay in streams by aquatic hyphomycetes (section 9.2.3(b)).

Studies by Kirby (1984) have shown that *Pythium* species are important early colonizers of the leaves of *Ranunculus penicillatus*, both of attached leaves and killed leaf segments placed in mesh bags in streams. The early samples from killed leaves were dominated by *Pythium* spp., but from segments left longer in the river, other fungi such as aquatic hyphomycetes and species of *Fusarium* were more abundant (Figure 9.6).

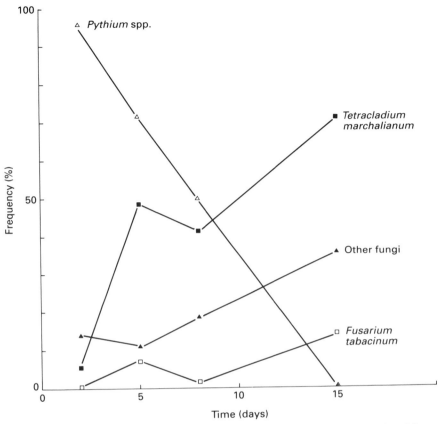

Figure 9.6 Percentage frequency of isolation of *Pythium* spp. and other fungi from sieved particles of dead *Ranunculus penicillatus* leaves at different periods after submergence in the River Exe, Devon, England. (Reproduced from Kirby, 1984.)

9.2.3 Ingoldian aquatic hyphomycetes

This group of freshwater fungi was named in honour of Professor C.T. Ingold, a pioneer on their taxonomy. About 300 species are known, with a worldwide distribution. Their characteristic habitat is in rapidly flowing, i.e. turbulent, well-aerated, non-polluted streams, although they have also been reported from lakes and terrestrial habitats. Typical substrata are leaves and branches of deciduous trees which have fallen into the water, but they also grow on submerged macrophytes, and as endophytes in the healthy roots of riparian trees such as *Salix* and *Alnus* which extend into water.

The conidia of Ingoldian fungi are unusually large, many spanning 50–100 μm or more. They also have very distinctive shapes, being

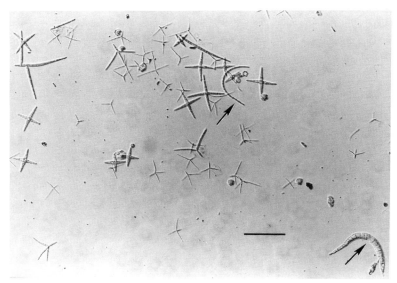

Figure 9.7 Spores of Ingoldian aquatic hyphomycetes from a foam sample collected from the River Teign, Devon, England. Most of the spores in this sample are branched (tetraradiate) but two sigmoid spores are also present (arrowed). Scale bar = 100 μm.

branched, often with four arms (**tetraradiate**) or worm-shaped, in the form of a steep helix (**sigmoid**) (Figures 9.7 and 9.8). These unusual spores, produced in great abundance, are adapted to underwater trapping, and germination is stimulated by contact with a surface (Webster, 1959, 1987; Webster and Davey, 1984; Read, Moss and Jones, 1991, 1992).

In their conidial state, these fungi are classified in the Hyphomycetes in the Deuteromycotina. However, an increasing number of conidial forms (anamorphs) are being connected with teleomorphs which belong to unrelated genera of Ascomycotina and Basidiomycotina (Webster, 1992). Ingoldian hyphomycetes are thus not a group of closely related organisms, but represent fungi which have become adapted morphologically and physiologically to aquatic habitats. Their conidia are believed to have resulted from convergent evolution (Webster, 1987). The teleomorphs usually develop on twigs and branches previously submerged in streams which have become exposed on the banks or projecting from the water as the stream level falls. The ascospores and basidiospores show no unusual morphological adaptations. They may be discharged into air, and thus possibly play an important role in long-distance transport.

Because of their characteristic conidia, the identification of Ingoldian hyphomycetes can often be done from these spores alone, which is certainly not the case for most other groups of fungi. Moreover, their conidia are readily trapped in foam in streams, enabling surveys of

FRESHWATER FUNGI 243

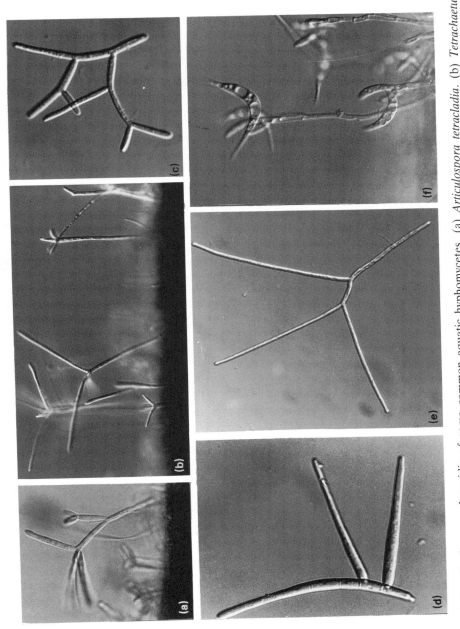

Figure 9.8 Conidiophores and conidia of some common aquatic hyphomycetes. (a) *Articulospora tetracladia*. (b) *Tetrachaetum elegans*. (c) *Varicosporium elodeae*. (d) *Tricladium splendens*. (e) *Tricladium chaetocladium*. (f) *Lunulospora curvula*.

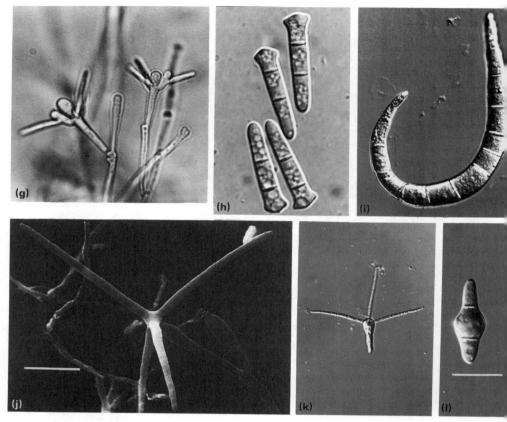

Figure 9.8 contd. (g) *Tetracladium marchalianum.* (h) *Heliscus lugdunensis.* (i) *Anguillospora crassa.* (j) *Lemonniera aquatica* conidiophore with two phialides. (Scanning electron micrograph by P.J. Sanders.) (k) *Clavariopsis aquatica.* (l) *Tumularia aquatica.* Scale bars = 10 μm (j, own scale), 40 μm (a,c,d,f,k); 20 μm (g,h,l); 80 μm (b,e).

distribution. A feature of ecological interest is the important role which this group of fungi plays in conditioning and processing leaf debris, making it palatable and nutritious to aquatic invertebrates. There have therefore been extensive ecological studies of Ingoldian aquatic hyphomycetes (for reviews, see Ingold, 1975a, 1976; Webster and Descals, 1981; Bärlocher, 1992a).

A guide to the taxonomy of these fungi has been provided by Ingold (1975b) and keys and descriptions by Descals, Marvanová and Webster (1995).

a) Techniques for study

Direct observation on randomly selected litter

A handful of deciduous tree leaves from a rapidly flowing stream in autumn will yield a rich harvest of aquatic hyphomycetes. If a single leaf is incubated at a temperature of about 15°C in a shallow dish of water, conidiophores develop overnight, often projecting from the leaf margin. Hanging drop cultures of leaf fragments may be used to follow conidial development, which is important in classification since tetraradiate and sigmoid spores may develop in a variety of ways.

Leaf packs and twig packs

Instead of using randomly selected leaf and twig material from streams, there are advantages in using packs of leaves or twigs of known tree species which can be attached to masonry bricks or placed in litter bags and immersed in streams. Amongst the advantages of using leaf and twig packs are that the place, time and duration of immersion can be controlled. Samples can be placed at different points along a river system and can also be moved about from one site to another within it. Leaf material is not scoured when the stream is in spate. The colonization of different tree leaf species under identical conditions can be compared. Leaf pack studies usually involve cutting out discs from the submerged leaves. The discs can then be incubated in water or dilute salt solution to permit sporulation and identification of the fungi which have colonized the leaf. Shearer and Webster (1985c) have compared the examination of randomly collected litter and leaf pack litter in the evaluation of community structure of aquatic hyphomycetes.

Twig packs are also used. Comparisons at the same site between leaf and twig colonization, between twigs of different tree species, and between corticated and decorticated twigs are possible. Direct observation can be done by scraping the twig surface. Incubating the twigs in aerated tubes also encourages the sporulation (and subsequent identification) of the fungi capable of fruiting, and such studies can be used to give quantitative estimates of conidial production (Shearer and Webster, 1991).

Leaf mapping

If a previously submerged leaf is cut up into small squares, 6 x 6 mm, which are then incubated separately in a dilute salt solution to permit sporulation, the distribution of aquatic hyphomycete colonies within the leaf can be mapped (Figure 9.9) (Shearer and Lane, 1983). Chamier, Dixon and Archer (1983) have used a similar technique in which 2 x 2 mm squares excised from repeatedly washed strips (transects) of *Alnus* leaves were plated on to dilute agar media. After sporulation the distribution of fungi in the transects could be reconstructed.

Particle plating

Identifications can be made from leaf fragments plated on to dilute non-selective media. The proportion of fragments which give rise to identifiable mycelia can be used as a measure of leaf colonization. The method used by Chamier, Dixon and Archer (1983) described above is effectively a particle plate technique although the particle size is rather large. Bärlocher and Kendrick (1974) used squares of about 1 mm side to follow colonization of ash, maple and oak leaves. Kirby, Webster and Baker (1990) used much smaller particles in the size range 212–700 μm obtained by passing homogenized material of the aquatic macrophyte *Ranunculus penicillatus* through sieves. The frequency of isolation of fungi increased with increasing particle size.

Cultures

Ingoldian hyphomycetes grow well in agar culture on a range of ordinary laboratory media. They appear to have no unusual nutritional requirements, many being able to utilize simple carbohydrates, and some can degrade cellulose and other polysaccharides. They can also utilize amino acids either as a carbon or a nitrogen source, and can immobilize dissolved nitrogen salts from stream water (Thornton, 1963, 1965; Suberkropp and Klug, 1981). In pure culture, sporulation normally only occurs when culture pieces are immersed in water, and spore production is greatly increased if culture pieces in water are agitated by compressed air or shaking (Webster and Towfik, 1972; Webster, 1975). Increasing the flow rate of water over cultures also stimulates sporulation in many cases (Sanders and Webster, 1980). These conditions simulate the turbulence of their normal habitats.

For many aquatic hyphomycetes from temperate streams, growth in culture is often better at low to moderate temperatures (around 15°C) than at higher temperatures (see further). Prolonged incubation of culture pieces in sterile water under diffuse light at low temperatures may result in the development of teleomorphs (Webster, 1992).

Spores in foam

Foam which accumulates near barriers in turbulent streams is a rich source of detached conidia (Figure 9.7) and from such material it is possible to make isolations (Descals, Webster and Dyko, 1977). Foam samples can be preserved for subsequent examination by the addition of fixative or by smearing on microscope slides and staining, and can be used to make comparisons between the aquatic hyphomycete spora of different river

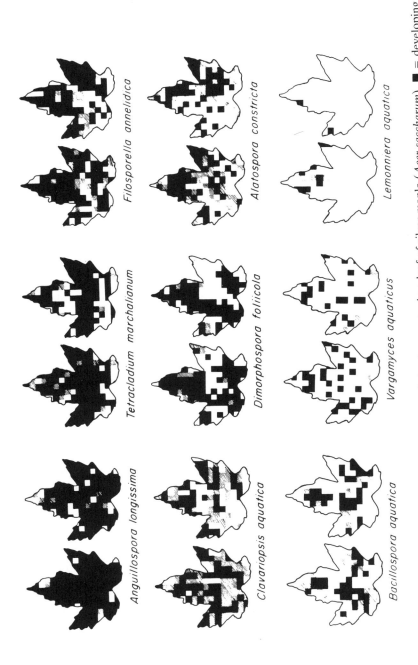

Figure 9.9 Leaf maps showing the distribution of aquatic hyphomycetes in a leaf of silver maple (*Acer saccharum*). ■ = developing conidia; ▨ = loose conidia. The left hand map of each pair represents the upper surface and the right hand map represents the lower surface. (Reprinted by permission, from Shearer and Lane, 1983, *Mycologia*, **75**, 498–508, The New York Botanical Garden.)

systems or to estimate species frequency of aquatic hyphomycetes throughout the year (Nilsson, 1964; Willoughby and Archer, 1973; Chauvet, 1991, 1992). However, foam samples may be unrepresentative compared with those obtained by Millipore filtration (see below). Branched spores are overrepresented as compared with spores of sigmoid and other shapes (Iqbal and Webster, 1973a).

Millipore filtration

Determinations of conidial concentrations can be made by filtering river water through a membrane filter (e.g. a Millipore filter) 5–8 μm pore size. A known volume of water is filtered in the field using a suction pump. The filter is stained with lactic acid cotton blue or lactic acid fuchsin which kills and stains the spores and renders the filter semitransparent (Iqbal and Webster, 1973b).

b) Ecological studies

When an autumn-shed leaf falls into a stream, it undergoes a series of changes resulting from physical, chemical, mechanical and biological action. Decomposition in the stream has been subdivided into three phases: **leaching**, **microbial colonization** and **invertebrate feeding** (Cummins, 1974). Leaching of soluble materials (carbohydrates, amino acids and phenolic compounds) may be quite rapid (Suberkropp, Godshalk and Klug, 1976), and 24–28 h immersion may result in the loss of up to 25% of the original weight (Webster and Benfield, 1986). If the leaf has been shed in the normal way by abscission after maturity, it will contain mycelia of several common terrestrial and phylloplane fungi such as species of *Aureobasidium, Cladosporium, Epicoccum, Alternaria*, etc. (Chapter 4). In the water, the leaf is rapidly colonized by the conidia of aquatic hyphomycetes which, in the autumn, may reach very high concentrations of over 1000 spores l^{-1} (see below). The original 'terrestrial' leaf inhabitants persist for a time, but are gradually replaced by aquatic fungi (Bärlocher and Kendrick, 1974; Chamier, Dixon and Archer, 1983). Within a few weeks, a mosaic of colonies of aquatic hyphomycetes is established (Figure 9.9). This early association is made up of a relatively small number of species (about 5 or 6), possibly resulting from competitive interactions (see later). It has been suggested that at an early stage of colonization, a leaf receives an 'imprint' from the stream spora which determines the dominant members of the fungal community throughout its decay (Bärlocher and Schweizer, 1983). Fungi account for 63–95% of the microbial biomass in decomposing submerged leaves of *Platanus, Quercus* and *Ulmus* (Findlay and Arsuffi, 1989). Bacteria play only a minor role in the early stages of decomposition. They increase in numbers with increasing time of submersion, possibly as a result of increased exposure of surface area of litter.

Aquatic hyphomycetes show little resource specificity, and many species can be collected on a wide range of different plant materials. However, certain species do show **resource preference**. Preference can be shown in a variety of ways, e.g. by more frequent occurrence on particular leaf species, on standard areas of different leaf baits in packs in the same stream, or by different levels of spore output. Inability to colonize a given kind of substratum may result from the presence of barriers to infection, such as a thick cuticle and epidermis, or to other anatomical features such as well-developed sclerenchyma or other lignified tissues, or to chemical factors such as the nitrogen content of the leaf or the presence of inhibitors such as phenolics. *Tetracladium marchalianum* (Figure 9.8) showed a higher colonization frequency and importance index on discs cut from leaf packs of hickory (*Carya glabra*) than on discs of oak (*Quercus alba*) submerged in Augusta Creek, Michigan, USA. *Triscelophorus monosporus* showed the opposite effect (Suberkropp, 1984). In comparisons between beech (*Fagus sylvatica*) and alder (*Alnus glutinosa*), it has been shown that *Tetracladium marchalianum* is abundant on alder leaves, but absent from beech leaves which were dominated by *Tricladium splendens* (Figure 9.8) (Bengtsson, 1983; Gönczol, 1989). The rate of mycelial growth of *T. marchalianum* inoculated onto beech leaves was very low, and no protease activity could be detected (Bengtsson, 1983). Possibly the metabolism of *T. marchalianum* is affected by inhibitors such as phenolics which complex with proteins, reducing the availability of leaf protein and inhibiting enzyme activity. Thomas, Chilvers and Norris (1992a) compared the frequency of occurrence of aquatic hyphomycetes on discs cut from randomly selected leaves of *Eucalyptus viminalis* and phyllodes of *Acacia melanoxylon* in an Australian stream. There was clear evidence of preference. *Alatospora acuminata* was recorded more frequently on *Acacia*, whilst *Tetrachaetum elegans* (Figure 9.8) was more common on *Eucalyptus*.

Coniferous leaves were thought to be poor substrata for aquatic hyphomycetes (Ingold, 1966), but later studies showed that they do become colonized, especially after prolonged submersion in a stream. The thick cuticle and epidermis are barriers to infection. If the cuticle is removed by steam or chemical treatment, the leaves become much more extensively colonized, as do leaves which are sliced longitudinally. Coniferous leaves also contain phenolic substances which inhibit the growth of aquatic hyphomycetes. These are eventually leached as the inner leaf tissues become exposed (Michaelides and Kendrick, 1978; Bärlocher and Oertli, 1978a,b; Bärlocher, Kendrick and Michaelides, 1978).

Woody materials, varying in size from small twigs to large branches or even whole tree trunks, fall into streams and become exposed to colonization by the aquatic hyphomycete spores. The fungal community which develops on woody substrata is similar, in part, to the leaf community, but is nevertheless distinctive (Révay and Gönczol, 1990). Shearer (1992) has reviewed the role of woody debris in the life cycles of aquatic hyphomycetes. One of the most important features is its longer persistence in a stream, sometimes extending over several years. In comparison, leaf material may only persist for a period of weeks, and by the time a crop of deciduous tree leaves is shed in the autumn, the previous year's crop has long since disappeared. Thus wood may represent an important reservoir of inoculum when other substrata are scarce. Living roots of riparian trees may also be another source of continuously available inoculum (Fisher, Petrini and Webster 1991; Sridhar and Bärlocher, 1992a). Another important property of woody materials is that they support the development of the teleomorphs of aquatic hyphomycetes, on which long-distance transport and sexual recombination may depend (Webster, 1992). Compared with the species known to be present in streams (for example, as spores in foam), the number of species which fruit on wood is smaller, i.e. wood selects only a fraction of the fungi available (Willoughby and Archer, 1973; Sanders and Anderson, 1979). Experimental studies based on submerged twig packs give little evidence of host preference, e.g. in a comparison between beech and alder (Révay and Gönczol, 1990) or oak and alder (Shearer and Webster, 1991). Shortly after immersion, corticated hardwood twigs develop pustules of *Heliscus lugdunensis, Cylindrocarpon* and *Fusarium*, but as the bark is shed, these early colonists disappear. It is likely that substances present in bark (e.g. of oak and alder) inhibit growth and development of many aquatic hyphomycetes, apart from the few species which develop in profusion on the bark of fresh twigs. Fungi fruiting late on wood include *Anguillospora crassa* (Figure 9.8) which is not common on leaves.

Table 9.3 Development of the fungal community on oak wood blocks of different sizes

Block size		Exposure period (weeks)					Total no. of species
mm	Relative area	2	4	6	8	13	
5 x 5	1	0[a]	1	3	6	6	6
10 x 10	4	0	4	4	7	6	9
20 x 20	16	2	3	5	10	–	12
40 x 40	64	4	9	8	11	–	16
80 x 80	256	6	15	15	13	–	19

[a] Number of species.
Source: Sanders and Anderson (1979).

Table 9.4 Distribution of species on oak blocks of different sizes (numbers in the table are number of blocks bearing each species, maximum 20)

Species	Block size (mm)					Total frequency (%)	Mean concentration of conidia $(l^{-1})^a$
	5	10	20	40	80		
Flagellospora curvula	11	11	11	9	17	59	396
Tricladium gracile	6	6	7	4	9	32	51
Dendrospora erecta	2	4	5	6	11	28	32
Casaresia sphagnorum	4	4	4	2	7	21	26
Articulospora tetracladia	2	1	1	6	11	21	410
Tricladium splendens	2	1	3	4	8	18	21
T. giganteum	–	3	7	5	9	24	13
Varicosporium elodeae	–	1	–	4	11	16	26
Lemonniera terrestris	–	1	1	2	6	10	0
Taeniospora gracilis	–	–	1	1	9	11	627
Anguillospora crassa	–	–	2	1	5	8	1
Alatospora acuminata	–	–	2	1	4	7	77
Tetrachaetum elegans	–	–	1	2	4	7	26
Anguillospora longissima	–	–	–	4	12	16	32
Volucrispora graminea	–	–	–	3	2	5	6
Clavatospora longibrachiata	–	–	–	2	1	3	621
Lemonniera aquatica	–	–	–	–	4	4	0
Flagellospora penicillioides	–	–	–	–	2	2	51
Tetracladium marchalianum	–	–	–	–	1	1	0
Mycocentrospora angulata	–	–	–	–	–	0	294

[a] 5 x 1 l samples over 8 weeks.
Source: Sanders and Anderson (1979).

Sanders and Anderson (1979) immersed square blocks of oak wood of varying size – 5, 10, 20, 40 or 80 mm sides – to study the effect of resource size on colonization by aquatic hyphomycetes. The blocks were removed from the stream at intervals from 2 to 13 weeks, and incubated in water for 3–7 days to allow aquatic hyphomycetes to sporulate. Filtration of the stream water during the exposure period provided figures for the concentration of spores in suspension. The numbers of species found on the different sizes of block are shown in Table 9.3.

Three interesting points emerge from these results. First, there is a delay in appearance of fungi on the smaller blocks. Secondly, there is an increase in the number of species appearing with increasing exposure. Thirdly, the number of species present is related directly to block size. This last conclusion indicates that resource size has a strong influence on the establishment of the species of aquatic fungi on it. Bärlocher and Schweizer (1983) reached a similar conclusion in studying the colonization of squares of different size cut from *Quercus* and *Acer* leaves. The species identified on the blocks of different size are listed in Table 9.4. The frequency of colonization of the blocks bears no relation to the measured concentration of the spores of different species in the stream. This is well shown by the fact that *Lemonniera terrestris*, which was found in a low proportion of blocks of 10 mm side or larger, was not detected by filtration, whilst *Mycocentrospora angulata*, which was detected at concentrations of 294 conidia l^{-1}, did not fruit on the block, i.e. it may be taxon-specific for reasons already described. The most frequently recorded fungus, *Flagellospora curvula*, which has sigmoid conidia, is probably less efficiently trapped than the tetraradiate conidia of some other fungi. The reason for the low number of species on the smaller blocks may be that interspecific competition in the smaller volume of wood may occur early, and may limit colony development and sporulation. This speculation, however, needs experimental investigation (see later).

The studies reported above show that there is a succession of fungi fruiting on submerged leaves. The detailed picture is dependent on the kind of leaf involved, the time (season) of submersion in the stream, which will in turn be related to the range of temperature to which the submerged leaves are subjected, and the abundance and species composition of spores available in the stream during the period of submergence. On leaves which are rapidly decomposed and consumed by invertebrates, such as those of alder, successional changes are more difficult to detect than on more persistent leaves such as those of oak (Chamier and Dixon, 1982a). Fungi which fruited early on leaf pack material of oak and hickory were *Flagellospora curvula*, *Lemonniera aquatica* and *Alatospora acuminata* (Suberkropp and Klug, 1976). Similar findings are reported by Gessner *et al.* (1993) from alder leaf packs submerged in autumn in the River Touyre in the French Pyrenees. Within 2 weeks of submergence, the leaves supported densely sporulating colonies of *Flagellospora curvula*, along with four other species, *Lemonniera aquatica*, *L. centrosphaera*, *L. terrestris* and *Tetrachaetum elegans*. These species persisted throughout the decay of the leaves, but after 4 weeks the community included 13 abundant species (with a frequency greater than 10%). Species which appeared late in the succession included *Clavatospora longibrachiata*, *Heliscella stellata* and *Goniopila monticola*. The succession was thus characterized by three stages: an assemblage of five pioneer species, a mature community of about 13 species and an impoverished successional stage with a reduced species diversity.

The arrival of additional species into the leaf and wood community, and the displacement of one dominant group of fungi by others, coupled with the reported patchy distribution of fungi colonizing leaves (Shearer and Lane, 1983; Chamier, Dixon and Archer, 1983) indicate some form of competition. It has also been claimed that fungi fruiting on submerged wood 'occur as discrete, usually dense, monospecies patches' (C.A. Shearer, unpublished). Khan (1987) examined the interactions of eight aquatic hyphomycetes when paired together on agar media, but found no evidence of interspecific competition inhibition at a distance or of hyphal interference. There were, however, differences in growth rate as the colonies approached each other, and it thus seems likely that more rapidly growing species might deprive slower growing forms of nutrients. Competitive interactions between 25 species of aquatic fungi normally occurring on leaves or on wood (or sometimes both) have been studied on agar by Shearer and Zare-Maivan (1988). The fungi included ascomycetes and fungi imperfecti. A wide range of response was reported. Aquatic hyphomycetes which ranked high in ability to inhibit the growth of other aquatic fungi were *Clavariopsis aquatica* (23 species inhibited), *Anguillospora* cf. *gigantea* and *Tetracladium marchalianum* (22). Lower ranking species included *Margaritispora aquatica* and *Triscelophorus monosporus* (20), *Heliscus lugdunensis* (16) and *Clavatospora longibrachiata* (9). These results are difficult to extrapolate to the field situation, but the least inhibitory fungi were those characteristic of leaves, whilst the more inhibitory species were wood inhabitants. These findings support the idea that persistent and late-colonizing fungi on long-lasting substrata are more likely to produce antagonistic substances than those on less persistent substrata.

Massarina aquatica, an aquatic loculoascomycete which fruits on wood and submerged tree roots, is the teleomorph of *Tumularia aquatica* (see Figure 9.8). This conidial state grows on submerged wood, but also on submerged leaves. Fisher and Anson (1983) tested the ability of diffusates from oak wood blocks colonized by *M. aquatica* to inhibit germination and growth of other fungi, by placing colonized blocks on water agar to allow leachates to diffuse into the agar. Germ tube production of *Tricladium giganteum* was significantly reduced close to the edge of the block and the growth of other fungi was also inhibited. It is thought that *M. aquatica* can produce one or more antifungal antibiotics.

In temperate and cool-temperate latitudes, streams bordered by deciduous trees show enormous variations in seasonal abundance of conidia of aquatic hyphomycetes directly related to the addition of autumn-shed leaves. In areas where litter input is spread over a longer period, the seasonal variations in abundance are less pronounced, e.g. in parts of Australia where litterfall in *Eucalyptus* forest may reach a maximum in the late-summer to early-autumn months, December–March (Thomas, Chilvers and Norris, 1989; Thomas, 1992b). In a stream in which needles of spruce (*Picea abies*), which are shed evenly throughout the year, dominated the terrestrial input, there was no clear peak in conidial concentration during the year (Bärlocher and Rosset, 1981).

The fall of tree leaf litter into a stream is quickly followed by a rapid rise in spore concentration in the water as shown by filtration. Chamier and Dixon (1982a) have estimated that the numbers of spores which can be produced per gram oven dry weight of leaf tissue are c. 140 000 for *Quercus* and *Alnus*. Even higher values have been claimed by Bärlocher (1982) who estimated that leaves of *Quercus* and *Larix* could produce 5–6 x 10^6 conidia g^{-1} within a 2-day period. Suberkropp (1991) has estimated that *Lunulospora curvula* allocates 60–80% of its biomass to sporulation, and *Anguillospora filiformis* 30–45%. Such prolific production is a ruderal trait, probably correlated with growth on relatively ephemeral substratum and the low chance that an individual spore has of being trapped on a suitable underwater substratum in a rapidly moving water current. Possibly rapid and prolific reproduction is an adaptation to minimize the effects of predation by aquatic invertebrates which graze on leaves colonized by aquatic hyphomycetes.

Spore concentrations between 10^3 and 10^4 l^{-1} have been reported by Iqbal and Webster (1973b) for a lowland stream in Devon, England, during October–December (Figure 9.10), and these autumnal values have been matched or exceeded by other workers elsewhere. In the following summer, as the leaf litter is decomposed, consumed by invertebrates, fragmented and swept downstream, the spore concentration may fall too low to be detected by filtration. Nevertheless, sufficient inoculum is available to infect leaves in the autumn, and it is possible that materials such as bud-scales, catkins, etc., falling in spring and early summer, contribute, as do colonies growing on woody material and riparian tree roots.

The frequency pattern of the total spore production masks variations in frequency of individual species. The seasonal abundance of the spores of *Clavariopsis aquatica* mirrors the total abundance of spores. Here the availability of leaf litter is probably the major factor controlling its abundance. For some other species, other factors are also important. Figure 9.10 shows that the peak concentrations of conidia of *Tricladium chaetocladium* are in the winter months from December to March in the River Creedy. A contrasting pattern of abundance is shown by *Lunulospora curvula*, detected in the River Creedy only from August to November. The suggested explanation that these different periods of spore production are controlled by temperature was explored by Webster, Moran and Davey (1976), who found differences in temperature optima for growth and sporulation, *L. curvula* having higher optima.

Experiments designed to separate the effects of temperature and litter availability have been reported by Suberkropp (1984), who immersed leaf

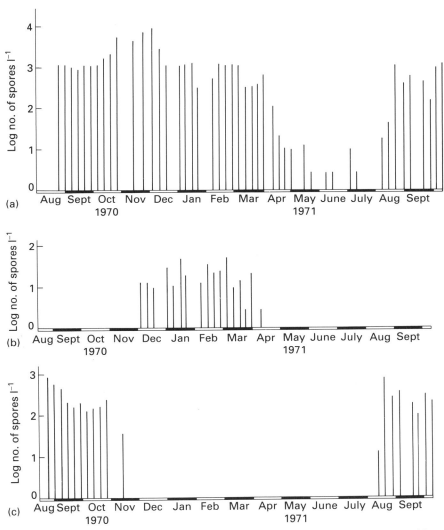

Figure 9.10 Changes in spore concentrations of aquatic hyphomycetes detected by Millipore filtration of the River Creedy, Devon, England, 1970–71. (a) Total spores; (b) spores of *Tricladium chaetocladium*; (c) spores of *Lunulospora curvula*. (Based on data from Iqbal and Webster, 1973b.)

packs of oak (*Quercus*) and hickory (*Carya*) for 4-week periods at different points in the stream characterized by different temperature ranges. Interchanges of packs colonized for a short period at one point in the stream before transfer to another point were also made. A summer assemblage of species dominant on leaf packs during the warmer months could be recognized: *Lunulospora curvula*, *Flagellospora penicillioides*, *Triscelophorus monosporus* and *Heliscus tentaculus*. *Flagellospora curvula* and *Lemonniera aquatica* were dominant on both leaf types in the colder months. *Alatospora acuminata* and *Tetracladium marchalianum* increased in importance as the water temperature increased. Transfer of packs briefly submerged at a warmer site to allow them to be colonized by members of the summer assemblage, *L. curvula* and *F. penicillioides*, then transferred to a cooler site, showed that both species could survive and sporulate in the cooler site.

Temperature affects different phases of the life cycle of a fungus: germination and infection, mycelial growth and sporulation (Koske and Duncan, 1974). Laboratory experiments show that species more abundant in cooler months have different temperature characteristics from species from the summer assemblage. For example, *L. curvula*, *H. tentaculus* and *F. penicillioides* only grow at temperatures above 5°C, whilst *F. curvula* and *Lemonniera aquatica* are able to grow at 1°C (Suberkropp, 1984; Suberkropp and Klug, 1981).

The San Marcos River in Texas is unusual in having a relatively constant temperature of 22°C ±1°C. Here some, but not all, of the fungi present are the same as those reported to belong to the summer assemblage of Suberkropp, e.g. *Lunulospora curvula* and *Triscelophorus monosporus* (Akridge and Koehn, 1987).

By contrast, a relatively constant winter temperature of 0.1°C occurs in the Njakajokk stream in the Swedish Arctic (68°N) from the end of October to May. For some of this period, the stream is covered with ice. Three species dominate the spore output during the year: *Flagellospora curvula*, *Lemonniera aquatica* and *Alatospora acuminata*, all of which are members of the cold water assemblage of Suberkropp (1984).

Temperature is also related to latitudinal and altitudinal distribution. A group of tropical and subtropical species has been recognized (Nilsson, 1964; Webster and Descals, 1981; Sridhar, Chandrashekar and Kaveriappa, 1992). Nawawi (1985) has listed 25 species which are tropical or subtropical in distribution, including *Brachiosphaera tropicalis*, *Campylospora chaetocladia*, *Heliscus tentaculus*, *Flagellospora penicillioides*, *Lunulospora curvula*, *Triscelophorus acuminatus*, *T. monosporus* and *Ingoldiella hamata*. All these species occur in Malaysia but some also occur in temperate climates. It is interesting that the following, all common in temperate latitudes, have not been found in Malaysia: *Tricladium splendens*, *Anguillospora*, *Culicidospora*, *Dendrospora* and *Lemonniera*.

Chauvet (1991) has surveyed the distribution of spores of aquatic hyphomycetes at 27 stations in south-western France based on foam samples collected at intervals throughout the year. The altitude of the stations varied between 9 and 935 m, with a pH range of 5–8.5 and a water temperature range of 2–20°C. Correspondence analysis was used to examine the relationship between distribution patterns and abiotic variables such as altitude, pH, temperature and season. A group of five species associated with lowland streams, high pH and high temperature was identified: *Campylospora chaetocladia*, *Campylospora* sp., *Heliscus tentaculus*, *Lunulospora curvula* and *Triscelophorus monosporus*. Two species, *Clavatospora longibrachiata* and *Tetrachaetum elegans*, appeared characteristic of acid water (pH <6), low altitude and autumn months. *Tetracladium marchalianum* and *Tricladium angulatum* were typical of lowland streams, whilst *Taeniospora gracilis* and *Tricladium chaetocladium* were found in mountain streams. Water pH appeared to be of secondary importance as compared with altitude.

Generally, where the gathering grounds for river systems are in mountainous or upland areas, as the streams descend into lowland and the tributaries merge to form larger rivers, there are associated changes in the instream and riparian vegetation and in chemical and physical characters of the water (Hynes, 1970; Chamier, 1992). These include changes in pH, conductivity, water hardness, dissolved oxygen, dissolved organic matter, flow characteristics, and amount and quality of trapped litter and sediment. Water quality is markedly affected by flow through forests of different type (coniferous litter results in low pH), and flow through agricultural areas may result in increase in dissolved nutrients through fertilizer run-off and leaching. Flow through populated areas may result in industrial or sewage pollution. As rivers approach the sea, they come under the influence of tides. There are also changes in the range and abundance of different animals within the length of a river. Bärlocher (1992b) has described variations in physical, chemical and biological parameters along river systems, all of which may affect community structure and activity of aquatic fungi.

Shearer and Webster (1985a,b) studied variation in spore concentration and in the colonization of alder leaf packs submerged at different points along the River Teign in Devon, England. The upper reaches arise in moorland where the water is acidic (pH 5.4–6.0). There are hardly any riparian trees, although there are several instream macrophytes. The main source of allochthonous litter is culms of a rush (*Juncus effusus*). The lower sites are characterized by higher pH (7.0–7.2), and are bordered by various trees whose detached leaves and branches are trapped amongst boulders. The upper (moorland) stretch of the river had much lower conidial concentrations, probably related to the scarcity of allochthonous litter. Species numbers increased in a downstream direction, and this might be associated with the greater variety of available substrata, or to features of water quality including higher pH. The community of aquatic hyphomycetes at the upland site was distinctly different from the downstream communities and less than 20% of all species were found on leaves at all the sites. *Fontanospora eccentrica* and *Varicosporium elodeae* were both common at certain times in the upland site, but were either absent or infrequent lower down. In contrast, *Tricladium splendens* and *Dimorphospora foliicola* showed the opposite pattern of distribution. The conidia of *F. eccentrica* and *V. elodeae* were present only in low concentration at the downstream sites. For example, the conidia of *F. eccentrica* made up 10–15% of the conidial pool at the upstream site and only 0–0.1% lower down. This suggests local production at the upstream site, and rapid extinction (i.e. loss due to trapping or dilution) as the spores are carried downstream. The idea is supported by the work of Metwalli and Shearer (1989), who studied species distribution and spore concentration in an Illinois stream whose banks were bordered by alternately clear-cut or wooded sections, using oak leaf packs. The concentration of conidia was high in the wooded areas but fell in the clear-cut areas downstream. The mean number of species was also lower in the clear-cut areas.

Thomas, Chilvers and Norris (1991a) have discussed factors related to changing spore concentrations in a stream and have provided a mathematical model (Thomas, Chilvers and Norris, 1991b) to explain the dynamics of spore populations in a body of moving water. The model enabled them to calculate the half-life of populations in terms of the distance downstream that the population would be halved in concentration. For three fungi tested, the half-lives gave similar values: *Alatospora acuminata* 0.69 km, *Clavariopsis aquatica* 0.78 km, and *Tetrachaetum elegans* 0.81 km. A good fit between simulated data and real data was obtained.

The above reports suggest that in a given river system, species number may be correlated with rising pH. When data from streams in Europe and Canada are correlated, however, this simple relationship becomes less

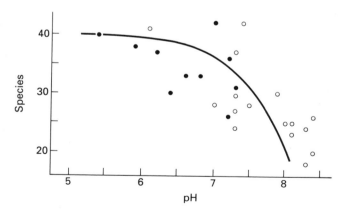

Figure 9.11 Plot of numbers of species of aquatic hyphomycetes recorded from streams of different pH based on data from Bärlocher (1987) ● and Wood-Eggenschwiler and Bärlocher (1983) ○. (Redrawn from Bärlocher, 1987 with permission.)

clear (Bärlocher and Rosset, 1981; Wood-Eggenschwiler and Bärlocher, 1983; Bärlocher, 1987). Figure 9.11 is a plot based on records from Europe and Canada, and shows a slow, possibly negligible, decrease in species in circumneutral waters (pH 5.7–7.2) with a much more rapid decline in alkaline waters (pH > 7.2) (Bärlocher, 1987).

The effects of pH on the abundance and activity of aquatic fungi are complex since H+ concentration is related to the abundance of plants (as possible substrata) and animals (as possible consumers). There are also features of water chemistry such as water hardness, alkalinity and the solubility of Ca and Al which are pH-dependent (Chamier, 1992). At low pH (<5.5), ionized forms of Al^{3+} predominate in solution, and some of these are toxic. In acid streams, submerged litter significantly accumulates Al more rapidly than in circumneutral streams and decomposition is slower (Chamier, Sutcliffe and Wishman, 1989).

Aquatic hyphomycetes possess an array of enzymes, including pectinases, capable of macerating leaf tissues (Chamier and Dixon, 1982a,b; Chamier, 1985; Zemek et al., 1985; Abdullah and Taj-Aldeen, 1989). The pectolytic enzymes are made up of several species of enzyme with different pH optima. It is therefore possible that the degradative activity of aquatic hyphomycetes is directly related to the pH of streamwater. An indirect effect is through the increasing availability of Ca^{2+} above pH 6.5. Increasing Ca^{2+} concentration increases pectolytic activity of *Tetrachaetum elegans* on a sodium polypectate medium at pH 6.5 and (in ecologically probable amounts) is also directly related to strength loss in alder leaf strips inoculated with the fungus. Stimulation of pectic enzyme activity of aquatic hyphomycetes by Ca^{2+} may in part explain the more rapid maceration of leaf tissues recorded in streams above pH 6.5 (Chamier and Dixon, 1983).

Where rivers flow into the sea, the organisms which thrive there must be able to tolerate variations in salinity. Observations of the distribution of aquatic hyphomycetes on wood blocks show that they do occur, even if they are not abundant in estuaries (Jones and Oliver, 1964). Byrne and Jones (1975a,b) have studied spore germination, vegetative growth and sporulation in two species, *Heliscus lugdunensis* and *Tetracladium setigerum*, in sea water at various dilutions or on agar to which sea water was added, at varying temperatures. Germination and growth occurred over a range of concentrations of sea water but sporulation was severely inhibited at quite low salinities, i.e. 30% sea water for *H. lugdunensis* and 10% for *T. setigerum*. These results indicate that freshwater aquatic hyphomycetes might be able to colonize substrata in brackish water, but would probably be unable to reproduce there.

The abundance of aquatic hyphomycetes in rapidly flowing, well-aerated streams suggests that they might be adapted physiologically to water with a high concentration of dissolved O_2. In stagnant habitats such as ponds and ditches, they are much less frequent, although *Tricladium splendens* is an exception. The mud surface of stagnant ponds, especially when carpeted by a layer of leaves, may rapidly become oxygen-deficient, and below the mud surface anaerobic conditions may prevail, as evidenced by the smell of hydrogen sulphide, which would rapidly combine with available oxygen. Even in rapidly flowing streams, accumulations of leaf and branch litter may become anaerobic if overlaid by sediments. Experimental studies indicate that different species of aquatic hyphomycetes vary in their capacity to survive anaerobic conditions. For example, the mycelium of *Articulospora tetracladia* failed to survive for 3 months, whilst a small proportion of colonies of *Tricladium splendens* and *Anguillospora rosea* survived for 1 year (Field and Webster, 1983). In similar studies in which beech leaf discs inoculated with aquatic hyphomycetes were incubated in buffer solutions at different concentrations of H_2S for 24 h, variation in survival was found, *A. tetracladia* being more sensitive than *T. splendens*. The aquatic hyphomycetes tested were more sensitive to H_2S than species of aero-aquatic fungi which are common in stagnant habitats (Field and Webster, 1985).

Bandoni (1981) has reviewed evidence that the conidia of some fungi commonly considered to be aquatic hyphomycetes may be found in terrestrial habitats. Bandoni has collected conidia from leaf litter remote from water and in water flowing down the stems of trees. There are also records of isolation from soil and from root surfaces, and as endophytes from roots in soil. The presence of spores in the habitats mentioned, or the isolation of a fungus from soil, does not prove that the fungus has active mycelium. However, more convincing evidence is available that some are active in terrestrial situations. Park (1982b) has shown that *Varicosporium elodeae*, which can be regularly found in the litter layer above soil, can colonize leaves or even filter paper buried in soil. Sanders and Webster (1978) have also shown that *V. elodeae* and *Articulospora tetracladia* can extend from colonized leaf discs sandwiched between sterile oak leaves buried in woodland litter, to infect the oak leaves. When leaves of oak which had been artificially or naturally colonized with certain aquatic hyphomycetes were hung in nylon bags above ground level, several species were capable of surviving for up to 13 weeks.

Since some of these so-called aquatic hyphomycetes are active and can survive in terrestrial habitats, and because some of them form teleomorphs which can release air-borne ascospores or basidiospores, it is probably best to regard them as amphibious fungi.

As indicated earlier, a series of chemical changes follows the immersion of tree leaves in streams, beginning with the rapid loss of soluble materials by leaching. Microbial colonization is accompanied by an absolute increase in protein content, first demonstrated by Kaushik and Hynes (1971) and later confirmed by many studies (see Bärlocher, 1985; Chauvet, 1987). The increase in protein largely results from fungal growth. There are also increases in other compounds, e.g. phosphate (Meyer, 1980). Physical changes also occur, notably the softening of leaf tissues as the result of the macerating activity of fungal enzymes (Suberkropp and Klug, 1980, 1981; Suberkropp, Arsuffi and Anderson, 1983; Chamier, 1985). These changes in the chemical and physical properties of submerged leaf litter have been termed 'conditioning'. There is much evidence that conditioned leaves are consumed in preference to non-conditioned leaves by leaf-shredding invertebrates such as the larvae of aquatic insects and *Gammarus* spp. Patches of leaf material colonized by fungi are preferred to uncolonized areas (Arsuffi and Suberkropp, 1985), and it is known that leaves colonized by certain fungal species are more attractive than others (Suberkropp, 1992). Aquatic invertebrates thus exert some control over the populations of aquatic hyphomycetes. In experiments in which leaf packs enclosed in fine- or coarse-mesh litter bags were immersed in streams, Bärlocher (1980) showed that the leaves in coarse-mesh bags, permitting the entry of invertebrates which fed on the leaves, had a lower cumulative species count of aquatic hyphomycetes than the leaves in the fine-mesh bags from which invertebrates were excluded. The rate of decay of the leaves in the coarse-mesh bags was greater than for the leaves enclosed in fine mesh. Thus the leaf-eating invertebrates are potent competitors of aquatic hyphomycetes. As well as eating fungal mycelium, they also remove substratum that might otherwise be used by fungi characteristic of later stages of succession.

There are several reasons why aquatic invertebrates preferentially ingest conditioned leaf material. Most do not possess gut enzymes capable of degrading plant polymers such as cellulose and lignin. Fungal enzymes transform plant polymers into digestible matter and their activity may continue in the invertebrate gut. The enrichment of the protein and probably the lipid content of leaf material associated with increase in fungal biomass renders conditioned leaf material more nutritious, whilst the macerating activity brought about by fungal enzymes makes it easier to ingest. For these reasons, animals fed on conditioned leaf material grow more rapidly than those fed on non-conditioned leaves, and expend less energy in foraging for food. Thus, although in most streams the bulk of the carbon and energy inputs is derived from allochthonous riparian tree leaf litter, this material is largely unavailable to aquatic invertebrates unless first colonized by microorganisms of which the aquatic hyphomycetes appear to be the most important (Bärlocher and Kendrick, 1976, 1981).

9.2.4 Aero-aquatic hyphomycetes

If leaves and twigs from the mud surface of a stagnant pond or slow-running ditch are washed and incubated in a moist chamber for a few days at room temperature, a remarkable array of large conidia develop. From the mycelium within the substratum, conidiophores extend into the air, giving rise to various kinds of buoyant, non-wettable propagules. In most propagules, air is entrapped. In *Helicoon* and *Helicodendron* (Figure 9.12), a cylindrical or barrel-shaped propagule is formed by the tight coiling of the developing conidium, whilst in *Clathrosphaerina* a hollow, clathrate sphere develops from a system of repeatedly dichotomous branches whose tips bend towards each other and touch (Figure 9.12). The propagules of *Aegerita candida* resemble a raspberry fruit, and are made up of inflated, clamped segments, between which air is trapped (Figure 9.12). *Cancellidium applanatum* has a flattened, gas-filled bag made up of parallel hyphae, and *Fusticeps bullatus* has a club-shaped, septate spore whose outer wall bears stud-like scales between which air is held.

Propagules of this kind are not developed in water, but only in air. In nature, they develop at the margins of drying ponds and ditches. When a dried-up pond is flooded again, the conidia float off and are dispersed. Premdas and Kendrick (1991) showed that leaves of *Acer saccharum* gently lowered onto the surface of a Canadian pond quickly acquired propagules of aero-aquatic fungi which had been floating in the surface film. The time between initial propagule attachment, colonization and subsequent sporulation can be as little as 1 week. These fungi are not confined to stagnant water habitats: they can be found on wood submerged in streams and rivers, moist leaf litter, and soil. The teleomorphs, where known, are inoperculate Discomycetes, Pyrenomycetes and Basidiomycotina with relatively simple fructifications formed on wet wood. It is obvious that the aero-aquatic way of life has evolved independently in unrelated fungal groups. Accounts of their morphology and ecology have been given by Webster and Descals (1981) and Fisher and Webster (1981). Goos (1987) has given an account of helicosporous forms.

Techniques for studying the ecology of aero-aquatic fungi in the laboratory and the field have been devised by Fisher (1977). The propagules germinate readily on suitable media, and sporulation occurs on dilute laboratory media such as 0.1% malt extract agar. Spores can be wetted by detergent and can then be used as inoculum. A valuable technique in studying survival or colonization is to inoculate large numbers of small discs cut out from leaves of beech (*Fagus sylvatica*). Such discs can be colonized and placed amongst leaf litter at the mud surface in ponds or in the laboratory. When the discs are removed from the field site, they are placed on moist filter paper, and the proportion of them which gives rise to sporulating colonies can be assessed. The use of freshly abscised leaves (brown) or leaves which had been submerged in water for some time (black), from which to prepare discs, enabled Fisher (1979) to compare the capacity of different aero-aquatic fungi to colonize fresh and partially decayed leaf discs in a eutrophic and an oligotrophic site. In general, the

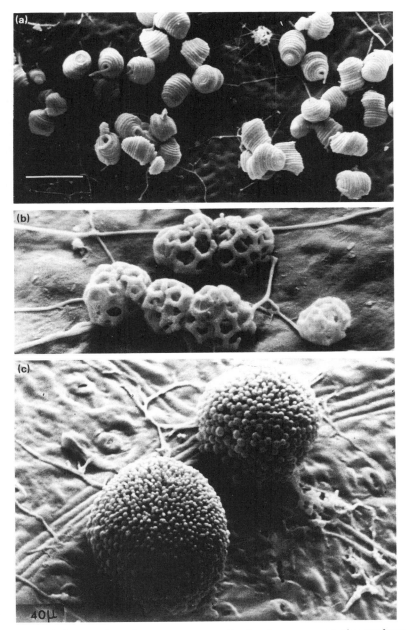

Figure 9.12 Propagules of some aero-aquatic fungi as seen on the surfaces of leaves. (a) *Helicodendron giganteum*. (b) *Clathrosphaerina zalewskii*. (c) *Aegerita candida*. Scale bar = 100 μm. (Scanning electron micrographs by P.J. Fisher.)

brown leaves were colonized more readily than the black ones, but in the oligotrophic habitat which was fairly well aerated, two species, *Helicodendron giganteum* and *Clathrosphaerina zalewskii*, were able to colonize brown or black leaf material equally well. Premdas and Kendrick (1991) found variations in the preference of different species of aero-aquatic fungi for freshly abscised (3–4 weeks) and 1-year-old leaves. In most cases, the fungi studied colonized freshly abscised leaves preferentially.

The mud surface of a pond containing large amounts of leaf litter can frequently become anaerobic, especially during the summer months, resulting from respiratory oxygen demand and from the lowered solubility of oxygen at higher temperatures. The saturation concentration of oxygen dissolved in water is about 9 mg l^{-1} (or ppm), but values in the range 0.1–0.6 mg l^{-1} are often detected immediately above the mud throughout the year, falling to zero in summer. Only a few millimetres beneath the surface, the mud may be anaerobic for much of the time, indicated by the black colour, largely due to iron sulphide (Fe_2S), and by the smell of hydrogen sulphide (H_2S). The growth and survival of aero-aquatic fungi under conditions of low oxygen availability is therefore of interest. Observations on leaves recovered from mud which had been anaerobic for several months show that after a few days' incubation in moist air conidia of aero-aquatic fungi develop. Similar results have been obtained after submerging beech leaf discs in anaerobic mud for 1 month during the summer, showing that colonization can occur under these conditions, probably from mycelium (Fisher, 1977). Incubation of the recovered discs in aerated water for a few days results in an increase in the number of species sporulating after incubation in air. This suggests that a period of vegetative growth under more aerobic conditions may permit the sporulation of some fungi which would otherwise not have been detected.

The survival of aero-aquatic fungi on beech leaf discs maintained in the laboratory under strictly anaerobic conditions has been studied by Field and Webster (1983). Five species of *Helicodendron* showed almost 100% survival for 6 months, and all species showed some survival for 12 months. The capacity of aero-aquatic fungi to survive anaerobiosis was significantly better than that of the Ingoldian hyphomycetes or saprolegniales tested. There is no evidence that aero-aquatic fungi can actually grow, i.e. increase the weight of mycelium, under anaerobic conditions. In experiments in which mycelia of four aero-aquatic fungi were grown in liquid beech leaf decoction equilibrated to different partial pressures of gases, the best mycelial growth was found in liquids equilibrated with air (160 mm O_2). Growth at lowered partial pressures of O_2 was significantly reduced (Fisher and Webster, 1979). Survival under anaerobic conditions probably occurs by means of thick-walled hyphae. The occurrence of sclerotia or chlamydospores in this group of fungi is rare.

Similar comparative studies have also been made of survival, under anaerobic conditions, at low concentrations of H_2S, a gas which is toxic to many organisms in low concentrations. The aero-aquatic fungi proved to be the most tolerant (Field and Webster, 1985).

Aero-aquatic fungi are found primarily in freshwater habitats, but in estuaries they may be subjected to variations in salinity. Several species can germinate and grow at high salinity levels, so that their relative scarcity in the lower reaches of tidal rivers may possibly be due to their inability to compete with other organisms in the colonization of suitable substrata, rather than intolerance of salinity (Field, 1983).

Fisher (1978) has investigated the ability of mycelium and spores of aero-aquatic fungi to survive desiccation. Homogenized, sterilized beech leaf debris was inoculated with mycelium and, after 7 weeks, the colonized debris was air-dried. The dried material was placed on garden soil. Of the eight species tested, all could survive in this form for over a month, and three species survived for 10 months, despite colonization of the leaf debris by other organisms. In most species, similar material survived drying in a desiccator for 1 month, and *Helicodendron triglitziense* survived for 9 months at a relative humidity of 4% and a moisture content of around 3%. The conidia were much less tolerant of desiccation. Conidia of six of the eight species retained the capacity to germinate after being in a desiccator for 10 days, but none survived for 20 days.

The aero-aquatic fungi are ubiquitous in appropriate habitats, although possibly more abundant in temperate and colder regions than in the tropics. Long-distance transport between isolated water bodies is possibly brought about by ascospores and basidiospores in those species which have teleomorphs. The buoyant propagules may aid dispersal during periods of flooding following a period of dry weather sufficient to allow mycelia from previously submerged substrata to sporulate. They may also help in dispersal and colonization after wetting and sinking. Since conidia and teleomorphs do not develop underwater, it is possible that spread underwater is by mycelial fragments or by leaf-to-leaf contact. Growth is most rapid in well-aerated water, but prolonged survival is possible at low oxygen levels or under completely anaerobic conditions. Some of these fungi can also survive prolonged drought and, under these conditions, dispersal as mycelium in wind-blown leaf litter might be possible, but it is unlikely that the conidia are dispersed independently in air.

9.3 MARINE FUNGI

Compared to the many thousands of fungal species known from terrestrial habitats, only about 500 have been described from oceans and estuaries which make up most of the world's surface (Kohlmeyer and Kohlmeyer, 1979; Kohlmeyer and Volkmann–Kohlmeyer, 1991). Although it was at first doubted that there were indigenous marine fungi, the publication by

Barghoorn and Linder (1944) entitled 'Marine fungi: their taxonomy and biology' provided a foundation and a stimulus to the study of these organisms. Kohlmeyer and Kohlmeyer (1979) have distinguished between **obligate marine fungi** which grow and sporulate exclusively in a marine or estuarine habitat and **facultative marine fungi** which are derived from freshwater or terrestrial sources and which are able to grow and possibly sporulate in the marine environment. There is also a fungal flora of inland salt lakes which resembles that of the sea.

9.3.1 The marine environment

Within the sea there is a range of chemical, physical and biological factors which control the distribution, activities and abundance of fungal inhabitants. The chief feature distinguishing the sea from fresh water is, of course, its salt content. The total dissolved solids per litre of a sample of sea water may vary, depending on the source, especially in relation to evaporation and nearby freshwater inflows. The values can vary from below 5‰ in estuaries, to 37‰ or more. A typical value for the salinity of ocean water is between 33 and 37‰, and the average salinity is usually stated to be 35‰. The proportion of the different salts in sea water is reasonably constant, but may, of course, be altered by pollution, and by proximity to land, rivers and terrestrial run-off. The pH of sea water lies in the range 7.5–8.4, but is usually between 8.1 and 8.3 at the surface. The photosynthetic activity and respiration of the phytoplankton affect the concentration of dissolved CO_2 so that, when photosynthesis is high, the CO_2 content decreases and the pH may rise to 8.3–8.5.

The temperature range of sea water is related to depth, latitude, insolation, season and time of day. It is also affected by such factors as proximity to ice, cooling by wind, and the upwelling of currents. Close to the poles, especially near to ice, the surface sea water temperature may be below 0°C. In the tropics, the surface temperature may never fall below 20°C. With increasing depth, the sea water temperature can also fall to low values.

All these factors, i.e. the salinity, pH and temperature of the water, influence the activity, abundance and distribution of marine fungi.

There are several groups of fungi or fungus-like organisms well-represented in the sea. The so-called lower fungi, reproducing by zoospores, include organisms of diverse affinity. Some members of the Chytridiales and the Lagenidiales are parasitic, e.g. on marine algae, whilst some grow saprotrophically. The Thraustochytriales and Labyrinthulales are both obligately marine. Although previously classified as fungi, they are now placed in a separate phylum Labyrinthomycota (Porter, 1989). The ecology of these organisms is not considered further (but for references, see Moss, 1986b).

The higher filamentous marine fungi include about 300 species, although more remain to be described. Most are ascomycetes and deuteromycetes and a few basidiomycetes. There are also marine yeasts. Some of the higher filamentous fungi are parasitic on marine algae or marine angiosperms, or grow symbiotically with brown algae, e.g. *Mycosphaerella ascophylli* on *Ascophyllum* and *Pelvetia* (Garbary and Gautam, 1989; Kingham and Evans, 1986). Marine lichens in which ascomycetes grow symbiotically with green algae or cyanobacteria include genera such as *Arthropyrenia, Verrucaria* and *Lichina*. Some marine fungi are associated with corals (for references, see Kohlmeyer and Volkmann-Kohlmeyer, 1989). However, the majority are saprotrophs on algae, wood, mangrove roots or marine angiosperms ('sea grasses').

Details of techniques for studying different groups of marine fungi have been given by Johnson and Sparrow (1961), Jones (1971), Hughes (1975) and Hyde, Farrant and Jones (1987).

9.3.2 Higher marine fungi

Several taxonomic groups of higher fungi have marine representatives, for example:

Ascomycotina
Eurotiales: Amylocarpus, Eiona
Sphaeriales: Halosphaeriaceae:
    Antennospora, Ceriosporopsis, Corollospora, Halosphaeria, Lindra, Lulworthia
Loculoascomycetes: *Aigialus, Halotthia, Leptosphaeria, Massarina, Mycosphaerella, Pleospora*

Basidiomycotina
Aphyllophorales: *Digitatispora, Halocyphina*
Gasteromycetes: *Nia*
Deuteromycotina *Asteromyces, Dendryphiella, Orbimyces, Sigmoidea, Varicosporina, Zalerion*

A fuller list of higher marine fungi will be found in Kohlmeyer and Volkmann-Kohlmeyer (1991).

Marine ascomycetes are often characterized by having asci whose walls deliquesce and ascospores which bear appendages. A few typical marine ascospores are illustrated in Figure 9.13. Ascospore appendages can develop in a great variety of ways, and their structure and developments have been much studied (Kirk, 1976; Jones and Moss, 1978; Jones, Johnson and Moss, 1983; Johnson, 1980). The marine basidiomycetes *Digitatispora marina* and *Nia vibrissa*, although not closely related, form basidiospores with radiating arms, so that the spore is tetraradiate or pentaradiate (Figure 9.14).

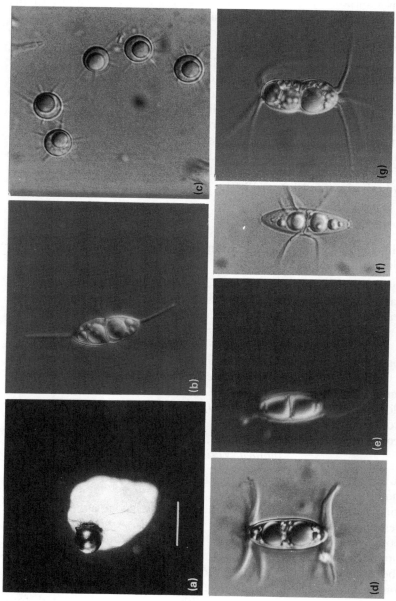

Figure 9.13 Marine ascomycetes. (a) Perithecium of *Corollospora maritima* attached to a sand grain. (b) *Corollospora maritima* ascospore with polar and equatorial appendages. (c) *Amylocarpus encephaloides* ascospores. (d) *Halosphaeria quadriremis* ascospore. (e) *Ceriosporopsis halima* ascospore. (f) *Halosphaeria mediosetigera* ascospore. (g) *Arenariomyces trifurcata* ascospore.

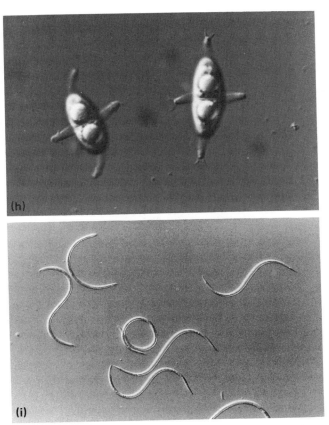

Figure 9.13 (h) *Ceriosporopsis calyptrata* ascospores. (i) *Lulworthia* sp. sigmoid ascospores. Scale bar = 200 μm (a); 20 μm (b–h); 50 μm (g,i).

The role of ascospore appendages or of branched basidiospores has been studied experimentally. Removal of the arms or appendages by sonication resulted in more rapid sedimentation than for undamaged spores (Rees, 1980), and Rees concluded that the extensions helped to keep the spores in suspension. It is also likely that they aid trapping to substrata such as wood (Hyde and Jones, 1989; Hyde, Moss and Jones, 1989).

Apart from *Varicosporina, Orbimyces, Robillarda* and *Dinemasporium*, which have conidia bearing appendages, the conidia of most marine deuteromycetes appear not to have distinctive morphological adaptations to flotation or trapping. The distribution and ecology of conidial marine fungi have been reviewed by Kohlmeyer (1981a) and Subramanian (1983).

There have been numerous studies on the physiology of higher marine fungi in culture (for references, see Jones and Byrne, 1976; Jennings, 1983,

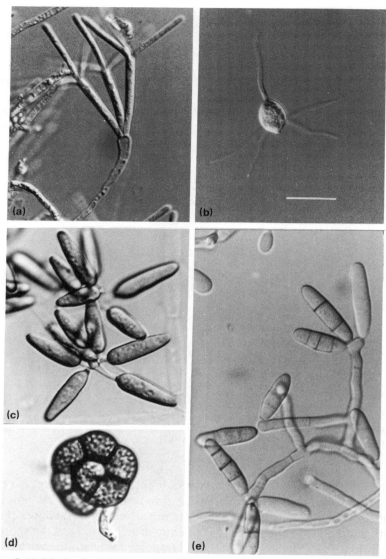

Figure 9.14 Marine basidiomycetes and deuteromycetes. (a) *Digitatospora marina* basidium showing three attached tetraradiate basidiospores. (b) *Nia vibrissa* pentaradiate basidiospore. (c) *Asteromyces cruciatus* conidia attached to conidiophores. (d) *Zalerion maritimum* detached conidium. (e) *Dendryphiella salina* conidiophores and conidia. Scale bar = 20 μm (a,b,c); 30 μm (d); 30 μm (e).

1986). Unfortunately, most studies of marine fungi in culture have provided little information of ecological significance (Jennings, 1986). This is partly because the commonly used batch culture technique does not match natural conditions and may result in large pH changes. A second reason is that in many experiments in culture, the fungi have been provided with abnormally high concentrations of carbohydrates, enabling them to synthesize organic solutes in the cytoplasm, thus to maintain a high solute potential, and consequently a turgor potential sufficient to allow growth to continue. Jennings (1986) has pointed out that for a fungus growing in the sea, sea water has three unusual properties: its low water potential, its high concentration of ions and its alkaline pH. The ability to grow in a medium of low water potential is correlated with a high internal osmoticum, generated by the synthesis of polyols such as glycerol, mannitol and arabitol. Ions may also make a major contribution to the solute potential of the protoplasm (Wethered, Metcalf and Jennings, 1985; Clipson and Jennings, 1990).

Many physiological studies have centred on the effects of salinity. It is surprising that in pure culture many marine fungi grow equally well on media made up with sea water or with distilled water. It is important, however, to distinguish between the effects of sea water on spore germination, vegetative growth and on sporulation, because, for successful life in the sea, a fungus must be able to germinate, grow and fruit in competition with other marine organisms.

a) Spore germination

The spores of most marine fungi have no constitutive dormancy, and are capable of immediate germination. There is, however, evidence that sea water contains a mycostatic factor that inhibits spore germination of some species of marine lignicolous fungi but not others. Kirk (1980) studied germination of conidia of *Orbimyces spectabilis*, *Trichocladium achrasporum*, *Dendryphiella salina* and *Zalerion maritimum*, and ascospores of *Halosphaeria mediostigera* in Millipore-filtered sea water with a salinity range of 15.5 to 27.5‰. *O. spectabilis*, *T. achrasporum* (the conidial state of *H. mediostigera*) and *Z. maritimum* showed no mycostatic inhibition, but the spore germination of *D. salina* and *H. mediosetigera* was inhibited in filtered natural sea water. The mycostatic effect was nullified by the addition of nutrients such as 0.1% glucose, 0.1% yeast extract or 0.1% $(NH_4)_3PO_4$. The requirement for nutrients to stimulate germination may be an adaptation preventing premature germination of spores not in contact with a suitable substratum. Kohlmeyer (1981b) has suggested that the mycostatic factor is effective for spores of obligate intertidal species, but does not operate on spores of subtidal or deep-sea species of marine fungi.

In many marine fungi, spore germination is relatively insensitive to variations in salinity. Byrne and Jones (1975a) studied spore germination in *Corollospora maritima* (ascospores), *Asteromyces cruciatus*, *Dendryphiella salina* and *Zalerion maritimum* (conidia) in sea water concentrations from 0 to 100%. The germination response in all was 'flat-topped', with spore germination between 50 and 100% within the range of salinities tested. This behaviour contrasts with that shown by ascospores of the marine pyrenomycete *Lindra thallassiae*, which grows on leaves of turtle grass (*Thalassia testudinum*), a submerged marine angiosperm. The spores of this fungus fail to germinate on distilled water media (Meyers and Simms, 1965).

b) Vegetative growth

As stated above, mycelial growth of many higher marine fungi can occur over a wide salinity range (Jones and Jennings, 1964). Jones, Byrne and Alderman (1971) have shown that mycelia of *Cremasteria cymatilis*, *Sporidesmium salinum* and *Lulworthia floridana* can grow over the range of 10–100% sea water. The rate of growth of *Cremasteria* was little affected by salinity variations in this range, whilst the other two species showed improved growth at the higher salinities. The marine pyrenomycetes *Lindra thalassiae*, *Lulworthia floridana* and *Halosphaeria mediosetigera* are also capable of growth over the range 0–100% sea water (Meyers and Simms, 1965), whilst the basidiomycete *Halocyphina villosa* can grow in water with a salinity range of 1–200% sea water (Rohrmann and Molitoris, 1986).

Davidson (1974) has compared the growth rate and respiration of two pyrenomycetes (*Gäumannomyces graminis*, a terrestrial cereal pathogen, and *Lulworthia medusa*, a lignicolous marine fungus) at different salinities from 0 to 28‰. Although both can grow within the salinity range tested, the growth of *G. graminis* is more rapid on media lacking sea water, whilst the opposite is true for *L. medusa*. The respiration rate of *G. graminis* is reduced in sea water, whilst the respiration rate of *L. medusa* was approximately the same in fresh and in sea water. It was postulated that in fresh water *L. medusa* devotes a high proportion of its respiratory energy to functions other than biomass increase, which may place the fungus at a competitive disadvantage in a freshwater environment.

Interactions between salinity and temperature have been noted by Ritchie (1957). He isolated a species of *Phoma* from submerged pine panels near Panama and tested its radial growth rate on media containing artificial sea salts up to a concentration of 9% w/v (i.e. up to nearly 2.5 times that of normal sea water) within the temperature range 7–37°C. The results (Figures 9.15 and 9.16) indicate that the optimum salinity for growth rises with increasing temperature. This so-called '**Phoma pattern**' of growth response to salinity and temperature has been found also in a species of *Pestalotia* isolated from the same substratum, in the marine

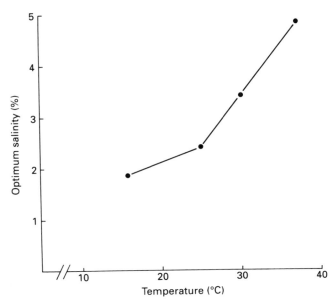

Figure 9.15 Growth response of *Phoma* sp. to increasing salt concentration over a range of temperatures *in vitro*. (Reproduced from Ritchie, 1957.)

basidiomycetes *Digitatispora marina* (Doguet, 1964), *Halocyphina villosa* (Rohrmann and Molitoris, 1986), and the hyphomycete *Zalerion maritimum* (Ritchie and Jacobsohn, 1963).

Table 9.5 Fruiting of marine ascomycetes in diluted sea water

	Seawater %					
	0	20	40	60	80	100
Lulworthia floridana	0	3	3	3	3	3
Lulworthia sp.	3	3	3	3	3	3
Lindra thalassiae	1	3	3	3	3	3
Halosphaeria mediosetigera	3	3	3	3	3	3
Torpedospora sp.	3	3	3	3	3	3

0 = Perithecia absent; 1 = perithecia immature; 3 = perithecia with asci and ascospores.
Source: Jones, Byrne and Alderman (1971).

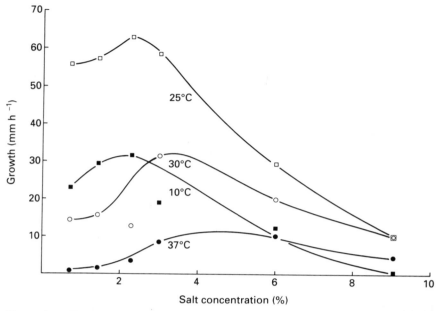

Figure 9.16 Relationship between optimum salinity for growth and temperature in *Phoma* sp. (Reproduced from Ritchie, 1957.)

c) Reproduction

The effects of salinity variations on the reproduction of higher marine fungi are variable. Some species are able to sporulate within the salinity range of 0–100% sea water, whilst others fail to develop fructifications, or have only immature fructifications in the absence of sea water (Jones, Byrne and Alderman, 1971) (Table 9.5). Doguet (1964) has shown that the marine basidiomycete *D. marina* can sporulate in diluted sea water within the range 5–25‰ (i.e. up to about 70% sea water). Sporulation is most prolific at 15–20°C, which may be related to the known distribution of the fungus in temperate waters, and may also explain why in nature it appears to fruit best only in the cooler months. *H. villosa* produces basidiocarps in culture over the range 25–100% sea water at temperatures between 22 and 27°C. This may reflect its natural habitat on the wood of mangroves, which are tropical and may be coastal or estuarine in their distribution (Ginns and Malloch, 1977; Hyde, 1986; Rohrmann and Molitoris, 1986).

Although vegetative growth of the marine deuteromycetes *Orbimyces spectabilis* and *Varicosporina ramulosa* takes place within the range 0–100% sea water, the amount of growth rising with increasing salinity, typical conidial development only occurs above a sea-water concentration of 20%. Below this concentration, conidia either fail to develop or the cells making up the body of the conidium may encyst. Chlamydospores may also occur in the mycelium (Meyers and Hoyo, 1966).

The ability of many higher marine fungi to germinate, grow and fruit over a range of 0–100% sea water is puzzling. It is known that some species can extend into estuaries, but what factors prevent their further extension into fresh water? Possibly competition from terrestrial or aquatic fungi is involved.

Jones (1974) has compared the behaviour of some different ecological groups of fungi in relation to salinity tolerance (Figure 9.17). The diagram shows that the spore germination, growth and reproduction of terrestrial ascomycetes are deleteriously affected by increasing salinity, and that aquatic hyphomycetes are even more drastically affected.

There is a wide range of substratum and mode of nutrition for higher marine fungi. Some are parasites of seaweeds, angiosperms and marine animals. Saltmarsh plants such as *Spartina* and *Halimione* bear an extensive flora of fungi, many of which are adapted to sea water. However, most ecological work has been done on the saprophytic fungi which colonize wood in or around the sea. Some tropical coasts are fringed by woody plants, e.g. *Cocos*, or by mangroves like *Rhizophora* and *Avicennia*, which are adapted to live in mud bathed by the sea. The leaves, branches, roots, seedlings and fruits of these plants which fall into the sea are colonized by marine fungi or, in the case of prop-roots of mangroves, are colonized *in situ*. Detritus from mangroves can provide the major carbon input into some estuarine ecosystems and microbial colonization increases the protein content of the detritus which provides food for invertebrates which are

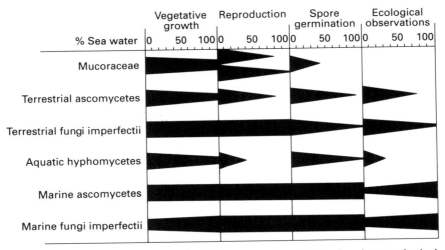

Figure 9.17 Physiological and ecological responses to salinity of various ecological groups of fungi. The scale represents percentage sea-water concentration. (After Jones, 1974.)

in turn consumed by crustacea and fish. There have been many studies of fungi colonizing mangrove detritus, including leaves and whole seedlings (Kohlmeyer, 1969, 1986; Anastasiou and Churchland, 1969; Fell and Master, 1975; Newell, 1976; Newell *et al.*, 1987; Hyde and Jones, 1988; Jones and Kuthubutheen, 1989; Newell and Fell, 1992). These studies will not be considered here, but reference will be made below to fungi colonizing woody mangrove debris.

9.3.3 Lignicolous marine fungi

Apart from woody fragments produced by maritime plants, there are considerable amounts of driftwood brought down to the sea by rivers, or brought there as a result of human activity. Studies on the higher fungi that colonize driftwood can be made by collecting samples which can be examined directly or after incubation in moist chambers. The fungi colonizing wood *in situ* in the sea can be studied by examining wooden piles or breakwaters. Alternatively, wood panels can be immersed in the sea at different depths and left in place for varying times before collection and examination, usually after moist-chamber incubation.

By submerging a string of panels of different kinds of wood, an indication of host preference by different marine fungi has been obtained. Meyers and Reynolds (1960) found that panels of basswood (*Tilia americana*) bore fruit-bodies of *Lulworthia*, *Ceriosporopsis* and *Halosphaeria*, but panels of yellow pine (*Pinus palustris*) did not. Similarly, Byrne and Jones (1974) showed that panels of beech (*Fagus sylvatica*) were more frequently colonized than those of Scots pine (*Pinus sylvestris*). Certain fungi show a preference for beech, e.g. *Halosphaeria appendiculata*, *H. hamata* and *Nautosphaeria cristaminuta*, whilst others prefer Scots pine, e.g. *Ceriosporopsis circumvestita* (Jones, 1976b).

Although there is an apparent succession of fungi fruiting on incubated wood panels, the sequence of fungi may reflect the time taken for fruit-bodies to develop. Amongst the first fungi to fruit on incubated wood panels are the hyphomycetes *Piricauda*, *Zalerion* and *Humicola*. Some panels may be dominated by *Lulworthia* sp. to the virtual exclusion of other fungi; this is possibly an example of antagonism.

a) Fruiting of marine lignicolous fungi in nature

Schaumann (1968) has studied the fruiting of marine fungi on incubated fragments of wood removed from posts driven into the sea-bed in the estuary of the River Weser in northern Germany. The salinity range varied from 26–31‰ at the seaward end to around 0.7‰ in the area dominated by fresh water. Schaumann has distinguished between four salinity zones, and has listed the characteristic fungi which he found in them.

- **Holeuryhaline species**: These occur in fresh, brackish and sea water, spanning the whole range from 0.5 to 30‰. They include the deuteromycetes *Cirrenalia macrocephala* and *Piricauda pelagica*.
- **Euryhaline species**: The distribution of this group resembles the former, but they do not colonize wood in fresh water. They include the deuteromycete *Dictyosporium pelagicum*, and the ascomycetes *Halosphaeria appendiculata*, *H. mediosetigera*, *Lignincola laevis*, *Ceriosporopsis calyptrata*, *Remispora hamata* and *R. maritima*.
- **Genuine brackish water species**: These inhabit regions with fluctuating salinity, including *Remispora pilleata* and the deuteromycete *Humicola alopallonella*.
- **Stenohaline species**: These live in salt water with a relatively narrow and decidedly high concentration. These typically marine forms include the ascomycete *Corollospora maritima* and the deuteromycete *Zalerion maritima*.

Vertical distribution of lignicolous fungi in the intertidal (eulittoral) zone on wooden pilings in the sea in the Weser Estuary and around the island of Helgoland in the North Sea has been compared (Schaumann, 1969, 1975). Around Helgoland, the salinity range is around 30–34‰, which is higher than in the estuary. Wood samples were removed from the pilings, and the heights above chart datum of the sampling points were recorded. Selected results are shown in Figure 9.18.

Fungi colonizing the pilings below the level of low water spring tides (LWS) would remain permanently submerged, but above this point they would be subjected to periodic emersion, and the effects of sun, wind, rain, etc. At the upper end, in the supralittoral zone, they would be uncovered by the sea for most of the time, but subjected to splash and spray. It is notable that some fungi, e.g. *Corollospora maritima*, *Lignincola laevis* and *Zalerion maritima*, were recovered from virtually the whole of the tidal range, whilst others, such as *Remispora maritima*, *Monodictys pelagica* and *Dictyosporium pelagicum*, were only recovered from the upper part of the eulittoral zone. The reasons for these differing patterns of distribution are not understood.

Differences in vertical distribution within the intertidal zone of fungi growing on overhanging branches and on prop-roots and subterranean exposed roots of mangroves (*Rhizophora* spp.) have been noted by Hyde (1988), working in Brunei. Of the 41 species collected, most (26 species) occurred in the middle level. Some species, such as *Halocyphina villosa*, *Lulworthia grandispora* and *Lulworthia* sp., formed fruit-bodies at all levels. Others, such as *Humicola allopallonella* and *Tricladium* sp., were restricted to lower levels, whilst *Aigialus* spp. fruited at the mid- and upper-littoral levels where the surface of the wood dries out. Whilst the distribution of fruit-bodies may be related to the degree of exposure to the

278 AQUATIC FUNGI

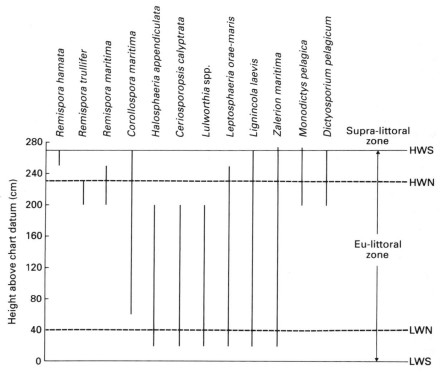

Figure 9.18 Vertical distribution of fruit-bodies of lignicolous marine fungi on posts submerged in the sea at Helgoland in the North Sea. HWS = High water at spring tide; HWN = high water at neap tide; LWN = low water at neap tide; LWS = low water at spring tide. (Redrawn from Schaumann, 1969.)

air, or to the extent of covering by mud, biotic factors may also be involved. For example, wood in the middle levels is more prone to attack by marine borers.

The distribution of wood-colonizing fungi in estuaries has been studied by Gold (1959) and Shearer (1972). Gold, working in the Newport River, North Carolina, USA, immersed wood panels at stations with a salinity gradient of 0–35‰. Four lignicolous fungi, *Ceriosporopsis halima*, *C. cambrensis*, *Lulworthia floridana* and *Trichocladium opacum*, were recovered within virtually the entire salinity range. There were indications of an interaction between higher temperatures, higher salinity and distribution, i.e. with increasing temperature the fungi were able to fruit better in water of higher salinity. This is reminiscent of the '*Phoma* pattern' of mycelial growth reported by Ritchie (1957) (section 9.3.2(b)). Shearer submerged balsa wood (*Ochroma*) blocks in the Putuxent River, Maryland, USA, at stations covering a salinity gradient of 0.05–29.0‰. Not all the fungi which fruited were marine; many were of 'terrestrial' and some of freshwater origin. Increasing salinity was correlated with an increase in the proportion of ascomycetes and a decrease in the proportion of deuteromycetes recovered, so that, at the most saline station, about two-thirds of the fungi found were ascomycetes. No evidence of a '*Phoma* pattern' of temperature–salinity relationships was found. Hughes (1975) has reviewed other studies on the distribution of fungi in estuaries.

Marine lignicolous fungi can exist at considerable depth in the sea, extending well beyond the sublittoral. Kohlmeyer and Kohlmeyer (1979) have claimed that 'fungal mycelia have been found in every piece of wood collected in the deep sea, unless a low oxygen content of the surrounding water had impeded development'. Down to a depth of several hundred metres, the fungi recorded include some familiar representatives from intertidal wood such as *Zalerion maritimum* and *Corollospora maritima*. At greater depths (over 1000 m), a few specialized fungi exist at high pressures, low temperatures and in constant darkness. They include the ascomycetes *Bathyascus vermisporus* and *Oceanitis scuticella*, and the deuteromycetes *Allescheriella bathygena* and *Periconia abyssa*. Samples of wood recovered from the deep sea rarely bear fungal fructifications, and Kohlmeyer (1977) has speculated that sporulation may only occur on small wood fragments broken away as a result of the activity of wood-boring animals.

The so-called lignicolous marine fungi are not confined to woody materials, and many have been reported from 'sea grasses' such as *Cymodocea*, *Posidonia*, *Thallasia* and *Zostera* (Hughes, 1975). Species of *Corollospora*, *Lindra*, *Lulworthia*, *Halotthia* and *Pontoporeia*, *Varicosporina* and *Dendryphiella* have been found on debris derived from such plants (Cuomo et al., 1985).

The decay of wood immersed in the sea is of great economic significance. It is brought about by bacteria (including actinomycetes), fungi and animals, especially wood-boring molluscs and crustacea. In most cases, the decay of submerged wood by ascomycetes and deuteromycetes is of the 'soft-rot' type. Fungal hyphae grow in the lumina of the wood cells, and from these penetrate to the lightly lignified S_2 layer of the secondary cell wall, often following the spiral orientation of the microfibrils making up the wall. Leightley and Eaton (1977) have given a detailed description of the decay of wood by the ascomycete *Halosphaeria mediosetigera* which causes soft-rot decay. The hyphae in this fungus are ensheathed in a mucilaginous layer which possibly contains wall-degrading enzymes. The basidiomycete *Nia vibrissa* causes white-rot decay in which a well-marked erosion zone surrounds the track of hyphae penetrating host cell walls (Leightley and Eaton, 1979; Jones, 1982).

Much interest has been devoted to the suggestion that marine fungi and bacteria may 'condition' wood and make it more susceptible to attack by marine borers (Kohlmeyer and Kohlmeyer, 1979). The gribble, *Limnoria tripunctata* (Crustacea, Isopoda) can live on fungus-free wood, but its lifetime can be extended if it is grown on softwood colonized by marine fungi. Apparently *L. tripunctata* does not reproduce unless marine fungi are included in its diet. The reason for this is unknown, but there have been speculations that the marine fungi might supply protein, vitamins or oils which may be required for reproduction.

Wood which has been in the sea for some time may be cast ashore bearing marine fungi visible at the time of collection or after incubation in moist chambers. On sandy shores, the driftwood can become partially or completely buried in sand, especially at the foot of sand dunes along the upper shore. Marine fungi may fruit on wood which has lain in the moist sand, and some can extend from the substratum to form ascocarps on the sand grains. It is also known that the appendaged spores of marine fungi may be trapped in foam and can then be carried on shore by wind, possibly to be caught onto suitable substrata or into the sand (the so-called **arenicolous fungi**) (Kohlmeyer, 1966; Wagner-Merner, 1972). Common spores in sea foam include ascospores *Corollospora maritima, Arenariomyces trifurcata* (Figure 9.13) and, in warmer seas where appropriate substrata such as *Zostera* and seaweeds are available, the conidia of *Varicosporina ramulosa*.

Ecological studies on the fungi characteristic of sand dunes have been made by Koch (1974) and Rees, Johnson and Jones (1979). Koch has distinguished between three zones on sandy shores in Denmark:

Zone 1: the wet, turbulent water's edge
Zone 2: the dry, windswept beach
Zone 3: the undisturbed area at the base of the dune and the gullies leading up to the inner dunes.

The condition of the wood in each zone is influenced by its water content, and probably by its salt content. In zone 1, the wood is often eroded, presumably by abrasion, whilst in zone 2, the wood is usually much drier and is capable of floating. In zone 3, the wood is often moist and its lower side often bears hyphae which extend into the sand. Only a small proportion of the wood collected in zone 1 bears fruit-bodies at the time of collection, and most of the fungi which do fruit have perithecia immersed in the wood. After incubation in moist chambers, perithecia of *Corollospora* spp. develop at the surface, indicating the presence of mycelium in the outer layers of the wood. The absence of fruit-bodies on fresh collections is probably due to abrasion. Wood from zone 2 similarly bears few fruit-bodies at the time of collection, and the effect of heat and drying may damage the mycelium or prevent development of fruit-bodies. Wood from zone 3 pressed into or partially covered by moist sand may develop perithecia of *Corollospora maritima, Arenariomyces trifurcata* and *Carbosphaerella leptosphaerioides* actually attached to the sand grains (Figure 9.13a). Perithecia of some of these fungi also develop on other hard materials such as barnacle tests. The physiological reasons for this preference for hard substrata for fruit-bodies are not known, but it is possible that dispersal of sand grains bearing perithecia may occur.

The marine deuteromycete *Varicosporina ramulosa* forms sclerotia attached to sand grains. Kohlmeyer and Charles (1981) have used the term **sclerocarp** for such structures which they interpret as propagules, and have presented anatomical evidence of homology between sclerocarps in this fungus with perithecia of members of the Halosphaeriaceae.

Isolations of fungi from sand near driftwood from Danish sand dunes by Rees, Johnson and Jones (1979) yielded different species depending on the technique used. If sand was plated directly on to agar media made up with sea water, the most commonly encountered fungi were terrestrial hyphomycetes, and some marine hyphomycetes such as *Asteromyces cruciatus*, *Monodictys pelagica* and *Dendryphiella salina* (Figure 9.14). When sand was added to sterile balsa-wood strips moistened by sea water, a number of marine ascomycetes (*Corollospora* spp. and *Carbosphaerella*) were identified. Presumably the spores of these fungi were trapped within the sand.

b) Geographical distribution of marine lignicolous fungi

In surveying the worldwide distribution of marine lignicolous fungi, there are several difficulties. The distribution of suitable substrata is obviously related to proximity to land with trees whose remains can reach the sea, although wood can drift for great distances. Some marine fungi are restricted to the prop-roots of mangroves, whilst others are host-specific parasites of marine algae, so that the distribution of these fungi is dependent on the distribution of the host plant. However, the majority of wood-inhabiting marine fungi are less specific and are not host-dependent in their distribution (Kohlmeyer, 1983). The activities of biologists with interest in and competence to identify marine fungi are not evenly distributed. For example, the polar regions are probably under represented because of difficulty of access and facilities for study. The world distribution of marine fungi has been discussed by Hughes (1974, 1975, 1986), Kohlmeyer (1983) and Booth and Kenkel (1986). Hughes (1974) has mapped the records of distribution of marine lignicolous fungi and has related these to five biogeographical regions defined in terms of surface-water isotherms. The surface-water temperature regimes used to delimit these five regions are:

- **Tropical**: The regions between the isotheres for 20°C water temperature

at the coldest time of the year. This region corresponds with that of reef-building corals.
- **Subtropical**: The region between the 17°C isocryme for August in the southern hemisphere and February in the northern hemisphere.
- **Temperate**: A region bounded towards the equator by the 17°C isocryme for the coldest calendar month and towards the pole by the 10°C isothere for the warmest calendar month.
- **Arctic and Antarctic**: Regions separated from temperate regions by a line following the 10°C isothere for August in the north and for February in the south.

Each of these regions has its characteristic flora of marine fungi.

Ceriosporopsis halima has a wide distribution, but is not common in the tropical zone. *Halosphaeria hamata* 'is perhaps best considered as a species of temperate regions', but there are problems in identification of this and similar fungi which make for uncertainty in mapping its distribution. *Antennospora quadricornuta* 'is the most common marine ascomycete in tropical waters and appears to be virtually restricted to this zone', an opinion confirmed by Hyde (1986). Ascospore germination occurs readily at 28°C within a few hours, but not at 20°C (Kohlmeyer, 1968).

Corollospora maritima (Figure 9.13a,b) is a ubiquitous and common fungus of sandy beaches with a worldwide distribution (Kohlmeyer, 1984). However, five cultures from widely separated localities, although morphologically similar, showed significantly different growth responses to temperature variations (Bebout *et al.*, 1987), suggesting that there are physiological races adapted to different temperatures.

The relationship between temperature, seasonal occurrence and distribution of three marine deuteromycetes has been investigated by Boyd and Kohlmeyer (1982). *Asteromyces cruciatus* is an arenicolous fungus, and *Sigmoidea marina* grows on seaweeds and marine angiosperms. Both are distributed in the temperate zone. *Varicosporina ramulosa* also grows on seaweeds and marine angiosperms, but has a subtropical distribution. Conidia of all three fungi can be collected in sea foam, and regular sampling of foam can indicate seasonal abundance. Some correlation was found between the effects of temperature on growth and survival and distribution. For *V. ramulosa*, the optimum temperature for linear growth was 30–40°C, and for dry weight increase 20–30°C. This fungus can also survive temperatures as low as 10°C but makes little or no growth at this temperature. It possesses temperature-resistant propagules (**sclerocarps**) which are attached to sand. They can endure prolonged exposure to a wide temperature range (45 to −70°C), which would enable *V. ramosa* to survive the extreme temperature ranges found on sandy beaches (Kohlmeyer and Charles, 1981). The upper temperature limit for vegetative growth for *A. cruciatus* and *S. marina* was 35°C, but prolonged exposure of both fungi to this temperature resulted in an inability to grow even at room temperature.

It is thus apparent that temperature is an important factor in relation to the distribution of marine fungi. The studies of Booth and Kenkel (1986), in which ordination analysis was used to group fungi with similar distributions together and relate them to environmental variables, showed that temperature gradient is the dominant factor, whilst the combined effect of temperature–salinity variation is of secondary importance.

10
NEMATOPHAGOUS FUNGI

An interesting ecological group of fungi are those which capture or grow parasitically on nematodes (eelworms), their cysts or eggs. Over 150 species are known (Gray, 1988) belonging to diverse taxonomic groups, as the following list shows.

Chytridiomycetes:	*Catenaria*
Oomycetes:	*Haptoglossa*, *Myzocytium*, *Nematophthora*, *Protascus*
Zygomycetes:	*Acaulopage*, *Macrobiotophthora*, *Meristacrum*, *Rhopalomyces*, *Stylopage*
Ascomycetes:	*Atricordyceps* (teleomorph of *Harposporium anguillulae*)
Basidiomycetes:	*Hohenbuehelia* (teleomorph of *Nematoctonus*), *Hyphoderma*, *Pleurotus*
Deuteromycetes:	*Arthrobotrys*, *Dactylaria*, *Dactylella*, *Drechmeria*, *Harposporium*, *Monacrosporium*, *Verticillium*.

The foundation work on this group of fungi was done by Drechsler, starting in 1941. Accounts have also been given by Soprunov (1958), Duddington (1955, 1957), and Duddington and Wyborn (1972), Barron (1977, 1978, 1981), Subramanian (1983), Dowe (1987) and Gray (1988). Cooke and Godfrey (1964) have provided keys to identification.

Nematodes are very varied in their habitats, feeding habits and general biology. They are abundant in soil and are especially numerous around plant roots. They are common in dung, in rotting wood, and also occur in the sea and in fresh water. Many nematodes ingest bacteria, whilst others penetrate fungal hyphae and consume the contents. Some are serious plant parasites. The plant-parasitic nematodes are of two types: free-living or sedentary. The sedentary nematodes include species of *Meloidogyne* which cause root-knot diseases, and the cyst nematodes such as species of *Heterodera* and *Globodera*. Sedentary nematodes spend most of their life within plant roots and form adult stages there. They feed by penetrating plant cells using mouthparts modified as stylets. The females of cyst nematodes enlarge and become packed with eggs. Their body wall hardens to form a resistant cyst, within which the eggs remain alive for many years. There are also serious animal-pathogenic nematodes with motile stages released into the soil from faeces.

Three kinds of nematophagous fungi have been distinguished depending on the manner in which they attack their hosts.

Predatory fungi have hyphae which extend into the soil or other substratum and capture their prey by a variety of structures: sticky hyphae, sticky knobs, adhesive networks, non-constricting rings or constricting rings (Figure 10.1). Most predatory fungi only form traps in the presence of nematodes in response to specific substances released from the nematode (Section 10.3). However, in a few cases traps may form spontaneously, in the absence of nematodes, and may even develop directly on spore germination – the so-called **conidial traps** (Dackman and Nordbring-Hertz, 1992). Form-genera with species that capture nematodes by means of traps include *Arthrobotrys*, *Dactylaria* and *Monacrosporium*. After capture, the body of the nematode is penetrated enzymatically, and trophic hyphae extend through the body and digest the contents.

The second group of nematode-destroying fungi are termed **endoparasitic** or **endozoic**. In contrast to the predatory group, endoparasitic predacious fungi probably have no extensive mycelial state in the soil away from their hosts and are regarded as ecologically obligate parasites. They produce conidia which may be ingested (in *Harposporium*) or may become attached to the host cuticle. Penetration of a germ tube through the gut wall or the cuticle is followed by the development of trophic hyphae and the formation of conidia external to the host. The form-genera *Verticillium*, *Harposporium*, *Drechmeria* and *Nematoctonus* include endoparasitic species (Figure 10.2). The chytridiomycete fungus *Catenaria* and the Oomycete *Myzocytium* form zoospores which are attracted to nematodes and encyst on the cuticle before penetration occurs.

A third group of nematophagous fungi are **parasites of eggs or cysts**. Egg parasites include *Rhopalomyces* (Barron, 1973) and *Dactylella oviparasitica*, a parasite of *Meloidogyne* (Stirling and Mankau, 1978, 1979). Parasites of cysts include the Oomycete *Nematophthora gynophila* on females and cysts of *Heterodera* spp. (Kerry and Crump, 1980), *Verticillium chlamydosporium* parasitic on *Heterodera* and *Meloidogyne* (Gams, 1988) and *Paecilomyces lilacinus* (Stirling, 1988).

10.1 TECHNIQUES FOR STUDYING NEMATOPHAGOUS FUNGI

Techniques for studying predatory nematophagous fungi have been described by Duddington (1955), Wyborn, Priest and Duddington, (1969),

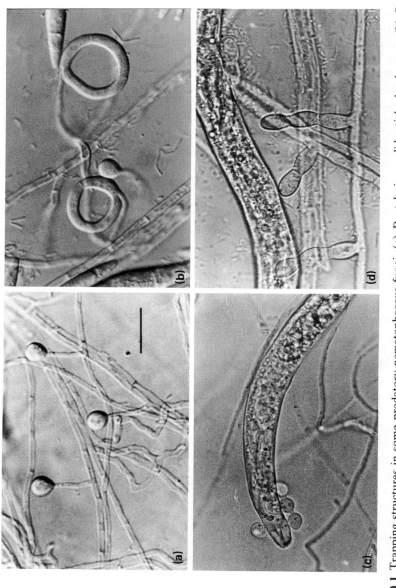

Figure 10.1 Trapping structures in some predatory nematophagous fungi. (a) *Dactylaria candida* sticky knob traps. (b) *Dactylaria candida* sticky knob traps and non-constricting ring traps formed on the same mycelium. (c) Nematode with several sticky knob traps attached to its anterior end and which it has detached from the mycelium. Such traps are still capable of infection. (d) *Monacrosporium ellipsosporum*: short adhesive branch traps.

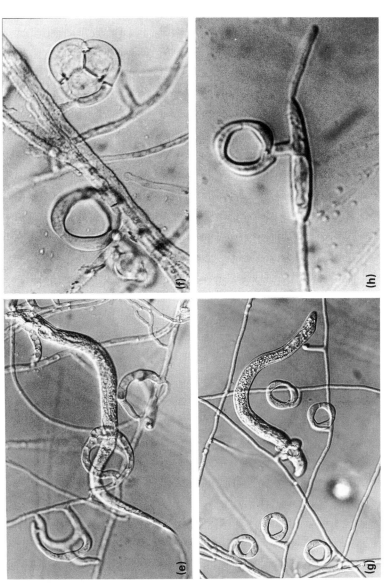

Figure 10.1 *cond.* (e) *Arthrobotrys oligospora* adhesive network traps with a nematode held by a sticky secretion formed on the inner face of the hyphae making up the network. (f) *Monacrosporium doedycoides* constricting ring traps: the trap on the left is still open but the one on the right has closed. (g) A nematode caught in a constricting ring trap. (h) *Arthrobotrys dactyloides* constricting ring trap formed by a germinating conidium. Scale bar = 20 μm (a,b,d,f); 25 μm (c); 50 μm (e); 40 μm (g).

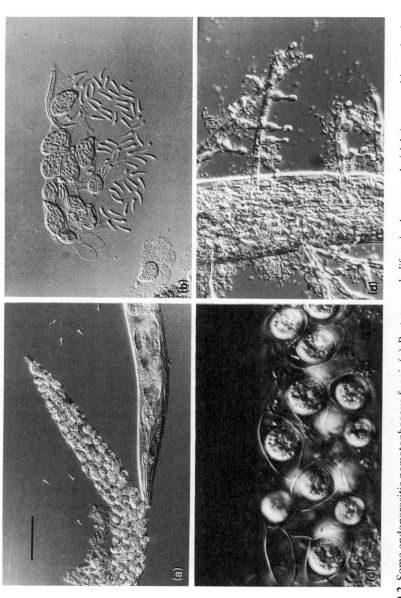

Figure 10.2 Some endoparasitic nematophagous fungi. (a) *Protascus subuliformis*. A nematode (right) approaching a dead nematode filled with sporangia of *Protascus*. (b) *Protascus subuliformis*. The remains of a nematode killed by *Protascus*. Sporangia have been released, some of which have dehisced and the awl-shaped sporangiospores have escaped. (c) *Protascus subuliformis*. Oospores in the cadaver of a nematode. (d) *Verticillium obovatum* conidiophores projecting from the cadaver of a nematode. The conidiogenous cells are phialides.

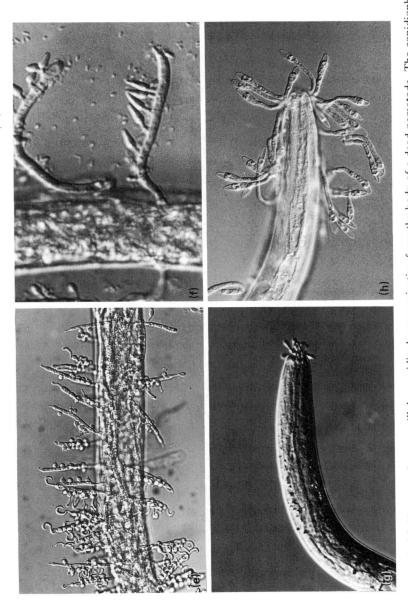

Figure 10.2 *contd.* (e) *Harposporium anguillulae* conidiophores projecting from the body of a dead nematode. The conidiophores form globose phailides from which coiled conidia develop. (f) *Drechmeria coniospora* conidiophores with phialides and conical phialoconidia. (g) Nematode with conidia of *Drechmeria coniospora* around its mouth. The conidia are attached by adhesive buds. (h) *Nematoctonus leiosporus* conidia attached to the cuticle of *Panagrellus redivivus*. Attachment is by a mucilage-covered trap at the tip of a short germ tube. (Photograph h by E. Poloczek.) Scale bar = 40 μm (a,g); 20 μm (b,c,d,f); 30 μm (h).

Barron (1977, 1982), Gray (1988) and Dackman, Jansson and Nordbring-Hertz (1992). In the **sprinkle plate** method, a small quantity (0.5–1.0 g) of soil or other substratum is sprinkled onto a Petri dish containing a weak agar medium such as dilute cornmeal agar or tap-water agar, and the plate is incubated for several weeks at room temperature. Eelworms present in the soil crawl out on to the agar, feeding on bacteria, and increase in number. Cultures of nematodes can be used to provide additional hosts and improve the chance of isolating nematophagous fungi, but this is not essential.

There is little evidence of host preference among predatory nematophagous fungi. However, the stylet mouthparts of plant-parasitic nematodes are unsuitable for ingesting the conidia of the endozoic fungus *Harposporium*. The presence of nematodes encourages the development of mycelia and trapping organs of predatory hyphomycetes and nematodes are attracted to both. Conidiophores and conidia develop around the trapped nematodes (Figure 10.2). Cultures are easily prepared from conidia, and most predatory nematophagous fungi grow well saprotrophically.

Quantitative or semi-quantitative methods for demonstrating the relative abundance of propagules of nematophagous fungi in soil have been devised (see Barron, 1982). Eren and Pramer (1965) used a soil dilution technique in which an aliquot of a soil suspension was added to water agar in a range of dilutions. After 3 days' incubation, nematodes were added. The plates were incubated for a further 3-week period and scored for the presence or absence of trapped nematodes. The results were analysed to give an estimate of most probable number (MPN). However, when the method was tested by the addition of known numbers of conidia of *Arthrobotrys conoides*, only about 15% were recovered, possibly because the number of spores exceeded the numbers of potential host nematodes. Dackman *et al.* (1987) have also used a soil dilution technique based on MPN estimations. This technique gave higher estimates of propagule number than the sprinkle-plate method. The numbers of nematodes, bacteria and nematophagous fungus propagules were all greatly affected by soil treatment. In heavily manured soil, 20 propagules g^{-1} dry wt of soil were estimated, whilst in soils treated with fertilizer but without manure only about 1 propagule g^{-1} was found. Stirling, McHenry and Mankau (1979) estimated that in soil from a Californian peach orchard there were about 5–50 propagules of *Arthrobotrys dactyloides* and *Monacrosporium ellipsosporum* g^{-1} soil. Mankau (1975) used membrane filters to filter a soil suspension of sieved soil. The membrane filter was then placed on dilute cornmeal agar. Nematodes and nematophagous fungi migrated into the agar and were observed within 6–12 days. Counting the number of parasitized nematodes allowed a semi-quantitative comparison between soil samples.

These indirect techniques suggest that predatory nematophagous fungi are active in soil, and this has been confirmed by direct observations on soils using epi-illumination (Kliejunas and Ko, 1975) or on nematodes extracted from soil by sieving. A proportion of the extracted nematodes may bear traps of predatory fungi or conidiophores of endozoic fungi (Capstick, Twinn and Waid, 1957). Buried Cellophane recovered from soil may carry trapped nematodes (Tribe, 1957). Cooke (1961) followed the activity of predatory fungi growing on discs of weak cornmeal agar attached to microscope slides buried in soil. At intervals, the slides were recovered and the agar surface scored for the presence or absence of different kinds of trapping structure as an indication of activity.

Endoparasitic nematophagous fungi may also be found in cultures prepared by soil-sprinkling techniques, but since the predatory fungi grow more rapidly and are very effective in ensnaring nematodes, eelworms infected with endoparasitic fungi may not be obvious, and alternative techniques have been developed to demonstrate their presence. Barron (1969) has made use of the fact that the conidia of most predatory fungi are larger than those of endozoic fungi to separate conidia by differential centrifugation. Giuma and Cooke (1972) have concentrated endozoic fungi by allowing infected nematodes to migrate through tissue paper from saturated soil held in filter funnels (the **Baermann funnel technique**). After centrifugation and washing, sluggish or immobile nematodes containing fungal thalli were transferred to cornmeal agar. Barron (1978) simplified this technique by omitting centrifugation and pouring the contents of the tube in which the nematodes had collected directly on to water agar plates. A comparison of the effectiveness of different techniques for isolating predatory and endozoic nematophagous fungi has been made by Bailey and Gray (1989).

Parasitized females of cyst-forming nematodes are detected by direct microscopic examination of females from the surface of infected plant roots or by culturing the fungi which grow out from them. Nematode cysts can be removed from soil by sieving, and when the cysts are ruptured, egg masses are released. These can be spread on to water agar plates, and subcultures from hyphae which grow out can be transferred to other media to permit sporulation and identification (Dackman, Jansson and Nordbring-Hertz, 1992).

10.2 DISTRIBUTION AND ABUNDANCE

There have been numerous surveys of the occurrence of nematophagous fungi (for references see Gray, 1987; Saxena and Mukerji, 1991). They occur throughout the world from the Equator to the Poles. There appears to be little difference between the species found in tropical, temperate or Antarctic soils. Gray (1983) surveyed 161 samples of soil and plant material from Ireland classified into broad habitat types such as deciduous or coniferous tree litter, temporary or permanent pasture, peatland and coastal vegetation. Nematophagous fungi were found in all habitats. Ninety per cent of sites with coniferous litter or coastal vegetation sites produced samples containing these fungi, whilst permanent pasture and composts had lower percentages, around 50%. Highest species diversity was found in composts and in peat (4–5 species). The commonest endoparasitic fungi were *Myzocytium* spp. (9.3%), *Verticillium balanoides* (6.8%) and *Harposporium anguillulae* (6.2%). The most common predatory fungi were *Monacrosporium bembicodes* (8.7%), *M. mamillatum* (6.8%) and *M. ellipsosporum* (6.2%). Two species of *Arthrobotrys*, *A. musiformis* and *A. robusta*, were significantly more abundant in permanent pasture than in the other habitats surveyed, whilst *A. oligospora* was found only in permanent pasture and in coastal vegetation in this study. However, it is generally considered to be one of the most abundant nematophagous fungi of temperate soils.

In a more extensive survey, Gray (1985) surveyed the presence of nematophagous fungi in 206 Irish soil samples, and attempted to correlate the presence of particular types of fungus with organic matter, soil moisture and pH. Endoparasites (*Myzocytium*, *Harposporium*, *Drechmeria*) were present in samples with higher moisture and organic matter and lower pH, and were detected more frequently in samples with high densities of nematodes. The only single factor correlated with the presence of predatory nematophagous fungi was soil pH, more predators being present in samples with a mean pH of 5.5. Analysis of the occurrence of these fungi in relation to their trapping mechanism revealed more interesting associations. Predators with sticky hyphae, branches or knobs were not significantly correlated with the moisture or organic matter content of the soils, whilst fungi with adhesive nets or constricting rings were significantly associated with these factors. The adhesive net-formers (e.g. *Arthrobotrys musiformis*) were associated with soils of lower water and organic matter content.

Similar findings were reported by Gray and Bailey (1985) in a study of the vertical distribution of nematophagous fungi in a deciduous woodland soil to a depth of 35 cm. The greatest species diversity occurred in the upper litter and humus-rich zones. The species characteristic of this zone were predominantly forms with sticky branches or constricting rings. In this zone, nematodes were more numerous than in the underlying mineral layers. In these lower layers, the endoparasitic *V. balanoides* and the net-forming predatory fungi were more frequently isolated than in the upper soil layers.

Ayen and Lysek (1986), working in a sandy beechwood soil in Germany, isolated endoparasitic fungi more frequently in winter than in summer. The higher number of isolations was correlated with high soil moisture content. This is not surprising since the zoosporic fungus *Myzocytium* made up half the isolations.

These and other investigations have shown that predacious fungi are more abundant in soils containing decomposing organic matter. This is probably related to the abundance of free-living nematodes feeding on bacteria, fungal hyphae and organic particles. Cooke (1962a,b, 1963a,b, 1964, 1968) has used the buried agar disc technique to assess the activity of predacious fungi following the addition of organic matter to soil. The numbers of nematodes in the soil were estimated by counting the nematodes expelled from soil samples in a Baermann funnel (Peters, 1955; Barron, 1982).

In experiments in which 3 g of sucrose was added to 250 g of soil (Cooke, 1962a), predacious activity rose steeply, reaching a maximum after 4 weeks, and declining to non-detectable levels after 12 weeks. The nematode population showed a similar pattern of change, rising to a maximum value 15 times that of the control soil after 5 weeks, and then declining to low levels after 7 weeks (Figure 10.3). A successional pattern was detected. During weeks 2–6, fungi with adhesive reticulate traps, such as *A. oligospora*, *D. psychrophila* and *Trichothecium cystosporium*, were dominant. Later, during weeks 5–11, the discs were dominated by fungi with short adhesive branches, such as *Monacrosporium cionopagum*. In another experiment, different amounts of sucrose were added to the soil (Figure 10.3). When 1–3 g of sucrose was added, there was a rapid rise in activity, followed by a steep decline to zero at 7 weeks. At higher sucrose levels, only negligible predacious activity was detected, but the nematode population continued to rise over the 7-week period of the experiment. Similar experiments were conducted using chopped cabbage in amounts varying from 5 to 20 g per 250 g of soil as an amendment (Cooke, 1962b). The results are shown in Figure 10.4. The numbers of nematodes rose much higher than when sucrose was added. It is noteworthy that preda-

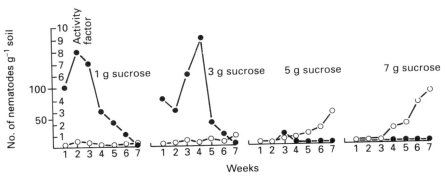

Figure 10.3 The relationship between predacious activity and nematode population during the decomposition of different amounts of sucrose. ○ = Nematode population; ● = activity factor. (Redrawn from Cooke, 1962a, with permission.)

cious activity rose rapidly, and fell to zero within 4–5 weeks. The nematode population also rose rapidly in response to organic matter amendment and then declined, but the nematode population was still at a high level after the predacious activity had fallen to zero. It can be concluded that there is no simple relationship between the population level of nematodes and the activity of predacious fungi.

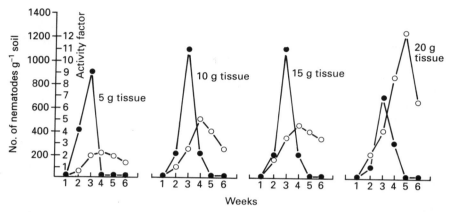

Figure 10.4 Relationship of predacious activity to nematode population during decomposition of different amounts of chopped cabbage leaf tissue. ○ = Nematode population; ● activity factor. (Redrawn from Cooke, 1962b, with permission.)

The consortium of bacteria, fungi, ciliates, rotifers, nematodes and other animals that inhabit sewage includes endozoic nematophagous fungi, especially the zoosporic fungus *Catenaria anguillulae*, whose zoospores can swim readily in the aquatic milieu, and the conidial fungus *Drechmeria coniospora*. Gray (1984) has shown that in small laboratory fermenters charged with mixed liquor samples from domestic sewage works, cyclical variations in the density of the nematode prey and these two parasitic fungi occurred over a time-span of about 20–30 days. The incidence of high percentage infection by *C. anguillulae* lowered the nematode population density from about 185 ml^{-1} to zero within a few days (Figure 10.5). This was followed by a recovery to a density of about 120 ml^{-1} after about 40 days. There was evidence of a similar density-dependent epidemic of *D. coniospora* (Figure 10.6). Such cyclical oscillations in populations are typical of many density-dependent predator–prey relationships.

10.3 ECOLOGICAL CHARACTERISTICS

The demonstration that predatory fungi with adhesive networks dominate early in the succession of such fungi on buried agar discs is probably

ECOLOGICAL CHARACTERISTICS 295

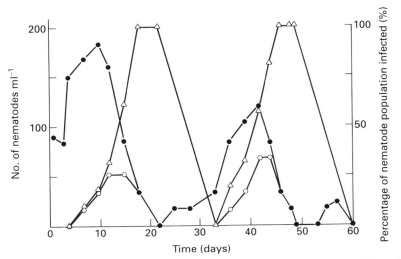

Figure 10.5 The numbers of nematodes infected with *Catenaria anguillulae* (○) and the percentage of the nematode population infected (△) compared with the total number of nematodes (●) in the mixed liquor from Naas Sewage Treatment Works during a 60-day experimental period. (Redrawn from Gray, 1984, with permission.)

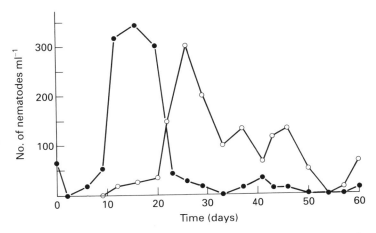

Figure 10.6 Changes in the numbers of nematodes infected with *Drechmeria coniospora* (○) compared with the uninfected (●) nematode population in the mixed liquor from Leixlip Sewage Treatment Works during the experimental period. (Redrawn from Gray, 1984, with permission.)

correlated with the fact that they grow more rapidly than other nematophagous fungi, and that they have the capacity to compete with other members of the soil microflora, i.e. they have a greater **competitive saprophytic**

ability than other types of predatory fungi (Cooke, 1963a; Garrett, 1951). Nordbring-Hertz and Jansson (1984) have classified nematophagous fungi into three groups as shown in Table 10.1. Fungi in group 1, represented by *A. oligospora*, are capable of rapid growth, but are not very effective in lowering the population level of nematodes in natural soil. It is interesting that *Arthrobotrys* spp. have also been shown to coil around the hyphae of some other soil fungi such as *Rhizoctonia*, resulting in their collapse and lysis (Tzean and Estey, 1978; Persson, Veenhuis and Nordbring-Hertz, 1985), showing that their activities in soil are not confined to nematodes. Group 2 nematophagous fungi, exemplified by *Dactylaria candida*, *D. gracilis* and *Monacrosporium cionopagum*, are rather slow-growing, relatively weak saprophytes, but with a considerable ability for lowering the population level of nematodes in soil. This capacity has been termed **predacity**. Group 3 fungi, consisting of the endoparasites *Drechmeria coniospora* and *Harposporium anguillulae*, are very slow-growing. It is likely that they are ecologically obligate parasites, i.e. they have no free-living existence in nature in the absence of their nematode hosts.

The three groups of nematophagous fungi also differ in their ability to attract nematodes. Jansson and Nordbring-Hertz (1979) used an assay in which discs of nematophagous and non-nematophagous fungi were placed in opposite quadrants of Petri-dishes containing dilute cornmeal agar, with control discs containing no fungus in alternate quadrants. A suspension of

Table 10.1 Ecological groups of nematophagous fungi

Characteristic	Nematode-trapping fungi		Endoparasitic fungi
	Group 1	Group 2	Group 3
Trapping organs	Adhesive networks: inducible by nematodes or with chemicals	Adhesive knobs, adhesive branches, (non)-constricting rings: spontaneously produced	Conidia, either adhering to nematode cuticles or ingested by nematodes
Growth pattern	Fast growing and relatively good saprophytes: weak predacious ability	Relatively slow-growing weak saprophytes: great predacious ability	Mostly obligate parasites; very slow growing
Type species	*Arthrobotrys oligospora* *A. conoides*	*Dactylaria candida* *Monacrosporium cionopagum* *D. gracilis*	*Drechmeria coniospora* *Harposporium anguillulae*

Source: Nordbring-Hertz and Jansson (1984).

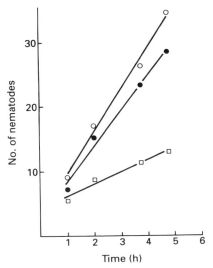

Figure 10.7 Attraction of *Panagrellus redivivus* to nematophagous fungi. The slope of each line indicates the attraction intensity: ○=*Harposporium anguillulae*, group 3; ●=*Monacrosporium ellipsosporum*, group 2; □ =*Arthrobotrys superba*, group 1. Each point is the mean from five replicate plates. (Redrawn from Jansson and Nordbring-Hertz, 1979, with permission.)

nematodes was placed in the centre of the dish, and the numbers of nematodes which had migrated into the different quadrants were counted at intervals up to 6 h. Some typical results are given in Figures 10.7 and 10.8.

The attractiveness (or attraction intensity) of the different groups of fungi can be compared from the slopes of the lines. The steeper the line, the greater the attractiveness. In a similar study, Jansson (1982) showed that the endoparasitic *D. coniospora* was highly attractive to *Panagrellus*.

The strategies of the different ecological groups of nematophagous fungi in attracting nematodes varies. Although, in many cases, the vegetative mycelium is attractive and may even provide a source of food for fungal-feeding nematodes, there is also evidence that the trapping organs themselves provide a further attraction (Field and Webster, 1977; Jansson, 1982). There is a sophisticated 'dialogue' between the nematode prey and the fungal predator. The presence of nematodes stimulates the development of trapping organs, i.e. a switch from a saprophytic to a predacious habit. The triggers or morphogens which induce trap formation are most probably small peptides (Nordbring-Hertz, 1973). The special attractants of the trapping organs have not been identified. There are, however, specific carbohydrates in the nematode cuticle, which bind with saccharide-specific proteins (lectins) contained in the adhesive lining the network

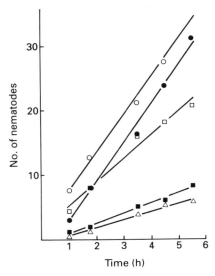

Figure 10.8 Attraction intensity curves for some members of groups 1 and 2 of nematophagous fungi. Group 2: *Dactylaria candida* (○), *Monacrosporium ellipsosporum* (●) and *M. cionopagum* (□). Group 1: *Arthrobotrys oligospora* (■) and *A. musiformis* (△). Each point is the mean from five replicate plates. (Redrawn from Jansson and Nordbring-Hertz, 1979, with permission.)

traps of *A. oligospora*, and there are indications of a similar mechanism for the traps of several other predatory and endozoic fungi (Nordbring-Hertz, 1984; Nordbring-Hertz, Friman and Mattiason, 1982; Tunlid, Jansson and Nordbring-Hertz, 1992). After contact with nematodes, *A. oligospora* mycelium produces a nematotoxin that renders its prey inactive. Filtrates from cultures of the fungus not in contact with nematodes have no apparent effect on the activity of *Rhabditis* species (Olthof and Estey, 1963).

As we have seen, the endoparasites *Harposporium anguillulae* and *Drechmeria coniospora* have mycelia which are highly attractive to nematodes. The conidia of *Drechmeria*, but not those of *Harposporium*, produce a further attractant (Jansson, 1982). Nematodes are attracted to feed in the vicinity of a *Harposporium* mycelium (possibly on a dead nematode) and may ingest the conidia, resulting in infection through the gut. The conidia of *Drechmeria* are stimulated by the presence of nematodes to produce adhesive buds at their narrow end (Dijksterhuis, Veenhuis and Harder, 1990). In the bacterial-feeding nematode *Panagrellus redivivus*, the specific infection sites to which the adhesive buds attach are the sites of chemoreception around the mouth, and there is evidence that a specific carbohydrate is localized at these points, resulting in firm attachment of the adhesive conidia by binding to a lectin (Jansson and

Nordbring-Hertz, 1983, 1984). Penetration then takes place through the cuticle.

Nematoctonus conidia are most frequently found attached by adhesive to the anterior end of the fungus-feeding nematode *Aphelenchus avenae* (Giuma and Cooke, 1971). In several species of *Nematoctonus*, it has been shown that germinating conidia and mycelia can produce nematotoxins. These compounds, which probably have a carbohydrate component, are potent at very low concentrations and are very effective in immobilizing their prey. Contact with one conidium can result in the immobilization of a vigorous nematode long before the cuticle has been penetrated (Giuma, Hackett and Cooke, 1973).

Examination of dead nematodes infected by predatory or endozoic fungi shows that they are very rarely attacked by saprophytic soil fungi. This may be the result of antifungal antibiotic activity against a range of soil fungi, as in the case of cultures of *Drechmeria* and *Harposporium* (Barron, 1977).

The conidia of *Nematoctonus* arise from clamped mycelia, indicating that they are basidiomycetes. Several connections have now been established between the *Nematoctonus* conidial state (anamorph) and teleomorps belonging to the genus *Hohenbuehelia* (Barron and Dierkes, 1977; Thorn and Barron, 1984, 1986). Species from both genera form basidiocarps on thoroughly rotted wood. Possibly the capture of nematodes provides a nitrogen supplement to their nutrition (Barron, 1992).

Another genus of wood-rotting fungi is *Hyphoderma* (Corticiaceae). Primary and secondary mycelia form stephanocysts, unusual two-celled conidia with a crown of hollow spines around the junction between the cells (Hallenberg, 1990). Conidia of this type, possibly of an unidentified species of *Hyphoderma*, attach firmly to the cuticles of the nematode *Aphelenchoides*. Germination, penetration and digestion ensue. Since death of the infected nematode takes up to 14 h, it is not thought likely that a toxin is involved (Liou and Tzean, 1992).

It is of interest that mycelium of several species of basidiomycete growing on rotting wood can rapidly inactivate nematodes, and can penetrate their dead bodies. *Pleurotus ostreatus* mycelium produces slender-stalked cells which secrete droplets of a toxin, ostreatin. Nematodes making contact with these droplets are rapidly paralysed. Fine hyphae from the *Pleurotus* then grow chemotropically towards the mouth and penetrate it. Digestion of the contents of the immobilized nematode then quickly follows (Barron and Thorn, 1987).

The form in which nematophagous fungi exist and survive in soil varies. There is some evidence that saprotrophs may be able to maintain their biomass even in the absence of nematodes, possibly, in the case of *Arthrobotrys* spp., as mycoparasites. Quantitative estimates of abundance suggest that some of the predatory fungi survive in the form of conidia, and it has also been suggested that conidial traps, which may develop in soil in response to fungistasis, may also function as survival structures (Dackman and Nordbring-Hertz, 1992). Chlamydospores develop in some endozoic fungi, e.g. *Verticillium* spp. (Gams, 1988), whilst oospores formed within the cadavers of nematodes or within their cysts may represent survival structures of endoparasitic Oomycetes (see Figure 10.2). Some extension of mycelium from dead infected hosts may occur. It is also known that infected nematodes to which the knob-like traps of *Dactylaria candida* have attached may break free from the supporting mycelium and carry the infection with them (Barron, 1975). Zoospores released from infected nematodes may also effect short-range dispersal.

10.4 BIOLOGICAL CONTROL OF PATHOGENIC NEMATODES

The possibility of using predatory nematophagous fungi in the control of nematode parasites of plants or mushroom mycelium is attractive, and the subject has been reviewed by Soprunov (1958), Mankau (1980, 1981), Tribe (1980), Kerry (1984) and Stirling (1988). Early attempts at control of eelworm populations was based on the observations of Linford, Yap and Oliviera (1938) that the addition of green manure in the form of chopped pineapple or grass leaves caused an increase in the population of free-living nematodes, followed by an increase in the activity of predacious fungi. These captured not only the free-living forms but also the parasitic nematode *Herodera marioni*, the cause of root-knot disease of pineapple. Linford and Yapp (1939) later showed that artificial inoculation of pineapple cuttings with predacious fungi, but especially *Monacrosporium ellipsosporum*, along with organic supplements, was associated with a reduction in the population level of *H. marioni* and a modest increase in growth of the host plant. They claimed that the inoculated fungi could be re-isolated from pots within which the plants were grown up to 15 months later. Experiments by other workers have rarely proved consistent in demonstrating effective biological control using predatory nematophagous fungi (Stirling, 1988). There are several reasons for this. First, it is notoriously difficult to introduce an organism into a soil to which it is not adapted in the face of competition from the native population. Introduced spores of predatory and endozoic nematophagous fungi are subjected to fungistatic effects and either fail to germinate or their germ tubes are quickly lysed (Mankau, 1962; Cooke and Satchuthananthavale, 1968; Giuma and Cooke, 1974). Secondly, to be effective, the predacious fungi must be capable of capturing nematodes at a time when the nematodes are multiplying rapidly. Thirdly, it may be difficult to sustain a high level of activity of a predacious fungus over the whole period of growth of a crop. The levels of organic supplement necessary to sustain a high level of predacious fungus activity may be uneconomic or impractical on a field scale. Since many of the plant-parasitic nematodes burrow deeply into their host tissues, they may not be vulnerable to predacious fungi or are only vulnerable in the brief juvenile stages. It is also likely that the choice of predatory fungus used has been based on rapid growth rate or ready availability, but as we have seen, there appears to be an inverse relationship between high growth rate and predacity. Until a fuller understanding of the complex interrelationships between the crop plant, the nematode, the predatory fungus and its competitors is available, it is premature to expect biological control of plant-parasitic nematodes. However, some success has been claimed following prior application of a commercial preparation of *Arthrobotrys superba* against tomato root-knot nematodes *Meloidogyne* (Cayrol, 1983). Successful commercial application of biological control of nematodes infesting mushroom composts has also been claimed. Mushroom mycelium is attacked by the nematode *Ditylenchus myceliophagus*. Introduction of *Arthrobotrys* on rye grain at the time of spawning (i.e. inoculation of) the mushroom compost has produced a significant increase in yield of mushrooms and a decrease in nematodes (Cayrol *et al.*, 1978).

Success has also been claimed in controlling root-knot of tomatoes caused by *Meloidogyne* using the endozoic fungus *Drechmeria coniospora* (Jansson, Jeyaprakash and Zuckerman, 1985). Reduction in root-galling was reported in the presence of *Drechmeria* but doubts have been expressed as to whether most of the nematodes used in the experiment were capable of infecting tomatoes (Stirling, 1988).

Amongst the egg and cyst parasites, *Paecilomyces lilacinus* shows promise for biological control, but although it has been claimed that the introduction of this fungus is associated with reduction of nematode populations, more convincing evidence is needed that the introduced fungus is responsible (Stirling, 1988). *Dactylella oviparasitica*, a parasite of *Meloidogyne* eggs, has also been shown in glasshouse trials to parasitize the relatively small egg masses of this nematode on peach rootstocks cv. Lovell. On Lovell peach, *Meloidogyne* forms egg masses containing only 250–400 eggs, but on grape rootstocks larger egg masses containing about 1500 eggs are laid. *D. oviparasitica* rarely parasitizes more than 50% of these larger egg masses, so that large numbers of viable eggs remain. Stirling, McHenry and Mankau (1978) believe that whilst *D. oviparasitica* may substantially reduce the egg population on peach roots, it is generally unable to do so on hosts such as grape.

11
PHOENICOID FUNGI

A specialized group of fungi (mostly ascomycetes and agarics) fruit amongst the ashes marking the sites of former fires. The term **phoenicoid fungi**, meaning 'arising from the ashes', has been proposed for them by Carpenter and Trappe (1985), but they are also referred to as **pyrophilous**, **anthracophilous** or **carbonicolous**.

Fires may be started deliberately to clear or manage vegetation (prescribed burning) or to remove slash and other debris following timber extraction. However, natural fires are normal, cyclic events in some types of vegetation, possibly due to lightning and the build-up of flammable materials. Some plant communities are adapted to periodic fire, e.g. stands of longleaf pine (*Pinus palustris*) in the southern USA, and some plant communities are maintained by periodic fires (Mutch, 1970; Ahlgren and Ahlgren, 1960; Kozlowski and Ahlgren, 1974). The ecology of phoenicoid fungi has been studied after fire in forests, prairies, and after volcanic eruptions.

The duration and severity of a fire may be variable. In 'natural' forest fires, although the surface vegetation and standing trees may be destroyed or damaged, the depth of penetration of the fire or its effects into the soil may not be very great (i.e. only a few centimetres). By contrast, where branches and litter have been accumulated into bonfires, the centre of the fire may reach very high temperatures, and the temperature of the soil surface may reach 500°C or more. At such bonfire sites (sometimes referred to as 'fireplaces' or 'burns'), there is often an accumulation of wood ash. Petersen (1970a) distinguished five layers at bonfire sites in Denmark:

1. White ash layer (0.5–4.0 cm)
2. Black ash layer, containing charred organic matter (1–4 cm)
3. Raw humus layer (1–7 cm)
4. Reddish-grey sand layer (4–13 cm)
5. Yellow sand layer.

11.1 CHEMICAL, PHYSICAL AND BIOLOGICAL CHANGES IN SOIL AFTER BURNING

Chemical changes associated with severe burning include an increase in pH and an initial increase is followed by differential leaching of salts. Values as high as pH 9.8–10.2 in the white ash layer have been recorded at bonfire sites in Denmark (Petersen, 1970a), and other workers have reported increases of 3–5 pH units following burning. The increase in pH in the white ash is due to the accumulation of salts which give an alkaline reaction in solution, such as carbonates, hydrogen carbonates and hydrogen phosphates. In the course of time as rainfall causes leaching of the surface layers, there is an increase in the pH in the raw humus layers beneath. A consequence of the increased pH of soil after burning is that bacterial populations may increase, including those of N_2-fixing bacteria such as *Azotobacter* and *Clostridium* (Ahlgren, 1974). The white ash layer is the result of more or less complete combustion of branches, so that it contains some elements in higher concentration than in the original material. As leaching proceeds, the concentrations of substances present in the ash will fall, but, because of differential solubility, the level of certain substances at the surface will fall more rapidly than others, accompanied by transfer to the raw humus layer. Some of these changes are illustrated in Figure 11.1 and in Table 11.1.

Figure 11.1 shows that although there is a steep decline in water-soluble

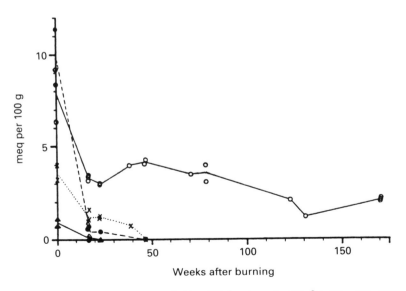

Figure 11.1 Content of water soluble CO_3^{2-} (x...x), SO_4^{2-} (●- -●), HCO_3^- (○–○), and Cl-(▲–▲) in the white ash at bonfire sites at Grib Skov II, Denmark. (After Petersen, 1970a, with permission.)

Table 11.1 Content of HCl-soluble and water-soluble metallic cations in the white ash of bonfire sites at Grib Skov II, Denmark (meq 100 g^{-1})

Weeks after burning	HCl-soluble				Water-soluble			
	Ca	Mg	K	Na	Ca	Mg	K	Na
0	525	130	78	22.5	0.42	2.22	30.0	3.98
17	525	118	54	23.3	0.28	0.38	4.68	0.87
47	480	93	40.5	15.0	0.45	1.47	2.03	0.56
79	465	87	34.5	15.0	0.64	2.03	2.63	0.89
171	157	79.5	15.7	7.05	0.17	1.06	1.17	0.32

Source: Petersen (1970a).

carbonate, sulphate and chloride, all of which have disappeared within 1 year, the hydrogen carbonate level remains high for over 3 years.

The white ash originally contains large amounts of calcium, but most of this is HCl-soluble rather than water-soluble, suggesting that it is mainly in the form of carbonate, phosphate or sulphate, rather than oxide or hydroxide. The level of water-soluble Ca was reduced to one-quarter after 171 weeks. The white ash also contains large amounts of HCl-soluble and water-soluble phosphorus immediately after burning, but the levels of each are approximately halved over a period of 171 weeks. Similar results have been found for magnesium. The salts of potassium and sodium, being readily soluble, are readily leached from the ash (Petersen, 1970a).

Physical changes also occur at bonfire sites. In an extensive fire, the canopy may be destroyed by the fire, or may have been opened by tree felling, exposing the soil to increased solar radiation. The ground vegetation and the surface litter may be destroyed and replaced by a layer of blackened ash which increases the absorbance and decreases the reflection of radiation. These effects combine to cause the burnt soil to have higher temperatures than soil from unburnt areas even after the fire has burnt out. Other changes in the properties of the burnt soil are increased exposure to direct rainfall, lower infiltration rates, and increased run-off.

Burning affects the living organisms present in the soil. The destruction of the soil fauna, especially animals which burrow in soil, may reduce soil porosity. It is important to distinguish between the effects of light burning or annual prescribed burning, and more severe fires resulting from periodic burning over longer intervals or slash burning. Annual prescribed burning, carried out to reduce the risk of natural fires, reduces the amount of organic matter in the F and H layers of the soil profile, and brings about a consequent reduction in the total numbers of fungi and bacteria. However, there appears to be little change in the numbers of organisms per gram of soil. No obvious qualitative differences were found in the fungi isolated by Jorgensen and Hodges (1970). After more severe slash fires in October, counts of bacteria the following May increased 25-fold, and the numbers of fungi decreased (Wright and Tarrant, 1957). In areas of sustained high temperature, bacteria, fungi, etc. are killed, but at the margin of the heated zone there is evidence of a selective effect in that certain organisms with resistant spores, such as the ascospores of certain soil fungi, may be stimulated to germinate by the effects of heat. The ecological effects of burning have been reviewed by Ahlgren and Ahlgren (1960, 1965), Ahlgren (1974), Bissett and Parkinson (1980), Gochenaur (1981), Christensen and Muller (1975) and Warcup (1981). In time, the burnt patches become re-vegetated, usually at first by mosses such as *Funaria* and *Ceratodon*, the liverwort *Marchantia* and by flowering plants such as *Chamerion angustifolium*.

In the moist tropics, the most conspicuous fungus to appear on charred vegetation after fire is *Neurospora crassa*, which forms profuse pink conidia. In temperate latitudes, the characteristic phoenicoid forest fungus flora is dominated by operculate discomycetes (Pezizales) and a few agarics. Edaphic factors such as the nature of the underlying rock, and the depth of soil or humus, and biotic factors such as whether the woodland is dominated by conifers or deciduous trees, and the development of moss cover over the burnt site, have their effects on the fungi which occur. Climatic effects based on the time of the fire in relation to season, rainfall, etc., also play an important role. Most ecological studies of fungi have been based on the presence and abundance or absence of fruit-bodies (ascocarps, basidiocarps), which in some species develop within a few days or weeks of burning, whilst in others only after several months. Fruit-bodies of some phoenicoids are illustrated in Figure 11.2.

11.2 ECOLOGICAL CHARACTERISTICS AND PHENOLOGY

From a survey of a large number of sites visited on several occasions, Petersen (1970a) has classified these fungi into four groups:

- **Group A**: Species which occur exclusively on burnt ground. These include *Anthracobia* spp., *Ascobolus carbonarius*, *Peziza anthracina*, *P. echinospora*, *P. petersii*, *Sphaerosporella hinnulea*, *Trichophaea abundans* (Discomycetes) and *Geopetalum carbonarium*, *Pholiota carbonaria*, *Tephrocybe carbonaria* (agarics).
- **Group B**: Species which, under natural conditions, occur exclusively on burnt ground, but may occur on disturbed unburnt ground. These include *Geopyxis carbonaria*, *Octospora* spp., *Peziza atrovinosa*, *P.*

Figure 11.2 Fruit-bodies of some common phoenicoid fungi. (a) *Peziza petersii* apothecia. (b) *Rhizina undulata* apothecia at the base of a pine tree damaged by fire.

praetervisa, *Trichophaea gregaria* and *T. hemisphaeridoides* (Discomycetes).
- **Group C**: Species which, under natural circumstances, are common on burnt ground, but which, under certain circumstances, occur on unburnt ground, e.g. *Rhizina undulata* (Discomycete) and *Coprinus angulatus*,

Figure 11.2 *contd.* (c) *Pholiota carbonaria* basidiocarps. (d) *Myxomphalia maura* basidiocarps. Scale bar = 2 cm (a); 4 cm (b, c, d).

Omphalia (Myxomphalia) maura (agarics). This group should probably also include *Pyronema* spp. which may occur on sterilized soil and on fresh plaster prepared from slaked lime.
- **Group D**: Species which occasionally occur on burnt ground, but which are more common on unburnt ground.

Moser (1949) has also classified these fungi into ecological groups. He has termed groups A and B **Anthrakobionte Pilze** (obligate fireplace fungi) and groups C and D **Anthracophile Pilze** (fungi favoured by burning). Ebert (1958) has produced a similar classification.

The **phenology** (time of appearance) of fruit-bodies has been studied by Petersen (1970a) by recording the frequency of fruit-body presence on a large number of sites over a 3-year period. The relative frequency (i.e. percentage of sites at which fruiting was recorded) plotted against month of occurrence, gives a good indication of seasonal abundance (Figures 11.3 and 11.4).

The relative frequency throughout the year of ascocarps of *Anthracobia melaloma* is shown in Figure 11.3 comparing two sites at which burning took place at the same time. The earlier fruiting at Magleby III (Figure 11.3b), where the fungus had reached maximum frequency even before it had appeared at Tureby, was ascribed to the fact that Magleby III was a moister site than Tureby.

Four agarics – *Myxomphalia maura*, *Geopetalum carbonarium*, *Ripartites tricholoma* and *Tephrocybe carbonaria* – fruited only in the autumn at all sites in three successive years (Figure 11.4). Other fungi were able to fruit in the spring and summer provided that sufficient moisture was available, e.g. *Coprinus angulatus*, *Geopyxis carbonaria*, *Trichophaea hemisphaerioides* and *Peziza praetervisa*. Petersen has concluded that most fireplace fungi are able to form fruit-bodies throughout the spring, summer and autumn, and that the usual summer minimum in fruit-body presence is due to unfavourable moisture conditions. This ability to use any period with suitable temperature and moisture conditions to form fruit-bodies must be advantageous to fungi which occur mainly or exclusively on such an unstable and ephemeral habitat as burnt ground.

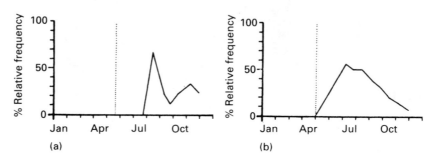

Figure 11.3 Seasonal variation graph showing the relative frequency of ascocarps of *Anthracobia melaloma* at two sites. (a) Eighteen bonfire sites at Tureby, Denmark. (b) Sixteen sites at Magleby III, Denmark. The vertical dashed line indicates the beginning of the investigation. (After Petersen, 1970a, with permission.)

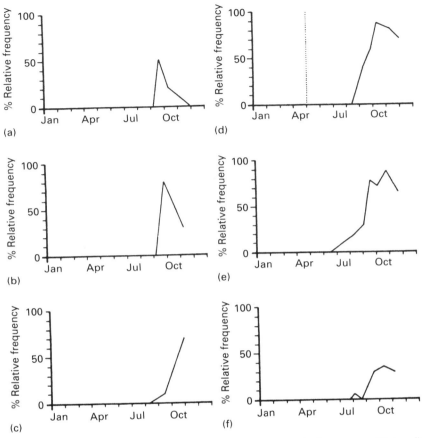

Figure 11.4 Seasonal variation graphs. (a)–(c) *Myxomphalia maura*; 10 bonfire sites at Grib Skov I. (a) 1963; (b) 1964; (c) 1965. (d)–(f) *Tephrocybe carbonaria*; 17 sites at Magleby II. (d) 1963; (e) 1964; (f) 1965. (After Petersen, 1970a, with permission.)

It is also of interest to know how long after burning the individual fungi take to fruit, and for how long after burning they are able to continue to fruit. Petersen (1970a) has presented information in the form of cumulative frequency curves based on the incidence of fruit-bodies at a number of sites. An example is given in Figure 11.5. The cumulative curves are shown in pairs: an **accumulative curve** and a **decumulative curve**. The accumulative curve shows, for a given locality, the number of bonfire sites where a species has been found up to a given time after burning as a percentage of the total number of sites where it occurred. The decumulative curve shows

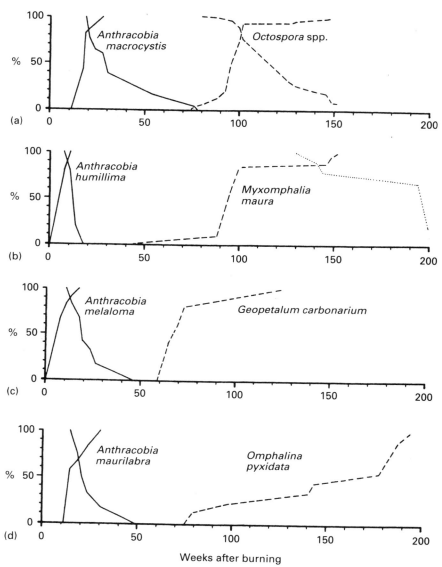

Figure 11.5 Cumulative fruiting frequency curves of phoenicoid fungi at bonfire sites in Denmark.(a) Solid lines: 23 sites at Tureby; dashed lines: 132 sites at Grib Skov II. (b) Solid lines: 5 sites at Hareskoven II; dashed line: 108 sites at Grib Skov II; dotted line: 10 sites at Grib Skov I. (c) Solid lines: 12 sites at Hareskoven II; dashed line: 5 sites at Horserod Hegn.(d) Solid lines: 12 sites at Tureby; dashed line: 9 sites at Grib Skov I. Ascending curves accumulative curves; descending curves decumulative curves. (After Petersen, 1970a, with permission.)

the number of bonfire sites where a species has been found after a given time after burning.

From information of this kind, the phoenicoid fungi have been arranged into four groups in relation to incidence of fruiting following burning:

- **Group I**: Species which, under favourable climatic conditions, appear on bonfire sites as early as 7 weeks after burning, but which do not fruit later than 80 weeks after burning. This group is exemplified by *Anthracobia* spp. and also by *Pyronema* spp. (Moser, 1949) and *Trichophaea abundans* (Turnau, 1984a).
- **Group II**: This is a heterogeneous group of species which fruit 10–15 weeks after burning. The duration of fruiting is very variable: for *Peziza trachycarpa* not later than 100 weeks after burning, for *P. praetervisa* and *Tephrocybe carbonaria* up to 150 weeks after burning, and for *Pholiota carbonaria* 190 weeks after burning. *Ascobolus carbonarius* and *Rhizina undulata* are also included in this group.
- **Group III**: This group comprises only *Trichophaea hemisphaerioides* and *Peziza endocarpoides*, which take 20–50 weeks to appear, and persist for 130–200 weeks.
- **Group IV**: This group is also heterogeneous and includes species which do not usually appear until 50 weeks after burning and may not disappear for 150 weeks, e.g. *Ripartites tricholoma*, or 200 weeks, e.g. *Myxomphalia maura*. Also included here are *Neotiella hetieri* and *Octospora* spp., whose apothecia are generally associated with the moss carpet.

Broadly similar fruiting patterns were found following a natural forest fire at Klosterhede Plantage in Denmark by Petersen (1971), although the severity of the fire was not as great as at bonfire sites, and there was less accumulation of ash and a smaller increase in pH.

In considering this grouping, it is important to stress that it is based solely on the appearance of fruit-bodies, and that we have no information about the extent of mycelial development in the period preceding fruiting.

The distribution of fruit-bodies of certain phoenicoid fungi may give a clue to the trophic relationships of the mycelia. *Geopyxis carbonaria* and *Rhizina undulata* are generally restricted to coniferous forests. Petersen (1971) noted that following a natural forest fire in Denmark, the fructifications of *Anthracobia maurilabra* occurred in hollows in the ground surrounding the trunks of charred *Picea*, and often followed the distribution of the horizontally disposed roots lying at the boundary between the raw humus layer and the mineral soil. The fructifications of *A. macrocystis*, *G. carbonaria* and *P. praetervisa* were distributed in the same way, and it is presumed that all these fungi are saprotrophs involved in the decay of roots. Moser (1949) has reported that *G. carbonaria* is also associated with charred needles of *Picea* (see also Ebert, 1958). Others may be associated with living root systems. A good example is *Rhizina undulata*, which is especially common at bonfire sites in pine plantations growing on acid soil. The characteristic yellow mycelium can be found in the raw humus layer, but extends on to living conifer roots where it may grow parasitically and spread to other trees, causing group death (Jalaluddin, 1967a,b; Gremmen, 1971; Ginns, 1968, 1974). Tests in pot culture of the possible pathogenicity of post-fire discomycetes to seedlings of lodgepole pine (*Pinus contorta*) have shown that whilst most are non-pathogenic, some could cause significant reduction in seed germination, e.g. *Pyropyxis rubra* and *Peziza endocarpoides*. Disease symptoms and reductions in radicle length were also associated with infection by *P. rubra*, *R. undulata* and *A. carbonarius* (Egger and Paden, 1986a). Associations of a different kind have also been demonstrated. *Sphaerosporella brunnea* forms ectomycorrhizae with *Pinus*, *Picea*, *Larix* and *Populus* (Danielson, 1984). *Anthracobia maurilabra* and *A. tristis* grow on the surface of living roots of pine, and may also invade the cortex (Egger and Paden, 1986b). Similarly, Warcup (1990) has shown that several species of post-fire discomycetes can form sheathing mycorrhizas with *Eucalyptus* and *Melaleuca*.

Trophic relationships of post-fire discomycetes have been studied by testing their enzymic activities when grown on a range of chemically defined substrata (Egger, 1986). Egger has classified the decomposer community into:

1. Species that utilize litter and (or) fine roots. This community includes *Anthracobia* spp., *Ascobolus carbonarius* and *Trichophaea abundans*. They are weak producers of phenol oxidases and do not degrade lignin. They appear to use the major non-lignified components of the soil.
2. Species that utilize wood and (or) woody roots. This group was exemplified by *Peziza* spp. They produce phenol oxidases and a wide range of hydrolytic enzymes including cellulase, indicating a generalized pattern of substrate utilization.
3. In addition, there is a community whose mycelia are associated with living roots, including *Pyropyxis rubra*, *Rhizina undulata* and *Sphaerospora brunnea*, which produce phenol oxidases but do not degrade lignin.
4. A fourth group of species did not fit into these three categories, e.g. *Geopyxis carbonaria* and *Trichophaea hemisphaerioides*, which formed associations with living roots but also degraded lignin. They may be weak parasites with the capacity to infect moribund roots.

Regular fires may be used in the management of agricultural land, as in the spring burning of prairies in North America or in the removal of cereal straw following combine-harvesting. Wicklow (1973, 1975) and Zak and Wicklow (1978a,b) have shown that there is a characteristic flora of ascomycetes associated with burnt prairie soil. By contrast with the community of operculate discomycetes which dominates forest bonfire sites, grassland sites are dominated by pyrenomycetes (Sphaeriales and Loculoascomycetes) including such genera as *Coniochaeta*, *Chaetomium*, *Podospora*, *Sordaria* and *Sporormiella*, many of which are also coprophilous and fruit on the dung of herbivores (Chapter 8). The ascocarps of these carbonicolous ascomycetes are inconspicuous, and their presence is best demonstrated by incubating burnt or heat-treated soil in Petri dishes with small amounts of straw or filter paper spread on the surface, on which the ascocarps can develop. Examination of the underside of the Petri dish lid for the deposits of discharged ascospores is a useful way of confirming sporulation.

Wicklow (1975) has shown that there is a fruiting succession of ascomycetes when burnt prairie soil is incubated in the laboratory, and this is shown in Table 11.2

11.3 EXPERIMENTAL STUDIES

There are a number of observations and experimental studies which throw light on the ecology of phoenicoid fungi. Stimulation of growth and fruiting may be the result of a number of different post-fire modifications to the soil.

Availability of nutrients

Heat treatment of soil, e.g. by steaming or autoclaving, but without burning, results in increased solubility of both inorganic and organic substances. There may be not only quantitative, but also qualitative changes, e.g. an increase in the amount of low molecular weight organic

Table 11.2 Total and % occurrence (in parentheses) of ascocarps on soil collected from a burnt prairie stand

	Length of incubation (days)				Total occurrence
	7	14	25	50	30 dishes
Coniochaeta discospora	13				13 (43)
C. tetraspora	12				12 (40)
Gelasinospora calospora	7				7 (23)
Podospora glutinans		9	10	10	29 (97)
P. vesticola		14	5	8	27 (90)
Ascobolus carbonarius				4	4 (13)
Podospora curvispora				18	18 (60)
Sporormiella pilosella			1	11	12 (40)

Source: Wicklow (1975).

compounds (Christensen and Muller, 1975). Evidence that heating alone, in the absence of burning, can stimulate the fruiting of certain pyrophilous fungi, is shown by the appearance on steamed glasshouse soils of ascocarps of *Pyronema* and of other fungi such as *Peziza ostracoderma*.

Stimulation of germination by heating

Heating may stimulate the germination of ascospores. El-Abyad and Webster (1968a) showed that a short exposure to 50°C enhanced the germination of a number of pyrophilous discomycetes. In some cases, e.g. *Pyronema domesticum, Ascobolus carbonarius* and *Peziza praetervisa*, ascospore germination occurred readily at 20°C, but was further stimulated by heat treatment. For pyrophilous pyrenomycetes, Zak and Wicklow (1978a) showed enhanced germination of several species following exposure to steam at 35–110°C for up to 3 mins. In burnt prairie soils, the heat from burning may stimulate the germination of ascospores not only of the phoenicoid fungi, but also of coprophilous fungi lying dormant in the soil (Wicklow, 1975). For *Rhizina undulata*, Jalaluddin (1967a) showed that heating ascospores at 37°C for 3 days stimulated germination and that treatment at 45°C for 8 h was also effective. Heat-treated ascospores suspended in Seitz-filtered exudates from heated pine root segments showed increased germination over treatment in which exudates from untreated segments were used. The activation of ascospore germination of *Neurospora* spp. by heat is well-known. Those of *N. crassa* and *N tetraspora* survive activation after periods of 5–30 min at temperatures of 50–70°C (Sussman and Halvorson, 1966). The ascospores of *Neurospora* have black, thick-ribbed walls, but the conidia (*Monilia*) have thin walls, and are capable of germination without heat treatment, but can survive elevated temperatures of 45–50°C for 5 min (*N. tetraspora*) or 35–37°C for 1–3 h (*N. sitophila*).

c) Reduced competition

Heating may reduce competition from other soil microorganisms. One effect on soil of the heat generated by burning is to kill most of the organisms present or to reduce their activity, and it is possible that phoenicoid fungi take advantage of the reduced competition to colonize and fruit. El-Abyad and Webster (1968b) showed that *Pyronema domesticum* would fruit readily on autoclaved burnt soil if inoculated by ascospores before or at the same time as non-sterile soil. If, however, non-sterile soil was added to the sterilized soil 3 days before the *Pyronema* ascospores, no fruiting was observed. *Pyronema* spp. and *Trichophaea abundans* are capable of very rapid mycelial growth, and this may also enable them to colonize burnt soil in advance of competitors, and the same may also be true of *Neurospora* spp. Not all phoenicoid fungi grow rapidly, however, and *Peziza praetervisa* is outgrown by a number of common soil fungi.

Similar findings for prairie phoenicoid fungi have been made by Wicklow (1975) and Wicklow and Zak (1979), who inoculated heat-activated ascospores of several of these fungi onto sterilized (autoclaved) prairie soil. Here they fruited readily, but when heat-activated ascospores were added to non-sterile soil, no fruiting occurred, presumably as a consequence of competition.

Phoenicoid fungi must also compete amongst themselves (Turnau, 1984b). Amongst the community of fungi which colonize spring-burnt tallgrass prairie soil, Wicklow and Hirschfield (1979b) established that there was a competitive hierarchy. Two species which fruited late in the succession (after 6 weeks on sterilized soil), *Sporormiella pilosella* and *Podospora pilosa*, were capable of limiting colony growth in culture of five other ascomycetes which fruited earlier in the succession. An antibiotic was obtained from culture filtrates of *P. pilosa* which inhibited the growth of some of the earlier-fruiting fungi. Whether such an antibiotic is produced or is active in burnt soil is not known.

It therefore seems likely that the restriction of phoenicoid fungi to burnt ground is a result of their inability to compete with other microorganisms in non-burnt soil. The spores of most soil fungi fail to germinate when added to non-sterile soil – a phenomenon termed **fungistasis** or **mycostasis** (Lockwood, 1977) (section 7.3). Phoenicoid fungi are exceptional in that heat-activated ascospores, when added to non-sterile (fungistatic) soil are capable of germination (Wicklow and Zak, 1979; Lockwood and Filonow, 1981; Wicklow, 1989). It has been argued that, since phoenicoid fungi are at a competitive disadvantage in unburnt soil in relation to other soil fungi, once their spores have been triggered by heat to germinate, any further check to germination would be suicidal, since it would allow competing organisms to establish themselves.

Chemical changes

Chemical changes in the soil following burning may affect the growth and fruiting of phoenicoid fungi. Evidence that some phoenicoid fungi may be adapted to the unusual chemical composition of ashes has been obtained from two sources. First, it has been observed that several of these fungi are not confined to burnt soil, but can also fruit on disturbed ground, especially where lime-rich materials such as building rubble, mortar and old plaster have been tipped. Petersen (1970b, 1985) has listed several such occurrences, including the fruiting of *Peziza praetervisa*, *Trichophaea hemisphaerioides* and *Geopyxis carbonaria*. Analysis of soil samples shows that they almost always contain $CaCO_3$ and have high pH values. Secondly chemical treatment of soil in the absence of heating may result in the fruiting of fungi usually found only on burnt soil. Hora (1959) applied horticultural fertilizers or hydrated lime ($CaCO_3$) to experimental plots within Scots pine plantations 25, 30 and 80 years old, growing on an acid sandy soil of pH 3.3–4.5. Weekly counts were made of the numbers of basidiocarps. In the lime-treated plots which received 20 g m^{-2} in the winter months of 1956, there was an enormous increase in the number of basidiocarps of *Myxomphalia maura* in the autumn of the following seasons.

Peterson (1970b) has made similar, but more extensive, observations. In one series of observations at Nyrup Hegn on a sandy moraine in Denmark, separate small experimental plots of 1 m^2 in a *Picea* plantation were treated at monthly intervals with 2.5 kg $CaCO_3$. This is a very high rate of application. It resulted in a pH in the surface layers of the soil of c. 8. The plots were visited at 1–2 week intervals and examined for the fruit-bodies of phoenicoid fungi. Results showing the time in weeks between the application of lime and the first appearance of ascocarps and basidiocarps are given in Table 11.3.

It is clear that many phoenicoid fungi are stimulated to fruit by the application of lime alone, i.e. without heating of the soil. The time of appearance is related to the date of liming. Treatments in August and September are associated with the shortest time intervals for fruiting (5–7 weeks). The lime-treated plots developed dense moss carpets of *Ceratodon* and *Funaria*, and the fruiting of the discomycetes *Lamprospora* and

Table 11.3 Number of weeks from the treatment with $CaCO_3$ until the appearance of the species on the plots, in relation to the time of the treatment

Species	Plot treated in the beginning of											
	Nov	Dec	Jan	Feb	Mar	Apr	May	Jun	Jul	Aug	Sep	Oct
Ascobolus pusillus	41	40	34	35	40	23	18	17	9	11	33	31
Myxomphalia maura	105		97	93	85	83	78	72	65	61	59	
Trichophaea hemisphaerioides			97							62	59	
Iodophanus carneus	53		37	41	25		17		8	5	7	
Lamprospora dictydiola	105			92	88			74				
Octospora spp.	79	72	71	66	60	57	50	46	41	40	58	54
Omphalina pyxidata			95									
Peziza endocarpoides	52	44	43	40			18	18	11	8	9	
P. granulosa	49		43									
P. praetervisa	42	47	43	34	28	25	20	17	8	5	37	

Source: Petersen (1970b).

Octospora and the basidiomycete *Omphalina pyxidata* were associated with the mosses. Petersen concluded that the main reason for the fruiting of the fungi was the increase in pH rather than the presence of calcium. In other plots in which calcium was presented in the form of $CaCl_2$ or $CaHPO_4$, there was no increase in the fruiting of phoenicoid fungi.

Natural burning of soil produces a combination of chemical and physical changes, and Zak and Wicklow (1980) have conducted elegant experiments using prairie soils to investigate the effects of steam heating and the presence of ash on the ascomycete community which develops in such soils in the presence or absence of competing microorganisms such as would be present in underlying soil unaffected by temperature changes during a fire. Samples of prairie soil were collected in the spring, prior to spring burning, and small subsamples (4 g) were subjected to the following treatments:

1. Steam treatment by aerated steam at up to 80°C for 60 s. These samples were regarded as controls.
2. Steam treatment followed by the deposition of a thin layer of the ashed remains of dead prairie plants.
3. Steam treatment, following which the steamed soil was placed over a layer of untreated (i.e. non-sterile) prairie soil.
4. A combination of treatments 2 and 3, i.e. steamed soil underlaid by non-sterile soil and overlaid by ashes. This treatment, which simulates that which surface soils would suffer during a natural fire, is termed **'simulated burn'**.

The treated soils were incubated in Petri dishes (six replicates). The surface of the soil was scanned for ascocarps (mostly small apothecia, perithecia and pseudothecia), and the undersides of the lids were scanned for deposits of projected ascospores. The number of species fruiting was recorded, and a species diversity index calculated. Results are shown in Tables 11.4 and 11.5.

It is notable that many of the fungi identified, with the exception of *Dasyscyphus virgineus* and *Ascobolus carbonarius*, are coprophilous. Column 1 shows that steaming alone stimulated the fruiting of 18 ascomycetes. The addition of ash to steamed soil reduced the number of species to 10, a reduction of 44%, but two species, *Podospora curvispora* and *Gelasinospora calospora*, responded to the addition of ash by fruiting in all six dishes. The presence of a layer of non-sterile soil below the steamed soil also reduced the number of species fruiting, possibly as a result of competition. When the treatments were combined (simulated burn), the total number of species (12) and some individual species frequencies were higher than when such treatments were applied singly.

Further analysis of the data is given in Table 11.5. The greatest number of species per sample, the total number of species and the highest species

Table 11.4 The effect of ash and layering with untreated soil on the frequency (%) of occurrence of post-fire ascomycetes among soil samples steamed for 60 s

Species	Treatment			
	Control (steamed)	Ash deposition	Untreated soil layer	Simulated burn
Arnium leporinum	17			
Chaetomium brevipilum	17			
C. spirale	67			
Delitschia araneosa	17			
Sordaria macrospora	33			
Coniochaeta discospora	50	17		
Dasyscyphus virgineus	17	17		
Podospora curvispora	50	100		83
Chaetomium longirostre	17		17	
Podospora sp. 1	17		17	17
Podospora dakotensis	17			100
Gelasinospora calospora	50	100	17	83
Podospora curvicolla	100	50	17	100
P. glutinans	50	50	50	17
P. pilosa	100	50	67	67
Sporormiella pilosella	83	17	33	50
S. subtilis	83	33	33	67
Ascobolus carbonarius	17	33		
Chaetomium indicum			17	
Podospora sp. 2			17	
Sordaria fimicola			17	50
Chaetomium murorum				17
Podospora collapsa				17

Source: Zak and Wicklow (1980)

diversity were found in soils heated at the intermediate temperatures of 55 and 70°C, and with the simulated burn treatment.

Zak and Wicklow (1980) have concluded that the organization and structure of the post-fire ascomycete community is, in part, determined by abiotic factors such as elevated temperatures, the deposition of an ash layer and biotic factors associated with subsurface (untreated) soil layers. Temperatures of 55–70°C promote ascospore germination and, at the same time, cause a significant reduction in the biomass of competing microbes. In the absence of competition, the ascomycetes are stimulated to fruit. The high pH and chemical composition of the ash may also limit the growth of

Table 11.5 Response of ascomycete community developing from prairie soil steamed at five different temperatures to ash and layering with untreated soil

Treatment	Temperature (°C)					Significance		
	35	40	55	70	85	Temperature	Treatment	Interaction
Number of species per sample[a]								
Control (steamed)	2.9	0.8	4.0	2.1	1.4	***	***	***
Ash deposition	1.5	1.1	2.2	2.2	<0.1			
Untreated soil layer	1.2	<0.1	0.1	0.4	1.0			
Simulated burn	1.3	<0.1	4.6	4.2	0.5			
Total no of species[b]								
Control	4.5	3.0	6.5	4.0	3.5	***	***	NS
Ash deposition	3.5	1.5	3.5	4.5	0.5			
Untreated soil layer	4.0	0.5	2.0	2.0	3.0			
Simulated burn	4.0	0.5	7.0	7.0	2.5			
Species diversity[c]								
Control	0.49	0.25	0.57	0.39	0.32	***	***	NS
Ash deposition	0.32	0.13	0.42	0.38	0.00			
Untreated soil layer	0.34	0.00	0.18	0.16	0.26			
Simulated burn	0.37	0.00	0.63	0.58	0.18			

***, $P < 0.001$; NS, not significant.
[a] Mean of six samples.
[b] Mean number of species observed from two replicates (three samples per replicate) for each temperature–treatment combination.
[c] Brillouin diversity index based on two replicates (three samples per replicate) for each temperature–treatment combination.
Source: Zak and Wicklow (1980).

competitors. It seems possible that, as soil alkalinity decreases, recolonization of the soil by competitors may begin. These authors have summarized their ideas in the form of a diagram (Figure 11.6).

11.4 FRUITING OF PHOENICOID FUNGI FOLLOWING VOLCANIC ERUPTIONS

One of the most exciting recent discoveries in fungal ecology is that, following the volcanic eruption of Mount St Helens in Washington State, USA, large numbers of fungi normally associated with forest fire disturbance grew, often in great abundance, on the volcanic deposits. The main eruption occurred in May 1980, associated with clouds of hot volcanic ash (**tephra**), landslides and avalanches from the snow-covered upper slopes.

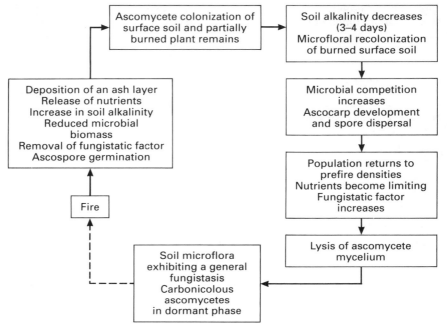

Figure 11.6 Suggested relationships between abiotic and biotic factors determining the structure and response of a post-fire ascomycete community. Processes within each box, except for the fire, are a consequence of the preceding events and represent key stages in the development and decline of the post-fire ascomycete community. (Redrawn from Zak and Wicklow, 1980, with permission.)

A large area was covered by a blanket of hot debris, and blast winds moved through forests. It has been estimated that within 10–20 km of the crater, temperatures ranged from 200 to 340°C, falling to *c.* 150°C within 10 min. Carpenter, Trappe and Ammirati (1987) have given a graphic account of the succession of phoenicoid fungi which fruited in the devastation zone in the following 3 years. Within 6 weeks of the eruption, the anamorph (conidial state) and apothecia of *Anthracobia melaloma* were found in wet depressions in the tephra, or in moist areas shaded by fallen trees. Perithecia of *Gelasinospora reticulospora* were also found at this time. One year later, apothecia of the larger phoenicoid discomycetes were found, including *Peziza* spp., *Trichophaea hemisphaerioides* and *Rhizina undulata*. Agarics such as *Coprinus plicatilis*, *Pholiota carbonaria*, *Psathyrella carbonicola* and *Schizophyllum commune* were also common. In September and October 1981, the numbers of basidiomycetes fruiting had increased, *Myxomphalia maura* being abundant on wet tephra. The numbers of ascomycetes had decreased, but *Aleuria aurantia* was abundant on tephra along stream banks. The succession was thus remarkably similar to that found following forest fires.

In summary, it becomes obvious that although phoenicoid fungi share similar habitats, they may do so for a variety of reasons. Some are stimulated to fruit by heat, others by increased pH or altered soil chemistry, others by disturbance. Many are apparently intolerant of competition. Others are associated with the moss carpet which develops on bonfire sites. One of the puzzling features of their biology is the form in which they survive in the time between successive fires, as spores (ascospores, basidiospores, conidia), mycelium or sclerotia.

12
FUNGI OF EXTREME ENVIRONMENTS

12.1 THERMOTOLERANT AND PSYCHROTOLERANT FUNGI

The majority of fungi are **mesophiles**, and grow at temperatures in the range of 5–35°C with optimum temperatures for growth between 25 and 30°C. Fungi exist, however, which grow very well outside the mesophilic temperature range; some are cold-tolerant (**psychrotolerant**) and are capable of growth near or below 0°C, while others are heat-tolerant (**thermotolerant**) and grow above 40°C. Many thermotolerant fungi grow over most of the mesophilic range but others are so highly adapted to growth in hot conditions that they can grow only at relatively high temperatures. These are the so-called **thermophiles** (heat-lovers) with high optimum temperatures for growth, and defined by Cooney and Emerson (1964) as those fungi unable to grow below 20°C. Similarly, among psychrotolerant species **psychrophiles** exist with low maximum and optimum temperatures for growth (Table 12.1).

While dividing fungi into different groups with respect to tolerance of temperature extremes is useful to distinguish between the physiological behaviour of different species, these divisions are rather arbitrary and are, of course, quite artificial since psychrophiles, mesophiles, etc. are all part of a continuum of response to temperature by fungi.

12.1.1 Thermotolerant and thermophilic fungi

Thermotolerant and thermophilic fungi are common in large, well-insulated, damp heaps of plant materials where the waste heat from microbial metabolism becomes trapped, causing the internal temperatures of the heap to rise in a self-heating (**thermogenetic**) process. A commonplace situation where heat-tolerant fungi thrive is in garden compost heaps. Piling plant debris into heaps is a simple but effective way of raising

Table 12.1 Cardinal temperatures (°C) for growth of some mesophilic, heat-tolerant and cold-tolerant fungi

Species	Minimum	Optimum	Maximum
Thermophiles			
Rhizomucor (Mucor) miehei	25	42–45	57
Rhizomucor (Mucor) pusillus	20	40–45	55
Chaetomium thermophile	27	50–52	58
Talaromyces thermophilus	27	47.5	59
Humicola insolens	23	45–47	55
Thermomyces lanuginosa	30	47.5–52	60
Thermotolerants			
Aspergillus fumigatus	12	40	52
Coprinus cinereus	13	–	45
Neurospora sitophila	4	36	44
Mesophiles			
Mucor mucedo	8	20	30
Mucor hiemalis	8	20–25	35
Mortierella ramanniana	10	23–25	35
Penicillium chrysogenum	4	23	36
Psychrophiles			
Cylindrocarpon magnusianum	0	c.5	c.15
Mucor strictus	0	c.10	20
Mucor oblongisporus	0	c.10	20
Candida scottii	0	c.10	15–16
Psychrotolerants			
Phacidium infestans	−3	15	27
Phytophthora infestans	2	18–21	26
Cryptococcus sp.	3	20–22	25
Chrysosporium pannorum	−5	18	28

Sources: Data mainly from Lilly and Barnett (1951); Di Menna (1960); Hagan and Rose (1961); Sumner, Morgan and Evans (1969); Flanagan and Scarborough (1974); Rosenberg (1975); Mislivec and Tuite (1970); Hurst and Pugh (1983).

temperature in order to accelerate decay, and if compost heaps are well made and large enough, temperatures at peak heating can easily reach 60°C, providing ideal conditions for the growth of heat-tolerant fungi.

Successions of fungi in composts have been followed in wheatstraw compost by Chang and Hudson (1967). Their results are probably typical for most composts. Freshly made wheatstraw compost heaps contain a variety of fungi with populations of soil-inhabiting and leaf-inhabiting mesophiles predominating. As the temperature rises, heat-tolerant species increase in number and soon become dominant, replacing the mesophiles from all but the cooler outer surfaces of the heap as the internal temperature becomes lethal. High temperatures may last for several weeks, depending on the size of the heap, but eventually the heap begins to cool, as decay slows down, at which point the decomposed interior is slowly recolonized by mesophiles which re-invade from the persisting surface populations. The actively growing populations of thermophiles die out but other thermotolerant species, capable of growth at lower temperatures, remain active in large numbers and continue to play a part in any further decomposition.

Heat-tolerant fungi show a whole spectrum of physiological behaviour with regard to their ability to decompose plant polymers, and some quite complex interspecific relationships evidently exist in this respect. Some can hydrolyse pure cellulose in culture, notably *Humicola insolens*, *Chaetomium thermophile* and *Aspergillus fumigatus*, and many hydrolyse carboxymethylcellulose, wheatstraw xylan and arabino-xylan (Chang, 1967; Flannigan and Sellars, 1972). The ability to hydrolyse hemicellulose seems on the whole to be stronger and more widespread than the ability to hydrolyse cellulose. A few fungi, such as *Talaromyces thermophilus* (*Penicillium dupontii*), may be weakly lignolytic (Tansey *et al.*, 1977; Jain, Kapoor and Mishra, 1979) and *Allescheria* species, *Thielavia terrestris* and *Paecilomyces* species are soft-rots of wood that degrade some lignin (Eslyn *et al.*, 1975). *Phanerochaete chrysosporium*, among thermotolerant species, degrades lignin extensively in lignocellulose (Rosenberg, 1978).

The majority of isolates of *Rhizomucor miehei* (*Mucor miehei*), *Rhizomucor pusillus* (*Mucor pusillus*) and *Thermomyces lanuginosus* (*Humicola lanuginosa*) degrade no cellulose (Chang, 1967; Kane and Mullins, 1973; Jain, Kapoor and Mishra 1979). On wheatstraw composts *Rhizomucor pusillus* disappears early in the succession and seems to be a primary sugar fungus. *Thermomyces lanuginosus* persists as a secondary sugar fungus in mutualistic relationships with some of the true cellulose decomposers of composts (Chang, 1967; Hedger and Hudson, 1974). While this view of the status of these fungi seems generally correct, it is clear that variation exists between isolates from within the same species. Flannigan and Sellars (1972) reported, for example, that isolates of *Rhizomucor pusillus* and *Thermomyces lanuginosus* hydrolysed carboxymethylcellulose suggesting that these isolates could hydrolyse some forms of cellulose, while Somkuti *et al.* (1966) found that an isolate of *Rhizomucor pusillus* possessed a number of cellulolytic enzymes and could degrade native cellulose. Mishra *et al.* (1981) isolated a strain of *Thermomyces lanuginosus* which decomposed native cellulose in wheat straw. Pure cellulose was not attacked by the fungus unless wheatstraw extract was added suggesting that cellulolytic activity in some straws may depend upon the presence of growth factors.

The heat-tolerant fungi of composts are common to a variety of warm habitats and liable to be present wherever organic matter heats up for any reason. They are common in mushrooms beds and composts, where they play an important part in the preparation of the compost prior to inoculation with the mushroom spawn (Fergus, 1969). The partial decomposition of the raw materials of the compost is essential for the successful cultivation of mushrooms (*Agaricus* species) and it is important that self-heating is encouraged during this process, not only to promote decay to condition the material but also partially to sterilize the compost. When composting is finished and after the temperature falls, the mushroom spawn is inoculated into the compost. If the composting has been carried out correctly, the mushroom mycelium rapidly colonizes the compost in the presence of very much reduced competition from other mesophiles.

Heat-tolerant fungi have also been found in more exotic warm habitats such as birds' nests (Apinis and Pugh, 1967), coal tips (Evans, 1971), volcanic hot springs (Tansey and Brock, 1971; Hedger, 1975) and power plant cooling pipes and effluents (Ellis, 1980). Where protective measures are inadequate, heat-tolerant fungi can become involved in the spoilage of agricultural products during storage. They can cause moulding and deterioration of hay which has been insufficiently dried (Gregory et al., 1963) and may attack grain stored in a moist condition in sealed silos if these become aerobic for any reason (Flannigan, 1969; Mulinge and Apinis, 1969; Lacey, 1971). *Rhizomucor pusillus* and *Aspergillus fumigatus* are known animal pathogens and pose a potential health hazard to those who handle mouldy grain or hay (Lacey, 1975; Ogundero, 1980a).

In the tropics where heating up is considerably aided by insolation, heat-tolerant fungi can cause deterioration of groundnut and palm oil. Fungal lipases promote the oxidation of the oils to fatty acids, which impart rancid flavours and render the oils unfit for manufacture into margarines etc. (Ogundero, 1980b, 1981).

In spite of the high temperatures which they require for growth, thermophiles occur in temperate soils where they grow as temperature allows (Apinis, 1963; Tansey and Jack, 1976). Even in Britain the sun can easily warm the soil above 20°C on sunny days. Contrary to possible expectations, however, there is no evidence that thermophiles are commoner than mesophiles in tropical soils, where it seems that lack of moisture in the dry season may be a serious limiting factor (Gochenaur, 1975; Hedger, 1975).

Some attention has been given to the possibility of treating town refuse, which contains a lot of cellulosic materials in the form of wastepaper, plant debris, etc., by composting it, and in certain circumstances this can be an efficient and sanitary method of disposal (Kane and Mullins, 1973). It is essential that high temperatures are generated during the composting of such waste, not only to promote rapid decay, but also to reduce bacterial populations to levels acceptable to public health standards. Thermotolerant fungi occur naturally in these situations and may have an important role to play in future developments (Stutzenberger, Kaufman and Lossin, 1970).

Thermotolerant fungi have great potential for other applied uses, including the safe and cheap disposal, by composting, of the enormous quantities of animal waste from intensive farming (Seal and Eggins, 1972), the conversion of animal wastes to fungal protein for recycling as food for cattle and poultry and the bioconversion of lignocellulosic crop and industrial wastes.

Efficient bioconversion of lignocellulosic wastes requires that lignin be broken down as an essential first step in order to allow cellulase enzymes access to the cellulose. This can be accomplished by physical and chemical means but biodegradation would be preferred since energy costs could be lower and some of the pollution and waste disposal problems associated with chemical methods can be avoided. Possibilities exist to use the basidiomycete *Phanerochaete chrysosporium*, a fast-growing thermotolerant causing white-rot, for bioconversions that require delignification. Pilot studies show that white-rot delignification of lignocellulosic wastes can upgrade the quality of these materials, enabling them to be used as food for cattle, improving palatability and *in vitro* digestibility by rumen fluids (Kirk and Moore, 1973). The efficiency of these bioconversions depend upon the plant material used and the choice of fungus (Zadrazil, 1980; Platt, Hadar and Chet, 1983). *Phanerochaete* has also been grown on the lignin and cellulosic wastes of paper pulp manufacture producing useful fungal protein, and if this were to be combined with the purification of the waters of pulp manufacture, which contain abundant amounts of these wastes, the process might be made economically attractive (Ek and Eriksson, 1980). Yields of fungal protein from such wastes are about 14% of the final product and have about a 70% efficiency of digestion by rats and pigs (Eriksson and Larsson, 1975).

Some success has been obtained using cellulase-less mutants of *Phanerochaete* with the aim of delignifying wood in order to reduce demand for energy in the production of pulp for paper manufacture without affecting the strength of cellulose in the final product (Eriksson, Grunewald and Vallander, 1980; Eriksson, 1985).

Thermotolerants are particularly valuable for use in bioconversion technology because they reduce the need for cooling during growth. This cuts costs compared with the use of mesophiles where growth would stop if the heat of metabolism were not removed. An additional benefit of working at high temperature is that the end-product is virtually free of contamination by mesophiles.

For a useful review of lignin biodegradation in biotechnology, see Buswell and Odier (1987).

12.1.2 Psychrotolerant fungi

Many fungi occur in both Antarctic and Arctic tundras where for a month to two in the summer the temperature rises above 0°C for long enough to allow the growth of a few plant species and an associated mycoflora. These fungi are capable of surviving the very low temperatures of these regions, which can reach −40°C. All the major subdivisions of the Eumycota are represented. Typically in Antarctica lists show that members of the fungi imperfecti and sterile mycelia are the dominants (Flanagan and Scarborough, 1974; Dowding and Widden, 1974). Pugh and Allsop (1982) found *Chrysosporium pannorum* and *Mortierella* species dominant on buried cellulosic and keratinous substrata.

In the Antarctic, macrofungi are confined to the Antarctic peninsula, and to the islands of the sub-Antarctic zone where woody plants are present. In general, the higher fungus flora is very similar to that of northern temperate and Arctic regions and is dominated by bryophilous agaric and ascomycete genera such as *Galerina, Omphalina* and *Bryosphaeria* (Pegler, Spooner and Lewis Smith, 1980).

However, hardly any of these fungi are confined to these regions and for the most part they are psychrotolerant strains of mesophiles adapted to grow at temperatures down to about −5°C. Many have evidently been introduced by human activity.

True psychrophiles include some *Mucor* species and certain yeasts (Table 12.1).

Yeasts dominate the spring and autumn mycoflora in Antarctic soils. Seemingly they are well-adapted to grow in the freeze–thaw conditions that occur during these seasons. Freeze–thawing of plant tissues releases soluble sugars etc. into the soil, on which the yeasts multiply. Yeasts isolated from Antarctic soils include species in the genera *Candida, Rhacodium* and *Cryptococcus* (Tubaki, 1961; Di Menna, 1960, 1966). In the summer, when the temperatures rise above 0°C the yeast populations crash due to replacement by filamentous fungi (Wynn-Williams, 1980).

Cold-tolerant fungi also occur as contaminants and spoilage organisms of food in cold storage. Many *Penicillium* species are capable of growth below −3°C (Panasenko and Tartarenko, 1940) and occur as common contaminants of fruit and vegetables in refrigerated stores. *Thamnidium, Mucor, Penicillium*, yeasts and especially *Cladosporium* species are known spoilage microorganisms of meat in cold stores (Jay, 1987). Cold-tolerant strains of *Botrytis cinerea, Mucor mucedo, M. piriformis, M. circinelloides, Rhizopus stolonifer* and *R. sexualis* can cause serious spoilage and loss of strawberries, peaches, tomatoes and other soft fruits in store at 0–5°C

(Dennis and Cohen, 1976; Smith, Moline and Johnson, 1979).

A number of fungal diseases of plants develop at low temperatures. The optimum temperature for the growth of most of these fungi is below 20°C and most have a lower limit of growth at about −5°C. Some well-known examples are *Typhula* species (*Typhula idahoensis*, a pathogen of overwintering cereals and ley grasses, is at its most virulent at about −1.5°C), *Fusarium nivale* (snow mould), *Phacidium infestans* (a pathogen of conifers) and *Coprinus psychromorbidus* (cottony snow mould of grasses and cereals in Canada). For further information on these low-temperature pathogens see Deverall (1968), Jamalainen (1974), and Traquair and Hawn (1982).

12.1.3 Physiology of adaptation to temperature extremes

Modern theories as to why some microorganisms are tolerant of high temperatures propose that tolerance of high temperature resides in the special physical and chemical properties of the cell macromolecules, membranes and organelles, which allow them to remain stable at high temperatures. Gathering evidence, however, to support these ideas has been very slow and in some cases the results of experimental work have not proved to be incontrovertible.

Studies on the stability of macromolecules of microorganisms growing at high temperatures indicate that many proteins are produced which are thermostable. Thermophilic bacteria, for example, produce numerous enzymes and structural proteins with demonstrably high degrees of thermostability (for references see Crisan, 1973). Some confusion has arisen, however, since some enzymes from mesophilic microorganisms are also highly thermostable. This is perhaps not entirely unexpected since mesophiles need to be able to withstand abnormal or exceptional increases in ambient temperature to ensure survival. Perhaps more puzzling is the fact that it was also discovered that the reverse also applied, that is, not all enzymes of thermophiles are thermostable. This led to the suggestion that the physiological basis of thermophily is the ability to replace rapidly important thermolabile macromolecules as they are lost. The rapid synthesis hypothesis, as it was called, has now however largely fallen out of favour. The mechanism of protein thermostability is not fully understood. It may be due to subtle changes in the amino acid complement or to binding with certain ions: Zn, Ca or Co. Heat-shock proteins could be involved (see below).

Little work has been done on thermostability of fungal proteins but what little information is available suggests that they are very similar to other microorganisms in this respect. It has been shown (Crisan, 1969) that proteins extracted from *Thermomyces lanuginosus* and *Talaromyces thermophilus* are not all thermostable at 60°C but that the proportion of thermostable proteins present increased as the temperature of incubation increased. This is consistent with the observation that some enzymes isolated from thermophilic fungi are thermostable and have optimum temperatures above 60°C, while others have lower optima and rapidly lose activity at 60°C (Craveri and Colla, 1966; Loginova and Tashpulatov, 1967; Somkuti and Babel, 1968; Somkuti and Somkuti, 1969; Broad and Shepherd, 1971). Several heat-shock proteins have been implicated in the survival of fungi at elevated temperatures (Plesofsky-Vig and Brambl, 1993). These are proteins that appear or increase as the temperature rises in both thermotolerant and mesophilic species. Among other functions, some are concerned with facilitating protein folding during normal growth and may help to reform proteins denatured at high temperatures.

The nucleic acids of thermotolerant microorganisms do not appear to be exceptionally thermostable, although this evidence is drawn strictly from work on bacteria (see Crisan, 1973).

Of other possible special properties of thermotolerant microorganisms most can be said about adaptations of membranes where recent advances in our understanding of membrane structure and function allow some reasonable hypotheses to be put forward. The functional integrity of membranes depends upon maintaining structure in a liquid crystalline state. In this condition the membrane is in the correct state of fluidity for membrane-bound enzymes to function and it can selectively control the passage of ions and molecules in and out of the cell.

On theoretical grounds, maintaining membranes in good physical condition at ambient temperature would appear to depend on having lipids with the correct fatty acid composition. Lipids tend to form gels when the temperature falls below their transition temperature while the higher the temperature rises above the transition temperature the more fluid they become.

The transition temperature is influenced by the proportions of saturated and unsaturated fatty acids that make up the lipid. Saturated fatty acids, because of their low numbers of double bonds, raise the transition temperature, while unsaturated fatty acids lower the transition temperature. If membranes are composed of lipids with a high proportion of saturated fatty acid, they will be better able to maintain functional structure at higher temperatures than those composed of a high proportion of unsaturated fatty acid, which will tend to become too fluid as the temperature rises. Conversely, membranes composed of lipids with a high proportion of unsaturated fatty acid will be able to accommodate to lower temperatures better than those composed of high proportions of saturated fatty acid. The latter will tend to form gels and become too rigid and consequently very leaky as the temperature falls. Logically, therefore, it would be expected that the membranes of microorganisms living at low ambient temperatures would contain a high proportion of unsaturated fatty acids in membrane lipids while those living at higher ambient temperatures would contain a greater proportion of the saturated form.

What evidence is available suggests that this is so. Generally speaking, thermophilic fungi contain significantly lower levels of unsaturated fatty acid in membrane lipids compared with mesophiles (Mumma, Sehura and Fergus, 1971). Thermophilic *Mucor* spp. contain higher proportions of oleic acid (one double bond), less linoleic acid (two double bonds) and much lower quantities of linolenic acid (three double bonds) than mesophiles in the same genus (Sumner, Morgan and Evans, 1969). In thermophilic and thermotolerant *Mucor* and *Rhizopus* species the degree of unsaturation tends to decrease as temperature rises and increase as temperature falls (Sumner, Morgan and Evans, 1969; Sumner and Morgan, 1969; Hammond and Smith, 1986). To some extent these species show homeoviscous adaptation (Sinensky, 1974), that is they show a tendency to change the proportion of saturated to unsaturated fatty acids in response to changes in temperature. Thermophilic fungi, however, tend to produce lipids with a higher saturated fatty acid content at all temperatures compared with mesophiles (Sumner, Morgan and Evans, 1969). This last point is especially interesting since it may provide a clue as to why thermophilic fungi cannot grow below 20°C. It may simply be that they cannot produce enough unsaturated fatty acid at lower temperatures. They may lack specific desaturases or more likely lower temperatures may limit desaturase function in these fungi.

Significantly, increase in temperature was found to make no difference to the degree of fatty acid saturation of psychrotolerant *Mucor* spp. or mesophilic yeasts (Sumner, Morgan and Evans, 1969; Hunter and Rose, 1972). Failure to increase the degree of saturation in these species with increase in temperature may explain why psychrotolerant and mesophilic species cannot grow at high temperatures.

Predictably, the fatty acid composition of the lipids of thermotolerant fungi has been found to resemble that of mesophiles at lower temperatures and that of thermophiles at higher temperatures (Sumner, Morgan and Evans, 1969).

The capacity of membranes to remain functional over a range of temperatures also appears to be due to a buffering effect exerted by sterols (Harrison and Lunt, 1980). Sterols have a liquefying effect on lipids tending to the gel condition and a condensing effect on those tending to liquefy (Demel and de Kruyff, 1976). Sterols are always present in high levels in thermophiles, the commonest being ergosterol (Weete, 1980).

In spite of the conviction of these arguments and the supporting experimental evidence, some doubt has been placed on these ideas. Esser (1979), working with *Bacillus stearothermophilus*, discovered a mutant in which the level of saturated fatty acids in the cell lipids did not rise as it grew at increased temperatures. This and other evidence led to the suggestion that the chemical properties of the fatty acids of lipids are less important than their physical properties in adaptation to high temperatures. Esser (1979) speculated that growth at high temperatures may depend on the ability of microorganisms to produce phospholipids which can form clusters or domains in the membranes by phase separation.

As yet there is very little information available on the effects of high temperatures on the functioning of other cell components in fungi or other microorganisms except in the case of ribosomes, where work with bacteria has shown that in thermophilic species these are exceptionally thermostable (Saunders and Campbell, 1966; Stenesh and Yang, 1967).

If little is known about the physiology of thermal adaptation in fungi, then it must be said that even less seems to be known about the physiology of those that grow at low temperatures. Presumably, the enzymes of these fungi must be synthesized and remain active at low temperatures, although the former seems to be the more important. At least in mesophilic bacteria poor enzyme synthesis, rather than low enzyme activity, appears to be the critical factor which limits growth at low temperatures (Ingraham, 1962).

Adaptations to growth at low temperatures must also include mechanisms to increase the solute concentration of the hyphal fluids, a step that would be necessary to prevent freezing and the loss of water to the saline environment that will arise as water freezes out of solutions. Prevention of hyphal shrinkage is vital to the preservation of membrane function. Increase in internal solute concentration is probably achieved through the synthesis of polyols and the accumulation of ions from the external bathing solutions in much the same way that xerotolerant fungi osmoregulate (section 12.2.3).

The membranes of psychrotolerant fungi probably contain lipids with high levels of unsaturated fatty acids to accommodate growth at low temperatures. The lipids of psychrotolerant yeasts have been found to contain more unsaturated fatty acids than mesophiles and these increase when the temperature falls (Kates and Baxter, 1962; Brown and Rose, 1969; McMurrough and Rose, 1973; Kerekes and Nagy, 1980). Studies on *Thamnidium elegans* and *Mucor strictus* suggest that mycelial psychrophiles are probably similar in this respect (Manocha and Campbell, 1978; Dexter and Cooke, 1984). On the other hand, Sumner, Morgan and Evans (1969), working with *Mucor strictus* and *M. oblongisporus*, found no greater degree of unsaturation of their lipids when they were grown at 10°C compared with growth at 20 or 25°C. Hammond and Smith (1986), separating changes in total lipid content from changes occurring specifically in membrane lipids, detected a fall in unsaturation in *M. psychrophilus* as temperature increased. Higher levels of membrane lipid unsaturation were detected in the psychrophile compared with a mesophile and a thermophile at optimum growth temperatures.

Low maximum temperatures for the growth of psychrophiles may be because of the presence of highly sensitive enzymes which are affected by moderate increases in temperature. In *Cryptococcus* spp. apparently there are such enzymes among those concerned with amino acid synthesis and respiration (Rose, 1962). There is evidence too that membranes become increasingly leaky above optimum temperatures for growth, indicating that the failure to synthesize sufficient saturated fatty acid to maintain the integrity of membranes at higher temperatures may also be important. In *M. strictus* the K efflux measured at 25°C was found to be 2–3 times that of closely related mesophilic species at the same temperature (Dexter and Cooke, 1985).

12.2 XEROTOLERANT AND OSMOTOLERANT FUNGI

While there are some good ecological reasons for distinguishing between these two groups, the physiological differences between them are not very great and some fungi are both xerotolerant and osmotolerant. The characteristic which these fungi share is an ability to grow on substrata where the water potential is very low due to low moisture content and/or a high solute concentration. Xerotolerant fungi can grow on very dry materials with low matric potential while osmotolerant fungi may be found growing where the osmotic potential is very low.

Tolerance of fungi to water stress can be conveniently assessed by measuring growth or germination while exposed to high concentrations of salts or non-electrolytes in agar gels. Under these conditions water potential is controlled by the osmotic potential and the effects of matric potential can be considered to be negligible. In the past water activity (a_w) has been used to express the availability of water under these circumstances. Alternatively, cultures may be held in low relative humidities above solutions of low osmotic potential. The relationship between water activity, relative humidity and water potential is explained in section 3.2.3 (d).

Xerotolerant and osmotolerant fungi are by definition those capable of growth at water activities below $0.85a_w$ (Schmiedeknecht, 1960) but many grow at $0.75a_w$ and below (Table 12.2). A solution with a water activity of $0.75a_w$ would be in equilibrium with a relative humidity of 75% and at 25°C would have a water potential of -39.8 MPa. The lowest limit for the growth of non-xerotolerant fungi is about $0.85a_w$ at 25°C (-24.5 MPa) or higher.

Species capable of growing at very low water activities are not numerous and the majority of those that do are 'lower' ascomycetes in the Hemiascomycetes and Plectomycetes. Some common examples are listed in Table 12.2, but for a more comprehensive list see Corry (1987).

Table 12.2 Lowest reported water activity (a_w) and water potential equivalents (ψ) for the growth and germination of some common osmotolerant and xerotolerant fungi (at 25°C unless otherwise indicated)

Species	a_w	ψ (MPa)
Monascus (Xeromyces) bisporus	0.62	−66.2
Saccharomyces rouxii (30°C)	0.61	−69.7
Debaryomyces hansenii	0.65	−59.6
Aspergillus restrictus group		
A. conicus (22°C)	0.70	−48.9
A. restrictus	0.70	−49.4
Aspergillus glaucus group		
A. amstelodami	0.70	−49.4
A. chevalieri (33°C)	0.71	−48.7
A. repens (24°C)	0.71	−47.3
A. sejunctus (A. ruber) (24°C)	0.71	−47.3
A. candidus	0.75	−39.8
A. ochraceus	0.78	−34.4
A. flavus (33°C)	0.78	−35.3
Penicillium chrysogenum	0.78	−34.4
P. implicatum	0.78	−34.4
P. martensii (23°C)	0.79	−32.4
P. spinulosum	0.80	−30.8
P. expansum	0.82	−27.5
P. citrinum	0.82	−27.5
P. cyclopium	0.84	−24.1
P. frequentens	0.84	−24.1
Saccharomyces baillii	0.80	−30.8

Sources: Based on data from Ayerst (1969); Pitt (1975); Pitt and Hocking (1977); Hocking and Pitt (1979).

12.2.1 Osmotolerant fungi

These are the yeasts, aspergilli, etc. of confectionery, honey, jams, syrups, nectars, dried fruits and other foods with a high solute concentration. Some can grow at exceedingly low water activities. Growth at about $0.60 a_w$ (equivalent to almost −70 MP at 25°C) has been recorded for *Monascus (Xeromyces) bisporus* (Table 12.2). Some of the yeasts in this group thrive in both sugary and salty environments and the genera *Debaryomyces*, *Hansenula* and *Pichia* include several facultative marine species (Kohlmeyer and Kohlmeyer, 1979). *Saccharomyces rouxii*, and other osmophilic yeasts, play an important part in the ripening of salty soy sauces and miso pastes during fermentation (Onishi, 1963). Others may be found causing serious losses of sugary foods in stores.

12.2.2 Xerotolerant fungi

These are the fungi of dry environments. The genus *Aspergillus* is collectively the most xerotolerant group of fungi and contains many of the most xerotolerant forms. Some *Penicillium* species are also xerotolerant and these constitute another large important group. Generally speaking, however, *Pencillium* species are less well-adapted to low water activities when compared with *Aspergillus* species. The lowest water activities which will permit germination of *Penicillium* species in culture lie in the range 0.78–$0.84a_w$, compared with below $0.78a_w$ for the majority of *Aspergillus* species (Table 12.2). In temperate climates xerotolerant *Penicillium* species are commoner than *Aspergillus* species in soil and tend to replace them on substrata with higher moisture contents. In general, *Aspergillus* species can only compete successfully with *Penicillium* species at higher temperatures and low water activity (Magan and Lacey, 1984b). Xerotolerant fungi are best known to microbiologists as storage fungi, the spoilage microorganisms of agricultural products and dried foods in store. The principal fungal contaminants of seeds, cereals and hay at harvest are field fungi, *Cladosporium* species, and the other inhabitants of plant phylloplanes (described in section 4.2), together with some specific parasites in the case of seeds and cereals (*Fusarium, Drechslera* species, etc.). Field fungi require minimum water activities of at least $0.85a_w$ for growth (Magan and Lacey, 1984a) and therefore tend to die out at the low water activities maintained in the storage of dried foods and crops. Normally the damage they do to crops before storage is minimal.

Species of most storage fungi are represented at harvest, although in low numbers (Flannigan, 1970a). They can compete well with most field fungi (Magan and Lacey, 1984b) and may increase under post-harvest conditions. Provided, however, the moisture content of the stored materials remains at a minimum of about 12% of the fresh weight, the growth of storage fungi will be prevented. The corresponding water activity for non-oily grains stored at this moisture content is below $0.60a_w$ but sunflower seeds, groundnuts and other oily seeds need to be stored somewhat drier because of their corresponding higher water activity (Figure 12.1). If, however, storage conditions are bad and the moisture content rises, or the materials have a moisture content above the critical 12% when placed in store, then contaminating storage fungi will slowly colonize (Clarke, Niles and Hill, 1967; Clarke *et al.*, 1968). For this to happen when grain or hay has been deliberately dried before storage is rather unusual and when it does come about, it is usually because the materials have not been uniformly dried to the correct level. If moisture pockets persist, even if these are only very slightly above the critical level, there will be enough to allow some of the more xerotolerant fungi such as *Aspergillus halophilicus* and those in the *A. glaucus* and *A. restrictus* groups to grow (Table 12.2). Once colonization has begun, provided the temperature is high enough, progressive deterioration of the stored materials will proceed with increasing rapidity as moisture is released by respiration. Colonization follows a succession in which species occupy distinct positions that are very strictly determined by moisture content. Marked population changes occur as the moisture content increases as species more competitive at higher moisture contents develop. Between about 14 and 15% moisture content the increase is sufficient to cause the succession to move from *A. halophilicus* through *A. restrictus*, *A. glaucus*, to *A. candidus*. At higher moisture contents *A. flavus* predominates (Christensen and Sauer, 1982).

The optimum temperature for the growth of many storage fungi tends to be relatively high (about 30°C), while a number such as *A. fumigatus*, *A. flavus* and *A. candidus* are thermotolerant and grow above 40°C (Ayerst, 1969). Growth will therefore be favoured by the rise in temperature of the materials caused by the heat released during metabolism. This heat tends to become trapped by the sheer bulk of the stored material so that self-heating occurs.

Some *Penicillium* species are capable of germinating at 0.78–$0.80a_w$ (Table 12.2); however, serious contamination of stored grain by these species only occurs at moisture contents above 18% (about $0.85a_w$ at 20°C).

At ambient temperature, *Penicillium brevicompactum* usually appears first, followed by *P. verrucosum* at slightly higher water activities. Above $0.90a_w$ these fungi are replaced by *P. hordei* and others and as the temperature rises these are in turn replaced by *P. capsulatum* at about 30°C and by *P. piceum* at 35°C and at water activities above $0.92a_w$ (Hill and Lacey, 1983, 1984; Magan and Lacey, 1984a,b)

Interaction between temperature and water activity has a very significant bearing on the problem of preservation of dried crops in store. Generally, for xerotolerant fungi the greatest tolerance to low water activity is close to the optimum temperature for growth (Ayerst, 1969). Generation of heat during moulding will therefore tend to compensate for low water activities in the stored materials and permit growth at near the absolute minimum water activity value for individual species, while conversely, lower temperatures will tend to raise the water activity value at which growth can begin. In fact, most storage fungi grow very slowly below 12°C whatever the moisture content of the substratum. In theory, agricultural products could be safely stored at moisture contents above 12% (up to about 16% in practice) provided the temperature is maintained below about 10°C. Above about 20% moisture content, field fungi, which are generally more

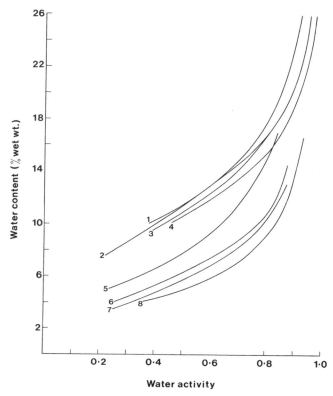

Figure 12.1 Relationship between water activity and moisture content in oily and non-oily seeds. 1. English maize; 2. wheat; 3. barley; 4. American maize; 5. soya bean; 6. linseed; 7. sunflower; 8. groundnuts. (Reprinted from *Journal of Stored Products Research*, 7, S.W. Pixton and S. Warburton, Moisture content relative humidity equilibrium, at different temperatures, of some oilseeds of economic importance, 261–9, copyright 1971, with kind permission from Elsevier Science Ltd, The Boulevard, Langford Lane, Kidlington OX5 1GB, UK.)

tolerant of lower temperature, will begin to grow. Experiments have shown that maize and wheat can be stored for years without deterioration at moisture contents between 15 and 16% at temperatures between 5 and 10°C (Papavizas and Christensen, 1958; Qasem and Christensen, 1958). Whether the cost of cooling to below ambient temperatures is economically sensible compared with the cost of drying is, of course, another matter.

Agricultural products attacked by storage fungi may be completely destroyed ultimately, but before that stage is reached serious devaluation of the crop will have occurred. Grain will become discoloured, and some of the food stores will be respired so reducing the economic value. Fatty acids with rancid flavours will accumulate from the oxidation of the seed oils and the crop will generally be rendered unfit for human consumption. The value of the crop as seed will also be seriously affected as embryos are penetrated and die, reducing overall viability (Christensen and Kaufman, 1969; Warnock, 1971).

Investigations of the enzymatic activities of the principal storage fungi show that cellulase production is generally poor. Members of the *Aspergillus glaucus* group are active producers of lipase and β1-4 glycosidase but produce little amylase or xylanase compared with *A. flavus* and other species. This correlates well with observations that the former, as initiators of decay in grain at low water activity, attack the embryos, where there are sugars and oils, rather than the endosperm with its carbohydrate food stores and slightly lower water potential (Flannigan 1970b; Flannigan and Sellars, 1972).

In addition to these problems mouldy grain may become completely useless because of the accumulation of mycotoxins. Several storage fungi are known to produce highly dangerous mycotoxins which can pose serious health problems for man and his livestock if consumed (Christensen and Kaufman, 1965; Christensen, 1971). **Aflatoxins** are well-known examples. These are powerful carcinogens produced by some strains of *Aspergillus flavus* when growing on nuts and cereal grains. This danger first came to light when young turkeys died after they were fed contaminated groundnuts (Asplin and Carnaghan, 1961; Spensley, 1963). Fortunately, *A. flavus* is not an extreme xerotolerant and does not begin to grow vigorously until the moisture content reaches about 18%.

Other less well-known but no less important mycotoxins produced by storage fungi are **ochratoxin** from *A. ochraceus* and **rubratoxin** from *Penicillium rubrum*. Burnside *et al.* (1957) showed experimentally that when the latter was allowed to grow on corn and subsequently fed to pigs, death followed within a few days from kidney and liver damage.

Other health hazards are associated with storage fungi. *Aspergillus fumigatus* is a human pathogen causing **aspergillosis**, an invasive disease of the lungs. Strictly speaking, it is not xerotolerant but it is commonly associated with such fungi growing under damper conditions. Even at optimum temperature (40°C) the minimum water activity for growth is in excess of $0.88a_w$ (Magan and Lacey, 1984a).

Under certain conditions exposure to storage fungi can lead to the development of allergic diseases akin to **farmers' lung disease** (Lacey, 1975; Neergaard, 1977). A case in point is **malthouse workers' lung disease** caused by inhaling dust contaminated with *Aspergillus clavatus*. Constant exposure to spores of this fungus leads to the development of hypersensitivity to the fungal glycoproteins and the development of allergic alveolitis (alveolar filling, haemorrhaging and inflammation). In severe cases workers become extremely debilitated (Grant *et al.*, 1976; Blyth *et al.*, 1977; Blyth, 1978).

The current practice of storing grain intended for animal feed moist in sealed or unsealed silos, under anaerobic conditions, in order to control mould growth and to avoid costly artificial drying and rewetting when needed, may increase the risk of invasion by storage fungi and crop losses, with attendant health hazards for those employed in unloading the grain. Sealed silos may leak and become aerobic or storage fungi may invade from the top of unsealed silos if the grain is not removed quickly enough (Lacey, 1971). Similar troubles may arise in moist hay and grain which has been preserved chemically with propionic acid or other fatty acids. These are fungistatic and inhibit spore germination in many fungi but moulding may occur if untreated pockets persist or if the fatty acids are metabolized by resistant fungi (Lord, Cayley and Lacey, 1981; Lord *et al.*, 1981).

12.2.3 Physiology of adaptation to growth at low water potentials

This subject has been reviewed by Brown (1976, 1978) and Griffin (1981b). Most of our original knowledge of the physiological mechanisms which allow fungi to grow at low water potentials came from the study of yeasts, notably *Saccharomyces rouxii*. Most of the yeasts which are tolerant of low water potentials are osmotolerant rather than osmophilic, i.e. they have no absolute requirement for low water potentials and grow successfully over a wide range of water potentials.

Tolerance of microorganisms to low water potentials seems to depend fundamentally upon an ability to modify the internal environment of the cell. Osmoregulatory substances accumulate from the environment or are synthesized, which depresses the internal water potential below that of the environment and ensures that the microorganism does not lose water and is able to maintain turgor.

In fungi a major role has been proposed for **glycerol** and other polyhydric alcohols (**polyols**) in countering the adverse effects of water stress. They accumulate in osmotolerants and xerotolerants at low water potentials (Adler, Pedersen and Thunbald-Johansson, 1982; Luard, 1982a; Hocking and Norton, 1983) and reach a high proportion of the dry weight of cells growing under severe water stress. In *Saccharomyces rouxii* measurements of 9–15% of the dry weight for glycerol production (Brown, 1974) are not exceptional for fungi growing at very low water potentials.

Polyols have been termed **compatible solutes** since they do not interfere with normal cell metabolism and high concentrations can be tolerated within cells. Glycerol meets all the requirements for such a substance. It has a low affinity for protein and so does not bind with enzymes or inhibit their function (Brown, 1976, 1978; Adler, 1978). Most fungi synthesize polyols and store them as food reserves (Lewis and Smith, 1967) and synthesis of polyols may increase in all fungi in response to water stress. However, a major physiological difference between fungi able to grow at low water potentials and those which cannot is that the latter show an increasing tendency to leak polyols from hyphae as the internal concentration increases. This does not happen in fungi tolerant of water stress and seems to indicate that the explanation of the difference in behaviour is that fungi tolerant of water stress can maintain high internal concentrations of polyols without undue expenditure of energy and so avoid the need to divert resources to polyol production at the expense of growth (Edgley and Brown, 1978, 1983).

Ions and non-electrolytes accumulating in hyphae from the external osmoticum also serve to lower the internal water potential when hyphae are exposed to low water potentials. This is a rapid method of adjustment and at only slightly reduced water potentials may be sufficient to enable the fungus to osmoregulate without significant synthesis of polyol (Luard, 1982a,b,c).

In the marine fungus *Dendryphiella salina* the accumulation of specific ions from the external osmoticum seems to stimulate or inhibit metabolic pathways leading to the synthesis of specific polyols. Glycerol synthesis is associated with sodium and tends to be replaced by other polyols when magnesium accumulates (Wethered, Metcalf and Jennings, 1985).

Polyols probably have other physiological functions in hyphae growing at low water potentials. The enzymes of xerotolerant fungi are not adapted to function at low water potentials (Brown, 1976) and a function of polyols in hyphae under water stress seems to be to protect enzymes from the effects of low internal water potentials. If bound water were removed from the hydrophobic groups of the cell polymers at low water potentials to make up for the deficit of solvent water in the cytoplasm, this would result in a disturbance of the structure of the polymers and cause their precipitation. Schobert (1977) and Schobert and Tschesche (1978) suggested that polyols could prevent this by substituting water-like R–OH groups for water at the hydrophobic site.

The amino acid proline, which is an osmoregulatory solute found in some of the lower fungi (Luard, 1982a), may fulfil the same function by transforming hydrophobic sites, with high numbers of water molecules per site, into hydrophilic sites with lower numbers of water molecules per site.

Polyols are also thought to protect enzymes from the inhibitory effects of the accumulation of high concentrations of Na and Mg ions at low osmotic potentials.

Osmotolerant yeasts that grow on both salty and sugary substances are generally much more tolerant of low water potentials in the latter situation (Onishi, 1963). *Saccharomyces rouxii* tolerates water activities down to about $0.60 a_w$ when sugars are used to adjust osmotic potential as opposed to a lower limit of about $0.85 a_w$ when NaCl is used. This seems to be a simple inhibitory effect: at high concentrations of NaCl, sodium and chloride ions probably accumulate in cells at a faster rate than they can be pumped out.

Among xerotolerant moulds there are some, such as *Monascus* (*Xeromyces*) *bisporus*, *Aspergillus amstelodami* and *A. repens*, which are xerophilic or nearly so and grow very poorly, if at all, at water activities above about $0.98a_w$ (Pitt and Hocking, 1977; Flannigan and Bana, 1980). These fail to grow in the usual range of laboratory media and can easily be overlooked in routine investigations unless media heavily fortified with sugars are used. At the moment we can only speculate about the unique physiology of these species. High polyol concentrations may be needed to maintain the functional integrity of membranes and organelles which may disintegrate if the cell solution becomes too dilute. In the case of *Aspergillus sejunctus* there is evidence that failure to grow at high water activities is due to a general reduction in protein synthesis and a decline in the production of respiratory enzymes (Stevens *et al.*, 1983).

Another feature of the fungi which inhabit dry environments is that their spores are often adapted to withstand desiccation. This is believed to be due to protection afforded to membranes by **trehalose**, which is synthesized during the ripening and drying of the spore. Dehydration of membranes in the absence of trehalose brings morphological change; phospholipids form cystalline complexes in which the density of the polar head groups of the lipids increase as water, which normally separates them, is removed. Membranes which become gels in this fashion are damaged and cannot reform properly and function normally on rehydration. Membranes dried in the presence of trehalose at concentrations similar to those present in protected cells suffer no damage and function normally on rehydration (Crowe, Crowe and Chapman, 1984). Trehalose is thought to protect membranes by substituting its hydroxyl groups for those of water between the polar head groups in the membrane, much as glycerol is thought to protect enzymes in cells held at low water potentials.

13
TERRESTRIAL MACROFUNGI

13.1 INTRODUCTION

The ecology of fungi that form macroscopic fruit-bodies is, in many respects, better known than the ecology of microfungi. This is because the presence of fruit-bodies enables information to be readily gathered about distribution, seasonal abundance, habitat, substratum preference, etc. Although certain ascomycetes have large ascocarps, the best-known macrofungi are the basidiomycetes, especially the Hymenomycetes. Macroscopic basidiocarps are also formed by jelly fungi (Heterobasidiomycetes) and by puffballs, stinkhorns, bird's nest fungi and others, arbitrarily classified together as Gasteromycetes. Fungi with macroscopic basidiocarps thus belong to very diverse taxonomic groups and it should not surprise us that there is also diversity in their nutritional relationships (Rayner, Watling and Frankland, 1985). Some are saprotrophs, playing an important role in litter and wood decay, whilst others are symbionts. Some symbionts, especially the fungi which form mycorrhizal relationships with trees, may have only a limited free-living existence away from their living hosts. Watling (1982) has related the major taxa of basidiomycetes to their ecology, and has listed the nutritional relationships of some of the major orders and their families (Table 13.1).

Although there are variations in the life cycles of basidiomycetes, the basidiocarps which produce the basidiospores following nuclear fusion in sexual reproduction are generally borne on dikaryotic mycelia. Basidiospores are produced in vast numbers and are widely dispersed by wind over large distances (Kramer, 1982). The concentration of basidiospores in air may reach high levels. For example, Gregory and Hirst (1952) showed that in August and September at Rothamsted, UK, the concentration of coloured basidiospores rarely fell below 1000 m^{-3} (see also Gregory, 1973). Basidiospores germinate to form homokaryotic mycelia which, compared with dikaryotic mycelia, are usually relatively short-lived, being converted to dikaryons by fusion with compatible monokaryons, or by

Table 13.1 Nutritional relationships of basidiomycete macrofungi

	Family		Family		Family
Hymenomycetes		**Hymenomycetes** *contd.*		**Hymenomycetes** *contd.*	
Agaricales	Agaricaceae	Polyporales	Schizophyllaceae[w]	Hymenochaetales	Hymenochaetoideae
	Bolbitiaceae		Polyporaceae[w]		Vararioideae
	Coprinaceae		Corticiaceae[w]		Lachnocladiaceae
	Cortinariaceae[l,m]		Podoscyphaceae		
	Lepiotaceae		Stereaceae[w]	**Gasteromycetes**	
	Strophariaceae		Punctulariaceae[w]		Phallales[l]
	Hygrophoraceae[part m]		Auriscalpiaceae[w]		Sclerodermatales[m]
	Pleurotaceae[l]		Lentinellaceae[w]		Lycoperdales
Tricholomatales	Tricholomataceae[l,m]		Clavicoronaceae[w]		Geastrales[l]
Amanitales	Amanitaceae[m]		Hericiaceae[w]		Tulostomatales
Pluteales	Pluteaceae[l]		Amylariaceae[w]		Nidulariales[l]
	Entolomataceae		Bondarzewiaceae[w]		Hymenogastrales[m]
Russulales	Russulaceae[m]		Echinodontiaceae[w]		Melanogastrales[m]
	Boletaceae[m]		Fistulinaceae[w]		Gautieriales[m]
	(including		Ganodermataceae[w]		
	Strobilomycetaceae)		Sparassidaceae	l= Families that are dominantly wood-rotters and litter-decomposers.	
Boletales	Gomphidiaceae[m]		Aphelariaceae	m= Primarily mycorrhiza-formers.	
	Paxillaceae[l,m]		Clavulinaceae	w= Families composed primarily of wood-rotters.	
	(Coniophoraceae)[l]		Pterulaceae	Source: Watling (1982).	
Thelephorales	Bankeraceae		Clavarioid alliance		
	Thelephoraceae[some m]		Clavariaceae		
Gomphus alliance	Gomphaceae		Hymenochaetaceae[w]		
	Ramariaceae		Asterostromatoideae		

fusion with adjacent dikaryons. Asexual reproduction may also occur in some basidiomycetes by the formation of conidia of various kinds on monokaryotic or on dikaryotic mycelia. Dry conidia are generally aerially disseminated, but sticky conidia are also known which are probably dispersed by mites, insects, etc.

About 10% of basidiomycetes are **homothallic**, basidiocarps developing on homokaryotic mycelia developing from single basidiospores. In the 90% that are **heterothallic**, basidiocarps only develop from dikaryotic mycelia formed by fusion between compatible monokaryons. In **bipolar** (or **unifactorial**) species mating is controlled by a single gene, whilst in **tetrapolar** (or **bifactorial**) species two non-linked gene loci are involved which favours outbreeding. This has a considerable bearing on their population biology and ecology.

The genetic control of mating in basidiomycetes is an example of **homogenic incompatibility** – i.e. when identical alleles are present at one or both of the mating-type loci in paired monokaryons, they are incompatible. A different kind of incompatibility, which is expressed best in dikaryotic mycelia, is termed **heterogenic incompatibility**, **vegetative incompatibility** or **somatic incompatibility**. When two genetically identical dikaryotic mycelia of the same species are paired in culture the hyphae may anastomose without adverse effects since the genotypes are homogenic. If, however, two genetically different dikaryotic strains are paired together, although anastomosis may occur adverse cytoplasmic reactions and death of the individual hyphae rapidly ensues (Rayner and Todd, 1979; Thompson and Rayner, 1983).

In culture, the pairing of heterogenic dikaryotic strains of the same species usually leads to the development of pigments and abnormally shaped cells in the contact zone. These correspond to the development of zone lines in decaying wood that separate colonies of genetically different strains of the same and also those of different species within the wood. The ability to distinguish between genetically identical and dissimilar populations of the same species provides a very useful tool with which to study basidiomycete population biology: it has been used to map and study the distribution of cord-formers in woodland (Thompson and Rayner 1982a,b; Thompson, 1984) and of mycelia of ectomycorrhizal fungi (Dahlberg, 1992) and to analyse populations of wood-rotting basidiomycetes in dead tree stumps (Rayner and Todd, 1977; Todd and Rayner, 1978, 1980; Rayner, 1991).

Attempts have also been made to identify and locate the distribution of basidiomycete mycelia in litter and soil using antibody techniques. These are at present somewhat limited and still need refinement (section 7.1). Other techniques make use of characteristic microscopic features (clamp connections, hyphal diameter, wall thickness, crystalline deposits on or in the cells) to identify mycelia (Stalpers, 1978; Rayner and Todd, 1979;

Rayner and Boddy, 1988). Many of the mycorrhizal fungi which form sheathing mycorrhizas with trees develop characteristic, branching root systems which can be recognized by colour, the presence of mycelial cords, etc.

The form of the mycelium may vary throughout the life of a fungus. The **primary homokaryotic mycelium**, which develops from a germinating basidiospore, has simple, transverse septa, but on conversion to the **secondary dikaryotic mycelium**, clamp connections are usually formed at the septa. Hyphae may develop into strands, or branch and radiate out to form mycelial fans. Gregory (1984) has referred to some of the alternative forms which mycelia can take as '**modes**'. Rayner, Boddy and Dowson (1987b) and Rayner and Coates (1987) have developed the idea of differing modes of mycelial growth in relation to strategies of colonization. **Mycelial fans** result from divergent growth of branch hyphae, a wide angle being subtended between a main hypha and a lateral. Convergent growth, associated with narrow angles of branching, and cohesion between hyphae leads to the formation of **mycelial cords** (section 2.2.3). Due to increased rates of conduction in cords compared with individual hyphae, the rate of extension growth of mycelial cords may be several times greater than that of individual hyphae (Rishbeth, 1968).

The form that a mycelium may take as it extends through soil or litter will depend on many abiotic variables (e.g. amount and quality of food resources) and biotic variables such as the activities of competitors (Boddy, 1984, 1993). In elegant experimental studies of the wood-decaying, cord-forming *Hypholoma fasciculare*, Dowson, Rayner and Boddy (1986) and Rayner, Boddy and Dowson (1987b) have shown how a sparse, exploratory mycelium may extend from a colonized wood block into surrounding soil. Contact between the mycelium and a fresh woody resource unit, possibly associated with directional growth of mycelium or mycelial cords, is followed by the formation of a fan-shaped, effuse mycelium over the fresh resource unit. Meanwhile, there is also a thickening of the mycelial cords connecting the two blocks and the regression of any non-connecting hyphae. It is possible that directional growth of mycelium and cords towards fresh resource units is mediated by volatile substances (Mowe, King and Senn, 1983; Thompson and Rayner, 1983; Thompson, 1984). For a time, following contact with a new woody resource, the outward extension of the mycelium may be checked, but eventually new fans of mycelium extend outwards from the newly colonized blocks. Rayner, Boddy and Dowson (1987b) comment on 'the remarkable ability of the mycelium to behave as a co-ordinated unit . . .: as an army it sends out scouting parties, establishes lines of communication, brings in reinforcements by redirecting the movements of its troops, conquers domain, establishes bases and moves on.' Cords are most commonly associated with wood-decay fungi because of the need to locate scattered resources.

However, they may be produced by other basidiomycetes such as *Suillus bovinus*, a mycorrhizal associate of *Pinus* (Read, 1984).

The mycelia of basidiomycetes vary greatly in their longevity. Those which colonize small ephemeral substrata may survive for weeks only. By contrast, the mycelia which colonize the heartwood of large trees may survive for years, whilst some mycelia of fairy ring fungi have been estimated to survive for centuries. The parasitic wood-rotting *Armillaria bulbosa* spreads by means of rhizomorphs and by basidiospores. It is bifactorial. Using genetical finger-printing techniques and analysis of mating-type alleles, comparisons of isolates from rhizomorphs and from basidiocarp tissue have been made in a Michigan hardwood forest (Smith, Bruhn and Anderson, 1992). It was found that a single clone (i.e. a genetically identical individual mycelium) could extend over an area of 15 ha. Estimates of possible age (based on known rates of rhizomorph extension) indicated that this clone was about 1500 years old and that its mass exceeded 100 t, making it one of the largest and oldest living organisms.

13.2 WOOD-DECAY MACROFUNGI

Ecologically, basidiomycetes with macroscopic fruit-bodies can be broadly classified into those inhabiting wood and litter, and those which form mycorrhiza. There is, inevitably, in such a simplified classification, some overlap of roles: for example, wood-decaying fungi may also have mycelia which decompose litter, and some mycorrhizal fungi may also be involved in litter decay (Trojanowski, Haider and Hutterman, 1984).

13.2.1 Decay of standing trees

The ecology of fungi associated with the decay of wood has been extensively described by Rayner and Boddy (1988). The attack on wood by basidiomycetes often begins in the standing tree. The branches of mature trees are frequently colonized, especially the lower branches and any others whose growth is suppressed by shade. Boddy and Rayner (1981, 1983a) have investigated the communities of basidiomycetes involved in the decay of attached branches of oak (*Quercus petraea* and *Q. robur*). Entire branches were removed and cut at intervals into sections 1–2 cm thick. Isolations were made from the decay zones, and some of the sections were incubated in polythene bags to encourage mycelial growth. The fungi were identified and their distributions plotted. Isolates were paired opposite each other on agar plates to investigate interspecific antagonisms. Somatic compatibility/incompatibility testing of isolates of the same species enabled different genotypes to be identified and the limits of individual mycelia plotted. In *Q. robur*, 12 basidiomycetes, all causing white-rot, were found in 16 branches; all were members of the Polyporales with the

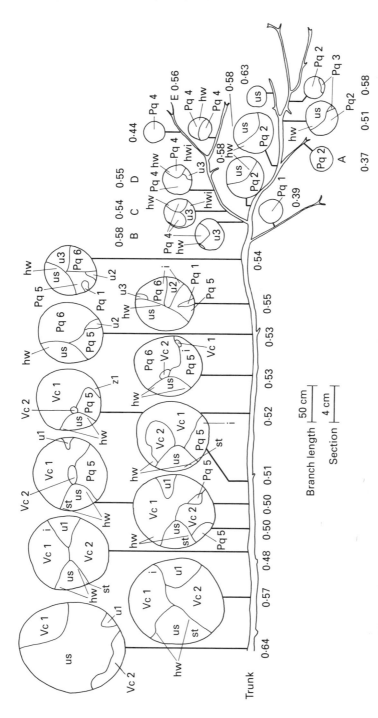

Figure 13.1 Decay community in an oak branch at a relatively early stage of decay. Different basidiomycete individuals were demarcated by interactive zone lines (i) in the wood. The predominant species were *Vuilleminia comedens*, represented by two individuals Vc1 and Vc2, and six individuals of *Peniophora quercina* Pq1–6. Three unidentified basidiomycetes, u1–3, occupied smaller volumes. The symbol hw represents heartwood wings and the symbol us represents living tissue. The numerical values are for the relative density of the wood (g cm^{-3}), an index of decay. (Redrawn from Boddy and Rayner, 1983a, with permission.)

exception of the heterobasidiomycete *Exidia glandulosa*. The frequency of species found is listed in Table 13.2.

Boddy and Rayner (1983a) studied the distribution of basidiomycetes in an oak branch measuring over 8 m in length at a relatively early stage of decay (Figure 13.1). They found that early stages of decay were brought about by *Stereum gausapatum, Phlebia rufa, Phellinus ferreus* and *Vuilleminia comedens*. Sometimes a single individual mycelium can extend for several metres. These fungi, together with *Peniophora quercina* and *Exidia glandulosa* (Figure 13.2) are regarded as pioneer species which can grow on living or recently dead wood. The points of entry are probably through small branch wounds. As the living tissues are invaded, the oak tree responds by forming premature heartwood, in which the vessels are packed with tyloses, and wings of prematurely formed heartwood may extend along the interface of the infected and uninfected sapwood. Such wings can readily be seen on attached dead lower branches of oak trees. *Coriolus versicolor*, which was only found infrequently in the sample studied, is regarded as a secondary invader of already dead or decayed wood. In paired culture with others it is able to replace *Phlebia rufa, Schizopora paradoxa, Stereum hirsutum* and several of the pioneer species (Table 13.3). There was no evidence that *C. versicolor* was ever replaced by other basidiomycetes. By contrast, when the pioneer fungus *Stereum gausapatum* is confronted by another pioneer, e.g. *Exidia glandulosa*, there is a deadlock – i.e. neither fungus replaces the other (Table 13.3).

Table 13.2 Frequency of basidiomycetes identified in 16 attached branches of oak (*Quercus robur*)

Species	Frequency
Stereum gausapatum	11
Exidia glandulosa	6
Vuilleminia comedens	5
Peniophora quercina	4
Phellinus ferreus	4
Phlebia radiata	3
Schizopora paradoxa	3
Phlebia rufa	2
Hyphoderma setigerum	2
Coriolus versicolor	1
Stereum hirsutum	1
Peniophora lycii	1

Sources: Boddy and Rayner (1983a).

Figure 13.2 Basidiocarps of some common wood-rotting basidiomycetes. (a) *Peniophora quercina*. (b) *Exidia glandulosa*. (c) *Coriolus versicolor*. (d) *Stereum hirsutum*. Scale bar = 2 cm.

Table 13.3 Interactions by basidiomycetes in attached oak branches

	Replaced	Deadlock	Replaced by
Coriolus versicolor	Exidia glandulosa Peniophora quercina Stereum gausapatum Vuilleminia comedens		
Stereum gausapatum		Exidia glandulosa Schizopora paradoxa Vuilleminia comedens	Coriolus versicolor Schizopora paradoxa

Source: Boddy and Rayner (1983a).

S. gausapatum may extend from the lateral branches into the heartwood of the main trunk, causing 'pipe rot', a form of decay reducing the economic value of the timber. In attached oak branches, it is often found that there is only a single individual mycelium present, which can extend for several metres. Boddy and Rayner (1982) have reported single individual mycelia which extended for up to 3.6 m along branches which had been dead for only one growing season. This implies infection by a mechanism other than a single infection court, possibly by the dissemination of mycelial fragments or other propagules (modules) in water columns within the host wood. This kind of behaviour is interpreted by Rayner and Boddy (1986) as an example of a **latent invasion strategy**. Spread of inoculum from the point of infection may not be followed by extensive mycelial development, which is possibly held in check by stress imposed by such factors as high water content with consequent low oxygen levels in the sapwood. Alleviation of stress, e.g. by drying of the sapwood, might then lead to proliferation of the mycelium.

Some of the fungi characteristic of attached branches are capable of extending into the upper part of the trunks of oak trees, especially when these are suppressed by overcrowding and shade. These include *Exidia glandulosa*, *Phellinus ferreus*, *Peniophora quercina* and *Vuilleminia comedens* (Boddy and Thompson, 1983).

Studies of the fungal communities growing on attached branches of ash (*Fraxinus excelsior*) show similarities to and differences from the fungal communities on oak (Boddy, Gibbon and Grundy, 1985; Boddy, Bardsley and Gibbon, 1987). Three ascomycetes, seven basidiomycetes and two deuteromycetes were found in a sample of 18 branches. The most common species were *Daldinia concentrica* and *Hypoxylon rubiginosum* (ascomycetes) and *Peniophora lycii* and *P. limitata* (basidiomycetes). The role of the pioneer colonizers, i.e. fungi growing in partly living and recently dead branches, was taken by the ascomycetes, in contrast with the basidiomycetes characteristic of oak. As in oak branches, there was evidence of extensive colonization by individual mycelia (in this case by ascomycetes) and latent invasion has again been suggested.

Pairings on malt agar between cultures of *Daldinia* and of fungi common on ash branches show that *Daldinia* is strongly combative against most of them. For example, *Hypoxylon rubiginosum*, *Mycoacia uda*, *Peniophora lycii*, *P. quercina* and *Phlebia radiata* were completely or partially replaced, or were subjected to deadlock by *Daldinia*. An interesting exception was *Coriolus versicolor*. On malt agar *Coriolus* and *Daldinia* displayed deadlock or partial replacement of *Daldinia* by *Coriolus*. When the water potential was lowered to -1.3 MPa, the reaction changed to partial replacement or complete overgrowth of *Daldinia* by *Coriolus*. When pairings were made under atmospheres of increased CO_2 concentration (e.g. 20% O_2, 60% CO_2), *Daldinia* replaced *Coriolus*. If similar behaviour occurs on wood in the field, it is clear that abiotic variables such as water potential and gaseous concentrations could affect the outcome of competition between these fungi.

Not all the fungi present as mycelium in attached branches fruit on them *in situ*, although some certainly do. On attached oak branches, basidiocarps of *Peniophora quercina* and *Phellinus ferreus* are fairly frequent.

Other colonists of standing trees, mostly members of the Polyporales, are largely confined to the heartwood, growing at first on living trees, and sometimes continuing to survive and fruit on dead trees and stumps. They include *Ganoderma adspersum* common on beech (*Fagus*) and less frequently on other hosts, *Rigidoporus ulmarius* on elm (*Ulmus*), *Inonotus hispidus* usually on ash (*Fraxinus*), *I. dryadeus* on oak (*Quercus*) and *Phaeolus schweinitzii* on conifers. There are several possible reasons for the restriction of certain fungi to heartwood. Heartwood has, in general, a lower water content than the sapwood of living trees, and a high water content is known to inhibit mycelial growth of many fungi and to limit decay because of low O_2 diffusion rates. Heartwood may also have a high CO_2 content, and certain of the specialist heartwood colonizers appear to be tolerant of high CO_2 levels while other less tolerant species are excluded (Hintikka, 1982). Heartwood is also rich in 'extractives' (Hillis, 1962; Highley and Kirk, 1979; Pearce, 1987; Kuć and Shain, 1977; Rayner and Boddy, 1988), including alkaloids, resins and phenolics, many of which are toxic to most fungi (section 6.1). However, the fungi which colonize heartwood are clearly able to grow despite the presence of these extractives, and there is evidence that certain fungi that are selective for particular host trees are tolerant to the extractives from those heartwoods. Examples of this are *Stereum sanguinolentum* on balsam fir (*Abies balsamea*) (Etheridge, 1962) and *Polyporus* (*Tyromyces*) *amarus* on incense cedar (*Libocedrus decurrens*) (Wilcox, 1970).

The fact that heartwood has no ability to produce defence reactions because it is dead may also be significant. Cooke and Rayner (1984) have suggested that heart-rot fungi are S-selected, i.e. they are adapted to a habitat which excludes most other fungi by virtue of stress factors such as the lack of easily assimilable substrates, the presence of inhibitory extractives and unusual aeration conditions.

13.2.2 Decay of fallen timber

When dead branches of trees fall to the ground, conditions change. Instead of the wide fluctuations in moisture content to which they were subjected when attached, the moisture content of branches in contact with the ground is usually higher. Variation in temperature may also be less extreme. The branches may also be subjected to colonization by fungi from the soil and to attack by a range of soil animals.

The fungal communities which develop on piles of branches on the ground are affected greatly by the degree of exposure, contact with other wood or with the ground. In a study of fungi colonizing piles of oak and ash branches, it was found that *Stereum hirsutum* occurred almost exclusively on aerially exposed branches, whilst *Phallus impudicus* and *Armillaria mellea* were restricted to branches in contact with the ground. *Hypholoma fasciculare* was most common on branches in contact with other wood or with the ground (Carruthers and Rayner, 1979). The reasons for these differences may have to do with the way in which colonization takes place: by mycelial cords in *Phallus*, *Hypholoma* and *Armillaria*, or by spores in the case of *Stereum*. It is very likely that differences in the water content of the branches also exert an influence.

The effects of temperature variations on the incidence of fungi on cut branches (slash) of lodgepole pine (*Pinus contorta* var. *latifolia*) in Alberta, Canada, have been studied by Loman (1962, 1965). A survey of the fungi growing on the slash showed that four basidiomycetes were most frequently isolated: *Peniophora phlebioides* and *Lenzites saepiaria* (*Gloeophyllum saepiarium*) were the dominant decay fungi in the upper 5 cm of exposed slash, whilst *Stereum sanguinolentum* and *Coniophora puteana* were mainly active at depths of more than 5 cm within the slash pieces. Temperature variations were much greater in the surface slash, with maximum temperatures as high as 50–55°C on clear sunny days. The differential distribution of the four fungi was correlated with the optimum temperatures for their growth and wood-decay rates, and with the maximum temperature at which the mycelium could survive. *P. phlebioides* and *L. saepiaria* are 'high-temperature' fungi, capable of growth and survival over a wide temperature range and with high temperature optima. By contrast, *S. sanguinolentum* and *C. puteana* grow within a narrower temperature range and have lower optimal temperatures, and are killed at the higher temperatures at which *P. phlebioides* and *L. saepiaria* can survive.

Many basidiomycetes fruit on the stumps of felled or fallen trees and on cut or fallen timber lying on the forest floor. The cut stump or felled trunk represents a largely vacant resource available for colonization and the process of felling is an example of enrichment disturbance (Cooke and Rayner, 1984; Rayner and Boddy, 1988; Rayner and Todd, 1979). The cut face is exposed to colonization by spores of many different fungi, whilst the below-ground portion of the stump is available for colonization from spores in the soil, from mycelium or from mycelial cords. The early phase of colonization of the newly exposed resource, i.e. the phase of primary resource capture, is characterized by species with ruderal-like behaviour which depend for their success on rapid arrival (in turn dependent on abundance of propagules and effective means of dispersal), rapid germination and rapid mycelial growth. The species present during this stage of decomposition may be quite numerous and diverse, but, as the colonies expand, they come into contact with each other, and begin to compete for resources and interact with each other in various ways (Rayner and Webber, 1984). The outcome of competition is that the ruderals are gradually displaced by a more limited range of slower growing combative and stress-tolerant species.

Changes in the fungal populations and community structure have been studied by Coates and Rayner (1985a,b,c) and Chapela, Boddy and Rayner (1988) using cut logs of beech (*Fagus sylvatica*) placed upright, partially buried in the ground in deciduous woodland. The patterns of distribution of mycelia were followed over a period of 4½ years. Fundamental differences were found between the colonization of surfaces exposed to the arrival of air-borne spores and those buried in the ground. Some of the common fungi identified from exposed surfaces are shown in Table 13.4, and include ascomycetes, such as *Coryne*, *Hypoxylon*, *Melanomma* and *Xylaria*, and the deuteromycete *Nodulisporium*, an anamorph of a member of the Xylariaceae. The decline in frequency with increasing time of the pioneer colonizers *Chondrostereum purpureum* and *Coryne sarcoides* should be noted. The presence near the aerially exposed surfaces of the logs of *Armillaria bulbosa*, *Phallus impudicus* and *Tricholomopsis platyphylla* is probably the result of colonization from the soil.

Isolations made from small blocks of decayed wood below the cut surface showed the presence of homokaryotic and heterokaryotic mycelia of three basidiomycetes: *Bjerkandera adusta*, *Coriolus versicolor* and *Stereum hirsutum*.

The lower ends of the logs buried in the soil were quickly colonized by *Xylaria hypoxylon* and a number of basidiomycetes forming cords, e.g. *Armillaria bulbosa*, *Phallus impudicus*, *Phanerochaete velutina* and *Tricholomopsis platyphylla*. All these fungi were capable of spread by

Table 13.4 Overall occurrence of fungi at or near the aerially exposed cut surface of beech logs 24, 52 and 97 weeks after exposure, as detected by direct incubation and observation following destructive sampling

Species	Total no. logs colonized (%)		
	24 weeks	52 weeks	97 weeks
Armillaria bulbosa	13.8	42.2	14.1
Bjerkandera adusta	12.1	17.8	14.1
Chondrostereum purpureum	10.3	0	0.9
Coriolus versicolor	20.7	31.1	35.8
Corticium evolvens	0	0	14.1
Coryne sarcoides	10.3	8.9	1.9
Hypoxylon fragiforme	5.2	8.9	24.5
H. nummularium	56.9	37.8	29.2
Melanomma pulvis-pyrius	84.5	84.4	74.5
Nodulisporium sp.	8.6	17.8	8.5
Phallus impudicus	0	0	2.8
Phanerochaete velutina	0	6.7	4.7
Stereum hirsutum	13.8	35.5	14.1
Tricholomopsis platyphylla	0	0	2.8
Xylaria hypoxylon	63.8	93.3	78.3
Total no. logs sampled	58	45	106

Source: Coates and Rayner (1985a).

superficial or subcortical mycelium spreading over or under the bark. Some of these cord-forming fungi had penetrated to the upper face of the logs (30–40 cm in length) by subcortical spread within 97 weeks. The percentage occurrence of fungi on the buried parts of the beech logs is shown in Table 13.5.

Early stages of colonization were indicated by mottled discoloration spreading into the logs from both the upper and buried cut surfaces. This was followed by the resolution of the decay into a mosaic of columns, mostly of white-rot, each occupied by a single genotype. Coates and Rayner (1985c) have classified the fungi colonizing beech logs into six ecological classes, each with a distinctive colonization strategy. Figure 13.3 represents the changing spatial distribution of decay fungi in beech logs in each of the ecological classes up to 97 weeks after exposure.

- **Class 1**: This comprises air-borne basidiomycetes causing white-rot, e.g. *Bjerkandera adusta, Coriolus versicolor* and *Stereum hirsutum*. They

Table 13.5 Occurrence of various fungi on the buried parts of beech logs

Species	Total no. logs colonized (%)		
	24 weeks	52 weeks	97 weeks
Armillaria bulbosa	76.5	28.6	44
Bjerkandera adusta	0	7.1	8
Coriolus versicolor	0	14.3	16
Coryne sarcoides	11.8	7.1	0
Cristella sulphurea	17.6	0	0
Hypholoma fasciculare	0	21.4	0
Mucor sp.	17.6	0	4
Ozonium sp.	5.9	0	4
Phallus impudicus	17.6	21.4	24
Phanerochaete velutina	5.9	7.1	28
Stereum hirsutum	23.5	0	4
Trichoderma spp.	17.6	14.3	28
Tricholomopsis platyphylla	5.9	21.4	28
Xylaria hypoxylon	94.1	43	52
Total no. logs sampled	17	14	25

Source: Coates and Rayner (1985b).

established numerous decay columns separated by interaction zone lines.
- **Class 2**: This includes the ascomycetes *Hypoxylon fragiforme* and *H. nummularium*, which formed wedge-shaped decay columns with the broad edge adjacent to the cambium. The columns were extensive after 24 weeks, despite slow growth in culture, and it is possible that this pattern of distribution follows a latent invasion strategy.
- **Class 3**: This group is exemplified by the ascomycete *Xylaria hypoxylon*, which formed white decay columns spreading from both surfaces, but more extensively from the lower surface. Possibly this is due to unfavourable factors near the upper surface.
- **Class 4**: This is exemplified by *Armillaria bulbosa*, which developed rhizomorphs and was restricted to the periphery of the logs.

Comparison of Tables 13.4 and 13.5 shows clearly the increased colonization from the soil of *Armillaria*, *Phanerochaete* and *Tricholomopsis*, and reduced colonization by *Bjerkandera* and *Coriolus*. This suggests that the colonization of the latter species is usually by air-borne basidiospores.

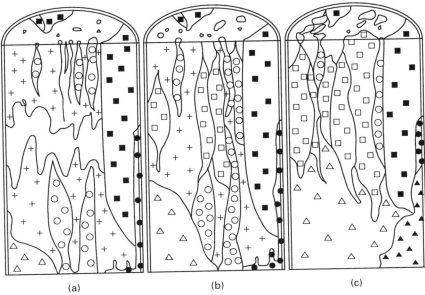

Figure 13.3 Idealized diagrams illustrating typical spatial patterns of colonization of beech logs by different ecological classes of fungi invading naturally from the aerial and buried cut surfaces. (a) 24 weeks, (b) 52 weeks, (c) 97 weeks after exposure. □ = **Class 1**: air-borne white-rot-causing basidiomycetes arriving early but not normally establishing clearly demarcated decay columns until between 24 and 52 weeks; often several or more genotypes of the same species invading; e.g. *Bjerkandera adusta*, *Coriolus versicolor*, *Stereum hirsutum*. ■ = **Class 2**: *Hypoxylon* spp. especially *H. mummularium* establishing rapidly in peripheral wedge-shaped decay columns, possibly via a latent invasion mechanism. ○ = **Class 3**: *Xylaria hypoxylon*, rapidly forming white-rot-decay columns spreading from both aerial and buried cut surfaces. ● = **Class 4**: *Armillaria bulbosa*, rapidly invading peripheral wood via subcortical growth and direct penetration by rhizomorphs. △ = **Class 5** (early colonizers): e.g. *Phallus impudicus* and *Tricholomopsis platyphylla*: mycelial-cord-forming basidiomycetes invading progressively from the base, often as single genotypes, and establishing intense, homogeneous white-rot decay. ▲ = **Class 5** (late colonizers): e.g. *Phanerochaete velutina*. + = **Class 6** (ruderals): basidiomycetes, e.g. *Chondrostereum purpureum*, *Corticum evolvens* and non-basidiomycetes, e.g. *Coryne sarcoides*, fungi imperfecti, *Mucor* spp.; predominant at early stages of community development in discoloured wood lacking clear demarcation into decay columns. (Reproduced from Coates and Rayner, 1985c, with permission.)

- **Class 5**: This group contains the cord-forming basidiomycetes invading from the base, e.g. *Phallus impudicus*, *Tricholomopsis platyphylla* and *Phanerochaete velutina*. These fungi are usually present as a single genotype, which extends upwards and occupies much of the available cross-section of the log.

- **Class 6**: This group includes ruderal fungi such as the ascomycete *Coryne sarcoides* and the basidiomycete *Chondrostereum purpureum*, which occurred regularly and often extensively, but became progressively less prevalent with increasing time.

13.2.3 Interactions between colonies of decay fungi

The occupation of large volumes of wood by a single genotype over a period of time suggests that in some cases there may be deadlock when competing colonies meet. In other cases, changes in occupation indicate that there is replacement of the mycelium of one fungus by another (Rayner and Webber, 1984). Table 13.6 shows examples of combative interactions between some of the fungi colonizing beech logs as shown by reactions on the logs themselves or on agar. From results of this kind, the six ecological classes were ranked from most to least combative in the following order:

$$5 > 1 > 2 = 3 = 4 > 6$$

This ranking is in agreement with the observed course of community development in which the ruderals (class 6), although initially dominant, were replaced by the stress-tolerant classes 2, 3 and 4. Class 1 and class 5 fungi ultimately became dominant.

Similar interactions between wood-rotting basidiomycetes have been reported for other hardwoods. Rayner (1978) analysed the interactions of over 200 different pairs of fungi isolated from hardwood stumps and showed a good correlation between the interactions observed on 3% malt extract agar and those observed on wood. An interesting example was the demonstration that *Bjerkandera adusta* was replaced by *Pseudotrametes gibbosa*. It is known that in deciduous wood, *P. gibbosa* selectively colonizes wood previously occupied by *B. adusta*. In agar culture, the hyphae of *P. gibbosa* wrap around those of *B. adjusta*, and this is associated with death of the latter (Rayner and Todd, 1979; Rayner and Webber, 1984). Similarly, decay columns occupied by *Coriolus versicolor* are sometimes invaded by *Lenzites betulina*. It has been shown that *L. betulina* is mycoparasitic specifically on the mycelium of *Coriolus*, and it has been suggested that the ability to parasitize *Coriolus* enables *Lenzites* to capture large volumes of wood which it can then decompose (Rayner, Boddy and Dowson, 1987a,b; Chapela, Boddy and Rayner 1988).

13.2.4 Intraspecific antagonism

Analysis of the populations of *Coriolus versicolor* in dead tree stumps has provided interesting information about the population biology of dikary-

Table 13.6 Examples of combative interactions observed in beech logs after 97 weeks' exposure

	Replaced by	Deadlock with
Armillaria bulbosa	Hypholoma fasciculare Phallus impudicus Phanerochaete velutina Tricholomopsis platyphylla	Coriolus versicolor Trichoderma viride
Chondrostereum purpureum[a]	Bjerkandera adusta Coriolus versicolor	
Hypoxylon nummularium	Phallus impudicus Phanerochaete velutina Trichoderma viride Tricholomopsis platyphylla	Armillaria bulbosa Bjerkandera adusta Coriolus versicolor Nodulisporium sp. Stereum hirsutum Xylaria hypoxylon
Phanerochaete velutina	Coniophora puteana Tricholomopsis platyphylla	Coriolus versicolor Phallus impudicus
Stereum hirsutum	Hypholoma fasciculare Phallus impudicus Phanerochaete velutina Tricholomopsis platyphylla	Bjerkandera adusta Coriolus versicolor Hypoxylon nummularium
Xylaria hypoxylon	Armillaria bulbosa Coriolus versicolor Hypholoma fasciculare Phallus impudicus Phanerochaete velutina Stereum hirsutum Trichoderma viride Tricholomopsis platyphylla	Hypoxylon nummularium Nodulisporium sp.

[a]Interaction observed in culture on malt agar only.
Source: Coates and Rayner (1985c).

ons occupying columns of decaying wood (Rayner and Todd, 1977; Todd and Rayner, 1978; Williams, Todd and Rayner, 1971, 1981). Sections of stumps of trees such as birch (*Betula*) or oak (*Quercus*) show that they may contain a mosaic of dikaryotic colonies separated by narrow dark zones – 'zone-lines'. Several separate colonies, sometimes in excess of 10, have been observed in a single stump. Sequential slices show that the colonies extend as vertical columns over considerable lengths of the stump (Figure 13.4).

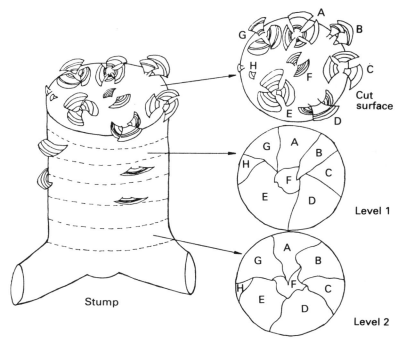

Figure 13.4 Diagram illustrating procedure for analysis of a population of *Coriolus versicolor* in a stump. The stump is first cut into transverse slices as indicated by the dotted lines. The position of the different decay columns, delimited by narrow dark zones, on separate slices is noted, and correlated with that of the fruit-bodies at the surface. The decay columns and their corresponding fruit-bodies are labelled accordingly, and isolates made from them as follows: (i) from fruit-body tissue; (ii) from single basidiospores from fruit-body used for tissue isolation; (iii) from the decayed wood at different levels. (Reproduced from Todd and Rayner, 1978, with permission.)

Basidiocarps, often with distinctive morphology, may form at the cut surface and along the vertical face of the stump, and it is possible to relate the individual distinctive groups of basidiocarps with the dikaryotic colony from which they developed. Dikaryotic cultures can be isolated onto agar from the separate colonies and from the corresponding basidiocarps, and it is also possible to isolate monokaryotic cultures from single basidiospores. *Coriolus versicolor* shows bifactorial (tetrapolar) incompatibility with multiple alleles at each mating type locus, and crosses between monokaryotic isolates derived from basidiocarps of quite distinctive morphology are generally completely intercompatible, forming dikaryons readily. This shows that the basidiocarps of distinctive morphology are conspecific, i.e. that *C. versicolor* is polymorphic. By appropriate techniques, it is also possible to

dedikaryotize (i.e. obtain monokaryotic isolates from) the dikaryons derived from separate decay columns, and these are capable of crossing with a proportion of the monokaryons derived from basidiospores. When the separate dikaryotic cultures are paired together, their mycelia usually anastomose, but anastomosis is quickly followed by an antagonistic response. This is shown by death of the anastomosing hyphae and by dark brown pigmentation at the interface between the two dikaryons (Figure 13.5), the antagonistic response indicating genetic dissimilarity between the interacting dikaryons. The dikaryons in the wood retain their separate identity over several seasons, and may form annual crops of basidiocarps which also retain their separate genetic identity.

The phenomenon of **intraspecific antagonism** appears to be widespread amongst ascomycetes and basidiomycetes, and is especially obvious in dikaryotic wood-rotting basidiomycetes. It serves to protect the integrity of dikaryons from exchange of nuclei (and cytoplasm) from dissimilar dikaryons, although, as we have seen, the dikaryons are free to transfer nuclei to contiguous monokaryons and convert them to new dikaryons (**di–mon mating**). The paradox of the conversion of a monokaryon to a dikaryon by an adjacent dikaryon is that it results in the formation of adjacent dikaryotic colonies which are mutually antagonistic.

The exposure of a cut stump or trunk to a rain of genetically diverse spores has, as a consequence, the development of a community of several or many genetically distinct dikaryotic mycelia, sometimes composed of one or more species. This is probably an artefact caused by human intervention. In intact trees, the number of separate individual mycelia may be much more reduced, in the extreme case to a single individual mycelium occupying a whole tree. Barrett and Uscuplic (1971) showed that trees of Sitka spruce (*Picea sitchensis*) infected with the polypore *Phaeolus schweinitzii* each contained a single individual mycelium distinct from that present in adjacent trees. Colonization by *P. schweinitzii* of the base of the trunk is normally through a single root, and it is likely that the first strain to colonize the base may monopolize development through the main trunk. *Piptoporus betulinus*, the birch polypore (Figure 13.6) provides another example. Multiple infections of standing *Betula* trunks are rare, and usually there are only one or two dikaryons per trunk (Adams, Todd and Rayner, 1981). It is believed that this may be due either to a low level of infection, or to competition between dikaryons growing into the trunk from broken branches.

13.2.5 Distribution of cord-forming individuals

By contrast with the situation in which a single tree trunk may contain many dikaryons of one fungus or in which a population of infected trees may each contain a single distinct individual dikaryon, *Tricholomopsis*

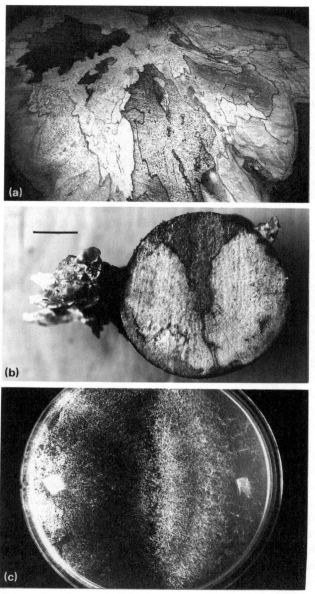

Figure 13.5 Interactions between colonies of wood-rotting fungi. (a) Cut end of beech stump showing a mosaic of brown- and white-rotting basidiomycete colonies. The dark patches are colonies of Xylariaceae. (b) Transverse section of a birch branch containing two colonies of the white-rotting basidiomycete *Coriolus versicolor*, each of which has formed basidiocarps. Where the two colonies have met black interaction zones have developed. (c) Interaction in culture between two genetically dissimilar dikaryons of *Coriolus versicolor*. A dark-pigmented zone has formed where the two mycelia have met. Scale bar = 20 cm (a); 6 cm (b).

Figure 13.6 *Piptoporus betulinus* basidiocarp projecting from the dead trunk of a birch tree. Scale bar = 5 cm.

platyphylla (Figure 13.7), a cord-forming agaric, may have genetically similar individuals which extend over wide areas of woodland (Thompson and Rayner, 1982b; Thompson, 1984). In an area of thinned oakwood some 20 ha in extent, samples from mycelial cords, fruit-bodies and wood infected by *T. platyphylla* were taken in order to identify the individuals by means of antagonistic reactions in culture. From a random sample of 133 dikaryotic isolates, 22 individuals were recognized. The distribution of these individuals is shown in Figure 13.8.

A study of this distribution map will show that certain individuals, e.g. types 10 and 11, are distributed over large distances (80–100 m). The most likely explanation of this is that they have spread by migrating cords. No other individuals were identified within the boundary of type 11, although this individual apparently extended into territory occupied by type 10. The distribution of individuals of type 1 showed two discontinuous patches over 150 m apart. The mating system of *T. platyphylla* is bifactorial and it is highly improbable that separate individuals sufficiently similar to inter-mingle in culture could have developed independently. It seems probable, therefore, that type 1 earlier had a more extensive continuous distribution, and that the original mycelium has broken up into two separate colonies.

Similar distribution patterns have been recorded for other cord-formers, e.g. *Phanerochaete velutina* and *Armillaria* spp. (Thompson and Boddy, 1983; Anderson *et al.*, 1979).

Figure 13.7 *Tricholomopsis platyphylla* basidiocarps around a dead tree trunk. They are interconnected by extensive white mycelial cords which have been exposed by scraping aside the surface litter. Scale bar = 8 cm.

13.2.6 Phenology

The time of appearance of fruit-bodies of lignicolous basidiomycetes is variable. In some cases, e.g. *Ganoderma, Fomes* and *Rigidoporus* spp., the sporophores are perennial, and may survive on the living tree or dead stump over several years. In most other cases, the fruit-bodies are seasonal, a fresh crop being produced each year, possibly for several seasons, dependent on the survival of a food base sufficient to support mycelial growth and fruiting. The seasonal production of basidiospores has an effect on colonization and even species with perennial basidiocarps may not produce basidiospores throughout the year. Nuss (1975, 1986) made a study of the seasonal production of basidiospores from sporocarps of polypores over a 2-year period. *Laetiporus sulphureus, Polyporus ciliatus* and *Polyporus squamosus* formed basidiospores only in spring and early summer (May–June). *Phaeolus schweinitzii* formed spores from September to January. In *Ganoderma adspersum* and *G. applanatum*, spore production from the perennial sporophores was generally limited to March–November. In the winter, the hymenial tubes of the previous season were closed over by hyphae, and new hymenial tubes developed the following spring. Basidiospore production in *Heterobasidion annosum* could continue throughout the whole year. At 0°C, the sporulation of most fungi was interrupted (e.g. *Bjerkandera adusta*) or completely stopped (e.g. *Daedalia quercina, Ganoderma* spp.), but for certain fungi, e.g. *Daedaleopsis*

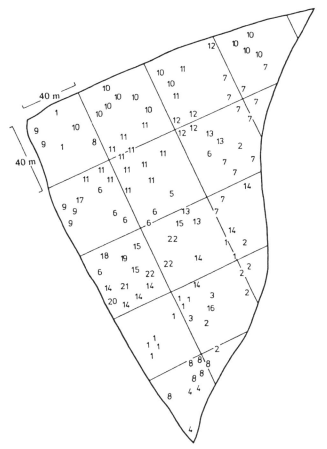

Figure 13.8 Distribution of mycelial types of *Tricholomopsis platyphylla* indicated by different numbers. (Reproduced from Thompson and Rayner, 1982b, with permission.)

confragosa, *Heterobasidion annosum* and *Piptoporus betulinus*, sporulation could continue even below temperatures of 0°C. Temperatures above 20°C inhibited sporulation, and above a temperature of 27°C sporulation stopped altogether.

Flammulina velutipes, a saprotrophic agaric which grows on dead hardwood stumps (Figure 13.9), is well-known for its ability to fruit in winter. It can continue to develop fresh crops of basidiocarps even during very cold winters. Vegetative growth of the mycelium can continue at 0°C, and sporulation occurs even at −5°C. Sporophores which have frozen hard can start sporulation again soon after they are unfrozen (Ingold, 1980,

Figure 13.9 *Flammulina velutipes* basidiocarps growing from a moss-covered dead tree trunk in winter. Scale bar = 4 cm.

1981). The ability of sporophores to revive and sporulate after being frozen at −18°C has been reported for *Auricularia* spp., *Exidia glandulosa* and *Stereum hirsutum* (Ingold, 1982).

The basidiocarps of *Schizophyllum commune* may survive in a brittle state for long periods on dry fallen timber, but on wetting they are capable of releasing basidiospores within a few hours. The same is true of *Auricularia auricula* and *A. mesenterica*.

Observations on dead tree trunks over several years show that a succession of different fungi fruit on them. Cooke and Rayner (1984) have cautioned against the use of the term 'succession' in such cases. What is generally being studied is a fruit-body succession, and this may give a misleading impression of the structure of the community of fungal mycelia within the rotting wood. It is obvious that there will be a period during the early colonization of the wood when the mycelium will not be large enough or will not have accumulated sufficient reserves to permit fruiting, and similarly, especially in the case of fungi forming seasonal sporophores, there may be long periods of the year in which there is little external evidence of their presence. Again there are cases known where an active mycelium produces no fruit-bodies at all. It is therefore important to distinguish between studies in which accounts of fungal successions are based on observations of presence or absence of fruit-bodies, and those in which attempts have been made to isolate mycelia from decaying wood.

13.3 LITTER DECOMPOSERS

The decomposition of litter is brought about by the combined activities of the bacterial, fungal and animal populations of the soil. Basidiomycetes are especially suitable organisms for the decomposition of forest litter because they possess a wide range of enzymes capable of degrading the lignocellulose complex of plant litter (Swift, 1982). It has been estimated that basidiomycetes account for 60% of the living microbial biomass in mull soils, whilst in mor soils the proportion may be much higher (Frankland, 1982). It is well-known that the fruiting of agarics is more prolific in mor soils than in mull soils. This is illustrated in Table 13.7, which shows that the total fresh weight of agaric sporophores collected from permanent plots over three successive years was over five times greater in comparable areas of a mor woodland than a mull woodland in Cumbria, UK (Hering, 1982).

The major basidiomycete genera decomposing litter are *Clitocybe*, *Collybia*, *Mycena*, *Marasmius*, *Hydnum*, *Agaricus* and *Tricholoma* (see also Table 13.1).

Table 13.7 Total numbers and fresh weights of decomposer agarics collected from two woodlands

Species	Mull site		Mor site	
	Number	Fresh weight (g)	Number	Fresh weight (g)
Clitocybe langei sensu Singer			41	30
Collybia peronata			50	139
Cystoderma amianthinum			17	10
Laccaria amethystea			34	43
L. laccata	44	12	352	171
Marasmius epiphyllus	190	12		
M. ramealis	31	1		
Mycena galericulata	28	27	203	210
M. galopus	226	36	292	34
M. metata	171	10		
M. polygramma	24	33		
M. speirea	211	4		
M. sanguinolenta	4	1	14	1
Psathyrella squamosa	4	1	62	85
Total	933	127	1065	723

Source: Hering (1982).

The individual components of forest litter are discontinuous; nevertheless in the mass they provide a continuous distinctive habitat (Rayner, Boddy and Dowson, 1987b). All the components are usually already colonized by microfungi before they fall. On the ground, especially as they become covered by later deposits of litter, the microclimate permits the colonization of other fungi (section 4.4). Although the habitat is continuous, the separate components may not be equally suitable for growth and fruiting of all litter-inhabiting fungi, some of which are restricted to particular components. Rayner, Watling and Frankland (1985) have reviewed the resource relations of higher fungi. Good examples of component-restricted litter decomposers are those which fruit on buried *Pinus* cones, e.g. *Auriscalpium vulgare* (Figure 13.10) and *Pseudohiatula tenacella*. Beech (*Fagus*) cupules also support a specialized group of fungi such as *Xylaria carpophila* (Figure 13.10) (Carré, 1964). Non-component-restricted fungi include *Clitocybe flaccida*, *C. nebularis* and *Collybia butyracea*. *Mycena galopus* grows on a wide range of materials in coniferous and angiosperm litter (Figure 13.11).

13.3.1 Autecology of *Mycena galopus*

The autecology of *M. galopus* has been investigated intensively (Frankland, 1984). It is widely distributed in north temperate woodlands and is not associated with any special soil type. The mycelium is perennial and may consist of separate hyphae or be aggregated into threads about 1 mm

Figure 13.10 Fruit-bodies of two resource-limited litter-decomposing fungi. (a) *Xylaria carpophila* conidial and perithecial stromata growing on overwintered fallen beech cupules. (b) *Auriscalpium vulgare* basidiocarps growing from a fallen pine cone. The hymenium is borne on spines on the underside of the eccentric cap. Scale bar = 2 cm.

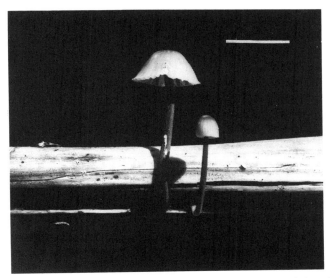

Figure 13.11 *Mycena galopus* basidiocarps growing on a petiole of bracken (*Pteridium aquilinum*). Scale bar = 1 cm.

in diameter. It is tolerant of low temperatures, is capable of decomposing *Quercus* litter at 4°C, can grow at −2°C and survives at −12°C (Hintikka, 1964; Frankland, 1984). It is regarded as a secondary colonist, i.e. colonizing plant materials which are already well-colonized by other fungi. Although capable of growth and fruiting on leaves of deciduous trees such as oak (*Quercus*) and conifers such as spruce (*Picea*), on petioles of bracken (*Pteridium*) and on small twigs, it appears to be incapable of colonizing bulky woody debris, possibly because of restricted aeration and the presence of inhibitory substances in wood. *M. galopus* utilizes most constituents of litter: soluble carbohydrates, hemicelluloses, cellulose, lignin and protein. It also produces polyphenol oxidases to detoxify litter phenols. The ability to break down lignin and cellulose enables it to function as a typical white-rot decay fungus.

The detection of mycelium of *M. galopus* by direct observation is difficult because it has few distinctive characteristics, but identification is possible by tracing the mycelium to the base of fruit-bodies. The use of specific antibodies conjugated with a fluorescent stain has also enabled the mycelium to be distinguished (Frankland *et al.*, 1981). Other techniques include the plating of litter particles on nutrient media and comparison of the cultures made with those obtained from basidiocarps. Litter (e.g. *Quercus* leaves) can be incubated to induce fruiting. Using this technique, it has been shown that *M. galopus* occurs on more than 80% of oak leaves

2 years after leaf-fall at Meathop Wood, Cumbria, UK, indicating that it is a key decomposer in this habitat (Frankland, 1984).

The colonization of litter by *M. galopus* may be by basidiospores or from mycelium. It is likely that colonization does not immediately occur on fresh litter, but only after this has undergone a period of decay by other organisms.

The abundance of *M. galopus* mycelium varies throughout the year. In the litter layer of Grizedale Forest, a spruce forest in Cumbria, it reaches a maximum between August and November. Fruiting was most prolific in September and October (Newell, 1984a). However, the abundance of fruit-bodies is very much dependent on suitable weather. In years in which the climate is unfavourable, basidiocarps may only be found in the lee of a tree stump or in similar places providing shelter from the prevailing wind (Frankland, 1984).

Studies by Newell (1984a,b) and Frankland (1984) have shown that the distribution and activity of the mycelium of *M. galopus* are profoundly affected by abiotic variables such as moisture content of litter, as well as by competition with another litter-decomposing agaric, *Marasmius androsaceus*, and selective grazing by a collembolan (springtail), *Onychiurus latus*. *Marasmius androsaceus*, the horse-hair fungus, grows amongst coniferous litter (and on dead heather), forming black hair-like rhizomorphs extending from one needle to another (Figure 2.4) (Gourbière, Pepin and Bernillon, 1987). By tracing the mycelia of these two fungi from their stipe bases into spruce litter, it was found that the mycelium of *Marasmius androsaceus* was mostly confined to the layer at the surface 3 mm (with a mean depth for the initiation of fruit-bodies of 1 mm). The mycelium of *Mycena galopus* generally occurred in deeper litter layers of the F_1, 5–9 mm (mean fruiting depth 6 mm) or, when in association with *Marasmius androsaceus*, even lower, from 6 to 12 mm (mean fruiting depth 10 mm) (Figure 13.12).

The collembolan *O. latus* is myceliophagous, and examination of its gut contents often reveals the presence of basidiomycete mycelium recognized by the presence of clamp connections. It is most abundant in the autumn, usually in the L and F_1 horizons of the litter. A lower abundance of *Onychiurus* in the L horizon is strongly correlated with a lower moisture content here compared with the F_1.

At temperatures above 2°C in the laboratory, *Marasmius androsaceus* grew more readily than *Mycena galopus*, and at the mean field temperature (11°C), it grew twice as fast. Probably as a result of its superior growth rate, *Marasmius androsaceus* showed a greater capacity to colonize spruce needles, whether taken from the L or the F_1 horizon, than *Mycena galopus*. It also caused more rapid decomposition of spruce litter than *Mycena galopus*. It might therefore be expected that *Marasmius androsaceus* would be dominant in the field, but in permanent quadrats in spruce

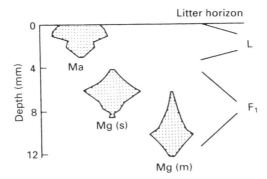

Figure 13.12 The depth of fruiting of *Marasmius androsaceus* (Ma) and *Mycena galopus* (Mg) in single (s) and mixed (m) species clumps in Sitka spruce litter at Grizedale Forest. The width of each 'kite' at any depth is proportional to the percentage number of fruit-body bases which were initiated at that depth. (Reproduced from Newell, 1984a, with kind permission from Elsevier Science Ltd, The Boulevard, Langford Lane, Kidlington, OX5 1GB, UK.)

plantations, 61% of the basidiocarps were of *Mycena galopus* and 38% were of *Marasmius androsaceus*.

The answer to this paradox appears to lie in the preference of *O. latus* for the mycelium of *Marasmius androsaceus*. When sterilized spruce needles were placed in contact with equal numbers of needles colonized by either *Marasmius androsaceus* or *Mycena galopus* in the presence or absence of *O. latus* the proportion of the sterile needles colonized by one or other of these fungi was dependent on grazing by the collembolan. As shown in Figure 13.13, in ungrazed cultures *Mycena galopus* colonized 28% and *Marasmius androsaceus* 72% of the needles, whilst in grazed cultures the proportion grazed was reversed to *Mycena galopus* 62%, *Marasmius androsaceus* 38% (Newell, 1984b).

Some confirmation of these findings was obtained from field experiments in which the *Onychiurus* was partially excluded from the litter by fine nylon mesh. In the areas from which the collembolan was excluded, *Mycena galopus* produced significantly fewer and *Marasmius androsaceus* significantly more fruit-bodies than in control areas from which the collembolan was not excluded.

Newell (1984b) concluded from her work:

These experimental results and observations suggest that the distribution of *M. galopus* and *M. androsaceus* in Grizedale Forest probably was influenced by *O. latus*. Selective grazing by *O. latus* of *M. androsaceus* mycelium in preference to that of *M. galopus* may have altered the outcome of competition in the F_1 horizon sufficiently to allow *M. galopus* to dominate this horizon.

370 TERRESTRIAL MACROFUNGI

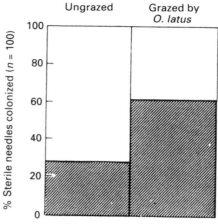

Figure 13.13 The colonizing abilities of *Marasmius androsaceus* □ and *Mycena galopus* ▨, with and without grazing by *Onychiurus latus* (20 per replicate), after 24 days at 11°C in mixed cultures which initially contained equal quantities of litter inoculated with each of the two fungi and 10 labelled sterile Sitka spruce needles per replicate. (Redrawn from Newell, 1984b, with kind permission from Elsevier Science Ltd, The Boulevard, Langford Lane, Kidlington, OX5 1GB, UK.)

Restricted grazing by *O. latus* due to lower densities of the animal probably explains why *Marasmius androsaceus* flourishes in the upper 2–3 mm of the litter. The complete absence of *Mycena galopus* here is probably largely explained by the contrasting minimum water potential requirements of the two fungi. *Mycena galopus* is intolerant of low water potentials and requires -1.0 to -3.6 MPa to make microscopically observable growth. Compared with this, *Marasmius androsaceus* can grow at much lower water potentials with a minimum requirement of between -6.0 and -7.2 MPa (Dix, 1984b).

The horizontal distribution of the mycelium of *Mycena galopus* has been followed over a number of seasons in permanent quadrats using the incidence of basidiocarps as an index of presence (Swift, 1982; Frankland, 1984). The perennial nature of the mycelium is indicated by the fairly close correspondence between the position of basidiocarps in successive years. There is, however, a suggestion of 'movement', i.e. displacement of basidiocarp production. In a plot measuring 10 x 4.5 m, at least 1000–2000 basidiocarps (i.e. 44 m^{-2}) have been recorded in productive seasons. Frankland (1984) has estimated that the ratio of production of basidiocarps to production of vegetation mycelium in *Quercus* litter was 1:10, indicating the presence of an abundance of mycelium possibly of the order of 700 mg m^{-2}.

Basidiocarps are not randomly distributed, but the factors responsible for localized production, such as variation in depth of litter or nutrients exported from the tree canopy above, are not understood. Somatic incompatibility/compatibility tests in culture on dikaryons obtained from different basidiocarps as a guide to the extent of the distribution of individual mycelia have shown genetical identity between isolates obtained from basidiocarps up to 2–5 m apart.

13.3.2 Fairy ring fungi

A number of basidiomycetes growing in woodland or in pasture form basidiocarps in arcs or circles, known as fairy rings. Over 60 species are recorded as being associated with fairy rings (Hawksworth, Sutton and Ainsworth, 1983). Gregory (1982) has distinguished between tethered and free rings. Tethered rings are formed by mycorrhizal associates of trees, and the arcs or circles of basidiocarps are centred on the tree to which they are attached (Ford, Mason and Pelham, 1980). Free rings are formed by several genera, e.g. *Agaricus*, *Clitocybe*, *Collybia*, *Marasmius* and *Lycoperdon*, which include litter decomposers. The mycelium develops as a centrifugally expanding annulus with the leading edge extending outwards into fresh terrain, and the trailing edge undergoing death and lysis. The regular geometry of the distribution of fruit-bodies and of the mycelium in

Figure 13.14 *Clitocybe nebularis* basidiocarps forming a fairy ring in woodland litter. (Photograph by R. Sexton.)

the litter has enabled studies to be made on the genetics, dynamics and the interactions between fairy rings of the same species or with mycelia of other fungi. The woodland litter decomposer *Clitocybe nebularis* (Figure 13.14) has been the subject of such a study (Dowson, Rayner and Boddy, 1989) in a deciduous woodland dominated by *Fagus*. The mycelial annulus of *C. nebularis* is a band 30–40 cm wide, with a leading edge composed of white mycelial cords extending into the litter and the mineral soil, and a central region of dense white mycelium ramifying through the bleached litter. Fruit-bodies developed from this central band. At the trailing edge, the mycelium becomes fragmented and lysed. Intact annuli 6–8 m in radius and arcs with calculated radius of 10–30 m were discovered. Estimates of the average rate of mycelial extension of 2.3 mm day^{-1} indicated that the annuli were 7–9 years old and the arcs 11–33 years old.

When sods were cut from rings, reorientated through 180° and replaced in the same hole the mycelium gradually disappeared, whereas sods replanted several metres in advance or behind a ring continued to grow at the same rate as the controls (sods cut and replaced in the same hole orientated in the original direction) and maintained growth in the same direction. Dowson, Rayner and Boddy (1989) interpreted these results as indicating a high degree of polarity of ring development, such as would occur if the growth front acted as a sink for resources supplied by the mycelium behind it. The advancing mycelial cords can be viewed as having a foraging function.

Interactions on agar media between the mycelium of *C. nebularis* and the mycelia of other litter-decomposing fungi such as species of *Clitocybe*, *Collybia*, *Tricholoma* and *Psathyrella* resulted either in deadlock or replacement of the *C. nebularis* mycelium by its opponent. This lack of combative ability in interspecific interactions, and the ability of the mycelium by rapid growth to exploit temporary abundance in the form of seasonal litter-fall, indicates that *C. nebularis* has ruderal characteristics.

Fairy rings are especially obvious in pastures, where they are associated

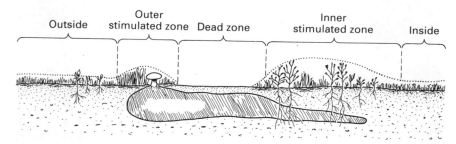

Figure 13.15 Section through turf and soil of a portion of a fairy ring due to *Marasmius oreades*. (Redrawn from Smith, 1957, with permission from the Sports Turf Research Institute.)

Figure 13.16 *Marasmius oreades*. (a) Ring of basidiocarps in a lawn. The darker patches of grass inside and outside the ring of basidiocarps represent darker green areas of grass. (b) Basidiocarps. Scale bar = 50 cm (a); 0.6 cm (b).

with the growth and fruiting of basidiomycetes including agarics, hydnums, clavarias and puffballs (Ramsbottom, 1953; Smith, Jackson and Woolhouse, 1989). They are particularly well-developed in open communities of poor grassland. The form of the ring may vary, depending on the species of the fungus and on the amount of rainfall. Fairy rings associated with *Marasmius oreades* (Figures 13.15 and 13.16) have three adjacent concentric rings of abnormal turf, an outer zone where the growth of the grass is stimulated (it is taller and greener), a middle zone where the grass may be dead and the ground bare (especially in dry seasons), and an inner zone of stimulated growth, often colonized by ruderal herbs which have recolonized the previously bare zone (Figure 13.15). Fruit-bodies of the causal fungus may be found in the outer ring, but may not develop every year.

Fairy rings extend outwards as an annulus, and measurements of the average annual increment of growth and of ring diameter provide means of estimating the age of some rings. Some estimates of radial growth rate are given in Table 13.8 (Dickinson, 1979).

It has been shown that large rings of *M. oreades* can be several centuries old, an *Agaricus tabularis* ring 60 m in diameter was thought to be 250 years old, and a *Calvatia cyathiformis* ring 200 m in diameter was thought to be 420 years old.

Somatic incompatibility testing in culture has enabled the genetical relationship of rings formed by *M. oreades* to be compared. Mallett and Harrison (1988) studied 23 rings in a small area and discovered 16 distinct individuals. Dikaryons from two separate rings grew together without confrontation, indicating that they were genetically identical and represented clonal propagation. How the two portions of the same clone became separated is not clear, although transfer by lawn-maintenance equipment cannot be eliminated.

The antagonistic confrontation of genetically dissimilar mycelia on agar

Table 13.8 Growth rates of fairy rings formed by several fungi in grasslands

Species	Radial growth rate (mm year^{-1})	Reference
Agaricus tabularis	120	Shantz and Piemeisel (1917)[a]
Calvatia cyathiformis	240	Shantz and Piemeisel (1917)[a]
Hydnum suaveolens	230	Thomas (1905)[b]
Lepiota sordida	152–204	Bayliss Elliott (1926)[c]
Marasmius oreades	99	Bayliss (1911)[d]
" "	140	Smith (1957)[e]
" "	165	Wolf (1971)[f]
" "	300–350	Ingold (1974)[g]
" "	135	Dickinson (1979)

Method of obtaining data based on:
[a] the occurrence of fruit-bodies in several rings over 10 years;
[b] measurements of one ring over 9 years;
[c] measurements of two rings over 1 year;
[d] measurements of three rings over 4 years;
[e] measurements of one ring over 11 months;
[f] an impression of the growth of 26 rings over several years;
[g] the occurrence of fruit-bodies in one ring over 3 years.

has its counterpart in nature when two different rings of the same species contact each other. It is generally found that the two rings do not intersect, but that growth stops at the point of contact (Smith, 1980). Parker-Rhodes (1955) uses the terms **bilateral extinction** for the situation in which neither opponent survives and **unilateral extinction** when one survives. By contrast, it is claimed that when rings of different species contact each other, they may continue to grow and intersect each other, a phenomenon termed **indifference**. Whether the mycelia of these different species are, in fact, indifferent to each other seems unlikely, and further microscopical observations of their behaviour as they approach and make contact with each other are needed.

Theoretical treatments of the kinetics (arrival and growth) of fairy rings have been given by Parker-Rhodes (1955) and by Stevenson and Thompson (1976). Assuming that there is a constant arrival rate σ (the number of new rings per unit area per unit time) and a constant growth rate ρ (radial extension per unit time), the factor determining survival in competitive juvenile systems is $\sigma\rho$, but in the longer term when equilibrium is established, the relevant factor is σ/ρ.

Parker-Rhodes (1955) has concluded that the ring habit is very inefficient ecologically on the grounds that a ring-forming fungus occupies a low proportion of the available ground. He regards the ring habit as typical of those plant associations which have a low rate of access to organic matter, such as poor grasslands. However, as we have seen, 'free' rings do occur in woodlands where litter fall and accumulation are appreciable, and *Clitocybe nebularis* provides an example.

The reason for the zone of stimulated grass growth at the outer margin of an advancing fairy ring is that in the region of the ring there is more rapid decomposition of soil organic matter in the mycelial zone. Weaver (1975) showed that the organic matter content of soil exploited by a *Marasmius oreades* fairy ring is lower than in unaffected soils. He concluded that decomposition rates are 2–3 times higher than in unaffected soil. The levels of extractable NO_3, NH_4 and P are also higher in soil under the dark green outer ring than in soil beyond this zone. When fruiting of the fungus occurs, however, the levels of these substances are depleted. Soil that has been previously occupied by the fungus has significantly lower concentration of extractable NO_3, NH_4 and P than unaffected soil (Fisher, 1977).

The death of grass and the resulting bare zone are probably due to three effects: parasitic attack by the fungus (Filer, 1965), drought, and toxin production. The mycelial zone is considerably drier than the soil outside it because of impeded water percolation (Smith, 1957). It has also been shown that culture filtrates from the mycelium of *M. oreades* contain toxins which damage wheat root tips (Traquair and McKeen, 1986). Other possible deleterious substances produced by the mycelium include HCN (Lebau and Hawn, 1963). The stimulation of grass growth in the inner ring is believed to be associated with increased fertility of the soil associated with the decomposition of the *Marasmius* mycelium, and this is associated with higher bacterial counts.

The mycelium of *M. oreades* is intermingled with grass roots and is usually confined to the surface of soil down to a depth of about 8–10 cm, but can extend to a depth of 30 cm. Warcup (1951b) has compared the numbers of species of fungi isolated by his soil plate technique (small crumbs of soil placed directly in Petri dishes and covered with agar) from soil samples taken within and outside the mycelial zone. The number of species in the mycelial zone (4–7) was impoverished as compared with the numbers immediately below it (13–18). The species isolated from the mycelial zone were often distinct from those outside it. The reasons for this restricted microflora are not fully known. Smith and Rupps (1978), using a dilution plate technique, were unable to provide clear confirmation of Warcup's findings. However, they isolated several soil fungi antagonistic to *M. oreades* mycelium from the mycelial zone, perhaps suggesting that only strong competitors can survive here.

The reasons for the continuing outward growth of *M. oreades* mycelium are complex, but it is possible that nutrient depletion of previously occupied soil and a build-up of microbial antagonists may play a part. It has also been suggested that the mycelium may produce a self-inhibitory metabolite (Smith, 1980). It is unusual to find one fairy ring inside another, suggesting that it is difficult for the fungus to re-establish itself in previously colonized soil.

13.4 MYCORRHIZAL MACROFUNGI

The roots of forest trees are invariably associated with fungal hyphae (Harley, 1989), and the feeder roots are often richly branched and encased in a sheath or mantle of fungal pseudoparenchyma. The ensheathing fungus penetrates between the cortical cells, forming single-layered flattened sheets of closely packed hyphae, the **Hartig net**. Infection does not extend into the stele, and the living cortical cells are not penetrated. This kind of association is an example of **mutualistic symbiosis**. The fungal partner or mycobiont acquires the bulk of its carbon supply from the chlorophyllous host or autobiont, whilst the mineral supply of the latter is enhanced as a result of infection by the fungus, which may form an extensive network of hyphae ramifying through the litter layers of the soil. The mycorrhizal root systems themselves are concentrated in the surface layers of the litter, and their mycelia are mainly in the F horizon where the resource quality of the litter and its decomposition rate are high. Infections of this type are termed **sheathing** or **ectomycorrhizae**, and they are characteristic of roots of conifers (e.g. Pinaceae) and broadleaved trees (e.g. members of the Fagaceae, Betulaceae and Myrtaceae). They are particularly well-developed in forests on moder, mull or brown earth soils in zones of moderate latitude and altitude, especially where the climate shows seasonal change, with some surface drying of the soil (Read, 1984, 1991). Mycorrhiza of this type develop very poorly in waterlogged soil (Read and Boyd, 1986).

Most of the ectomycorrhizal fungi belong to the Agaricales, Boletales and the Gasteromycetes. They may also be produced by members of the Tuberales (ascomycetes) and Endogone (zygomycetes). Some of the more common genera of ectomycorrhizal basidiomycetes are:

Agaricales and Boletales
Amanitaceae
Amanita
Boletaceae
Boletus
Leccinum
Suillus
Cortinariaceae
Cortinarius
Hebeloma
Gomphidiaceae
Gomphidius
Paxillaceae
Paxillus
Russulaceae
Russula
Lactarius
Tricholomataceae
Clitocybe
Laccaria
Leucopaxillus
Tricholoma

Aphyllophorales
Cantharellaceae
Cantharellus
Craterellus
Clavariaceae
Clavaria
Clavariadelphus
Hydnaceae
Hydnum
Thelephoraceae
Thelephora

Hysterangiales
Rhizopogonaceae
Rhizopogon

Sclerodermatales
Pisolithaceae
Pisolithus
Sclerodermataceae
Scleroderma

More extensive lists of mycorrhizal partners have been given by Trappe (1962), Harley and Smith (1983) and, for British flowering plants, by Harley and Harley (1987).

Fungi which form ectomycorrhizae differ in their host specificity. It is important to distinguish between the ability to cause the characteristic symptoms of mycorrhizal infection (root branching, sheath formation, development of Hartig net) and the ability to fruit (i.e. from basidiocarps) in association with a given host. Molina and Trappe (1982) have distinguished three different groups:

1. Fungi with wide ectomycorrhizal host potential, low specificity, whose

basidiocarps are usually associated in the field with diverse hosts. Examples are *Amanita muscaria, Boletus edulis, Laccaria laccata, Lactarius deliciosus, Paxillus involutus* and *Pisolithus arhizus* (= *P. tinctorius*). *A. muscaria* is well-known for its ability to form mycorrhizae with coniferous hosts (*Pinus, Picea*) and with deciduous hosts such as *Betula*. Similarly *B. edulis* and *L. laccata* have a wide range of possible associates. *P. involutus* forms ectomycorrhizae with conifers and with *Betula* (Laiho, 1970). *P. arhizus* is capable of infecting numerous hosts, and has been extensively used for artificial inoculation. It has a worldwide distribution and it has been claimed that it can infect any tree host which has ectomycorrhiza (Marx, 1977).
2. Fungi with intermediate host potential yet specific or limited in basidiocarp–host associations. The genus *Suillus* is associated in the field with conifers, especially species of *Pinus* and *Larix*. *Suillus grevillei* (= *Boletus elegans*) is almost invariably found fruiting under *Larix* species, but there are occasional reports of fruit-bodies under *Pinus*.
3. A more specialized group of fungi with a very narrow host potential. This group includes the gasteromycete *Alpova diplophloeus*, which forms mycorrhiza only with *Alnus* spp. *Cortinarius pistorius* and *Rhizopogon cokeri* are limited to pines in experimental inoculations, but *R. cokeri* can form mycorrhiza in the field with *Tsuga mertensiana*. *Alpova* and *Rhizopogon* both possess hypogeous basidiocarps. Molina, Massicotte and Trappe (1992a) have pointed out that most hypogeous basidiomycetes show stronger host restriction than epigeous species, emphazing the close co-evolution of the specialized hypogeous fungi with their hosts.

For a further discussion of ecological specificity see Harley and Smith (1983) and Molina, Massicotte and Trappe (1992b).

The reasons for the different degrees of host specificity by ectomycorrhizal fungi are not known. Duddridge (1987) has discussed how a plant root distinguishes between a potential mycorrhizal partner and the diverse population of saprophytic and parasitic fungi present in the rhizosphere. Compatible partners may be able to evade recognition as pathogens because of an ability to avert or an ability to tolerate the defence systems of the host. She has considered a number of steps in the infection process: arrival in the rhizophere, growth and attachment to the root surface, and formation of the Hartig net. This is similar to pathogen attack where virulence is associated with compatibility with the host.

13.4.1 Arrival in the rhizosphere

Colonization of a tree root may occur in several ways: by basidiospores germinating near to it; by hyphae, mycelial strands or rhizomorphs in the soil, possibly extending from established infections; from sclerotia; and from root-to-root contact. Lateral root primordia may also be infected at the time of emergence by mycelium in the parent root.

Studies on the germination of basidiospores of ectomycorrhizal basidiomycetes on agar have usually given erratic results with low percentage germination, but germination may be stimulated by the proximity of yeasts such as *Rhodotorula* and mycelia of the same or of different species of fungus (Fries, 1984, 1987b). The proximity to roots may stimulate basidiospore germination (Fries and Birraux, 1980) and exudates from roots may also be effective. When tree roots in soil are allowed to grow over microscope slides bearing agar films in which basidiospores have been suspended, spore germination may be considerably improved, e.g. 96% of spores of *Paxillus involutus* germinated close to birch roots in untreated soil from a birch wood and chemotropically directed growth of germ tubes towards the birch root occurs (Ali and Jackson, 1988). Similarly, when roots of *Pinus* seedlings are dipped into agar suspensions of basidiospores of the mycorrhizal gasteromycete *Rhizopogon luteolus* or of the boletes *Suillus luteus* and *S. granulatus*, and the seedlings grown on in agar or in soil, good spore germination occurs. Twenty-one-day-old seedlings were more effective in inducing germination of *R. luteolus* spores than 4-day-old seedlings (Theodorou and Bowen, 1987), probably reflecting increased leakage of exudates from the older seedlings.

The root surface and its immediate surroundings are zones of enhanced microbial activity where different kinds of soil microorganisms are stimulated to grow in response to the sloughing off of root cells and the exudation of a range of organic material (section 7.4). Mycorrhizal fungi thus have to compete with other members of the rhizosphere population. The substances present in the rhizosphere may stimulate spore germination, e.g. by suppressing soil fungistasis, and may also stimulate mycelial growth. Particular interest focuses on the specificity of such substances and their possible role in the selection of specific mycorrhizal partners to a particular tree host. Exudates from birch roots grown in soil are active in stimulating spore germination in *Hebeloma crustuliniforme*, but not in certain other fungi known to form mycorrhiza with birch (e.g. *Paxillus involutus*, *Amanita fulva*). In this case, fractionation of the root exudates showed that the active fraction was ninhydrin-positive, indicating the presence of amino groups (Ali and Jackson, 1988). Extracts from homogenates of pine roots stimulated germination of *Hebeloma mesophaeum*, and germination promotion activity is associated with lipidic fraction of these extracts (Fries and Swedjemark, 1986). Fatty acid fractions from exudates of axenically grown pine seedlings have also been shown to enhance the mycelial growth of *Laccaria amethystina* and *L. bicolor*, but not that of *Leccinum aurantiacum* (Fries, Bardet and Serck-Hanssen, 1985). Fries *et al.* (1990) have identified abietic acid as an activator of basidiospore germination in ectomycorrhizal species of *Suillus*.

Thus, although root exudates do stimulate germination and growth of mycorrhizal fungi, their chemical nature is variable, and does not yet provide an explanation for the specificity of relationships between fungi and their hosts.

13.4.2 Growth and attachment to the root surface

Growth of the fungus in the rhizosphere is followed by adhesion to the root surface, but adhesion is non-specific because it occurs on host as well as on non-host roots. On non-host roots, a sheath is usually not formed. Duddridge (1986) made experimental studies of infection by *Suillus grevillei*, normally associated with *Larix*, on host and non-host roots. No mycorrhizae were formed with *Betula*, *Allocasuarina*, *Picea*, *Alnus* or *Pinus nigra*. In the case of *P. nigra*, although a loose sheath was formed, the epidermal cells were penetrated and killed. Mycorrhizae were formed in association with *Pinus sylvestris* and *Pseudotsuga menziesii*, but ultrastructural changes were observed in these cases which suggested that the relationship was not fully compatible (see below).

13.4.3 Formation of the Hartig net

The formation of the Hartig net appears to be a specific phenomenon, and does not take place in non-hosts. In the examples quoted above, a true Hartig net was formed by *S. grevillei* on *Pseudotsuga* but not on *Pinus*. Increased production of phenolic materials was observed at the host–fungus interface on non-susceptible hosts, and the host cell walls were often thickened. Signs of degradation were also detected in the fungus. Such phenomena are similar to the interactions observed between plant pathogens and non-susceptible hosts.

The distribution of the mycelium of ectomycorrhizal fungi is clearly related to the distribution of the host and its root system. It is doubtful if the mycelia of most such fungi have a prolonged existence in soil or litter if not attached to host roots. The extent to which mycelium extends from the root varies with the host and also with the fungus. The surface of a mycorrhizal beech root is smooth and there is probably little exploration of the surrounding litter layer by hyphae of the mycorrhizal fungus. By contrast, the mycelium may extend from pine roots for over a metre (Schramm, 1966). Ogawa (1985) dissected soil layers in forests in Japan and discriminated between three types of mycelial distribution – the fairy ring, the irregular mat and the dispersed colony – but Read (1992) has doubted whether Ogawa's fairy ring mycelia (including such genera as *Tricholoma, Lepista, Agaricus, Clitocybe*) belong to mycorrhizal fungi because they lack a sheath or Hartig net. More likely they belong to fungi which utilize continuously distributed nutrient resources such as forest litter. Mycorrhizal fungi with irregular mats show localized areas of intensive mycelial development. They include species in the genera *Cortinarius, Rozites, Sarcodon* and *Suillus*. The hypogeous fungi *Hysterangium* and *Gautieria* form dense mycelial mats (Cromack *et al.*, 1979; Griffiths and Caldwell, 1992). In the case of *H. crassum* which forms ectomycorrhiza with Douglas fir (*Pseudotsuga menziesii*), the mats, up to 1 m in diameter, occur at a depth of about 4–9 cm and may occupy up to 16.7% of the soil volume. The bulk of the mycelium making up the mat occurs in the mineral soil or at its interface with the litter.

The dispersed colony pattern is characteristic of many of the well-known ectomycorrhizal fungi, including such genera as *Amanita, Boletus, Lactarius, Paxillus* and *Russula*. *Pinus* seedlings inoculated with *Suillus bovinus* bear fans of mycelium capable of extending into unsterile peat at rates of up to 3 mm day^{-1} (Read, 1984). The mycelium shows directional growth towards accumulations of nutrient-rich litter. It may also aggregate into rope-like differentiated strands, especially when in contact with a compatible root system (Duddridge, Malibari and Read, 1980). By means of such strands, several host plants may be interconnected, and water and other materials may be conducted.

In forest areas and plantations, it is possible to trace the extent of individual mycelia, making use of the fact that genetically identical dikaryotic mycelia (**genets** or **clones**) freely intermingle in pure culture, whilst genetically distinct dikaryons show demarcation lines indicating mutual aversion (**somatic incompatibility**). Isolations made from basidio-

Table 13.9 **Stand characteristics, sporocarp density and numbers of clones of** *Suillus bovinus* **in** *Pinus sylvestris* **stands in Sweden**

	Site 1	Site 3
Age of stand (years)	12	125
Area (m^2)	400	400
Number of trees ha^{-1}	3800	380
Number of sporocarps found	250	150
Number of clones detected	36	1
Estimated number of clones ha^{-1}	900	25
Mean distance between outermost sporocarps of the clones (m)	0.7	30.0
Maximum distance between two sporocarps of a clone (m)	4.2	30.0

Source: Dahlberg and Slendid (1990)

carps are paired in culture, and from the type of interaction the identity and the mycelia can be confirmed (Fries, 1987a; Dahlberg and Stenlid, 1990; Dahlberg, 1992). In a study of the distribution of clones of *Suillus bovinus* in Scots pine forests in Sweden, Dahlberg and Stenlid (1990) compared a disused sand-pit colonized some 12 years earlier by naturally regenerated pines, with older established stands up to 125 years old (Table 13.9). The recently colonized area (site 1 in Table 13.9) had a large number of small clones, whilst the older site (site 3) had a single large clone extending over 30 m. At a smaller site of approximately the same age, two clones were detected. It is obvious that a single clone may infect several trees. The maximum radial rate of expansion (estimated from site 1) was 20 cm year^{-1}, and from this estimate it has been calculated that the largest clone was at least 75 years old. This contrasts with the lifespan of a *Pinus* mycorrhizal short root which persists for only a few months (Harley and Smith, 1983). It seems likely that as pine stands mature, the number of clones is reduced by intraspecific competition. As in other studies of the extent of individual mycelia, there are indications that clones may become fragmented (see Thompson and Rayner, 1982b, and section 13.2.5).

Although ectomycorrhizal basidiomycetes are clearly specialized in their physiology and nutrition, the features which distinguish them from other basidiomycetes such as litter- or wood-decomposers are difficult to specify. Harley and Smith (1983) have summarized previous investigations. Most ectomycorrhizal fungi require simple carbohydrates such as sugars, although some can utilize polysaccharides if provided with a small amount of sugar. Some can degrade holocellulose, lignocellulose and lignin (Trojanowski, Haider and Hutterman, 1984). Some can utilize protein as a carbon and a nitrogen source and translocate breakdown products to their host (Abuzinadah and Read, 1986a,b; Abuzinadah, Finlay and Read, 1986).

The development of ectomycorrhizal infections on seedlings is related to seedling age, light intensity, fungal partner and soil conditions (Harley, 1969). The formation of mycorrhizal root tips usually corresponds in time to the development of true leaves (as distinct from cotyledons) and thus to the probable onset of active photosynthesis. Increasing light intensity increases the number of root apices and the proportion which are converted to mycorrhizae. Mycorrhizal development is most vigorous on relatively infertile soils, especially those in which there is a deficiency or imbalance in available minerals such as nitrogen and phosphorus, and is therefore most obvious in soils with an accumulation of raw humus. High levels of phosphorus and nitrogen suppress mycorrhizal infection.

The fungi which become established on seedling roots usually do not persist as the tree ages, but are displaced by others, so that there is a succession of mycorrhizal partners (Dighton and Mason, 1985; Mason *et al.*, 1987). Evidence for this comes from several sources. It is possible to identify from morphological, anatomical and microscopical characters the diagnostic features of many fungus root-tip associations (Agerer, 1987–91; Ingleby *et al.*, 1990) and dissection of root tips from soil cores taken at increasing distance from a tree trunk show differing proportions of fungi (Figure 13.17).

Fungi isolated from mycorrhizal root tips show changes associated with age (Chu-Chou and Grace, 1981). The incidence of fruit-bodies of known mycorrhizal fungi also changes in relation to distance from the base of the tree (Figure 13.17; Last *et al.*, 1983). As the roots extend radially outwards, annuli of fruit-bodies can be distinguished, with the fungi at the periphery differing from those at the centre, nearer the trunk (Ford, Mason and Pelham, 1980). The composition of the species lists of basidiocarps of mycorrhizal and non-mycorrhizal fungi also changes in relation to the age of tree stands (Table 13.10). There are dangers in using the incidence of fruit-bodies as an indicator of succession, because fungi present on mycorrhizal root tips may take some time to fruit, but a good correlation has been found between the incidence of fruit-bodies and the characteristic mycorrhizal tips belonging to the same species.

Confirmation of the changing pattern of colonization of root tips with increasing age of the root system has been provided by Gibson and Deacon (1988), who transplanted 2-year-old birch saplings from a coal spoil site to linear troughs in a glasshouse. The sapling roots were washed and then arranged in the troughs which were filled with a brown earth. The soil was repeatedly inoculated with the mycelia of three ectomycorrhizal fungi: *Laccaria proxima*, *Hebeloma crustuliniforme* and *Lactarius pubescens*. The

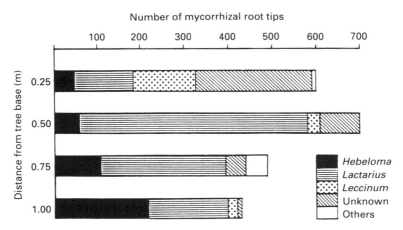

Figure 13.17 Numbers of mycorrhizas of different types in 15 soil cores taken at increasing distances from the base of a birch (*Betula pubescens*) tree. (Reproduced from Deacon, Donaldson and Last, 1983, in Dighton and Mason, 1985 by permission of Kluwer Academic Publishers and the British Mycological Society.)

Table 13.10 Succession of fruit-bodies of proven, or suspected, ectomycorrhizal fungi appearing in a stand of birches (*Betula* spp.) planted at Bush Estate, near Edinburgh, Scotland

Years after planting	Species
1	Nil
2	*Hebeloma crustuliniforme*
	Laccaria sp.
3	*Thelephora terrestris*
4	*Hebeloma fragilipes*
	H. sacchariolens
	H. mesophaeum
	Inocybe lanuginella
	Lactarius pubescens
6	*Cortinarius* sp.
	Hebeloma leucosarx
	Hymenogaster tener
	Inocybe petiginosa
	Leccinum roseofracta
	L. scabrum
	L. versipelle
	Peziza badia
	Ramaria sp.
7	Other *Cortinarius* spp.
	Other *Hebeloma* spp.
	Lactarius glyciosmus
	Leccinum subleucophaeum
10	*Hebeloma vaccinum*
	Russula betularum
	R. grisea
	R. versicolor

Source: Last *et al.* (1983)

whole root system was examined at intervals over three growing seasons and the distribution of the mycorrhizal root tips of different fungi was mapped and the location of basidiocarps noted. It was found that *L. pubescens* mycorrhiza only developed in the older regions of the root system, whilst *H. crustuliniforme* mycorrhiza developed in all the root regions. As a result of studies of this kind, early- and late-stage mycorrhiza-formers have been distinguished (Deacon, Donaldson and Last, 1983). Typical early-stage fungi are *Hebeloma* spp., *Laccaria* spp.

and *Thelephora terrestris* (Figure 13.19). On birch roots these early-stage fungi are succeeded by *Lactarius pubescens* (Figure 13.18 and Table 13.10). Late-stage fungi include *Cortinarius* spp., *Leccinum* spp. (Figure 13.20), *Russula* spp. and *Amanita rubescens*.

A distinction has to be made between seedlings growing well away from a mature tree and those close enough to develop root contact with it. In those seedlings close enough for root contact, or for mycelial or mycelial-cord contact, infection by late-stage fungi can occur earlier (Fleming, 1983).

In attempting to analyse differences in the behaviour of early- and late-stage fungi, experiments on seedling infection by basidiospores have been performed (Fox, 1983, 1986). A clear-cut difference was found between early-stage fungi (*Hebeloma* spp., *Inocybe* spp., *Laccaria* spp. and *Paxillus involutus*), all of which were capable of infecting birch seedlings from basidiospores, and late-stage fungi (*Amanita muscaria*, *Cortinarius* spp., *Lactarius* spp., *Leccinum* spp., *Russula* spp., *Scleroderma citrinum*) which could not. Similar findings have been made using seedlings of *Pinus banksiana* which formed ectomycorrhizae readily with mycelial slurries of *Thelephora*, *Laccaria*, *Hebeloma* and *Pisolithus*, but

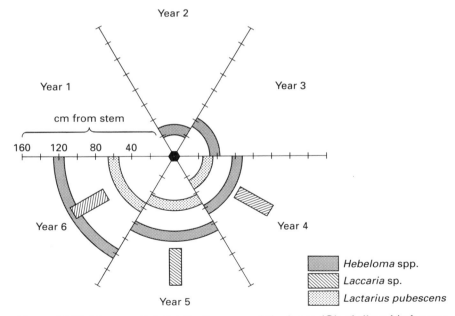

Figure 13.18 Mean spatial distribution around the bases (○) of silver birch trees (*Betula pubescens*) of fruit-bodies of sheathing mycorrhizal fungi during the first 6 years after planting. (Redrawn from Last *et al.*, 1983, by permission of Kluwer Academic Publishers.)

Figure 13.19 Fruit-bodies of two early-stage sheathing mycorrhiza fungi. (a) *Thelephora terrestris*. (b) *Laccaria laccata*. Scale bar = 2 cm (a); 1 cm (b).

failed to do so with *Lactarius*, *Suillus* and *Tricholoma* (Danielson, Visser and Parkinson, 1984).

Deacon and Fleming (1992) have suggested that early-stage fungi may be capable of infecting roots in the monokaryotic state in contrast with late-stage fungi where only dikaryotic mycelia can successfully infect. Assuming that most ectomycorrhizal fungi have tetrapolar (bifactorial) mating systems and a low number of basidiospores germinating near the host root surface, ability to infect in the monokaryotic state would greatly increase the chance of establishing a successful partnership.

As trees age, and their capacity to produce foliage increases, changes

Figure 13.20 Fruit-bodies of two late-stage sheathing mycorrhizal fungi. (a) *Leccinum scabrum*, a mycorrhizal associate of birch. (b) *Amanita rubescens*, a mycorrhizal associate of deciduous and coniferous trees. Scale bar = 4 cm.

occur in the soil and litter layers (Figure 13.21). Canopy closure tends to eliminate the ground flora. As the tree litter accumulates in quantity, its quality declines and the proportion of lignin and phenolic substances increases. The early-stage fungi are generally non-host-specific. With increasing time, the range of fungal species increases and then declines, especially in pure tree stands, to a small number of more host-specific late-stage fungi (Dighton, Poskitt and Howard, 1986). The N-nutrition of some late-stage mycorrhizal fungi may also be correlated with their greater capacity to utilize protein as compared with early-stage fungi (Abuzinadah and Read, 1986a,b; Read, Leake and Langdale, 1989). In cultures *in vitro*, late-stage fungi from birch showed a higher dependence on glucose for growth as compared with early-stage fungi, and it is believed that late-stage fungi have a higher energy demand (Gibson and Deacon, 1990).

The fruit-bodies of some of the early-stage fungi (e.g. those of *Laccaria* and *Hebeloma*) are smaller than those of the late-stage fungi (e.g. *Leccinum*) (Deacon, Donaldson and Last, 1983). It has been suggested (Last, Dighton and Mason, 1987) that the early-stage prolifically fruiting fungi with their wide host range are R-selected, i.e. show ruderal characteristics. The relatively few less-prolific fungi found on older trees following canopy closure, with their narrower host range, seem to be S-selected, i.e. have stress-tolerant characters. The early-stage fungi are able to colonize roots in soils with little or no tree litter, whilst the late-stage fungi colonize roots in soils with accumulations of plant litter. During the intermediate stages of forest stand development, when the species diversity of fungi is greatest, and when the early-stage fungi are being displaced by others, there is also an implication of C-selection, i.e. selection of fungi with combative characters (Figure 13.21).

A fuller discussion of the replacement of early- by late-stage fungi has

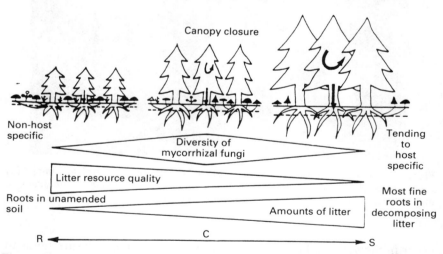

Figure 13.21 Scheme representing some of the many differnt aspects of forest succession including canopy development, changes in ground vegetation and alterations in the diversity of mycorrhizal fungi links with (i) R-, C- and S-selections and (ii) amounts and qualities of forest floor litter. (Reproduced from Last, Dighton and Mason, 1987, with permission from Elsevier Science Ltd, The Boulevard, Kidlington, OX5 1GB, UK.)

been provided by Deacon and Fleming (1992).

The fruiting of mycorrhizal basidiomycetes is affected not only by the species of host and the age of the host tree, but by its provenance, i.e. its source. This has been particularly well shown for birch (*Betula pendula* and *B. pubescens*). Seed collected from localities varying in latitude from 50°40′N (southern England) to 66°30′N (northern Sweden) were germinated and seedlings established in experimental plots near Edinburgh (55°52′N). Around the seedlings of more southerly provenance (between latitudes 50° and 54°59′N), fruit-bodies were found within 2 years, but around seedlings of more northerly provenance, fruit-body appearance was delayed, to 3 years for seedlings derived from between latitudes 55° and 64°59′N, or to 6 years around seedlings from the most northerly latitude 66°30′N (Mason *et al.*, 1982; Last and Fleming, 1985). Fruit-bodies were also much more abundant around the plants of southerly origin. These differences were correlated with the time of onset of autumnal leaf yellowing, which began earlier in the northern strains. The dependence of fruit-body production on photosynthetic activity of the tree host is also shown by experiments in which leaves were removed from birch trees, resulting in a drastic reduction in the number of sporophores (Last *et al.*, 1979). The severance of connections between tree roots and the tree by means of trenches cut around the tree limits basidiocarp production to the area inside the trench and to the edge of the trench (Romell, 1938; Laiho, 1970).

The incidence of mycorrhizal infection and of fruit-bodies is also affected by **edaphic factors**. Beech (*Fagus sylvatica*) can establish itself or has been planted in a range of different soil types. Tyler (1985, 1989) has analysed the macrofungal flora of Swedish beech forests in relation to the accumulation of soil organic matter and the chemical composition of soils under beech. Three hundred permanent observation plots of 5 m^2 in 30 beech forest sites were kept under observation over a 5-year period, and records were made of the incidence of sporophores. Chemical analyses (% metal ion saturation) and organic matter (%) of the surface soil layers after removal of the surface litter are illustrated in Figure 13.22. The observation plots have been classified into five groups: A–E.

Soils in groups A and B are well-developed mulls with relatively high metal ion saturation. Soils C–E are highly acid, with low metal ion saturation. Soils classified in group D are transitional between mull and mor, whilst soils in group E have a high organic matter content and a well-developed mor horizon.

The occurrence of species belonging to mycorrhizal genera is shown in Figure 13.23. There is a clear relationship between species diversity and soil type ranging from about one species per plot in the mull soils of type A to about six species per plot in the mor soils of type E. There are also differences in the frequency of sporophores, with a mean of about two per

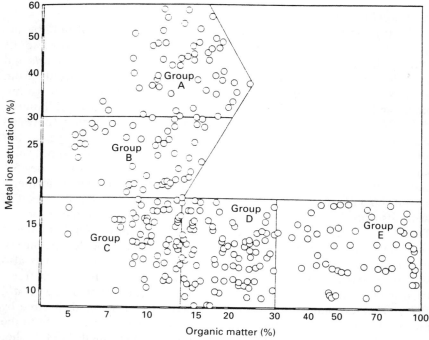

Figure 13.22 Distribution of the observation plots (5 m^2) in relation to metal ion saturation (%) (degree of neutralization) and organic matter (OM) content (%) of the humus layer (0–5 cm soil depth). Five edaphical groups are distinguished (A–E) and delimited by lines. The scales are logarithmic. (After Tyler, 1985.) (Figures 13.22–13.24 are reproduced by kind permission of Elsevier Science Ltd, The Boulevard, Kidlington, OX5 1GB, UK.)

plot in soils of type A and more than 20 per plot in soils of type E.

Figure 13.24 shows the distribution in the frequency of sporophores of five mycorrhizal species, and also of the non-mycorrhizal *Armillaria mellea*, in relation to the soil variables. The preference of *Cantharellus tubaeformis* and *Russula mairei* for base-poor soils with moderate to high organic matter accumulation contrasts with that of *Lactarius subdulcis* which extends into soils of types A and B.

In experiments in which beech leaf litter was removed for two consecutive seasons in a southern Swedish forest, Tyler (1991) found a considerable increase in sporophore production by *Russula* spp. (mostly *R. ochroleuca* and *R. fellea*) over untreated areas, but the underlying mechanism for this effect is not known.

There have been many other observations relating the occurrence of particular kinds of mycorrhizal fungi to soil type. The capacity of trees to establish on waste ground, such as on coal spoil or overburden from other

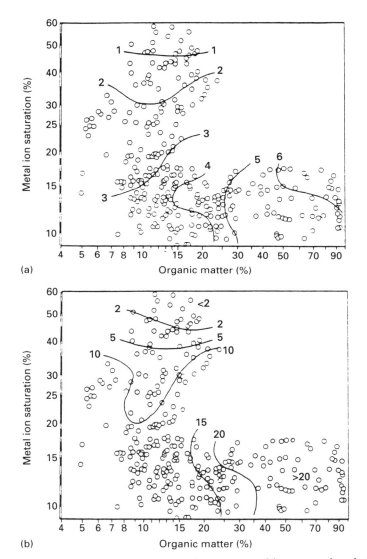

Figure 13.23 Distribution of species belonging to mycorrhiza genera in relation to metal ion saturation (%) and organic matter (OM) content (%) of the humus layer. (a) Mean number of species per observation plot. (b) Mean number of sporophores observed per observation plot. Logarithmic scales. Position of the plots along the axes indicated by circles. (After Tyler, 1985.)

mining activities, is dependent on their ability to form endo- (vesicular-arbuscular) or ectomycorrhizae. Ectomycorrhizal fungi known to establish on coal spoil and coking waste in association with *Betula*, *Quercus* or *Salix*

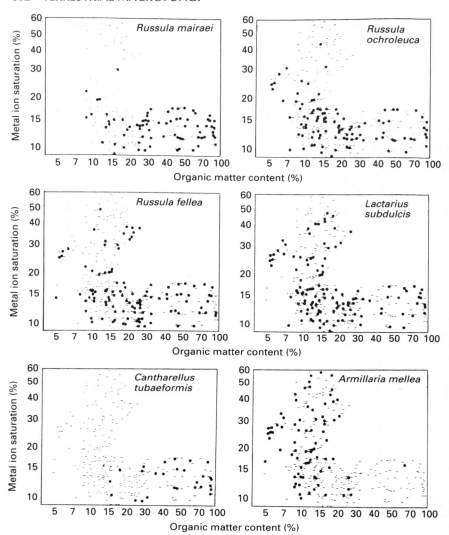

Figure 13.24 Distribution of five mycorrhizal species and the mainly parasitic *Armillaria mellea* in relation to metal ion saturation (%) and organic matter content (%) of the humus layer. Plots where a species was observed are indicated by a black dot; plots with no observation by a dash. (After Tyler, 1985.)

include *Hebeloma mesophaeum, Lactarius pubescens, L. rufus, L. turpis, Laccaria proxima, Paxillus involutus* and *Thelephora terrestris* (Watling and Lyon, 1988). *Laccaria laccata* is often the only agaric to develop in the pioneer colonization of shale by birch (Watling, 1984). In experimental inoculations of birch seedlings using basidiospores of *P. involutus*, it was found that mycorrhiza developed better in coal spoil waste than in a brown earth soil (Fox, 1986). At a restored bauxite mining site in which eucalyptus were planted on ridges of soil, *Laccaria* fruited on the ridges, whilst *Cortinarius* fruited in the furrows filled with accumulations of litter (Gardner and Malajczuk, 1985). The destruction by fire of litter accumulations in jarrah (*Eucalyptus marginata*) forest in Australia is associated with a reduction in the abundance of white and brown ectomycorrhizae of jarrah (caused by unidentified basidiomycetes) and an increase in the incidence of black ectomycorrhizae (probably caused by *Cenococcum geophilum*). Black ectomycorrhizae were dominant in the mineral horizons of the soil, whilst the white and brown ectomycorrhizae were mainly found in the litter layers (Reddell and Malacjzuk, 1984).

The fruiting of mycorrhizal basidiomycetes is affected by **soil amendments**. Hora (1959) applied fertilizers or lime to plots of plantation-grown *Pinus sylvestris*. Growmore (a general horticultural fertilizer), ammonium sulphate or superphosphate resulted in a striking increase in basidiocarp production by *Lactarius rufus* and *Paxillus involutus* as compared to control plots with no added fertilizer, whilst lime caused a reduction in basidiocarp production of both fungi. Experiments by Shubin, Ronkonen and Saukonen (1977) showed that the addition of nitrogenous fertilizers to a 30-year-old birch stand resulted in a decrease of basidiocarp production by *Amanita muscaria*, *Cortinarius* spp. and *Russula* spp., an increase by *Paxillus involutus*, and no change in some species of *Lactarius*, *Tricholoma* and *Xerocomus*. Menge and Grand (1978) also found that the addition of nitrogenous fertilizers (NH_4NO_3) to plots of *Pinus taeda* was associated with a reduction in the number of basidiocarps of some proven or potential ectomycorrhizal fungi. *Suillus hirtellus* produced many fewer basidiocarps in fertilized plots, but *Amanita citrina*, *Lactarius piperatus* and *L. subdulcis* were little affected by fertilizer treatment. Applications of a range of fertilizer treatments (CaO, P_2O_5, NH_4NO_3, K_2SO_4, singly or in mixtures) in a beech plantation resulted in striking differences in the fruiting of basidiomycetes (Garbaye *et al.*, 1979). Amongst known ectomycorrhizal fungi, *Hebeloma crustuliforme* was stimulated to fruit by the addition of calcium, but the further addition of N, P and K had a depressive effect. *Laccaria laccata* was also stimulated by calcium. *Boletus chrysenteron* was stimulated by treatments of phosphorus alone, or in combination with the other fertilizers.

The interpretation of results of this kind is difficult, because the addition of fertilizers may have effects not only on the fungus stimulated, but on other members of the soil microflora, and may also affect the physiology of the host, and the interrelationships between host and mycobiont.

Evidence is accumulating of a decline in the frequency of basidiocarps and the number of species of ectomycorrhizal fungi in Europe in recent years (Arnolds, 1988, 1991). It is suspected that the decline may be attributable, in part, to the deposition of acid rain. The gaseous pollutants include SO_2 and compounds of nitrogen (NO, NO_2, HNO_3, NO_3, NH_3 and NH_{4+}). The effects of this chemical mixture are complex because they affect not only the tree host but also the quality and quantity of assimilates available to the fungus. They also affect the understorey plants such as grasses, physical and chemical properties of the soil, and the activities of its biota, including earthworms. Experiments on forest plots fumigated with SO_2 or treated with nitrogenous fertilizers suggest that the addition of nitrogen rather than acidification is the main factor associated with the decline of ectomycorrhizal fungi (e.g. Shaw *et al.*, 1992). Fellner (1988) has distinguished three phases in the deterioration of ectomycorrhizal mycocoenoses: (1) inhibition of fructification – characteristic mycorrhizal fungi fruit only sporadically; (2) reduction in species diversity; and (3) total destruction of mycorrhizal mycocoenoses.

The **phenology** (time of appearance) and abundance of fruit-bodies of mycorrhizal fungi is controlled by edaphic and climatic factors, and by features of the physiology of the host. Some of the edaphic factors have already been discussed. Fruiting is in part dependent on the ability of the host to supply carbohydrates (and possibly other growth substances) in adequate amounts, and is thus linked to factors which regulate photosynthesis, such as expansion of foliage, daylength, light intensity, latitude, altitude, aspect, temperature and season. Agerer (1985) and Vogt *et al.* (1992) have reviewed the seasonal abundance of fruit-bodies of mycorrhizal fungi. Although it is well-known that sporophores are most abundant in the autumn, fruiting may occur over a much longer period, and in most years two separate peaks of fruit-body abundance can be distinguished: one in early summer and one in autumn.

The most important **climatic factors** affecting fruiting are temperature and rainfall. Low temperature in the litter layer may limit mycelial growth, and in cold regions where the litter layer is frozen, the ability of the mycelium to survive freezing may be a factor in the revival of mycelial growth in the spring. Moser (1958) showed that there were differences between upland and lowland strains of the same species of agaric in relation to temperature effects on mycelial growth and survival. A strain of *Suillus variegatus* associated with *Pinus mugho* at an altitude of 2100 m showed the first signs of mycelial damage only after 2 months freezing, in contrast to a lowland strain associated with *P. sylvestris* at 600 m which did not survive freezing for 5 days. There are also differences in the minimum temperatures at which mycelial growth occurs, the mountain strain being capable of growth between -2 and $+4°C$, whilst the lowland strain was not. Some species differences were also found: *S. granulatus* and *S. plorans* could grow between -2 and $+4°C$, whilst *S. aeruginascens* and *S. grevillei* could not. The ability to grow at lower temperatures effectively prolongs the growing season of certain fungi, and some correlation was noted between fungi which fruit early in the year and the ability of their mycelia to survive freezing and to grow at low temperature.

Some mycorrhizal fungi are capable of fruiting at low temperatures around 5°C, e.g. *Lactarius camphoratus* and *Russula fragilis* (Friedrich, 1940). Frost, especially if prolonged, kills most basidiocarps, and brings further fruiting to an end.

High temperature similarly has varied effects, e.g. it directly affects mycelial growth, fruiting, the drying of surface litter layers and the desiccation of fruit-bodies, which affects their ability to expand. It also affects the water relations and photosynthesis of the host. High soil temperature also affects the activities of other soil organisms, influencing competition. High temperature stimulates the mycelial growth of *Pisolithus arhizus*, which has an optimum temperature for mycelial growth of 34°C, and affects its ability to form ectomycorrhizae on loblolly pine (Marx and Davey, 1970).

The amount and distribution of rainfall have major effects on fruiting of most higher basidiomycetes, and mycorrhizal fungi are no exception. A critical minimum value for the water content of the litter layer is required for mycelial extension. According to Moser (1962, 1965), there are two critical steps in fruit-body development which are dependent on suitable conditions: the formation of primordia and their subsequent expansion. A dry period in the spring may prevent the development of primordia. Because the complete development of basidiocarps may take several weeks, there may be a delay between the onset of suitable conditions which precondition fruiting, and the emergence and expansion of fruit-bodies. The development of fruit-bodies is linked with their transpiration rate: generally those fungi in which fruit-body development is rapid have high rates of basidiocarp transpiration. Agerer (1985) found a good correlation between the relative abundance of fruit-bodies and the cumulative rainfall totals for the previous 7 weeks in a study of mycorrhizal fungi in the Schönbuch Nature Park in Germany (Figure 13.25).

Figure 13.25 shows the considerable variation in relative production which occurs between different years, as shown, for example, by comparing 1979 and 1982. It also indicates that there is variation in the time of maximum productivity, which was in week 44 in 1981 and week 35 in 1982. The twin peaks of production are also evident in all 4 years.

Similar bimodal peaks in numbers of fruit-bodies were observed in a beechwood plot and a pinewood plot near Oxford, UK, by Wilkins and Harris (1946). There were many times more fungal sporophores in the pinewood than the beechwood. Seasonal variations in fruit-body appearance were noted. *Hygrophoropsis aurantiaca* (a decomposer of coniferous litter) started fruiting early and finished early, whilst *Lactarius tabidus* (common in moist places in deciduous woods) started late but continued fruiting to the end. Hering (1966), however, noted different peak fruiting

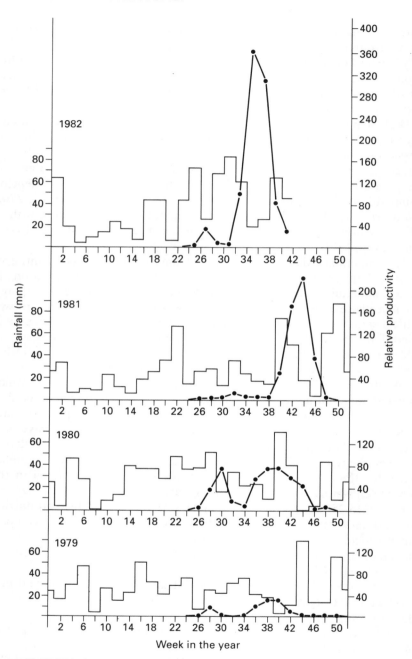

Figure 13.25 Relative productivity ●-● of mycorrhizal fungi in a beech–oak plot in Schönbuch Nature Park, Germany in four successive years, 1979–82. The histograms give rainfall totals for 2-week intervals. (After Agerer, 1985, with permission.)

times for mycorrhizal agarics in oak woods in the English Lake District, *Lactarius camphoratus* and *L. tabidus* appearing in early September, followed by *Russula emetica*, *Lactarius quietus*, *Laccaria laccata*, *Russula ochroleuca* and *Laccaria amethystea*.

Observations of the fruiting patterns of agarics and boleti in marked quadrats in woodlands near Tübingen, Germany, have indicated that the fruiting of certain fungi is negatively correlated with the fruiting of others (Agerer and Kottke, 1981). In quadrats in which *Collybia butyracea*, a non-mycorrhizal litter-decomposing fungus, fruited, *Russula fellea*, a mycorrhizal associate of *Fagus*, fruited relatively rarely. The fruiting of *Russula ochroleuca*, *R. vinosa*, *Lactarius rufus* and *L. necator* was likewise inhibited by the fruiting of *C. butyracea*. *R. ochroleuca* and *R. fellea* appeared to be mutually exclusive. The explanation of this phenomenon is not yet known, but antibiotic production has been suggested. Similar reports have been made from other countries. For example, there is little spatial overlap between the fruiting of the dominant *Russula* species in *Castanopsis cuspidata* and other broadleaved forests in Japan (Murakami, 1987). That interspecific competition is likely to occur within mycorrhizal root systems is shown by the detection of up to 15 species of ectomycorrhizal fungi in a 10 x 10 cm square cut from the organic layer of a beechwood soil (Brand, 1992).

REFERENCES

Abdullah, S.K. and Taj-Aldeen, S.J. (1989) Extracellular enzymatic activity of aquatic and aero-aquatic conidial fungi. *Hydrobiologia*, **174**, 217–23.

Abuzinadah, R.A., Finlay, R.D. and Read, D.J. (1986) The role of proteins in the nitrogen nutrition of ectomycorrhizal plants. II. Utilization of protein by mycorrhizal plants of *Pinus contorta*. *New Phytologist*, **103**, 495–506.

Abuzinadah, R.A. and Read, D.J. (1986a) The role of proteins in the nitrogen nutrition of ectomycorrhizal plants. I. Utilization of peptides and proteins by ectomycorrhizal fungi. *New Phytologist*, **103**, 481–93.

Abuzinadah, R.A. and Read, D.J. (1986b) The role of proteins in the nitrogen nutrition of ectomycorrhizal plants. III. Protein utilization by *Betula*, *Picea* and *Pinus* in mycorrhizal association with *Hebeloma crustuliniforme*. *New Phytologist*, **103**, 507–14.

Adams, P.B. (1971) *Pythium aphanidermatum* oospore germination as affected by time, temperature and pH. *Phytopathology*, **61**, 1149–50.

Adams, T.J.H., Todd, N.K. and Rayner, A.D.M. (1981) Antagonism between dikaryons of *Piptoporus betulinus*. *Transactions of the British Mycological Society*, **76**, 510–13.

Adebayo, A.A. Harris, R.F. (1971) Fungal growth responses to osmotic as compared to matric water potential. *Soil Science Society of America Proceedings*, **35**, 465–9.

Adebayo, A.A., and Harris, R.F. and Gardner, W.R. (1971) Turgor pressure of fungal mycelia. *Transactions of the British Mycological Society*, **57**, 145–51.

Adler, L. (1978) Properties of alkaline phosphatase of the halotolerant yeast *Debaryomyces hansenii*. *Biochimica et Biophysica Acta*, **522**, 113–21.

Adler, L., Pedersen, A. and Thunbald-Johansson, I. (1982) Polyol accumulation by two filamentous fungi grown at different concentrations of NaCl. *Physiologia Plantarum*, **56**, 139–42.

Agerer, R. (1985) Zur Ökologie der Mykorrhizapilze. *Bibliotheca Mycologica*, **97**, J. Cramer, Vaduz.

Agerer, R. (1987–91) *Colour Atlas of Ectomycorrhizae*, Einhorn Verlag, Schwäbish Gmünd.

Agerer, R.J. (1992) Ectomycorrhizal rhizomorphs: organs of contact, in *Mycorrhiza in Ecosystems*, (eds D.J. Read, D.H. Lewis, A.H. Fitter and I.J. Alexander), CAB International, Wallingford, UK, pp. 84–90.

Agerer, R. and Kottke, I. (1981) Sozio-Ökologische Studien an Pilzen von Fichten- und Eichen- Buchen- Hainbuchen-Waldern im Naturpark Schönbuch. *Zeitschrift für Mykologie*, **47**, 103–22.

Ahlgren, I.F. (1974) The effect of fire on soil organisms, in *Fire and Ecosystems*, (eds T.T. Kozlowski and C.E. Ahlgren), Academic Press, New York, pp. 47–72.

Ahlgren, I.F. and Ahlgren, C.E. (1960) Ecological effects of forest fires. *Botanical Review*, **26**, 483–533.

Ahlgren, I.F. and Ahlgren, C.E. (1965) Effects of prescribed burning on soil microorganisms in a Minnesota jack pine forest. *Ecology*, **46**, 304–10.

Ainsworth, A.M. and Rayner, A.D.M. (1986) Responses of living hyphae associated with self- and non-self fusions in the basidiomycete *Phanerochaete velutina*. *Journal of General Microbiology*, **132**, 191–201.

Akridge, R.E. and Koehn, R.D. (1987) Amphibious hyphomycetes from the San Marcos river in Texas. *Mycologia*, **79**, 228–33.

Alabi, R.O. (1971a) Factors affecting seasonal occurrence of Saprolegniaceae in Nigeria. *Transactions of the British Mycological Society*, **56**, 289–99.

Alabi, R.O. (1971b) Seasonal periodicity of Saprolegniaceae at Ibadan, Nigeria. *Transactions of the British Mycological Society*, **56**, 337–41.

Ali, N.A. and Jackson, R.M. (1988). Effects of plant roots and their exudates on germination of spores of ectomycorrhizal fungi. *Transactions of the British Mycological Society*, **91**, 253–60.

Alexander, J.V., Bourret, J.A., Gold, A.H. and Snyder, W.C. (1966). Induction of chlamydospore formation by *Fusarium solani* in sterile soil extracts. *Phytopathology*, **56**, 353–4.

Allen, E.A., Hoch, H.C., Steadman, J.R. and Stavely, R.J. (1991) Influence of leaf surface features on spore deposition and the epiphytic growth of phytopathogenic fungi, in *The Microbial Ecology of Leaves*, (eds J.H. Andrews and S.S. Hirano), Springer-Verlag, New York, pp. 87–110.

Allison, F.E., Murphy, R.M. and Klein, C.J. (1963) Nitrogen requirement for the decomposition of various kinds of finely ground woods in soil. *Soil Science*, **96**, 187–90.

Anastasiou, C.J. and Churchland, L.M. (1969) Fungi on decaying leaves in marine habitats. *Canadian Journal of Botany*, **47**, 251–7.

Ander, P. and Eriksson, K.E. (1976) The importance of phenol oxidase activity in lignin degradation by the white-rot fungus *Sporotrichum pulverulentum*. *Archives of Microbiology*, **109**, 1–8.

Ander, P. and Eriksson, K.E. (1978) Lignin degradation and utilization by microorganisms. *Progress in Industrial Microbiology*, **14**, 1–58.

Anderson, J.B., Ullrich, R.C., Roth, L.F. and Filip, G.M. (1979) Genetic identification of clones of *Armillaria mellea* in coniferous forests in Washington. *Phytopathology*, **69**, 1109–11.

Anderson, J.M. (1973) The breakdown and decomposition of sweet chestnut (*Castanea sativa* Mill.) and beech (*Fagus sylvatica* L.) leaf litter in two deciduous woodland soils. II. Changes in the carbon, hydrogen, nitrogen and polyphenol content. *Oecologia*, **12**, 275–88.

Anderson, J.M. and Ineson, P. (1984) Interaction between microbes and soil invertebrates in nutrient flux pathways of forest ecosystems, in *Invertebrate Microbial Interactions*, (eds J.M. Anderson, A.D.M. Rayner and D.W.H. Walton), *British Mycological Symposium* 6, Cambridge University Press, Cambridge, pp. 59–88.

Anderson, J.P.E. and Domsch, K.H. (1978) A physiological method for the qualitative measurement of microbial biomass in soils. *Soil Biology and Biochemistry*, **10**, 215–21.

Anderson, T.R. and Patrick, Z.A. (1978) Mycophagous amoeboid organisms from soil that perforate spores of *Thielaviopsis basicola* and *Cochliobolus sativus*. *Phytopathology*, **68**, 1618–26.

Andrews, J.H. (1992) Fungal life-history strategies, in *The Fungal Community: Its Organisation and Role in the Ecosystem*, 2nd ed, (eds C.G. Carroll and D.T. Wicklow), Marcel Dekker, New York, pp. 119–45.

Andrews, J.H., Kenerley, C.M. and Nordheim, E.V. (1980) Positional variation in phylloplane microbial populations within an apple tree canopy. *Microbial Ecology*, **6**, 77–84.

Angel, K. and Wicklow, D.T. (1975) Relationships between coprophilous fungi and fecal substrates in a Colorado grassland. *Mycologia*, **67**, 63–74.

Apinis, A.E. (1963) Occurrence of thermophilous fungi in certain alluvial soils near Nottingham. *Nova Hedwigia*, **5**, 57–8.

Apinis, A.E. and Pugh, G.J.F. (1967) Thermophilous fungi of birds' nests. *Mycopathologia et Mycologia Applicata*, **33**, 1–9.

Arnolds, E. (1988) The changing macromycete flora in the Netherlands. *Transactions of the British Mycological Society*, **90**, 391–406.

Arnolds, E. (1991) Mycologists and nature conservation, in *Frontiers of Mycology*, (ed. D.L. Hawksworth), CAB International, Wallingford, UK, pp. 243–64.

Arsuffi, T.L. and Suberkropp, K. (1985) Selective feeding by stream caddis fly (*Trichoptera*) detritivores on leaves with fungal-colonized patches. *Oikos*, **45**, 50–8.

Asplin, F.D. and Carnaghan, R.B.A. (1961) The toxicity of certain groundnut meals for poultry with special reference to their effect on ducklings and chickens. *Veterinary Recorder*, **73**, 1215–19.

Ausmus, B.S. (1973) The use of the ATP assay in terrestrial decomposition studies. *Bulletin Ecological Research Communication (Stockholm)*, **17**, 223–34.

Ayen, E. and Lysek, G. (1986) Endoparasitic nematode-destroying fungi in sandy soils of a beech wood in Berlin. *Microbiology Ecology*, **38**, 397–400.

Ayerst, G. (1969) The effects of moisture and temperature on growth and spore germination in some fungi. *Journal of Stored Product Research*, **5**, 127–41.

Bääth, E. (1988) Autoradiographic determination of metabolically active fungal hyphae in forest soil. *Soil Biology and Biochemistry*, **20**, 123–5.

Baigent, N.L. and Ogawa, J.M. (1960) Activity of the antibiotic produced by *Pullularia pullulans*. *Phytopathology*, **50**, 82.

Bailey, F. and Gray, N.F. (1989). The comparison of isolation techniques for nematophagous fungi from soil. *Annals of Applied Biology*, **114**, 125–32.

Bailey, J.A., Vincent, G.G. and Burden, R.S. (1974) Diterpenes from *Nicotiana glutinosa* and their effect on fungal growth. *Journal of General Microbiology*, **85**, 57–64.

Balis, C. (1976) Ethylene induced volatile inhibitors causing soil fungistasis. *Nature*, **259**, 112–14.

Balis, C. and Kouyeas, V. (1968) Volatile inhibitors involved in soil mycostasis. *Annales de l'Institute Phytopathologique Benaki (NS)*, **8**, 145–9.

Bandoni, R.J. (1981) Aquatic hyphomycetes from terrestrial litter, in *The Fungal Community: its Organisation and Role in the Ecosystem*, (eds D.T. Wicklow and G.C. Carroll), Marcel Dekker, New York, pp. 693–708.

Banerjee, A.K. and Levy, J.F. (1971) Fungal successions in wooden fence posts. *Material und Organismen*, **6**, 1–25.

Barber, D.A. and Martin, J.K. (1976) The release of organic substances by cereal roots into soil. *New Phytologist*, **76**, 69–80.

Barghoorn, E.S. and Linder, D.H. (1944). Marine fungi: their biology and taxonomy. *Farlowia*, **1**, 395–467.

Barley, K.P. (1970) The configuration of the root system in relation to nutrient uptake. *Advances in Agronomy*, **22**, 159–201.

Barley, K.P. and Sedgley, R.H. (1961) The blockage of soil pores by plant roots. *The Australian Journal of Science*, **23**, 279–98.

Bärlocher, F. (1980) Leaf eating invertebrates as competitors of aquatic hyphomycetes. *Oecologia*, **47**, 303–6.

Bärlocher, F. (1982) Conidium production from leaves and needles in four streams. *Canadian Journal of Botany*, **60**, 1487–94.

Bärlocher, F. (1985) The role of fungi in the nutrition of stream invertebrates. *Botanical Journal of the Linnean Society*, **91**, 83–94.
Bärlocher, F. (1987) Aquatic hyphomycete spora in 10 streams of New Brunswick and Nova Scotia. *Canadian Journal of Botany*, **65**, 76–9.
Bärlocher, F. (ed.) (1992a) *The Ecology of Aquatic Hyphomycetes*, Ecological Studies no. 94, Springer-Verlag, Berlin.
Bärlocher, F. (1992b) Recent developments in stream ecology and their relevance to aquatic mycology, in *The Ecology of Aquatic Hyphomycetes* (ed. F. Bärlocher), Springer-Verlag, Berlin, pp. 16–37.
Bärlocher, F. and Kendrick, B. (1974) Dynamics of the fungal population on leaves in a stream. *Journal of Ecology*, **62**, 761–791.
Bärlocher, F. and Kendrick, B. (1976) Hyphomycetes as intermediaries of energy flow in streams, in *Recent Advances in Aquatic Mycology*, (ed. E.B.G. Jones), Elek Science, London, pp. 435–46.
Bärlocher, F. and Kendrick, B. (1981) Role of aquatic hyphomycetes in the trophic structure of streams, in *The Fungal Community: its Organisation and Role in the Ecosystem*, (eds D.T. Wicklow and G.C. Carroll), Marcel Dekker, New York, pp. 743–60.
Bärlocher, F. and Oertli, J.J. (1978a) Colonization of conifer needles by aquatic hyphomycetes. *Canadian Journal of Botany*, **56**, 57–62.
Bärlocher, F. and Oertli, J.J. (1978b) Inhibitors of aquatic hyphomycetes in dead conifer needles. *Mycologia*, **70**, 964–74.
Bärlocher, F. and Rosset, J. (1981) Aquatic hyphomycete spora of two Black Forest and two Swiss Jura streams. *Transactions of the British Mycological Society*, **76**, 479–83.
Bärlocher, F. and Schweizer, M. (1983) Effects of leaf size and decay rate on colonization by aquatic hyphomycetes. *Oikos*, **41**, 205–10.
Bärlocher, F., Kendrick, B. and Michaelides, J. (1978) Colonization and conditioning of *Pinus resinosa* needles by aquatic hyphomycetes. *Archiv für Hydrobiologie*, **81**, 462–74.
Barnett, H.L. and Binder, F.L. (1973) The fungal host–parasite relationship. *Annual Review of Phytopathology*, **11**, 273–92.
Barnett, H.L. and Lilly, V.G. (1962) A destructive mycoparasite *Gliocladium roseum*. *Mycologia*, **54**, 72–9.
Barr, D.J.S. (1989). Phylum Chytridiomycota, in *Handbook of Protoctista*, (eds L. Margulis, J.O. Corliss, M. Melkonian and D.J. Chapman), Jones and Bartlett, Boston, pp. 454–66.
Barrett, D.K. and Uscuplic, M. (1971) The field distribution of interacting strains of *Polyporus schweinitzii* and their origin. *New Phytologist*, **70**, 591–8.
Barron, G.L. (1969) Isolation and maintenance of endoparasitic nematodes. *Canadian Journal of Botany*, **47**, 1899–902.
Barron, G.L. (1973) Nematophagous fungi: *Rhopalomyces elegans*. *Canadian Journal of Botany*, **51**, 2505–7.

Barron, G.L. (1975) Detachable adhesive knobs in *Dactylaria*. *Transactions of the British Mycological Society*, **65**, 311–12.
Barron, G.L. (1977) *The Nematode-Destroying Fungi. Topics in Mycobiology*, Canadian Biological Publications, Guelph.
Barron, G.L. (1978) Nematophagous fungi: endoparasites of *Rhabditis terricola*. *Microbial Ecology*, **4**, 157–163.
Baron, G.L. (1981) Predators and parasites of microscopic animals, in *Biology of Conidial Fungi*, II, (eds G.T. Cole and B. Kendrick) Academic Press, New York, pp. 167–200.
Barron, G.L. (1982) Nematode destroying fungi, in *Experimental Microbial Ecology*, 2nd ed, (eds R.G. Burns and J.H. Slater), Blackwell Scientific, Oxford, pp. 533–52.
Barron, G.L. (1992) Ligninolytic and cellulolytic fungi as predators and parasites, in *The Fungal Community: Its Organisation and Role in the Ecosystem*, 2nd edn (G.C. Caroll and D.T. Wicklow), New York, pp. 311–26.
Barron, G.L. and Dierkes, Y. (1977) Nematophagous fungi: *Hohenbuehelia*, the perfect state of *Nematoctonus*. *Canadian Journal of Botany*, **55**, 3054–62.
Barron, G.L. and Thorn, R.G. (1987) Destruction of nematodes by species of *Pleurotus*. *Canadian Journal of Botany*, **65**, 774–8.
Barton, R. (1957) Germination of oospores of *Pythium mamillatum* in response to exudates from living seedlings. *Nature*, **180**, 613–14.
Barton, R. (1960) Antagonism amongst some sugar fungi, in *The Ecology of Soil Fungi*, (eds D. Parkinson and J.S. Waid) Liverpool University Press, Liverpool, pp. 160–7.
Barton, R. (1961) Saprophytic activity of *Pythium mamillatum* in soils. II. Factors restricting *Pythium mamillatum* to pioneer colonization of substrates. *Transactions of the British Mycological Society*, **44**, 105–18.
Bashi, E. and Fokkema, N.J. (1977) Environmental factors limiting growth of *Sporobolomyces roseus*, an antagonist of *Cochliobulus sativus* on wheat leaves. *Transactions of the British Mycological Society*, **68**, 17–25.
Bavendamm, W. (1928a) Neue Untersuchungen über die Lebensbedingungen holzzerstörender Pilze. II Mitt. Gerbstoffversuche. *Zentralblatt für Bakteriologie, Parasitenkunde Infektionskrankheiten und Hygiene*, Abt 11, **76**, 172–277.
Bavendamm, W. (1928b) Über das Vorkommen und den Nachweis von Oxydasen bei holzzerstörenden Pilzen. *Zeitschrift für Pflanzenkrankheiten*, **38**, 257–76.
Bayliss, J.S. (1911) Observations on *Marasmius oreades* and *Clitocybe gigantea*, as parasitic fungi causing 'fairy rings'. *Journal of Economic Biology*, **6**, 111–32.

Bayliss Elliott, J.S. (1926) Concerning fairy rings in pastures. *Annals of Applied Biology*, **13**, 227–88.

Bebout, B., Schatz, S., Kohlmeyer, J. and Haibach, M. (1987) Temperature-dependent growth in isolates of *Corollospora maritima* Werdem. (Ascomycetes) from different geographical regions. *Journal of Experimental Marine Biology and Ecology*, **106**, 203–10.

Beckman, C.H., Mueller, W.C. and McHardy, W.E. (1972) The localization of stored phenols in plant hairs. *Physiological Plant Pathology*, **2**, 69–74.

Bell, A. (1983) *Dung Fungi. An Illustrated Guide to Coprophilous Fungi in New Zealand*, Victoria University Press, Wellington.

Bellotti, R.A. and Couse, N.L. (1980) Induction of mycelial strands in *Calvatia sculpta*. *Transactions of the British Mycological Society*, **74**, 19–25.

Bengtsson, G. (1983) Habitat selection in two species of aquatic hyphomycetes. *Microbial Ecology*, **9**, 15–26.

Benjamin, R.K. (1959) The merosporangiferous Mucorales. *Aliso*, **4**, 321–433.

Benoit, R.E. and Starkey, R.L. (1967) Inhibition of decomposition of cellulose and some other carbohydrates by tannin. *Soil Science*, **105**, 291–6.

Benoit, R.E., Starkey, R.L. and Basaraba, J. (1968) Effect of purified plant tannin decomposition of some organic compounds and plant material. *Soil Science*, **105**, 153–8.

Berdy, J. (1974) Recent developments of antibiotic research and classification of antibiotics according to chemical structure. *Advances in Applied Microbiology*, **18**, 309–402.

Bernhardt, E.A. and Grogan, R.G. (1981) Effect of soil matric potential on the formation and indirect germination of sporangia of *Phytophthora parasitica*, *P. capsici*, and *P. cryptogea*. *Phytopathology*, **72**, 507–11.

Berry, C.R. and Barnett, H.L. (1957) Mode of parasitism and host range of *Piptocephalis virginiana*. *Mycologia*, **49**, 374–86.

Betts, W.B., Dart, R.K. and Ball, M.C. (1988). Degradation of larch wood by *Aspergillus flavus*. *Transactions of the British Mycological Society*, **91**, 227–32.

Bharat, Rai. and Singh, D.B. (1980) Antagonistic activity of some leaf surface microfungi against *Alternaria brassicae* and *Drechslera graminea*. *Transactions of the British Mycological Society*, **75**, 363–9.

Bhatt, D.D. and Vaughan, E.K. (1962) Preliminary investigation on biological control of gray mold (*Botrytis cinerea*) of strawberries. *Plant Disease Reporter*, **46**, 342–5.

Bisset, J. and Parkinson, D. (1979a) Distribution of fungi in some alpine soils. *Canadian Journal of Botany*, **57**, 1609–29.

Bisset, J. and Parkinson, D. (1979b) Fungal community structure in some

alpine soils. *Canadian Journal of Botany*, **57**, 1630–59.
Bisset, J. and Parkinson, D. (1980) Long-term effects of fire on the composition and activity of the soil microflora of a subalpine coniferous forest. *Canadian Journal of Botany*, **58**, 1704–21.
Black, R.L.B. and Dix, N.J. (1976a) Utilization of ferulic acid by microfungi from litter and soil. *Transactions of the British Mycological society*, **66**, 313–17.
Black, R.L.B. and Dix, N.J. (1976b) Spore germination and germ hyphal growth of microfungi from litter and soil in the presence of ferulic acid. *Transactions of the British Mycological Society*, **66**, 305–11.
Black, R.L.B. and Dix, N.J. (1977) Colonization of Scots pine litter by soil fungi. *Transactions of the British Mycological Society*, **68**, 284–7.
Blaisdell, D.J. (1939) Decay of hardwoods by *Ustulina vulgaris* and other ascomycetes. *Phytopathology*, **29**, 2.
Blakeman, J.P. (1972) Effect of plant age on inhibition of *Botrytis cinerea* spores by bacteria on beetroot leaves. *Physiological Plant Pathology*, **2**, 143–52.
Blakeman, J.P. (1988) Competitive antagonism of air-borne fungal pathogens, in *Fungi in Biological Control Systems*, (ed. M.N. Burge) Manchester University Press, Manchester, pp. 141–60.
Blakeman, J.P. and Atkinson, P. (1976) Evidence for a spore germination inhibitor co-extracted with wax from leaves, in *Microbiology of Aerial Plant Surfaces*, (eds C.H. Dickinson and T.F. Preece), Academic Press, London, pp. 441–50.
Blakeman, J.P. and Atkinson, P. (1979) Antimicrobial properties and possible role in host-pathogen interactions of parthenolide, a sesquiterpene lactone isolated from the glands of *Chrysanthemum parthenium*. *Physiological Plant Physiology*, **15**, 183–92.
Blakeman, J.P. and Atkinson, P. (1981) Antimicrobial substances on the aerial surfaces of plants, in *Microbial Ecology of the Phylloplane*, **15**, 183–92.
Blakeman, J.P. and Brodie, I.D.S. (1976) Inhibition of pathogens by epiphytic bacteria on aerial plant surfaces, in *Microbiology of Aerial Plant Surfaces*, (eds C.H. Dickinson and T.F. Preece), Academic Press, London, pp. 529–57.
Blakeman, J.P. and Fokkema, N.J. (1982) Potential for biological control of plant diseases on the phylloplane. *Annual Review of Phytopathology*, **20**, 167–92.
Blakeman, J.P. and Fraser, A.K. (1971) Inhibition of *Botrytis cinerea* spores by bacteria on the surface of *Chrysanthemum* leaves. *Physiological Plant Pathology*, **1**, 45–54.
Blakeman, J.P. and Sztejnberg, A. (1973) Effect of surface wax on inhibition of germination of *Botrytis cinerea* spores on beetroot leaves. *Physiological Plant Pathology*, **3**, 269–78.

Blakeman, J.P. and Sztejnberg, A. (1974) Germination of *Botrytis cinerea* spores on beetroot leaves treated with antibiotics. *Transactions of the British Mcyological Society*, **62**, 537–45.

Blanchette, R.A. (1984) Screening wood decayed by white-rot fungi for preferential lignin degradation. *Applied Environmental Microbiology*, **48**, 647–53.

Blanchette, R.A. and Shaw, C.G. (1978) Association between bacteria, yeasts and basidiomycetes during wood decay. *Phytopathology*, **68**, 631–7.

Blyth, W. (1978) The occurrence and nature of alveolitis-inducing substances in *Aspergillus clavatus*. *Clinical Experimental Immunology*, **32**, 272–82.

Blyth, W., Grant, I.W.B., Blackadder, E.S. and Greenberg, M. (1977) Fungal antigens as a source of sensitization and respiratory disease in Scottish malt workers. *Clinical Allergy*, **7**, 549–62.

Boddy, L. (1983) Effect of temperature and water potential on growth rate of wood-rotting basidiomycetes. *Transactions of the British Mycological Society*, **80**, 141–9.

Boddy, L. (1984) The micro-environment of basidiomycete mycelia in temperate deciduous woodlands, in *The Ecology and Physiology of the Fungal Mycelium*, (eds D.H. Jennings and A.D.M. Rayner), *British Mycological Society Symposium 8*, Cambridge University Press, Cambridge, pp. 261–89.

Boddy, L. (1993) Saprotrophic cord-forming fungi: warfare strategies and other ecological aspects. *Mycological Research*, **97**, 641–55.

Boddy, L. and Griffith, G.S. (1989) Role of endophytes and latent invasion in the development of decay communities in sapwood of angiospermous trees. *Sydowia*, **41**, 41–73.

Boddy, L. and Rayner, A.D.M. (1981) Fungal communities and formation of heartwood wings in attached oak branches undergoing decay. *Annals of Botany*, **47**, 271–4.

Boddy, L. and Rayner, A.D.M. (1982) Population structure, intermycelial interactions and infection biology of *Stereum gausapatum*. *Transactions of the British Mycological Society*, **78**, 337–51.

Boddy, L. and Rayner, A.D.M. (1983a) Ecological roles of basidiomycetes forming decay communities in attached oak branches. *New Phytologist*, **93**, 77–88.

Boddy, L. and Rayner, A.D.M. (1983b) Origin of decay in living deciduous trees: the role of moisture content and a re-appraisal of the expanded concept of tree decay. *New Phytologist*, **94**, 623–41.

Boddy, L. and Thompson, W. (1983) Decomposition of suppressed oak trees in even-aged plantations. *New Phytologist*, **93**, 261–76.

Boddy, L., Bardsley, D.W. and Gibbon, O.M. (1987) Fungal comunities in attached ash branches. *New Phytologist*, **107**, 143–54.

Boddy, L., Gibbon, O.M. and Grundy, M.A. (1985) Ecology of *Daldinia concentrica*: effect of abiotic variables on mycelial extension and interspecific interactions. *Transactions of the British Mycological Society*, **85**, 201–11.

Boddy, L., Watling, R. and Lyon, A.J.E. (eds) (1988) *Fungi and Ecological Disturbance*. Proceedings of the Royal Society of Edinburgh, 94B.

Bondietti, E., Martin, J.P. and Haider, K. (1972) Stabilization of amino sugar units in humic type polymers. *Soil Science Society of America Proceedings*, **36**, 597–602.

Bonner, R.D. and Fergus, C.L. (1959) The fungus flora of cattle feeds. *Mycologia*, **51**, 855–63.

Booth, T. and Kenkel, N. (1986) Ecological studies of lignicolous marine fungi: a distribution model based on ordination and classification; in *The Biology of Marine Fungi*, (ed. S.T. Moss), Cambridge University Press, Cambridge, pp. 297–310.

Bowen, G.D. (1969) Nutrient status effects on loss of amides and amino acids from pine roots. *Plant and Soil*, **30**, 139–42.

Bowen, G.D. and Rovira, A.D. (1973) Are modelling approaches useful in rhizosphere biology? *Bulletins from the Ecological Research Committee (Stockholm)*, **17**, 443–50.

Boyd, P.E. and Kohlmeyer, J. (1982) The influence of temperature on the seasonal and geographical distribution of three marine fungi. *Mycologia*, **74**, 894–902.

Brady, B.L. (1960) Occurrence of *Itersonilia* and *Tilletiopsis* on lesions caused by *Entyloma*. *Transactions of the British Mycological Society*, **43**, 31–50.

Brame, C. and Flood, J. (1983) Antagonism of *Aureobasidium pullulans* towards *Alternaria solani*. *Transactions of the British Mycological Society*, **81**, 621–4.

Brand, F. (1992) Mixed associations of fungi in ectomycorrhizal roots, in *Mycorrhizas in Ecosystems*, (eds D.J. Read, D.H. Lewis, A.H. Fitter and I.J. Alexander) CAB International, Wallingford, UK, pp. 142–7.

Bray, J.R. and Gorham, E. (1964) Litter production in forests of the world. *Advances in Ecological Research*, **2**, 101–52.

Breeze, E.M. and Dix, N.J. (1981). Seasonal analysis of the fungal community on *Acer platanoides* leaves. *Transactions of the British Mycological Society*, **77**, 321–8.

Bremer, G.B. (1976) The ecology of marine lower fungi, in *Recent Advances in Aquatic Mycology*, (ed. E.B.G. Jones), Elek Science, London, pp. 313–33.

Breymeyer, A., Jakubczyk, H. and Olechowicz, E. (1975) Influence of coprophagous arthropods on microorganisms in sheep feces – laboratory investigations. *Bulletin de l'Academie Polonaise des Sciences. Série des Sciences Biologiques*, **23**, 257–62.

Brian, P.W. (1957) The ecological significance of antibiotic production, in *Microbial Ecology*, 7th Symposium of the Society of General Microbiology, (eds R.E. Williams and C.C. Spicer) Cambridge University Press, Cambridge, pp. 168–88.

Brierley, J.K. (1955) Seasonal fluctuations in the oxygen and carbon dioxide concentration in beech litter with reference to the salt uptake of beech mycorrhizas. *Journal of Ecology*, **43**, 404–8.

Bristow, P.R. and Lockwood, J.L. (1975) Soil fungistasis. Role of the microbial nutrient sink and of fungistatic substances in two soils. *Journal of General Microbiology*, **90**, 147–56.

Broad, T.E. and Shepherd, M.G. (1971) Purification and properties of glucose-6-phosphate dehydrogenase from the thermophilic fungus *Penicillium dupontii*. *Biochimica et Biophysica Acta*, **198**, 407–14.

Brodie, I.D.S. and Blakeman, J.P. (1975) Competition for carbon compounds by a leaf surface bacterium and conidia of *Botrytis cinerea*. *Physiological Plant Pathology*, **6**, 125–35.

Brondnitz, M.H. and Pascale, J.V. (1971) Thiopropanal-s-oxide a lachrymatory factor in onions. *Journal Agricultural Food Chemistry*, **19**, 269–72.

Brown, A.D. (1974) Microbial water relations: features in the intracellular composition of sugar-tolerant yeasts. *Journal of Bacteriology*, **118**, 769–77.

Brown, A.D. (1976) Microbial water stress. *Bacteriological Reviews*, **40**, 803–46.

Brown, A.D. (1978) Compatible solutes and extreme water stress in eukaryotic microorganisms. *Advances in Microbial Physiology*, **17**, 181–242.

Brown, A.D. (1990) *Microbial Water Stress, Principles and Perspectives*, John Wiley & Sons, Chichester, UK.

Brown, A.D. and Simpson, J.R. (1972) Water relations of sugar-tolerant yeasts: the role of intracellular polyols. *Journal of General Microbiology*, **72**, 589–91.

Brown, A.E., Finlay, R. and Ward, J. (1987). Antifungal compounds produced by *Epicoccum purpurascens* against soil-borne plant pathogenic fungi. *Soil Biology and Biochemistry* **19**, 657–64.

Brown, C.M. and Rose, A.H. (1969) Fatty acid composition of *Candida utilis* as affected by growth, temperature and dissolved oxygen tension. *Journal of Bacteriology*, **99**, 371–8.

Brown, D.E. and Halsted, D.J. (1975) The effect of acid pH on the growth kinetics of *Trichoderma viride Biotechnology and Bioengineering*, **17**, 1199–210.

Brown, G.E. and Kennedy, B.W. (1966) Effect of oxygen concentration on *Pythium* seed rot of soybean. *Phytopathology*, **56**, 407–11.

Brown, M.E. (1972) Plant growth substances produced by microorgan-

isms of soil and rhizosphere. *Journal of Applied Bacteriology*, **35**, 443–51.
Brown, M.E. (1975). Rhizosphere microorganisms – opportunists, bandits or benefactors, in *Soil Microbiology*, (ed. N. Walker), Wiley, New York, pp. 21–38.
Bruce, A. and King, B. (1983) Biological control of wood decay by *Lentinus lepideus* (Fr.) produced by *Scytalidium* and *Trichoderma* residues. *Material und Organismen*, **18**, 171–81.
Bruce, A., Austin, W.J. and King, B. (1984) Control of growth of *Lentinus lepideus* by volatiles from *Trichoderma*. *Transactions of the British Mycological Society*, **82**, 423–8.
Brückner, H. and Reinecke, C. (1989) Chromatographic assays for the rapid and sensitive detection of peptaibol mycotoxins (antibiotics) in filamentous fungi. *Journal of High Resolution Chromatography*, **2**, 113–16.
Bruehl, G.W. and Lai, P. (1966) Prior-colonization as a factor in the saprophytic survival of several fungi in wheat straw. *Phytopathology*, **56**, 766–8.
Bruehl, G.W., Millar, R.L. and Cunfer, B. (1969) Significance of antibiotic production by *Cephalosporium gramineum* to its saprophytic survival. *Canadian Journal of Plant Science*, **49**, 235–46.
Bull, A.T. (1970) Inhibition of polysaccharases by melanin. Enzyme inhibition in relation to mycolysis. *Archives of Biochemistry and Biophysics*, **137**, 345–56.
Bu'lock, J.D. (1961) Intermediary metabolism and antibiotic synthesis. *Advances in Applied Microbiology*, **3**, 293–342.
Burg, S.P. (1962) The physiology of ethylene formation. *Annual Review of Plant Physiology*, **13**, 265–302.
Burges, A. (1958) *Microorganisms in the Soil*, Hutchinson, London.
Burges, A. (1963) Some problems in soil microbiology. *Transactions of the British Mycological Society*, **46**, 1–14.
Burges, A. and Fenton, E. (1953). The effect of carbon dioxide on the growth of certain soil fungi. *Transactions of the British Mycological Society*, **36**, 104–10.
Burnside, J.E., Shippel, W.L., Forgacs, J. *et al.* (1957). A disease of swine and cattle caused by eating moldy corn. Experimental production with pure culture of molds. *American Journal of Veterinary Research* **18**, 817–824.
Burrage, S.W. (1971) Microclimate at the leaf surface, in *Ecology of Leaf Surface Microorganisms*, (eds T.F. Preece and C.H. Dickinson), Academic Press, London, pp. 91–102.
Buswell, J.A. and Odier, E. (1987) Lignin biodegradation. *Critical Reviews in Biotechnology*, **6**, 1–60.
Butcher, J.A. (1975) Nutritional factors affecting decay of softwoods by

soft-rot fungi. *Material und Organismen*, **3**, 277–86.
Byrne, P.J. and Jones, E.B.G. (1974) Lignicolous marine fungi. *Veröffentlichen des Instituts Meeresforschung Bremerhaven*, Supplement 5, 301–20.
Byrne, P.J. and Jones, E.B.G. (1975a) Effect of salinity on spore germination of terrestrial and marine fungi. *Transactions of the British Mycological Society*, **64**, 497–503.
Byrne, P.J. and Jones, E.B.G. (1975b) Effect of salinity on the reproduction of terrestrial and marine fungi. *Transactions of the British Mycological Society*, **65**, 185–200.
Cain, R.F. and Weresub, L.K. (1957) Studies of coprophilous ascomycetes. V. *Sphaeronaemella fimicola*. *Canadian Journal of Botany*, **35**, 119–31.
Cain, R.B., Bilton, R.F. and Darrah, J.A. (1968) The metabolism of aromatic acids by microorganisms. Metabolic pathways in the fungi. *Biochemical Journal*, **108**, 797–828.
Cairney, J.W.G., Jennings, D.H. and Veltkamp, C.J. (1989) A scanning electron microscope study of the internal structure of mature linear mycelial organs of four basidiomycete species. *Canadian Journal of Botany*, **67**, 2266–71.
Campbell, R. and Greaves, M.P. (1990) Methods for studying the microbial ecology of the rhizosphere, in *Methods in Microbiology*, (eds R. Grigorova and J.R. Norris), Vol. 22, Academic Press, London, pp. 447–77.
Campbell, R. and Porter, R. (1982) Low temperature scanning electron microscopy of microorganisms in soil. *Soil Biology and Biochemistry*, **14**, 241–5.
Canter, H.M. (1969) Studies on British chytrids. XXIX. A taxonomic revision of certain fungi found on the diatom *Asterionella*. *Journal of the Linnean Society (Botany)*, **60**, 85–97.
Canter, H.M. and Jaworski, G.H.M. (1979) The occurrence of a hypersensitive reaction in the planktonic diatom *Asterionella formosa* Hassall by the chytrid *Rhizophydium planktonicum* Canter emend., in culture. *New Phytologist*, **82**, 187–206.
Canter, H.M. and Jaworski, G.H.M. (1980) Some general observations of the zoospores of the chytrid *Rhizophydium planktonicum* Canter emend. *New Phytologist*, **84**, 515–31.
Canter, H.M. and Jaworski, G.H.M. (1981) The effect of light and darkness upon infection of *Asterionella formosa* Hassall by the chytrid *Rhizophydium planktonicum* Canter emend. *Annals of Botany London N.S.*, **47**, 13–30.
Canter, H.M. and Lund, J.W.G. (1948) Studies on plankton parasites I. Fluctuations in the numbers of *Asterionella formosa* Hass. in relation to fungal epidemics. *New Phytologist*, **47**, 238–61.

Canter, H.M. and Lund, J.W.G. (1951) Studies on plankton parasites III. Examples of the interaction between parasites and other factors in determining the growth of diatoms. *Annals of Botany London N.S.*, **15**, 359–71.

Canter, H.M. and Lund, J.W.G. (1953) The parasitism of diatoms with special reference to lakes in the English Lake District. *Transactions of the British Mycological Society*, **36**, 13–37.

Cantino, E.C. (1966) Morphogenesis in aquatic fungi, in *The Fungi An Advanced Treatise*, (eds G.C. Ainsworth and A.S. Sussman), Vol. II, Academic Press, London, pp. 283–337.

Capstick, C.K., Twinn, D.C. and Waid, J.S. (1957) Predation of natural populations of free-living nematodes by fungi. *Nematologica*, **2**, 193–201.

Carey, J.K. and Savory, J.G. (1979) Basidiospore germination on pine sapwood. *Bulletin of the British Mycological Society*, **13**, 84.

Carpenter, S.E. and Trappe, J.M. (1985) Phoenicoid fungi: a proposed term for fungi that fruit after heat treatment of substrates. *Mycotaxon*, **23**, 203–6.

Carpenter, S.E., Trappe, J.M. and Ammirati, J. (1987) Observations of fungal succession in the Mount St Helen's devastation zone 1980–1983. *Canadian Journal of Botany*, **65**, 716–44.

Carré, G.C. (1964) Fungus decomposition of beech cupules. *Transactions of the British Mycological Society*, **47**, 437–44.

Carroll, G.C. (1979) Needle micro-epiphytes in a Douglas fir canopy: biomass and distribution patterns. *Canadian Journal of Botany*, **57**, 1000–7.

Carroll, G.C. (1986) The biology of endophytism in plants with particular reference to woody perennials, in *Microbiology of the Phyllosphere*, (eds N.J. Fokkema and J. van den Heuvel), Cambridge University Press, Cambridge, 205–22.

Carroll, G.C. (1988) Fungal endophytes in stems and leaves: from latent pathogen to mutualistic symbiont. *Ecology*, **69**, 2–9.

Carroll, G.C. (1991a) Beyond pest deterrence – alternative strategies and hidden costs of endophytic mutualisms in vascular plants, in *Microbial Ecology of Leaves*, (eds J.H. Andrews and S.S. Hirano), Springer-Verlag, New York, pp. 358–75.

Carroll, G.C. (1991b) Fungal associations of woody plants as insect antagonists in leaves and stems, in *Microbial Mediation of Plant-Herbivore Interactions*, (eds P. Barbosa, V.A. Krischik and C.G. Jones), Wiley & Sons, New York, pp. 253–71.

Carroll, G.C. and Carroll, F.E. (1978) Studies on the incidence of coniferous needle endophytes in the Pacific Northwest. *Canadian Journal of Botany*, **56**, 3034–43.

Carroll, F.E., Müller, E. and Sutton, B.C. (1977) Preliminary studies on

the incidence of needle endophytes in some European conifers. *Sydowia*, **29**, 87–103.
Carruthers, S.M. and Rayner, A.D.M. (1979) Fungal communities in decaying hardwood branches. *Transactions of the British Mycological Society*, **72**, 283–9.
Cartwright, K.St.G. and Findlay, W.P.K. (1942) Principal decays of British hardwoods. *Annals of Applied Biology*, **29**, 219–53.
Cartwright, K.St.G. and Findlay, W.P.K. (1950) *Decay of Timber and its Preservation*, His Majesty's Stationery Office, London.
Caten, C.E. and Jinks, J.L. (1966) Heterocaryosis: Its significance in wild homothallic ascomycetes and fungi imperfecti. *Transactions of the British Mycological Society*, **49**, 81–93.
Cayrol, J.C. (1983) Lutte biologique contre les *Meloidogyne* au moyen d'*Arthrobotrys irregularis*. *Revue Nematologique*, **6**, 265–73.
Cayrol, J.C., Frankowski, J.P., Laniece, A. *et al.* (1978) Contre les nématodes en champignonnière. Mise au point d'une methode de lutte biologique à l'aide d'un Hyphomycète prédateur: *Arthrobotrys robusta* souche 'Antipolis' (Royal 300). *Pépiniéristes, Horticulteurs, Maraîchers – Revue Horticole*, **184**, 23–30.
Celio, D.A. and Padgett, D.E. (1989) An improved method of quantifying water mold spores in natural water columns. *Mycologia*, **81**, 459–60.
Chakraborty, S. and Old, K.M. (1982) Mycophagous soil amoebae: interactions with three plant pathogenic fungi. *Soil Biology and Biochemistry*, **14**, 247–55.
Chamier, A.-C. (1985) Cell-wall degrading enzymes of aquatic hyphomycetes: a review. *Botanical Journal of the Linnean Society*, **91**, 67–81.
Chamier, A.-C. (1987) Effect of pH on microbial degradation of leaf litter in seven streams of the English Lake District. *Oecologia*, **71**, 491–500.
Chamier, A.-C. (1992) Water chemistry, in *The Ecology of Aquatic Hyphomycetes*, (ed. F. Bärlocher), Springer-Verlag, Berlin, pp. 152–72.
Chamier, A.-C. and Dixon, P.A. (1982a) Pectinases in leaf degradation by aquatic hypomycetes I: the field study. The colonization pattern of aquatic hypomycetes on leaf packs in a Surrey stream. *Oecologia*, **52**, 109–15.
Chamier, A-C. and Dixon, P.A. (1982b) Pectinases in leaf degradation by aquatic hypomycetes: the enzymes and leaf maceration. *Journal of General Microbiology*, **128**, 2469–83.
Chamier, A-C. and Dixon, P.A. (1983) Effect of calcium-ion concentration on leaf maceration by *Tetrachaetum elegans*. *Transactions of the British Mycological Society*, **81**, 415–18.
Chamier, A-C. Dixon, P.A. and Archer, S.A. (1983) The spatial distribution of fungi on decomposing alder leaves in a freshwater stream. *Oecologia*, **64**, 92–103.
Chamier, A-C., Sutcliffe, D.W. and Wishman, J.P. (1989) Changes in

Na, K, Ca, Mg and Al content of submerged leaf litter, related to ingestion by the amphipod *Gammarus pulex* (L.). *Freshwater Biology*, **21**, 181–9.

Chang, Y. (1967) The fungi of wheat straw compost. II. Biochemical and physiological studies. *Transactions of the British Mycological Society*, **50**, 667–77.

Chang, Y. and Hudson, H.J. (1967) The fungi of wheat straw compost. I. Ecological studies. *Transactions of the British Mycological Society*, **50**, 649–66.

Chapela, I.H., Boddy, L. and Rayner, A.D.M. (1988) Structure and development of fungal communities in beech logs four and a half years after felling. *Microbial Ecology*, **53**, 59–70.

Chapela, I.H., Petrini, O. and Hagman, L. (1991) Monolignol glucosides as specific recognition messengers in fungus–plant symbioses. *Physiological and Molecular Plant Pathology*, **39**, 289–98.

Chard, J.M., Gray, T.R.G. and Frankland, J.C. (1985a) Purification of an antigen characteristic for *Mycena galopus*. *Transactions of the British Mycological Society*, **84**, 235–41.

Chard, J.M., Gray, T.R.G. and Frankland, J.C. (1985b) Use of an anti-*Mycena galopus* serum as an immunofluorescence reagent. *Transactions of the British Mycological Society*, **84**, 243–5.

Chauvet, E. (1987) Changes in the chemical composition of alder, poplar and willow leaves during decomposition in a river. *Hydrobiologia*, **148**, 35–44.

Chauvet, E. (1991) Aquatic hypomycete distribution in South-Western France. *Journal of Biogeography*, **18**, 699–706.

Chauvet, E. (1992) Dynamique saisonnière des spores d'hyphomycètes aquatiques de quatre rivières. *Nova Hedwigia*, **54**, 379–95.

Chen, A.W. and Griffin, D.M. (1966) Soil physical factors and ecology of soil fungi. V. Further studies in relatively dry soils. *Transactions of the British Mycological Society*, **49**, 419–26.

Chesters, C.G.C. (1948) A contribution to the study of fungi in soil. *Transactions of the British Mycological Society*, **30**, 100–17.

Chesters, C.G.C. and Bull, A.T. (1963) The enzymic degradation of laminarin. I., The distribution of laminarinase among microorganisms. *Biochemical Journal*, **86**, 28–31.

Chet, I., Harman, G.E. and Baker, R. (1981) *Trichoderma hamatum*: its hyphal interactions with *Rhizoctonia solani* and *Pythium* spp. *Microbial Ecology*, **7**, 29–38.

Chet, I., Zieberstein, Y. and Henis, Y. (1973) Chemotaxis of *Pseudomonas lachrymans* to plant extracts and water droplets collected from leaf surfaces of resistant and susceptible plants. *Physiological Plant Pathology*, **3**, 473–9.

Chinn, S.H.F. (1953) A slide technique for the study of fungi and actinomycetes with special reference to *Helminthosporium sativum*. *Canadian Journal of Botany*, **31**, 718–24.

Chinn, S.H.F. and Ledingham, R.J. (1957) Studies on the influence of various substances on the germination of *Helminthosporium sativum*. *Canadian Journal of Botany*, **35**, 679–701.

Christensen, C.M. (1971) Mycotoxins. *Critical Reviews in Environmental Control*, **2**, 57–80.

Christensen, C.M. and Kaufman, H.H. (1965) Deterioration of stored grains by fungi. *Annual Review of Phytopathology*, **3**, 69–84.

Christensen, C.M. and Kaufman, H.H. (1969) *Grain Storage, The Role of Fungi in Quality Loss*, University of Minnesota Press, Minneapolis.

Christensen, C.M. and Sauer, D.B. (1982) Microflora, in *Storage of Cereal Grains and their Products*, (ed. C.M. Christensen), American Association of Cereal Chemists Inc., St Paul, Minnesota USA, pp. 219–40.

Christensen, M. (1981) Species diversity and dominance in fungal communities, in *The Fungal Community*, (eds D.T. Wicklow and G.C. Carroll), Marcel Dekker, New York, pp. 201–32.

Christensen, M. (1989) A view of fungal ecology. *Mycologia*, **81**, 1–19.

Christensen, N.L. and Muller, C.H. (1975) Effects of fire on factors controlling plant growth in *Adenostoma* chaparral. *Ecological Monographs*, **45**, 29–55.

Chu-Chou, M. and Grace, L.J. (1981) Mycorrhizal fungi of *Pseudotsuga menziesii* in the North island of New Zealand. *Soil Biology and Biochemistry*, **13**, 247–9.

Ciegler, A., Kadis, S. and Ajl, S.J. (1971) *Microbial Toxins*, Vol. 6, Academic Press, London.

Clark, C.C., Miller, J.D. and Whitney, N.J. (1989) Toxicity of conifer needle endophytes to spruce budworm. *Mycological Research*, **93**, 508–12.

Clarke, J.H. (1966) Studies on fungi in the root region. V. The antibiotic effects of root extracts of *Allium* on some root surface fungi. *Plant and Soil*, **25**, 32–40.

Clarke, J.H., Niles, E.V. and Hill, S.T. (1967) Ecology of the microflora of moist barley. *Pest Infestation Research*, **1966**, 14–16.

Clarke, J.H., Hill, S.T., Niles, E.V. *et al.* (1968) Ecology of the microflora of moist barley. *Pest Infestation Research*, **1967**, 15–16.

Clausz, J.C. (1974) Periods of activity of water molds in a North Carolina lake, in *Phenology and Seasonality Modeling*, (ed. H. Leith), Springer, New York, pp. 191–203.

Clay, K. (1988) Fungal endophytes of grasses: A defensive mutualism between plants and fungi. *Ecology*, **69**, 10–16.

Clay, K. (1991) Fungal endophytes, grasses and herbivores, in *Microbial Mediation of Plant–Herbivore Interactions* (eds P. Barbosa, V.A. Krischik and C.G. Jones), Wiley & Sons, New York, pp. 199–226.

Clay, K. (1992) Endophytes as antagonists of plant pests, in *Microbial Ecology of Leaves*, (eds J.H. Andrews and S.S. Hirano), Springer-Verlag, New York, pp. 331–57.

Clay, K., Hardy, T.N. and Hammond, A.M. (1985a) Fungal endophytes of grasses and their effects on an insect herbivore. *Oecologia*, **66**, 1–6.

Clay, K., Hardy, T.N. and Hammond, A.M. (1985b) Fungal endophytes of *Cyperus* and their effects on an insect herbivore. *American Journal of Botany*, **72**, 1284–9.

Claydon, N., Allan, M., Hanson, J.R. and Avent, A.G. (1987) Antifungal alkyl pyrones of *Trichoderma harzianum*. *Transactions of the British Mycological Society*, **88**, 503–13.

Clipson, N.J.W. and Jennings, D.H. (1990) Role of potassium and sodium in generation of osmotic potential of the marine fungus *Dendryphiella salina*. *Mycological Research*, **94**, 1017–22.

Clubbe, C.P. (1980) The colonization and succession of fungi in wood. The International Research Group on Wood Preservation Document no. IRG/WP/1107.

Coates, D. and Rayner, A.D.M. (1985a) Fungal population and community development in cut beech logs. I. Establishment via the aerial and cut surface. *New Phytologist*, **101**, 153–71.

Coates, D. and Rayner, A.D.M. (1985b) Fungal population and community development in cut beech logs. II. Establishment via the buried cut surface. *New Phytologist*, **101**, 173–81.

Coates, D. and Rayner, A.D.M. (1985c) Fungal population and community development in cut beech logs. III. Spatial dynamics, interactions, and strategies. *New Phytologist*, **101**, 183–98.

Coggins, C.R., Hornung, U., Jennings, D.H. and Veltkamp, C.J. (1980) The phenomenon of point growth and its relation to flushing and strand formation in mycelium of *Serpula lacrimans*. *Transactions of the British Mycological Society*, **75**, 69–76.

Coker, W.C. (1923) *The Saprolegniaceae, with notes on other water molds*, University of North Carolina Press, Chapel Hill.

Collins, M.A. (1976) Colonization of leaves by phylloplane saprophytes and their interactions in this environment, in *Microbiology of Aerial Plant Surfaces*, (eds C.H. Dickinson and T.F. Preece), Academic Press, London, pp. 401–18.

Collins, M.A. (1982) Rust disease and the development of phylloplane microflora of *Antirrhinum* leaves. *Transactions of the British Mycological Society*, **79**, 117–22.

Collins, V.G. and Willoughby, L.G. (1962) The distribution of bacteria and fungal spores in Blelham tarn with particular reference to an experimental overturn. *Archiv für Mikrobiologie*, **43**, 294–307.

Cook, M.T. and Taubenhaus, J.J. (1911) The relations of parasitic fungi to the contents of the cells of host plants. I. The toxicity of tannin. *Bulletin*

Delaware College Agricultural Experimental Station, **91**, 1–77.
Cook, R.J. and Papendick, R.I. (1972) Influence of water potential of soils and plants on root disease. *Annual Review of Phytopathology*, **10**, 349–74.
Cook, R.J., Papendick, R.I. and Griffin, D.M. (1972) Growth of two root-rot fungi as affected by osmotic and matric water potentials. *Soil Science Society of America Proceedings*, **36**, 78–82.
Cooke, R.C. (1961) Agar disc technique for the direct observation of nematode-trapping fungi in the soil. *Nature, London*, **191**, 1411–12.
Cooke, R.C. (1962a) The ecology of nematode-trapping fungi in the soil. *Annals of Applied Biology*, **50**, 507-13.
Cooke, R.C. (1962b) The behaviour of nematode-trapping fungi during the decomposition of organic matter in the soil. *Transactions of the British Mycological Society*, **45**, 314–20.
Cooke, R.C. (1963a). The predaceous activity of nematode-trapping fungi added to soil. *Annals of Applied Biology*, **51**, 295–9
Cooke, R.C. (1963b) Ecological characteristics of nematode-trapping hyphomycetes. I. Preliminary studies. *Annals of Applied Biology*, **52**, 431–7.
Cooke, R.C. (1964) Ecological characteristics of nematode-trapping hyphomycetes. II. Germination of conidia in soil. *Annals of Applied Biology*, **54**, 375–9.
Cooke, R.C. (1968) Relationships between nematode-destroying fungi and soil-borne phytonematodes. *Phytopathology*, **58**, 909–13.
Cooke, R.C. (1983) Morphogenesis of sclerotia, in *Fungal Differentiation*, (ed. J.E. Smith), Marcel Dekker, New York, pp. 397–418.
Cooke, R.C. and Godfrey, B.E.S. (1964) A key to the nematode-destroying fungi. *Transactions of the British Mycological Society*, **47**, 61–74.
Cooke, R.C. and Rayner, A.D.M. (1984) *Ecology of Saprotrophic Fungi*, Longman, New York.
Cooke, R.C. and Satchuthananthavale, V. (1968) Sensitivity to mycostasis of nematode-trapping hyphomycetes. *Transactions of the British Mycological Society*, **51**, 555–61.
Cooney, D.G. and Emerson, R. (1964) *Thermophilic Fungi, an account of their Biology, Activities and Classification*, Freeman, San Francisco.
Corbett, N.H. (1965) Micro-morphological studies on the degradation of lignified cell walls by ascomycetes and fungi imperfecti. *Journal of the Institute of Wood Science*, **14**, 18–29.
Corry, J.E.L. (1987) Relationships of water activity to fungal growth, in *Food and Beverage Mycology* (ed. L.R. Beuchat), Van Nostrand Reinhold, New York, pp. 51–99.
Côté, W.H., Kutscha, N.P., Simson, B.W. and Timell, T.E. (1968)

Studies on compression wood. VI. Distribution of polysaccharides in the cell wall tracheids from compression wood of balsam fir (*Abies balsamea* L. Mill). *Tappi*, **51**, 33–40.
Coulson, C.B., Davies, R.I. and Lewis, D.A. (1960) Polyphenols in plant humus and soil. I. Polyphenols of leaves, litter and superficial humus from mull and mor sites. *Soil Science* **11**, 20–9.
Courtois, H. (1963) Beitrag zur Frage holzabbauender Ascomyceten und Fungi Imperfecti. *Holzforschung*, **17**, 176–83.
Covey, R.P. (1970) Effect of oxygen tension on the growth of *Phytophthora cactorum*. *Phytopathology*, **60**, 358–9.
Cowley, G.T. and Whittingham, W.F. (1961) The effect of tannin on the growth of selected soil microfungi in culture. *Mycologia*, **53**, 539–42.
Cowling, E.B. (1961) Comparative biochemistry of the decay of sweetgum sapwood by white-rot and brown-rot fungi. *Technical Bulletin 1285*, Forest Service, US Department of Agriculture, Washington, DC.
Craveri, R. and Colla, C. (1966) Ribonuclease activity of mesophilic and thermophilic molds. *Annals of Microbiology*, **16**, 97–9.
Crisan, E.V. (1969) The proteins of thermophilic fungi, in *Current Topics in Plant Science*, (ed. J.E. Gunckel), Academic Press, New York, pp. 32–3.
Crisan, E.V. (1973) Current concepts of thermophilism and thermophilic fungi. *Mycologia*, **65**, 1171–98.
Croft, J.H. and Jinks, J.L. (1977) Aspects of the population genetics of *Aspergillus nidulans*, in *Genetics and Physiology of Aspergillus*, (eds J.E. Smith and J.A. Pateman), Academic Press, London, pp. 339–60.
Cromack, K. and Caldwell, B.A. (1992) The role of fungi in litter decomposition and nutrient cycling, in *The Fungal Community: Its Organisation and Role in the Ecosystem*, 2nd edn, (eds G.C. Carroll and D.T. Wicklow), Marcel Dekker, New York, pp. 653–68.
Cromack, K., Sollins, P., Graustein, W.C. *et al.* (1979) Calcium oxalate accumulation and soil weathering in mats of the hypogeous fungus *Hysterangium crassum*. *Soil Biology and Biochemistry*, **11**, 463–8.
Crook, E.M. and Holden, M. (1948) Some factors affecting the extraction of nitrogenous materials from leaves of various species. *Biochemical Journal*, **43**, 181–3.
Crowe, J.H., Crowe, L.M. and Chapman, D. (1984) Preservation of membranes in anhydrobiotic organisms: the role of trehalose. *Science*, **223**, 701–3.
Cummins, K. (1974) Structure and function of stream ecosystems. *Bio Science*, **24**, 631–41.
Cuomo, V., Vanzanella, F., Fresi, E. (1985). Fungal flora of *Posidonia oceanica* and its ecological significance. *Transactions of the British Mycological Society*, **84**, 35–40.

Curl, E.A. and Truelove, B. (1986) *The Rhizosphere*, Springer-Verlag, Berlin.

Curry, J.P. (1969) The decomposition of organic matter in soil. I. The role of fauna in decaying grassland herbage. *Soil Biology and Biochemistry*, **1**, 253–66.

Curtis, E.J.C. (1969) Sewage fungus: its nature and effects. *Water Research*, **3**, 289–311.

Curtis, F.C., Evans, G.H., Lillis, V. *et al.* (1978) Studies on Mucoralean parasites. I. Some effects of *Piptocephalis* species on host growth. *New Phytologist*, **80**, 157–65.

Curtis, P.J. (1969) Anaerobic growth in fungi. *Transactions of the British Mycological Society*, **53**, 299–302.

Dackman, C. and Nordbring-Hertz, B. (1992) Conidial traps – a new survival structure of the nematode-trapping fungus *Arthrobotrys oligospora*. *Mycological Research*, **96**, 194–8.

Dackman, C., Jansson, H-B. and Nordbring-Hertz, B. (1992) Nematophagous fungi and their activities in soil, in *Soil Biochemistry*, vol. 7, (eds G. Stotzky and J-M. Bollag), Marcel Dekker, New York, pp. 95–130.

Dackman, C., Olsson, S., Jansson, H-B. *et al.* (1987) Quantification of predatory and endoparasitic nematophagous fungi in soil. *Microbial Ecology*, **13**, 89–93.

Dagley, S. (1971) Catabolism of aromatic compounds by microorganisms. *Advances in Microbial Physiology*, **6**, 1–42.

Dahlberg, A. (1992) Somatic incompatibility – a tool to reveal spatiotemporal mycelial structures of ectomycorrhical fungi, in *Mycorrhizas in Ecosystems*, (eds D.J. Read, D.H. Lewis, A.H. Fitter and I.J. Alexander), CAB International, Wallingford, UK, pp. 135–40.

Dahlberg, A. and Stenlid, J. (1990) Population structure and dynamics in *Suillus bovinus* as indicated by spatial distribution of fungal colonies. *New Phytologist*, **115**, 487–93.

Daniel, G., Volc, J. and Nilsson, T. (1992) Soft rot and multiple T-branching by the basidiomycete *Oudemansiella mucida* (Schrad ex Fr.) Hohn in pine wood. *Mycological Research*, **96**, 49–54.

Danielson, R.M. (1984) Ectomycorrhiza formation by the operculate Disomycete *Sphaerosporella brunnea* (Pezizales). *Mycologia*, **76**, 454–61.

Danielson, R.M., Visser, S. and Parkinson, D. (1984) The effectiveness of mycelial slurries of mycorrhizal fungi for the inoculation of container-grown Jack pine seedlings. *Canadian Journal of Forest Research*, **14**, 140–2.

Dart, P.J. and Mercer, F.V. (1964) The legume rhizosphere. *Archiv für Mikrobiologie*, **47**, 344–78.

Das, A.C. (1963) Ecology of soil fungi of rice fields. I. Succession of fungi on rice roots. 2. Association of soil fungi with organic matter. *Transac-*

tions of the British Mycological Society, **46**, 431–43.

Davidson, D.E. (1974) The effect of salinity on a marine and a freshwater ascomycete. *Canadian Journal of Botany*, **52**, 553–63.

Davidson, R.W., Campbell, W.A. and Blaisdell, D.J. (1938) Differentiation of wood-decaying fungi by their reactions on gallic or tannic acid medium. *Journal of Agricultural Research*, **57**, 683–95.

Davies, R.I., Coulson, C.B. and Lewis, D.A. (1964) Polyphenols in plant humus and soil. IV. Factors leading to increase in leaves and their relationship to mull and mor formation. *Journal of Soil Science*, **15**. 310–18.

Deacon, J.W. (1973) *Phialophora radicicola* and *Gaeumannomyces graminis* on roots of grasses and cereals. *Transactions of the British Mycological Society*, **61**, 471–85.

Deacon, J.W. (1974) Further studies on *Phialophora radicicola* and *Gaeumannomyces graminis* on roots and stems of grasses and cereals. *Transactions of the British Mycological Society*, **63**, 307–27.

Deacon, J.W. (1976) Studies on *Pythium oligandrum* an aggressive parasite of other fungi. *Transactions of the British Mycological Society*, **66**, 383–91.

Deacon, J.W. (1979) Cellulose decomposition by *Pythium* and its relevance to substrate-groups of fungi. *Transactions of the British Mycological Society*, **72**, 469–77.

Deacon, J.W. (1985) Decomposition of filter paper cellulose by thermophilic fungi acting singly, in combination and in sequence. *Transactions of the British Mycological Society*, **85**, 663–9.

Deacon, J.W. (1987) Programmed cortical senescence: a basis for understanding root infection, in *Fungal Infection of Plants*, (eds G.F. Pegg and P.G. Ayres), Cambridge University Press, Cambridge, pp. 285–97.

Deacon, J.W. and Fleming, L.V. (1992) Interaction of ectomycorrhizal fungi, in *Mycorrhizal Functioning: an Integrative Plant-Fungal Process*, (ed M.F. Allen), Chapman & Hall, New York and London, pp. 249–300.

Deacon, J.W., Donaldson, S.J. and Last, F.T. (1983) Sequences and interactions of mycorrhizal fungi on birch. *Plant and Soil*, **71**, 257–62.

De Boois, H.M. (1976) Fungal development on oak litter and decomposition potentialities of some fungal species. *Revue d'Ecologie et de Biologie du Sol*, **13**, 437–48.

Debout, B., Schatz, S., Kohlmeyer, J. and Haibach, M. (1987) Temperature-dependent growth in isolates of *Corollospora maritima* Werdem. (Ascomycetes) from different geographical regions. *Journal of Experimental Marine Biology and Ecology*, **106**, 203–10.

De Groot, R.C. (1972) Growth of wood-inhabiting fungi in saturated atmospheres of monoterpenoids. *Mycologia*, **64**, 683–870.

Dekker, R.F. and Richards, G.N. (1976) Hemicellulases: their occurrence, purification, properties and mode of action. *Advances in Carbohydrate Chemistry and Biochemistry*, **32**, 277–352.

Demel, R.A. and de Kruyff, B. (1976) The function of sterols in membranes. *Biochimica et Biophysica Acta*, **457**, 109–32.

Dennis, C. and Cohen, E. (1976) The effect of temperature on strains of soft fruit spoilage fungi. *Annals of Applied Biology*, **82**, 51–6.

Dennis, C. and Webster, J. (1971a) Antagonistic properties of species groups of *Trichoderma*. I. Production of non-volatile antibiotics. *Transactions of the British Mycological Society*, **57**, 25–39.

Dennis, C. and Webster, J. (1971b) Antagonistic properties of species groups of *Trichoderma*. II. Production of volatile antibiotics. *Transactions of the British Mycological Society*, **57**, 41–8.

Dennis, C. and Webster, J. (1971c) Antagonistic properties of species groups of *Trichoderma*. III. Hyphal interaction. *Transactions of the British Mycological Society*, **57**, 363–9.

Descals, E., Marvanova, L. and Webster, J. (1995) *Monograph of Ingoldian Aquatic Hyphomycetes* (preparation).

Descals, E., Webster, J. and Dyko, B.S. (1977) Taxonomic studies on aquatic hyphomycetes I. *Lemonniera* de Wildeman. *Transactions of the British Mycological Society*, **69**, 89–109.

Deschamps, A.M. and Leulliette, L. (1984). Tannin degradation by yeasts from decaying barks. *International Biodeterioration*, **20**, 237–41.

Deverall, B.J. (1968) Psychrophiles, in *The Fungi An Advanced Treatise*, vol. III, (eds G.C. Ainsworth and A.S. Sussman), Academic Press, London, pp. 129–34.

Dexter, Y. and Cooke, R.C. (1984) Fatty acids, sterols and carotenoids of the psychrophile *Mucor strictus* and some mesophilic *Mucor* spp. *Transactions of the British Mycological Society*, **83**, 455–61.

Dexter, Y. and Cooke, R.C. (1985) Effect of temperature on respiration, nutrient uptake and leakage in the psychrophile *Mucor strictus* Hagem. *Transactions of the British Mycological Society*, **84**, 131–6.

Dick, M.W. (1966) The Saprolegniaceae of the environs of Blelham Tarn: sampling techniques and the estimation of propagule numbers. *Journal of General Microbiology*, **42**, 257–82.

Dick, M.W. (1970) Saprolegniaceae on insect exuviae. *Transactions of the British Mycological Society*, **55**, 449–58.

Dick, M.W. (1971) The ecology of the Saprolegniaceae in lentic and littoral muds with a general theory of fungi in the lake exosystem. *Journal of General Microbiology*, **65**, 325–37.

Dick, M.W. (1973a) Saprolegniales, in *The Fungi: An Advanced Treatise*, vol. IVB, (eds G.C. Ainsworth, F.K. Sparrow and A.S. Sussman), Academic Press, New York and London, pp. 113–44.

Dick, M.W. (1973b) Leptomitales, in *The Fungi: An Advanced Treatise*, vol. IVB, (eds G.C. Ainsworth, F.K. Sparrow and A.S. Sussman),

Academic Press, New York and London, pp. 145–58.
Dick, M.W. (1976) The ecology of aquatic phycomycetes, in *Recent Advances in Aquatic Mycology*, (ed. E.B.G. Jones), Elek Science, London, pp. 513–42.
Dick, M.W. (1989) Phylum Oomycetes, in *Handbook of Protoctista*, (eds L. Margulis, J.O. Corliss, M. Melkonian and D.J. Chapman), Jones and Bartlett, Boston, pp. 661–85.
Dick, M.W. and Ali-Shtayeh, M.S. (1986) Distribution and frequency of *Pythium* species in parkland and farm soils. *Transactions of the British Mycological Society*, **86**, 49–62.
Dickinson, C.H. (1967) Fungal colonization of *Pisum* leaves. *Canadian Journal of Botany*, **45**, 915–27.
Dickson, C.H. (1971) Cultural studies of leaf saprophytes, in *Ecology of Leaf Surface Microorganisms*, (eds T.F. Preece and C.H. Dickinson), Academic Press, London, pp. 129–38.
Dickinson, C.H. (1973) Interaction of fungicides and saprophytes. *Pesticide Science*, **4**, 563–74.
Dickinson, C.H. (1979) Fairy rings in Norfolk. *Federation of British Plant Pathologists Newsletter*, **4**, 39–42.
Dickinson, C.H. (1981) Leaf surface microorganisms as pathogen antagonists and as minor pathogens, in *Strategies for the Control of Cereal Disease*, (eds J.F. Jenkyn and R.T. Plumb), Blackwell, Oxford, pp. 109–21.
Dickinson, C.H. (1986) Adaptations of micro-organisms to climatic conditions affecting aerial plant surfaces, in *Microbiology of the Phyllosphere*, (eds N.J. Fokkema and J. van den Heuvel), Cambridge University Press, Cambridge, pp. 78–100.
Dickinson, C.H. and Bottomley, D. (1980) Germination and growth of *Alternaria* and *Cladosporium* in relation to their activity in the phylloplane. *Transactions of the British Mycological Society*, **74**, 309–19.
Dickinson, C.H. and Macnamara, O.C. (1983) *In vitro* degradation of *Ilex aquifolium* leaf cutin by phylloplane fungi. *Transactions of the British Mycological Society*, **81**, 347–54.
Dickinson, C.H. and O'Donnell, J. (1977) Behaviour of phylloplane fungi on *Phaseolus* leaves. *Transactions of the British Mycological Society*, **68**, 193–9.
Dickinson C.H. and Pugh, G.J.F. (1965a) The mycoflora associated with *Halimione portulacoides*. I. The establishment of the root surface flora of the marine plant. *Transactions of the British Mycological Society*, **48**, 381–90.
Dickinson C.H. and Pugh, G.J.F. (1965b) The mycoflora associated with *Halimione portulacoides*. II. Root surface fungi of mature and excised plants. *Transactions of the British Mycological Society*, **48**, 595–602.

Dickinson C.H. and Pugh, G.J.F. (1965c) The mycoflora associated with *Halimione portulacoides*. III. Fungi on green and moribund leaves. *Transactions of the British Mycological Society*, **48**, 603–10.

Dickinson C.H. and Skidmore, A.M. (1976) Interactions between germinating spores of *Septoria nodorum* and phylloplane fungi. *Transactions of the British Mycological Society*, **66**, 45–56.

Dickinson, C.H. and Underhay, V.H.S. (1977) Growth of fungi in cattle dung. *Transactions of the British Mycological Society*, **69**, 473–7.

Dickinson, C.H. and Wallace, B. (1976) Effects of late application of foliar fungicide on activity of microorganisms on winter wheat flag leaves. *Transactions of the British Mycological Society*, **76**, 103–12.

Dickinson, D.J. (1982) The decay of commercial timbers, in *Decomposer Basidiomycetes: their Biology and Ecology*, (eds. J.C. Frankland, J.N. Hedger and M.J. Swift), *British Mycological Symposium 4*, Cambridge University Press, Cambridge, pp. 179–90.

Diem, H.G. (1971) Effect of low humidity on the survival of germinated spores commonly found in the phyllosphere, in *Ecology of Leaf Surface Microorganisms*, (eds T.F. Preece and C.H. Dickinson), Academic Press, London, pp. 211–20.

Diem, H.G. (1974) Microorganisms of the leaf surface: estimation of the mycoflora of the barley phyllosphere. *Journal of General Microbiology*, **80**, 77–83.

Dighton, J. and Mason, P.A. (1985) Mycorrhizal dynamics during forest tree development, in *Developmental Biology of Higher Fungi*, (eds D. Moore, L.A. Casselton, D.A. Wood and J.C. Frankland) *British Mycological Society Symposium 10*, Cambridge University Press, Cambridge, pp. 117–39.

Dighton, J., Poskitt, J.M. and Howard, D.M. (1986) Changes in occurrence of basidiomycete fruit bodies during forest stand development: with specific reference to mycorrhizal species. *Transactions of the British Mycological Society*, **87**, 163–71.

Dijksterhuis, J., Veenhuis, M. and Harder, W. (1990) Ultrastructural study of adhesion and initial stages of infection of nematodes by conidia of *Drechmeria coniospora*. *Mycological Research*, **94**, 1–8.

Dik, A.J. (1991) Interaction among fungicides, pathogens, yeasts and nutrients in the phyllosphere, in *Microbial Ecology of Leaves*, (eds J.H. Andrews and S.S. Hirano), Springer-Verlag, New York, pp. 412–29.

Di Menna, M.E. (1960) Yeasts from Antarctica. *Journal of General Microbiology*, **23**, 295–300.

Di Menna, M.E. (1966) Yeasts in Antarctic soils. *Antonie van Leeuwenhoek, Journal of Microbiology and Serology*, **32**, 29–38.

Dix, N.J. (1958) A comparative physiological study of the microfungi of *Dactylis glomerata*. PhD Thesis, University of Sheffield.

Dix, N.J. (1964) Colonization and decay of bean roots. *Transactions of the*

British Mycological Society, **47**, 285–92.
Dix, N.J. (1969) Further experimental studies of bean rhizosphere fungi. *Transactions of the British Mycological Society*, **52**, 451–7.
Dix, N.J. (1972) Effect of soil fungistasis on spore germination and germ tube growth in *Penicillium* species. *Transactions of the British Mycological Society*, **58**, 59–66.
Dix, N.J. (1974) Identification of a water-soluble fungal inhibitor in the leaves of *Acer platanoides*. *Annals of Botany*, **38**, 505–14.
Dix, N.J. (1979) Inhibition of fungi by gallic acid in relation to growth on leaves and litter. *Transactions of the British Mycological Society*, **73**, 329–36.
Dix, N.J. (1984a) Moisture content and water potential of abscissed leaves in relation to decay. *Soil Biology and Biochemistry*, **16**, 367–70.
Dix, N.J. (1984b) Minimum water potential for the growth of some litter-decomposing agarics and other basidiomycetes. *Transactions of the British Mycological Society*, **83**, 152–3.
Dix, N.J. (1985) Changes in the relationship between water content and water potential after decay and significance for fungal successions. *Transactions of the British Mycological Society*, **85**, 649–53.
Dix, N.J. and Cairney, J.W.G. (1985) Wood rotting by *Phallus impudicus*. *Transactions of the British Mycological Society*, **85**, 514–20.
Dix, N.J. and Christie, P. (1974) Changing sensitivity to soil fungistasis with age in *Drechslera rostrata* spores and associated permeability changes. *Transactions of the British Mycological Society*, **62**, 527–35.
Dix, N.J. and Simpson, A.P. (1984) Decay of leaf litter by *Collybia peronata*. *Transactions of the British Mycological Society*, **83**, 37–41.
Dobbs, C.G. (1971) Residual soil mycostasis. Resumé of a paper read at the first International Mycological Congress, Exeter, UK, 4 pp.
Dobbs, C.G. and Gash, M.J. (1965) Microbial and residual mycostasis in soils. *Nature*, **207**, 1354–6.
Dobbs, C.G. and Hinson, W.H. (1953) A widespread fungistasis in the soil. *Nature*, **172**, 197.
Doguet, G. (1964) Influence de la temperature et de la salinité sur la croissance et la fertilité du *Digitatéspora marina* Doguet. *Bulletin de la Société Francaise de Physiologie Vegetale*, **10**, 285–92.
Domsch, K.H. and Gams, W. (1969) Variability and potential to decompose pectin, xylan and carboxymethyl cellulose. *Soil Biology and Biochemistry*, **1**, 29–36.
Dowding, P. (1981) Nutrient uptake and allocation during substrate exploitation by fungi, in *The Fungal Community: its Organisation and Role in the Ecosystem*, (eds D.T. Wicklow and G.C. Garroll), Marcel Dekker, New York, pp. 621–63.
Dowding, P. (1986) Water availabilty, the distribution of fungi and their adaptation to the environment, in *Water, Fungi and Plants*, (eds P.G.

Ayres and L. Boddy), Cambridge University Press, Cambridge, pp. 305–20.
Dowding, P. and Widden, P. (1974) Some relationships between fungi and their environment in tundra regions, in *Soil Organisms and Decomposition in Tundra*, (eds A.J. Holding, O.W. Heal, S.F. Maclean and P.W. Flanagan), Tundra Biome Steering Committee, Stockholm, pp. 123–50.
Dowe, A. (1987) *Räuberische Pilze und andere Pilzliche Nematodenfeinde*, Ziemsen Verlag, Wittenberg Lutherstadt.
Dowson, C.G., Rayner, A.DM. and Boddy, L. (1986) Outgrowth patterns of mycelial cord-forming basidiomycetes from and between woody resource units in soil. *Journal of General Microbiology*, **132**, 203–11.
Dowson, C.G., Rayner, A.DM. and Boddy, L. (1989) Spatial dynamics and interactions of the woodland fairy ring fungus *Clitocybe nebularis*. *New Phytologist*, **111**, 699–705.
Drechsler, C. (1937) New Zoopagaceae destructive to soil rhizopods. *Mycologia*, **29**, 229–49.
Drechsler, C. (1941) Predaceous fungi. *Biological Reviews, Cambridge*, **16**, 265–90.
Duddington, C.L. (1951) The ecology of predaceous fungi. I. Preliminary survey. *Transactions of the British Mycological Society*, **34**, 322–31.
Duddington, C.L. (1955) Fungi that attack microscopic animals. *The Botanical Review*, **21**, 377–439.
Duddington, C.L. (1957) *The Friendly Fungi: A New Approach to the Eelworm Problem*, Faber & Faber, London.
Duddington, C.L. and Wyborn, C.H.E. (1972) Recent research on the nematophagous hyphomycetes. *The Botanical Review*, **38**, 545–65.
Duddridge, J.A. (1986) The development and ultrastructure of ectomycorrhizas. III. Compatible and incompatible interaction between *Suillus grevillei* (Klotzsch) Sing. and a number of ectomycorrhizal host *in vitro* in the absence of exogenous carbohydrate. *New Phytologist*, **103**, 457–64.
Duddridge, J. (1987) Specificity and recognition in ectomycorrhizal associations, in *Fungal Infection of Plants*, (eds G.F. Pegg and P.G. Ayres), *British Mycological Society Symposium, 13*, Cambridge University Press, Cambridge, pp. 25–44.
Duddridge, J.A., Malibari, A. and Read, D.J. (1980) Structure and function of mycorrhizal rhizomorphs with special reference to their role in water transport. *Nature*, **287**, 834–6.
Duncan, C.G. (1960) Wood-attacking capacities and physiology of soft rot fungi. US Department of Agriculture Forest Services. Forest Products Laboratory Report 2173, Madison, Wisconsin.
Duncan, C.G. and Eslyn, W.E. (1966) Wood decaying ascomycetes and fungi imperfecti. *Mycologia*, **5**, 642–5.
Duniway, J.M. (1976) Movement of zoospores of *Phytophthora cryptogea*

in soils of various textures and matric potentials. *Phytopathology*, **66**, 877–82.
Duniway, J.M. (1979) Water relations of water molds. *Annual Review of Phytopathology*, **17**, 431–60.
Durbin, R.D. (1959) Factors affecting the vertical distribution of *Rhizoctonia solani* with special reference to CO_2 concentration. *American Journal of Botany*, **46**, 22–5.
Durschner, U. (1983) Pilzliche Endoparasiten an beweglichen Nematodenstadien. *Mitteilungen aus der Biologischen Bundesanstalt für Land- und Forstwirtschaft*, **217**, 1–83.
Dwivedi, R.S. and Garrett, S.D. (1968) Fungal competition in agar plate colonization from soil inocula. *Transactions of the British Mycological Society*, **51**, 95–101.
Eamus, D. and Jennings, D.H. (1984) Determination of water, solute and turgor potentials of mycelium of various basidiomycetes causing wood decay. *Journal of Experimental Botany*, **35**, 1782–6.
Eamus, D. and Jennings, D.H. (1986) Turgor and fungal growth: studies on water relations of mycelia of *Serpula lacrimans* and *Phallus impudicus*. *Transactions of the British Mycological Society*, **86**, 527–35.
Eamus, D., Thompson, W., Cairney, J.W.G and Jennings, D.H. (1985) Internal structure and hydraulic conductivity of basidiomycete translocating organs. *Journal of Experimental Botany*, **36**, 1110–16.
Ebert, P. (1958) Das *Geopyxidetum carbonariae*, eine carbophile Pilzassociation. *Zeitschrift für Pilzkunde*, **24**, 32–44.
Edgley, M. and Brown, A.D. (1983) Physiological changes induced by solute stress in *Saccharomyces cerevisiae* and *Saccharomyces rouxii*. *Journal of General Microbiology*, **129**, 3453–63.
Edgley, M. and Brown, A.D. (1978) Response of xerotolerant and non-xerotolerant yeasts to water stress. *Journal of General Microbiology*, **104**, 343–5.
Edwards, C.A. and Heath, G.W. (1963) The role of soil animals in breakdown of leaf material, in *Soil Organisms* (eds J. Doeksen and J. van der Drift), North Holland Publishing Company, Amsterdam, pp. 76–84.
Egger, K.N. (1986) Substrate hydrolysis patterns of post-fire ascomycetes (Pezizales). *Mycologia*, **78**, 771–80.
Egger, K.N. and Paden, J.W. (1986a) Pathogenicity of post-fire ascomycetes (Pezizales) on seeds and germinants of lodgepole pine. *Canadian Journal of Botany*, **64**, 2368–71.
Egger, K.N. and Paden, J.W. (1986b) Biotrophic associations between lodgepole pine seedlings and post-fire ascomycetes (Pezizales) in monoxenic culture. *Canadian Journal of Botany*, **64**, 2719–25.
Ek, M. and Eriksson, K.E. (1980) Utilisation of the white-rot fungus *Sporotrichum pulverulentum* for water purification and protein produc-

tion on mixed lignocellulosic wastes. *Biotechnology and Bioengineering*, **22**, 2273–84.

El-Abyad, M.S.H. and Webster, J. (1968a) Studies on pyrophilous Discomycetes. I. Comparative physiological studies. *Transactions of the British Mycological Society*, **51**, 353–67.

El-Abyad, M.S.H. and Webster, J. (1968b) Studies on pyrophilous Discomycetes. II. Competition. *Transactions of the British Mycological Society*, **51**, 369–75.

Elad, Y., Chet, I. and Henis, Y. (1982) Degradation of plant pathogenic fungi by *Trichoderma harzianum*. *Canadian Journal of Microbiology*, **28**, 719–25.

Elad, Y., Barak, R., Chet, I. and Henis, Y. (1983a) Ultrastructure studies of the interaction between *Trichoderma* spp. and plant pathogenic fungi. *Phytopathologische Zeitschift*, **107**, 168–75.

Elad, Y., Chet, I., Boyle, P. and Henis, Y. (1983b) Parasitism of *Trichoderma* spp. on *Rhizoctonia solani* and *Sclerotium rolfsii* – scanning EM and fluorescence microscopy. *Phytopathology*, **73**, 85–8.

Eliasson, U. and Lundqvist, N. (1979) Fimicolous myxomycetes. *Botanisker Notiser*, **132**, 551–68.

Ellis, D.H. (1980) Thermophilic fungi isolated from a heated aquatic habitat. *Mycologia*, **72**, 1030–3.

Ellis, M.B. and Ellis, J.P. (1988) *Microfungi on Miscellaneous Substrates*, Croom Helm, London and Sydney.

Elmholt, S. and Kjøller, A. (1987) Measurement of the length of fungal hyphae by the membrane filter technique as a method for comparing fungal occurrence in cultivated field soils. *Soil Biology and Biochemistry*, **19**, 679–82.

Emerson, R. (1968) Thermophiles, in *The Fungi An Advanced Treatise*, vol. III, (eds G.C. Ainsworth and A.S. Sussman), Academic Press, London, pp. 105–25.

Emerson, R. and Held, A.A. (1969) *Aqualinderella fermentans* gen. et sp. n., a phycomycete adapted to stagnant waters II. Isolation, cultural characteristics and gas relations. *American Journal of Botany*, **56**, 1103–20.

Emerson, R. and Natvig, D.O. (1981) Adaptation of fungi to stagnant waters, in *The Fungal Community*, (eds D.T. Wicklow and G.C. Carroll), Marcel Dekker, New York, pp. 109–28.

Emerson, R. and Weston, W.H. (1967) *Aqualinderella fermentans* gen. et sp. nov., a phycomycete adapted to stagnant waters I. Morphology and occurrence in nature. *American Journal of Botany*, **54**, 702–19.

Eren, J. and Pramer, D. (1965) The most probable number of nematode trapping fungi in soil. *Soil Science*, **99**, 285.

Eriksson, K.E. (1981) Cellulases of fungi, in *Trends in the Biology of Fermentations for Fuels and Chemicals*, (ed. A. Hollaender), Plenum, New York and London, pp. 19–32.

Eriksson, K.E. (1985) Swedish developments in biotechnology related to the pulp and paper industry. *Tappi Journal*, **68**, 46–55.

Eriksson, K.E. and Larsson, K. (1975) Fermentation of waste mechanical fibers from a newsprint mill by the rot fungus *Sporotrichum pulverulentum*. *Biotechnology and Bioengineering*, **17**, 327–48.

Eriksson, K.E., Grunewald, A. and Vallander, L. (1980) Studies of growth conditions in wood for three white-rot fungi and their cellulase-less mutants. *Biotechnology and Bioengineering*, **22**, 363–76.

Eriksson, K.E., Pettersson, B. and Westermark, U. (1975) Enzymic mechanisms of cellulose degradation caused by the rot fungus *Sporotrichum pulverulentum*, in *Biological Transformation of Wood by Microorganisms*, (ed. W. Liese), Springer-Verlag, New York, pp. 143–52

Eslyn, W.E., Bultman, J.D. and Jurd, L. (1981) Wood decay inhibition by tropical hardwood extractives and related compounds. *Phytopathology*, **71**, 521–4.

Eslyn, W.E., Kirk, T.K. and Effland, M.J. (1975) Changes in the chemical composition of wood caused by six soft-rot fungi. *Phytopathology*, **65**, 473–6.

Esser, A.F. (1979) Physical chemistry of thermostable membranes in *Strategies of Microbial Life in Extreme Environments* (ed. M. Shilo) Verlag Chemie, Berlin, pp. 433–54.

Etheridge, D.E. (1958) The effect on variation in decay of moisture content and rate of growth in subalpine spruce. *Canadian Journal of Botany*, **36**, 187–206.

Etheridge, D.E. (1962) Selective action of fungus-inhibiting properties of balsam fir heartwood. *Canadian Journal of Botany*, **40**, 1459–62.

Evans, H.C. (1971) Thermophilous fungi of coal spoil tips. II. Occurrence distribution and temperature relationships. *Transactions of the British Mycological Society*, **57**, 255–66.

Eveleigh, D.E. (1985) Trichoderma, in *Biology of Industrial Microorganisms*, (eds. A.L. Demain and N.A. Solomon) Benjamin Cummings, Redwood City, CA, pp. 487–509.

Fell, J.W. and Master, I.M. (1975). Phycomycetes (*Phytophthora* spp. nov. and *Pythium* sp. nov.) associated with degrading mangrove (*Rhizophora mangle*) leaves. *Canadian Journal of Botany*, **53**, 2908–22.

Fellner, R. (1988) Effects of acid deposition on the ecotrophic stability of mountain forest ecosystems in Central Europe (Czechoslovakia), in *Ectomycorrhizas and Acid Rain* (eds A.E. Jansen, D. Dighton and A.H.M. Bresser) Berg en Dal, Bilthoven, pp. 116–121.

Fergus, C.L. (1969) The cellulolytic activity of thermophilic fungi and actinomycetes. *Mycologia*, **61**, 120–9.

Field, J.I. (1983) Physiological and ecological investigations on aero-aquatic hypomycetes. PhD thesis, University of Exeter.

Field, J.I. and Webster, J. (1977) Traps of predacious fungi attract

nematodes. *Transactions of the British Mycological Society*, **68**, 467–9.
Field, J.I. and Webster, J. (1983) Anaerobic survival of aquatic fungi. *Transactions of the British Mycological Society*, **81**, 365–9.
Field, J.I. and Webster, J. (1985) Effects of sulphide on survival of aero-aquatic and aquatic hypomycetes. *Transactions of the British Mycological Society*, **85**, 193–9.
Filer, T.H. (1965) Parasitic aspects of a fairy ring fungus, *Marasmius oreades*. *Phytopathology*, **55**, 1132–4.
Filip, Z., Haider, K., Beutelspacher, H. and Martin, J.P. (1974) Comparisons of IR-spectra from melanins of microscopic soil fungi, humic acids and model phenol polymers. *Geoderma*, **11**, 37–52.
Findlay, S.E.G. and Arsuffi, T.L. (1989) Microbial growth and detritus transformations during decomposition of leaf litter in a stream. *Freshwater Biology*, **21**, 261–9.
Findlay, W.P.K. (1950) The resistance of wood-rotting to desiccation. *Forestry*, **23**, 112–5.
Findlay, W.P.K. and Badcock, E.C. (1954) Survival of dry rot fungi in air-dry wood. *Timber Technology*, **62**, 137–8.
Fisher, P.J. (1977) New methods of detecting and studying the saprophytic behaviour of aero-aquatic hyphomycetes from stagnant water. *Transactions of the British Mycological Society*, **68**, 407–11.
Fisher, P.J. (1978) Survival of aero-aquatic hyphomycetes on land. *Transactions of the British Mycological Society*, **71**, 419–23.
Fisher, P.J. (1979). Colonization of freshly ascissed and decaying leaves by aero-aquatic hyphomycetes. *Transactions of the British Mycological Society*, 73, 99–102.
Fisher, P.J. and Anson, A.E. (1983) Antifungal effects of *Massarina aquatica* growing on oak wood. *Transactions of the British Mycological Society*, **81**, 523–7.
Fisher, P.J. and Petrini, O. (1990) A comparative study of fungal endophytes in xylem and bark of *Alnus* species in England and Switzerland. *Mycological Research*, **94**, 313–19.
Fisher, P.J. and Webster, J. (1979) Effect of oxygen and carbon dioxide on growth of four aero-aquatic hyphomycetes. *Transactions of the British Mycological Society*, **72**, 57–61.
Fisher, P.J. and Webster, J. (1981) Ecological studies on aero-aquatic hyphomycetes, in *The Fungal Community: its Organisation and Role in the Ecosystem*, (eds D.T. Wicklow and G.C. Carroll), Marcel Dekker, New York, pp. 709–30.
Fisher, P.J. Petrini, O. and Webster, J. (1991) Aquatic hyphomycetes and other fungi in living aquatic and terrestrial roots of *Alnus glutinosa*. *Mycological Research*, **95**, 57–61.
Fisher, R.F. (1977) Nitrogen and phosphorus mobilization by the fairy ring fungus *Marasmius oreades*. *Soil Biology and Biochemistry*, **9**, 239–41.

Fitter, A.H. and Hay, R.K.M. (1981) *Environmental Physiology of Plants*, Academic Press, London.

Flanagan, P.W. (1981) Fungal taxa, physiological groups and biomass: a comparison between ecosystems, in *The Fungal Community*, (eds D.T. Wicklow and G.C. Carroll), Marcel Dekker, New York, pp. 569–92.

Flanagan, P.W. and Scarborough, A.M. (1974) Physiological groups of decomposer fungi on tundra plant remains, in *Soil Organisms and Decomposition in Tundra*, (eds A.J. Holding, O.W. Heal, S.F. Maclean and P.W. Flannagan), Tundra Biome Steering Committee, Stockholm, pp. 159–81.

Flannigan, B. (1969) Microflora of dried barley grains. *Transactions of the British Mycological Society*, **53**, 371–9.

Flannigan, B. (1970a) Comparison of seed-borne mycofloras of barley, oats and wheat. *Transactions of the British Mycological Society*, **55**, 267–76.

Flannigan, B. (1970b) Degradation of arabinoxylan and carboxymethyl cellulose by fungi isolated from barley kernels. *Transactions of the British Mycological Society*, **55**, 277–81.

Flannigan, B. and Bana, M.S.O. (1980) Growth and enzyme production in aspergilli which cause deterioration in stored grain, in *Biodeterioration* (Proceedings of the 4th International Symposium Berlin), (eds T.A. Oxley, D. Allsopp and G. Becker), Pitman, London, pp. 229–36.

Flannigan, B. and Sellars, P.N. (1972) Activities of thermophilous fungi from barley kernels against arabinoxylan and carboxymethyl cellulose. *Transactions of the British Mycological Society*, **58**, 338–41.

Fleming, L.V. (1983) Succession of mycorrhizal fungi on birch: infection of seedlings planted around mature trees. *Plant and Soil*, **71**, 263–7.

Flodin, K. and Andersson, J. (1977) Studies on volatile compounds from *Pinus silvestris* and their effect on wood decomposing fungi. I. Identification of volatile compounds from fresh and heat dried wood. *European Journal of Forest Pathology*, **7**, 282–7.

Flodin, K. and Fries, N. (1978) Studies on volatile compounds from *Pinus silvestris* and their effect on wood decomposing fungi. II. Effects of some volatile compounds on fungal growth. *European Journal of Forest Pathology*, **8**, 300–10.

Floyd, R.A. and Ohlrogge, A.J. (1970) Gel formation on nodal root surfaces of *Zea mays*. *Plant and Soil*, **33**, 331–43.

Fogarty, W.H. and Kelly, C.T. (1979) Starch-degrading enzymes of microbial origin, in *Progress in Industrial Microbiology*, vol. 15, (ed. M.J. Bull), Elsevier, Amsterdam, pp. 87–150.

Fogel, R. (1985) Roots as primary producers in below ground ecosystems, in *Ecological Interactions in Soil*, (eds A.H. Fitter, D. Atkinson, D.J. Read and M.B. Usher), Blackwell, Oxford, pp. 23–36.

Fokkema, N.J. (1971) The effect of pollen in the phyllosphere of rye on colonization by saprophytic fungi and on infection by *Helminthosporium sativum*. *Netherlands Journal of Plant Pathology*, **77**, Suppl. 1, 1–60.

Fokkema, N.J. (1973) The role of saprophytic fungi in antagonism against *Drechslera sorokiniana* (*Helminthosporium sativum*) on agar plates and on rye leaves with pollen. *Physiological Plant Pathology*, **3**, 195–205.

Fokkema, N.J., Hauter, J.G., Kosterman, Y.J.C. and Nelis, A.L. (1979) Manipulation of yeasts on field grown leaves and their antagonistic effects on *Cochliobolus sativus* and *Septoria nodorum*. *Transactions of the British Mycological Society*, **72**, 19–29.

Foley, M.F. and Deacon, J.W. (1985) Isolation of *Pythium oligandrum* and other necrotrophic mycoparasites from soil. *Transactions of the British Mycological Society*, **85**, 631–9.

Ford, E.D., Mason, P.A. and Pelham, J. (1980) Spatial patterns of sporophore distribution around a young birch tree in three successive years. *Transactions of the British Mycological Society*, **75**, 287–96.

Forster, G.F. (1977) Effect of leaf surface wax on the deposition of airborne propagules. *Transactions of the British Mycological Society*, **68**, 245–50.

Foster, R.C. (1986) The ultrastructure of the rhizoplane and rhizosphere. *Annual Review of Phytopathology*, **24**, 211–34.

Foster, R.C. (1988) Microenvironments of soil micro-organisms. *Biology and Fertility of Soils*, **6**, 189–203.

Foster, R.C. and Rovira, A.D. (1973) The rhizosphere of wheat roots studied by electron microscopy of ultra-thin sections. *Bulletins from the Ecological Research Committee (Stockholm)*, **17**, 93–102.

Foster, R.C., Rovira, A.D. and Cook, T.W. (1983) *Ultrastructure of the Root Soil Interface*, American Phytopathological Society, St Paul, MN.

Fox, F.M. (1983) Role of basidiospores as inocula of mycorrhizal fungi of birch. *Plant and Soil*, **71**, 269–73.

Fox, F.M. (1986) Groupings of ectomycorrhizal fungi of birch and pine, based on establishment of mycorrhizas on seedlings from spores in unsterile soils. *Transactions of the British Mycological Society*, **87**, 371–80.

Frankland, J.C. (1966) Succession of fungi on decaying petioles of *Pteridium aquilinum*. *Journal of Ecology*, **54**, 41–63.

Frankland, J.C. (1969) Fungal decomposition of bracken petioles. *Journal of Ecology*, **57**, 25–36.

Frankland, J.C. (1974) Decomposition of lower plants, in *Biology of Plant Litter Decomposition*, (eds C.H. Dickinson and G.J.F. Pugh), Academic Press, London, pp. 30–6.

Frankland, J.C. (1975) Fungal decomposition of leaf litter in a deciduous wood, in *Biodegradation et Humification*, (eds G.K. Kilbertus, O. Reisinger, A. Mourey and J.A. Cancella da Fonseca), Pierron, Sarreguemines, pp. 33–40.

Frankland, J.C. (1981) Mechanisms in fungal successions, in *The Fungal Community* (eds D.T. Wicklow and G.C. Carroll), Marcel Dekker, New York, pp. 403–26.

Frankland, J.C. (1982) Biomass and nutrient cycling by decomposer basidiomycetes, in *Decomposer Basidiomycetes: their Biology and Ecology*, (eds J.C. Frankland, J.N. Hedger and M.J. Swift), *British Mycological Society Symposium 4*, Cambridge University Press, Cambridge, pp. 241–61.

Frankland, J.C. (1984) Autecology and the mycelium of a woodland litter decomposer, in *The Ecology and Physiology of the Fungal Mycelium*, (eds D.H. Jennings and A.D.M. Rayner), *British Mycological Society Symposium 8*, Cambridge University Press, Cambridge, pp. 241–60.

Frankland, J.C., Bailey, A.D., Gray, T.R.G and Holland, A.A. (1981) Development of an immunological technique for estimating mycelial biomass of *Mycena galopus* in leaf litter. *Soil Biology and Biochemistry*, **10**, 323–33.

Friedrich, K. (1940) Untersuchungen zur Ökologie der hoheren Pilze. *Pflanzenforschung*, **22**, 1–52.

Fries, L. (1956) Studies in the physiology of *Coprinus*. II. Influence of pH, metal factors and temperature. *Svensk Botanisk Tidskrift*, **50**, 47–96.

Fries, N. (1961) The growth-promoting activity of some aliphatic aldehydes on fungi. *Svensk Botanisk Tidskrift*, **55**, 1–16.

Fries, N. (1973) Effects of volatile organic compounds on the growth and development of fungi. *Transactions of the British Mycological Society*, **60**, 1–21.

Fries, N. (1984) Spore germination in the higher basidiomycetes. *Proceedings of the Indian Academy of Sciences (Plant Sciences)*, **93**, 205–22.

Fries, N. (1987a) Somatic incompatibility and field distribution of the ectomycorrhizal fungus *Suillus luteus* (Boletaceae). *New Phytologist*, **197**, 735–9.

Fries, N. (1987b) Ecological and evolutionary aspects of spore germination in the higher basidiomycetes. *Transactions of the British Mycological Society*, **88**, 1–7.

Fries, N. and Birraux, D. (1980) Spore germination in *Hebeloma* stimulated by living plant roots. *Experientia*, **36**, 1056–7.

Fries, N. and Swedjemark, G. (1986) Specific effects of tree roots on spore germination in the ectomycorrhizal fungus *Hebeloma mesophaeum* (Agaricales), in: *Physiological and Genetical Aspects of Mycorrhizae*, (eds V. Gianinazzi-Pearson and S. Gianinazzi), INRA, Paris, pp. 725–31.

Fries, N., Bardet, M. and Serck-Hanssen, K. (1985) Growth of ectomycorrhizal fungi stimulated by lipids from a pine root exudate. *Plant and Soil*, **86**, 287–90.

Fries, N., Sereck-Hanssen, K., Dimberg, H. and Theander, O. (1990)

Abietic acid, an activator of basidiospore germination in ectomycorrhizal species of the genus *Suillus* (Boletaceae). *Experimental Mycology*, **11**, 360–3.

Fuller, M.S. and Poyton, R.O. (1964) A new technique for the isolation of aquatic fungi. *Bioscience*, **14**, 45–6.

Gadgil, P.D. (1965) Distribution of fungi on living roots of certain Gramineae and the effect of root decomposition on soil structure. *Plant and Soil*, **22**, 239–59.

Gams, W. (1988) A contribution to the knowledge of nematophagous species of *Verticillium*, *Netherlands Journal of Plant Pathology*, **94**, 123–48.

Garbary, D.J. and Gautam, A. (1989) The *Ascophyllum*, *Polysiphonia*, *Mycosphaerella* symbiosis. I. Population ecology of *Mycosphaerella* from Nova Scotia. *Botanica marina*, **32**, 181–6.

Garbaye, J., Kabre, A., Letacon, F., Mousain, D. and Piou, D. (1979) Fertilization minerale et fructification des champignons supérieurs en hêtraie. *Annales des Sciences Forestières*, **36**, 151–64.

Gardner, D.E. and Hendrix, F.F. (1973) Carbon dioxide and oxygen concentrations in relation to survival and saprophytic growth of *Pythium irregulare* and *Pythium vexans* in soil. *Canadian Journal of Botany*, **51**, 1593–8.

Gardner, J.N. and Malajczuk, N. (1985) Succession of ectomycorrhizal fungi associated with eucalypts on rehabilitated bauxite mines in South Western Australia, in *Proceedings of the 6th North American Conference on Mycorrhizae*, (ed. R. Molina). Oregon State University Press, Corvallis, OR, p. 265.

Garrett, S.D. (1950) Ecology of the root-inhabiting fungi. *Biological Reviews, Cambridge*, 220–54.

Garrett, S.D. (1951) Ecological groups of soil fungi: a survey of substrate relationships. *New Phytologist*, **50**, 149–66.

Garrett, S.D. (1956) *Biology of Root-Infecting Fungi*, Cambridge University Press, Cambridge.

Garrett, S.D. (1963) *Soil Fungi and Soil Fertility*, Pergamon Press, Oxford.

Garrett, S.D. (1966) Cellulose decomposing ability of some cereal foot-rot fungi in relation to their saprophytic survival. *Transactions of the British Mycological Society*, **49**, 57–68.

Gessner, M.O., Thomas, M., Jean-Louis, A.-M. and Chauvet, E. (1993) Stable successional patterns of aquatic hypomycetes on leaves decaying in a summer cool stream. *Mycological Research*, **97**, 163–72.

Gibson, F. and Deacon, J.W. (1988) Experimental study of establishment of ectomycorrhizas in different regions of birch root systems. *Transactions of the British Mycological Society*, **91**, 239–51.

Gibson, F. and Deacon, J.W. (1990) Establishment of ectomycorrhizas in aseptic culture: effect of glucose, nitrogen and phosphorus in relation to

successions. *Mycological Research*, **94**, 166–72.
Gilbert, R.G., Menzies, J.D. and Griebel, G.E. (1969) The influences of volatiles from alfalfa upon growth and survival of microorganisms. *Phytopathology*, **59**, 992–5.
Gillespie, J., Latter, P.M. and Widden, P. (1988) Cellulolysis of cotton by fungi in three upland soils, in *Cotton Strip Assay: An Index of Decomposition in Soils*, (eds A.F. Harrison, P.M. Latter and D.W.H. Walton), Institute of Terrestrial Ecology, Grange-over-Sands, UK, pp. 60–7.
Gilliver, K. (1947) The effect of plant extracts on the germination of the conidia of *Venturia inequalis*. *Annals of Applied Biology*, **43**, 136–43.
Giltrap, N.J. (1982) Production of polyphenol oxidases by ectomycorrhizal fungi with special reference to *Lactarius* spp. *Transactions of the British Mycological Society*, **78**, 75–81.
Ginns, J.H. (1968) *Rhizina undulata* pathogenic on Douglas-fir seedlings in Western North America. *Plant Disease Reporter*, **52**, 579–80.
Ginns, J.H. (1974) *Rhizina* root rot: severity and distribution in British Columbia. *Canadian Journal of Forest Research*, **4**, 143–6.
Ginns, J.H. and Malloch, D. (1977) *Halocyphina*, a marine basidiomycete (Aphyllophorales). *Mycologia*, **69**, 53–8.
Giuma, A.Y. and Cooke, R.C. (1971) Nematotoxin production by *Nematoctonus haptocladus* and *N. concurrens*. *Transactions of the British Mycological Society*, **56**, 89–94.
Giuma, A.Y. and Cooke, R.C. (1972) Some endozoic fungi parasitic on soil nematodes. *Transactions of the British Mycological Society*, **59**, 213–18.
Giuma, A.Y. and Cooke, R.C. (1974) Potential of *Nematoctonus* conidia for biological control of soil-borne phytonematodes. *Soil Biology and Biochemistry*, **6**, 217–20.
Giuma, A.Y., Hackett, A.M. and Cooke, R.C. (1973) Thermostable nematotoxins produced by germinating conidia of some endozoic fungi. *Transactions of the British Mycological Society*, **60**, 49–56.
Gladders, P. and Coley-Smith, H.J.R. (1978) Interactions between *Rhizoctonia tuliparum* and soil microorganisms. *Annals of Applied Biology*, **89**, 131.
Glasare, P. (1970) Volatile compounds from *Pinus silvestris* stimulating the growth of wood-rotting fungi. *Archiv für Mikrobiologie*, **72**, 333–43.
Gleason, F.H. (1968) Nutritional comparisons in the Leptomitales. *American Journal of Botany*, **55**, 1003–10.
Glen, A.I., Hutchinson, S.A. and McCorkindale, N.J. (1966) Hexa-1–3–5-triyne a metabolite of *Fomes annosus*. *Tetrahedron Letters*, **35**, 4223–5.
Gochenaur, S.E. (1975) Distribution patterns of mesophilous and thermophilous microfungi in two Bahamian soils. *Mycopathologia*, **57**, 155–64.

Gochenaur, S.E. (1981) Responses of soil fungal communities to disturbance, in *The Fungal Community: its Organisation and Role in the Ecosystem*, (eds D.T: Wicklow and G.C. Carroll) Marcel Dekker, New York, pp. 459–79.
Godfrey, B.E.S. (1974) Phylloplane mycoflora of bracken. *Pteridium aquilinum*. *Transactions of the British Mycological Society*, **62**, 305–11
Godfrey, B.E.S. (1983) Growth of two terrestrial microfungi on submerged alder leaves. *Transactions of the British Mycological Society*, **81**, 418–21.
Godtfredson, W.O. and Vangedal, S. (1965) Trichodermin a new sesquiterpene antibiotic. *Acta Chemica Scandinavica*, **19**, 1088–102.
Gold, H.S. (1959) Distribution of some lignicolous ascomycetes and fungi imperfecti in an estuary. *Journal of the Elisha Mitchell Scientific Society*, **75**, 25–8.
Goldstein, J.L. and Swain, T. (1965) The inhibition of enzymes by tannins. *Phytochemistry*, **4**, 185–92.
Gönczol, J. (1989) Longitudinal distribution patterns of aquatic hyphomycetes in a mountain stream in Hungary. Experiments with leaf packs. *Nova Hedwigia*, **48**, 391–404.
Gooday, G.W. (1990) The ecology of chitin degradation. *Advances in Microbial Ecology*, **11**, 387–430.
Gooday, G.W., Green, D. and Shaw, G. (1974) Sporopollenin formation in the ascospore wall of *Neurospora crassa*. *Archives for Microbiology*, **101**, 145–51.
Gooday, G.W., Fawcett, P., Green, D. and Shaw, G. (1973) The formation fungal sporopollenin in the zygospore wall of *Mucor mucedo*: a role for the sexual carotenogenesis in the Mucorales. *Journal of General Microbiology*, **74**, 233–9.
Goos, R.D. (1987) Fungi with a twist: the helicosporous hyphomycetes. *Mycologia*, **79**, 1–22.
Gourbière, F., Pepin, R. and Bernillon, D. (1987) Microscopie de la microflora des aiguilles de sapin (*Abies alba*). III. *Marasmius androsaceus*. *Canadian Journal of Botany*, **65**, 131–6.
Grant, I.W.B., Blackadder, E.S., Greenberg, M. and Blyth, W. (1976) Extrinsic allergic alveolitis in Scottish malt workers. *British Medical Journal*, **1**, 490.
Grant, W.D. (1976) Microbial degradation of condensed tannins. *Science*, **193**, 1137–9.
Gray, N.F. (1983) Ecology of nematophagous fungi: distribution and habitat. *Annals of Applied Biology*, **102**, 501–9.
Gray, N.F. (1984) The effect of fungal parasitism and predation on the population dynamics of nematodes in the activated sludge process. *Annals of Applied Biology*, **104**, 143–50.
Gray, N.F. (1985) Ecology of nematophagous fungi: effect of soil mois-

ture, organic matter, pH and nematode density on distribution. *Soil Biology and Biochemistry*, **17**, 499–507.
Gray, N.F. (1987) Nematophagous fungi with particular reference to their ecology. *Biological Reviews*, **62**, 245–304.
Gray, N.F. (1988) Fungi attacking vermiform nematodes, in *Diseases of Nematodes*, II, (eds G.O. Poinar and H.B. Jansson) CRC Press, Boca Raton, Fl., pp. 3–37.
Gray, N.F. and Bailey, F. (1985) Ecology of nematophagous fungi: vertical distribution in a deciduous woodland. *Plant and Soil*, **86**, 217–23.
Greaves, M.P. and Darbyshire, J.F. (1972) The ultrastructure of the mucilaginous layer on plant roots. *Soil Biology and Biochemistry*, **4**, 443–9.
Greaves, M.P. and Webley, D.M. (1965) A study of the breakdown of organic phosphates by microorganisms from the root region of certain pasture grasses. *Journal Applied Bacteriology*, **28**, 454–65.
Greenwood, D.J. (1961) The effect of oxygen concentration on the decomposition of organic materials in soil. *Plant and Soil*, **14**, 360–76.
Greenwood, D.J. and Berry, G. (1962) Aerobic respiration in soil crumbs. *Nature*, **195**, 161–3.
Gregory, P.H. (1952) Presidential address. Fungus spores. *Transactions of the British Mycological Society*, **35**, 1–18.
Gregory, P.H. (1973) *The Microbiology of the Atmosphere*, 2nd edn, Leonard Hill Books, Aylesbury, UK.
Gregory, P.H. (1982) Fairy rings free and tethered. *Bulletin of the British Mycological Society*, **16**, 161–3.
Gregory, P.H. (1984) The fungal mycelium – an historical perspective, in *The Ecology and Physiology of the Fungal Mycelium*, (eds D.H. Jennings and A.D.M. Rayner), *British Mycological Society Symposium 8*, Cambridge University Press, Cambridge, pp. 1–22.
Gregory, P.H. and Hirst, J.M. (1952) Possible role of basidiospores as airborne allergens. *Nature*, **170**, 414.
Gregory, P.H., Lacey, M.E., Festenstein, G.M. and Skinner, F.A. (1963) Microbial and biochemical changes during the moulding of hay. *Journal of General Microbiology*, **33**, 147–74.
Gremmen, J. (1971) *Rhizina undulata* – a review of research in the Netherlands. *European Journal of Forest Pathology*, **1**, 1–6.
Griffin, D.M. (1960) Fungal colonization of sterile hair in contact with soil. *Transactions of the British Mycological Society*, **43**, 583–96.
Griffin, D.M. (1963) Soil moisture and the ecology of soil fungi. *Biological Reviews*, **38**, 141–66.
Griffin, D.M. (1965) The interaction of hydrogen ion, carbon dioxide and potassium ion in controlling the formation of resistant sporangia in *Blastocladiella emersonii*. *Journal of General Microbiology*, **40**, 13–28.
Griffin, D.M. (1966) Soil physical factors and the ecology of fungi. IV.

Influence of the soil atmosphere. *Transactions of the British Mycological Society*, **49**, 115–20.

Griffin, D.M. (1972) *Ecology of Soil Fungi*, Chapman & Hall, London.

Griffin, D.M. (1977) Water potential and wood decay fungi. *Annual Review of Phytopathology*, **15**, 319–29.

Griffin, D.M. (1981a) Water potential as a selective factor in the microbiology of soils, in *Water Potential Relations in Soil Microbiology. Soil Science Society of America Special Publication* 9, pp. 141–51.

Griffin, D.M. (1981b) Water and microbial stress. *Advances in Microbial Ecology*, **5**, 91–136.

Griffin, D.M. and Nair, N.G. (1968) Growth of *Sclerotium rolfsii* at different concentrations of oxygen and carbon dioxide. *Journal of Experimental Botany*, **19**, 812–16.

Griffin, G.J. (1969) *Fusarium oxysporum* and *Aspergillus flavus* spore germination in the rhizosphere of peanut. *Phytopathology*, **5**, 1214–22.

Griffin, G.J., Hale, M.G. and Shay, F.J. (1976) Nature and quantity of sloughed organic matter produced by roots of axenic peanut plants. *Soil Biology and Biochemistry*, **8**, 29–32.

Griffith, N.T. and Barnett, H.L. (1967) Mycoparasitism by basidiomycetes in culture. *Mycologia*, **59**, 149–54.

Griffiths, R.P. and Caldwell, B.A. (1992) Mycorrhizal mat communities in forest soils, in *Mycorrhizas in Ecosystems*, (eds D.J. Read, D.H. Lewis, A.H. Fitter and I.J. Alexander), CAB International, Wallingford, UK, pp. 98–105.

Grigorova, R. and Norris, J.R. (1990) *Methods in Microbiology*, vol. 22, Academic Press, New York.

Grime, J.P. (1977) Evidence for the existence of three primary strategies in plants and its relevance to ecological and evolutionary theory. *American Naturalist*, **111**, 1169–94.

Grime, J.P. (1979) *Plant Strategies and Vegetation Processes*, John Wiley & Sons, Chichester.

Gum, E.K. and Brown, R.D. (1977) Comparison of four purified extracellular 1, 4 β-D glucan cellobiohydrolase enzymes from *Trichoderma viride*. *Biochimica et Biophysica Acta*, **492**, 225–31.

Gunasekera, S.A. and Webster, J. (1983) Inhibitors of aquatic and aero-aquatic hyphomycetes in pine and oak wood. *Transactions of the British Mycological Society*, **80**, 121–5.

Gunasekera, S.A., Webster, J. and Legg, C.J. (1983) Effect of nitrate and phosphate on weight losses of pine and oak wood caused by aquatic and aero-aquatic hyphomycetes. *Transactions of the British Mycological Society*, **80**, 507–14.

Gundersen, K. (1961) Growth of *Fomes annosus* under reduced oxygen pressure and the effect of carbon dioxide. *Nature*, **190**, 649.

Gupta, A.K. and Mehrotra, R.S. (1989) Seasonal periodicity of aquatic

fungi in tanks at Kurukshetra, India. *Hydrobiologia*, **173**, 219–29.
Gupta, R.L. and Tandon, R.N. (1977) Growth inhibition of fungi by volatiles from *Streptomyces*. *Transactions of the British Mycological Society*, **68**, 438–9.
Haars, A., Chet, I. and Huttermann, A. (1981) Effects of phenolic compounds and tannin on growth and laccase activity of *Fomes annosus*. *European Journal of Forest Pathology*, **11**, 67–76.
Hagan, P.O. and Rose, A.H. (1961) A psychrophilic *Cryptococcus*. *Canadian Journal of Microbiology*, **7**, 287–94.
Haider, K. and Martin, J.P. (1967) Synthesis and transformation of phenolic compounds by *Epicoccum nigrum* in relation to humic acid formation. *Soil Science Society of America Proceedings*, **31**, 766–72.
Haider, K. and Martin, J.P. (1970) Humic acid-type phenolic polymers from *Aspergillus sydowi* culture medium, *Stachybotrys* spp. cells and autoxidized phenol mixtures. *Soil Biology and Biochemistry*, **2**, 145–56.
Haider, K. and Trojanowski, J. (1975) Decomposition of specifically ^{14}C-labelled phenols and dehydropolymers of coniferyl alcohol as models for lignin degradation by soft and white rot fungi. *Archives of Microbiology*, **105**, 33–41.
Hale, M.D. and Eaton, R.A. (1985) Oscillatory growth of fungal hyphae in wood cell walls. *Transactions of the British Mycological Society*, **84**, 277–88.
Hale, M.G., Moore, L.D. and Griffin, G.J. (1978) Interactions between non-pathogenic soil microorganisms and plants, in *Ecology of Root Pathogens* (eds Y.R. Dommergues and S.V. Krupa) Elsevier, New York, pp. 163–97.
Hall, G. (1987) Sterile fungi from roots of winter wheat. *Transactions of the British Mycological Society*, **89**, 447–56.
Hallenberg, N. (1990) Ultrastructure of stephanocysts and basidiospores in *Hyphoderma praetermissum*. *Mycological Research*, **94**, 1090–5.
Hallett, I.C. and Dick, M.W. (1981) Seasonal and diurnal fluctuations of Oomycete propagule numbers in the free water of a freshwater lake. *Journal of Ecology*, **69**, 671–92.
Halliwell, G. (1979) Microbial β glucanase, in *Progress in Industrial Microbiology*, vol. 15, (ed. M.J. Bull), Elsevier, Amsterdam, pp. 1–60.
Hamblin, A.P. (1981) Filter-paper method for routine measurement of field water potential. *Journal of Hydrology*, **53**, 355–60.
Hamlen, R.A., Lukezic, F.L. and Bloom, J.R. (1972) Influence of age and stage of development on neutral carbohydrate components in root exudases from alfalfa plants grown in a gnotobiotic environment. *Canadian Journal of Plant Science*, **52**, 633–42.
Hammond, D.P. and Smith, S.N. (1986) Lipid composition of a psychrophilic, a mesophilic and a thermophilic *Mucor* species. *Transactions of the British Mycological Society*, **86**, 551–60.

Hanlon, R.D.G. (1981) Influence of grazing by Collembola on the activity of senescent fungal colonies grown on media of different nutrient concentration. *Oikes*, **36**, 362–7.

Hanlon, R.D.G. and Anderson, J.M. (1979) The effects of Collembola grazing on microbiological activity in decomposing leaf litter. *Oecologia*, **38**, 93–100.

Harborne, J.B., Ingham, J.L., King, L. and Payne, M. (1976) The isopentenyl isoflavone luteone as a pre-infectional antifungal agent in the genus *Lupinus*. *Phytochemistry*, **15**, 1485–7.

Harley, J.L. (1969) *The Biology of Mycorrhiza*, Leonard Hill, London.

Harley, J.L. (1989) The significance of mycorrhiza. *Mycological Research*, **92**, 129–39.

Harley, J.L. and Harley, E.L. (1987) A check list of mycorrhiza in the British flora. *New Phytologist*, **105**, (Supplement), 1–102.

Harley, J.L. and Smith, S.E. (1983) *Mycorrhizal Symbiosis*, Academic Press, London.

Harley, J.L. and Waid, J.S. (1955a) A method of studying active mycelia on living roots and other surfaces in the soil. *Transactions of the British Mycological Society*, **38**, 104–18.

Harley, J.L. and Waid, J.S. (1955b) The effect of light upon the roots of beech and its surface population. *Plant and Soil*, **7**, 96–112.

Harper, J.E. (1962) A comparative ecological study of the fungi on rabbit dung. PhD thesis, University of Sheffield.

Harper, J.E. and Webster, J. (1964) An experimental analysis of the coprohilous fungus succession. *Transactions of the British Mycological Society*, **47**, 511–30.

Harrison, J.L. and Jones, E.B.G. (1971) Salinity tolerance of *Saprolegnia parasitica* Coker. *Mycopathologia et Mycologia Applicata*, **43**, 297–307.

Harrison, J.L. and Jones, E.B.G. (1974) Patterns of salinity tolerance displayed by the lower fungi. *Veröffentlichüngen des Instituts für Meeresforschung in Bremerhaven*, Supplement, 5, 197–220.

Harrison, J.L. and Jones, E.B.G. (1975) The effect of salinity on sexual and asexual sporulation of members of the Saprolegniaceae. *Transactions of the British Mycological Society*, **65**, 389–94.

Harrison, R. and Lunt, G.G. (1980) *Biological Membranes Their Structure and Function*, 2nd ed., Blackie, Glasgow.

Harrower, K.M. and Nagy, L.A. (1979) Effects of nutrients and water stress on growth and sporulation of coprophilous fungi. *Transactions of the British Mycological Society*, **72**, 459–62.

Hart, J.H. (1981) Role of phytostilbenes in decay and disease resistance. *Annual Review of Phytopathology*, **19**, 437–58.

Hart, J.H. and Hillis, W.E. (1972) Inhibition of wood rotting fungi by ellagitannins in the heartwood of *Quercus alba*. *Phytopathology*, **62**, 620–6.

Hawksworth, D.L. (1981) A survey of the fungicolous conidial fungi, in *Biology of Conidial Fungi*, vol. 1, (eds G.T. Cole and B. Kendrick) Academic Press, London, pp. 171–244.

Hawksworth, D.L. (1991) Presidential Address 1990. The fungal dimension of biodiversity: magnitude, significance and conservation. *Mycological Research*, **95**, 641–55.

Hawksworth, D.L., Sutton, B.C. and Ainsworth, G.C. (1983) *Ainsworth & Bisby's Dictionary of the Fungi*, 7th edn, Commonwealth Mycological Institute, Kew.

Hayes, A.J. (1982) Phylloplane micro-organisms of *Rosa* cv Piccadilly following infection by *Diplocarpon rosae*. *Transactions of the British Mycological Society*, **79**, 311–19.

Head, G.C. (1973) Shedding of roots, in *Shedding of Plant Parts*, (ed. T. Kozlowski), *Physiological Ecology*, Academic Press, New York, pp. 237–86.

Heath, G.W., Edwards, C.A. and Arnold, M.K. (1964) Some methods for assessing the activity of soil animals in the breakdown of leaves. *Pedobiologia*, **4**, 80–7.

Heath, G.W., Arnold, M.K. and Edwards, C.A. (1966) Studies in leaf litter breakdown. I. Breakdown rates of leaves of different species. *Pedobiologia*, **6**, 1–12.

Hedger, J.N. (1975) Ecology of thermophilic fungi in Indonesia, in *Biodégradation et Humification* (eds G. Kilbertus, O. Reisinger, A. Mourey and J.A. Cancella da Fonseca), Pierron, Sarreguemines, pp. 59–65.

Hedger, J.N. and Hudson, H.J. (1974) Nutritional studies of *Thermomyces laguginosus* from wheat straw compost. *Transactions of the British Mycological Society*, **62**, 129–43.

Held, A.A. (1970) Nutrition and fermentative energy metabolism of the water mold *Aqualinderella fermentans*. *Mycologia*, **62**, 339–58.

Held, A.A., Emerson, R., Fuller, M.S. and Gleason, F.H. (1969) *Blastocladia* and *Aqualinderella*, fermentative water molds with high carbon dioxide optima. *Science*, **165**, 706–9.

Helsel, E.D. and Wicklow, D.T. (1979) Decomposition of rabbit feces: role of the sciarid fly *Lycoriella mali* (Diptera: Sciaridae) in energy transformations. *Canadian Entomologist*, **111**, 213–18.

Henderson, M.E. (1960) Studies on the physiology of lignin decomposition by soil fungi, *The Ecology of Soil Fungi*, (eds D. Parkinson and J.S. Waid), Liverpool University Press, Liverpool, pp. 286–96.

Henderson, M.E.K. (1961) The metabolism of aromatic compounds related to lignin by some hyphomycetes and yeast like fungi of soil. *Journal of General Microbiology*, **26**, 155–65.

Henriksen, H.A. and Jørgensen (1952) *Fomes annosus* attack in relation to grades of thinning. An investigation on the basis of experiments.

Forstlige Forøksvaesen Denmark, **21**, 215–52.
Hering, T.F. (1965) Succession of fungi in the litter of a Lake District oakwood. *Transactions of the British Mycological Society*, **48**, 391–408.
Hering, T.F. (1967) Fungal decomposition of oak leaf litter. *Transactions of the British Mycological Society*, **50**, 267–73.
Hering, T.F. (1966) The terricolous higher fungi of four Lake District woodlands. *Transactions of the British Mycological Society*, **49**, 369–83.
Hering, T.F. (1982) Decomposing activity of basidiomycetes in forest litter, in *Decomposer Basidiomycetes: their Biology and Ecology*, (eds J.C. Frankland, J.N. Hedger and M.J. Swift), *British Mycological Symposium 4*, Cambridge University Press, Cambridge, pp. 213–25.
Hesseltine, C.W., Whitehill, A.R., Pidacks, C. *et al.*, (1953) Coprogen, a new growth factor present in dung required by *Pilobolus*. *Mycologia*, **45**, 7–19.
Highley, T.L. (1976) Hemicellulases of white and brown rot fungi in relation to host preference. *Material und Organismen*, **11**, 33–46.
Highley, T.L. (1977) Requirements for cellulose degradation by a brown-rot fungus. *Material und Organismen*, **12**, 25–36.
Highley, T.L. and Kirk, T.K. (1979) Mechanisms of wood decay and the unique features of heartrots. *Phytopathology*, **69**, 1151–9.
Highley, T.L., Bar-Lev, S.S., Kirk, T.K. and Larsen, M.J. (1983) Influence of O_2 and CO_2 on wood decay by heart-rot and sap-rot fungi. *Phytopathology*, **73**, 630–3.
Higuchi, T. (1980) Microbial degradation of dilignols as lignin models, in *Lignin Biodegradation: Microbiology, Chemistry and Potential Applications*, vol. 1, (eds T.K. Kirk, T. Higuchi and Hou-min Chang), CRC Press, Boca Raton, FL, pp. 171–94.
Hill, R.A. (1979) Barley grain microflora with special reference to *Penicillium* species. PhD thesis, Reading University, UK.
Hill, R.A. and Lacey, J. (1983) Factors determining the microflora of stored barley grain. *Annals of Applied Biology*, **102**, 467–83.
Hill, R.A. and Lacey, J. (1984) *Penicillium* species associated with barley grain in the UK. *Transactions of the British Mycological Society*, **82**, 297–303.
Hillis, W.E. (1962) *Wood Extractives and their Significance to the Pulp and Paper Industries*, Academic Press, New York and London.
Hintikka, V. (1964) Psychrophilic basidiomycetes decomposing forest litter under winter conditions. *Communicationes Instituti Forestalis Fenniae*, **59**, 1–20.
Hintikka, V. (1970) Selective effect of terpenes on wood-decomposing hymenomycetes. *Karstenia*, **11**, 28–32.
Hintikka, V. (1971a) *Mucor oblongisporus* as a psychrophilic secondary sugar fungus. *Karstenia*, **12**, 59–65.

Hintikka, V. (1971b) Tolerance of some wood-decomposing basidiomycetes to aromatic compounds related to lignin degradation. *Karstenia*, **12**, 46–52.

Hintikka, V. (1982) The colonisation of litter and wood by basidiomycetes in Finnish forests, in *Decomposer Basidiomycetes: their Biology and Ecology*, (eds J.C. Frankland, J.N. Hedger and M.J. Swift), British Mycological Society Symposium 4, Cambridge University Press, Cambridge, pp. 227–39.

Hintikka, V. and Korhonen, K. (1970) Effects of carbon dioxide on the growth of lignicolous and soil-inhabiting Hymenomycetes. *Communicationes Instituti Forestalis Fenniae*, **69**, 1–29.

Hirst, J.M. (1952) An automatic volumetric spore trap. *Annals of Applied Biology*, **39**, 257–65.

Hirst, J.M. (1965) Dispersal of soil microorganisms, in *Ecology of Soil-Borne Plant Pathogens*, (eds K.F. Baker and W.C. Snyder), J. Murray, London, pp. 69–81.

Hislop, E.C. and Cox, T.W. (1969) Effects of captan on the non-parasitic microflora of apple leaves. *Transactions of the British Mycological Society*, **52**, 223–35.

Hoch, H.C. and Provvidenti, R. (1979) Mycoparasitic relationships: cytology of the *Sphaerotheca fuliginea–Tilletiopsis* sp. interaction. *Phytopathology*, **69**, 359–62.

Hocking, A. and Norton, R.S. (1983) Natural-abundance ^{13}C nuclear magnetic resonance studies on the internal solutes of xerophilic fungi. *Journal of General Microbiology*, **129**, 2915–25.

Hocking, A.D. and Pitt, J.I. (1979) Water relations of some *Penicillium* species at 25°C. *Transactions of the British Mycological Society*, **73**, 141–5.

Hogg, B.M. (1966) Microfungi on leaves of *Fagus sylvatica*. II. Duration of survival, spore viability and cellulolytic activity. *Transactions of the British Mycological Society*, **49**, 193–204.

Hogg, B.M. and Hudson, H.J. (1966) Microfungi on leaves of *Fagus sylvatica*. I. The microfungal succession. *Transactions of the British Mycological Society*, **49**, 185–92.

Höhnk, W. (1935) Saprolegniales und Monoblepharidales aus der Umgebung Bremens, mit besonder Berucksichtigung der Oekologie der Saprolegniaceae. *Abhandlungen Naturwissenschaftlichen Verein zu Bremen*, **29**, 207–37.

Höhnk, W. (1939) Ein Beitrag zur Kenntnis der Phycomyceten des Brackwassers. *Kieler Meeresforschung*, **3**, 337–61.

Höhnk, W. (1952) Studien zur Brack- und Seewassermykologie I. *Veröffentlichen des Instituts für Meeresforschung Bremerhaven*, **1**, 115–25.

Höhnk, W. (1953) Studien zur Brack- und Seewassermykologie III. *Veröffentlichen des Instituts für Meeresforschung Bremerhaven*, **2**, 52–108.

Holter, P. (1979) Effect of dung beetles and earthworms on the disappearance of cattle dung. *Oikos*, **32**, 393–402.
Hora, F.B. (1959) Quantitative experiments on toadstool production in woods. *Transactions of the British Mycological Society*, **42**, 1–14.
Hora, T.S. and Baker, R. (1970) Volatile factor in soil fungistasis. *Nature*, **225**, 1071–2.
Hora, T.S. and Baker, R. (1972) Soil fungistasis: microflora producing a volatile inhibitor. *Transactions of the British Mycological Society*, **59**, 491–500.
Hsu, S.C. and Lockwood, J.L. (1971) Responses of fungal hyphae to soil fungistasis. *Phytopathology*, **61**, 1355–62.
Hubbard, J.P., Harman, G.E. and Hadar, Y. (1983) Effect of soil borne *Pseudomonas* spp. on the biological control agent *Trichoderma hamatum*, on pea seeds. *Phytopathology*, **73**, 655–9.
Hudson, H.J. (1962) Succession of micro-fungi on ageing leaves of *Saccharum officinarum*. *Transactions of the British Mycological Society*, **45**, 395–423.
Hudson, H.J. (1968) The ecology of fungi on plant remains above the soil. *New Phytologist*, **67**, 837–74.
Hudson, H.J. (1978) Introduction to the significance of interactions in successions in natural environments. *Annals of Applied Biology*, **89**, 155–8.
Hudson, H.J. and Webster, J. (1958) Succession of fungi on decaying stems of *Agropyron repens*. *Transactions of the British Mycological Society*, **41**, 165–77.
Hughes, G.C. (1962) Seasonal periodicity of the Saprolegniaceae in the South-Eastern United States. *Transactions of the British Mycological Society*, **45**, 519–31.
Hughes, G.C. (1974) Geographical distribution of the higher marine fungi. *Veröffentlichen des Instituts für Meeresforschung in Bremerhaven*, Supplement 5, 419–1.
Hughes, G.C. (1975) Studies of fungi in Oceans and estuaries since 1961 I. Lignicolous, caulicolous and foliicolous species. *Oceanography and Marine Biology Annual Review*, **13**, 69–180.
Hughes, G.C. (1986) Biogeography and the marine fungi, in *The Biology of Marine Fungi*, (ed. S.T. Moss), Cambridge University Press, Cambridge, pp. 275–95.
Hulme, M.A. and Shields, J.K. (1975) Antagonistic and synergistic effects for biological control of decay, in *Biological Transformation of Wood by Microorganisms*, (ed. W. Liese), Springer-Verlag, Berlin, pp. 52–63.
Humpherson-Jones, F.M. and Cooke, R.C. (1977) Induction of sclerotium formation by acidic staling compounds in *Sclerotinia sclerotiorum* and *Sclerotium rolfsii*. *Transactions of the British Mycological Society*, **68**, 413–20.

Hung, L.L. and Trappe, J.M. (1983) Growth variation between and within species of ectomycorrhizal fungi in response to pH *in vitro*. *Mycologia*, **75**, 234–41.
Hunter, K. and Rose, A.H. (1972) Lipid composition of *Saccharomyces cerevisiae* as influenced by growth temperature. *Biochemica et Biophysica Acta*, **260**, 639–53.
Hurst, J.L. and Pugh, G.L.F. (1983) Association between *Chrysosporium pannorum* and *Mucor hiemalis* in *Poa flabellata* litter. *Transactions of the British Mycological Society*, **81**, 151–3.
Hutchinson, S.A. (1971) Biological activity of volatile fungal metabolites. *Transactions of the British Mycological Society*, **57**, 185–200.
Hyde, H.A. and Williams, D.A. (1946) A daily census of *Alternaria* spores caught from the atmosphere at Cardiff in 1942 and 1943. *Transactions of the British Mycological Society*, **29**, 78–85.
Hyde, K.D. (1986) Frequency of occurrence of lignicolous marine fungi in the tropics, in *The Biology of Marine Fungi*, (ed. S.T. Moss), Cambridge University Press, Cambridge, pp. 311–22.
Hyde, K.D. (1988) Observations on the vertical distribution of marine fungi on *Rhizophora*, spp. at Kampong Danau mangrove, Brunei. *Asian Marine Biology*, **5**, 77–81.
Hyde, K.D. and Jones, E.B.G. (1988) Marine mangrove fungi. *P.S.Z.N.I. Marine Ecology*, **9**, 15–33.
Hyde, K.D. and Jones, E.B.G. (1989) Observations on ascospore morphology in marine fungi and their attachment to surfaces. *Botanica Marina*, **32**, 205–8.
Hyde, K.D., Farrant, C.A. and Jones, E.B.G. (1987) Isolation and culture of marine fungi. *Botanica Marina*, **30**, 291–303.
Hyde, K.D., Moss, S.T. and Jones, E.B.G. (1989) Attachment studies in marine fungi. *Biofouling*, **1**, 287–98.
Hynes, H.B.N. (1970) *The Ecology of Running Waters*, Liverpool University Press, Liverpool.
Ikediugwu, F.E.O. (1976) Ultrastructure of hyphal interference between *Coprinus heptemerus* and *Ascobolus crenulatus*. *Transactions of the British Mycological Society*, **66**, 281–90.
Ikediugwu, F.E.O. and Webster, J. (1970a) Antagonism between *Coprinus heptemerus* and other coprophilous fungi. *Transactions of the British Mycological Society*, **54**, 181–204.
Ikediugwu, F.E.O. and Webster, J. (1970b) Hyphal interference in a range of coprophilous fungi. *Transactions of the British Mycological Society*, **54**, 205–10.
Ikediugwu, F.E.O., Dennis, C. and Webster, J. (1970) Hyphal interference by *Peniphora gigantea* against *Heterobasidion annosum*. *Transactions of the British Mycological Society*, **54**, 307–9.
Ingleby, K., Mason, P.A., Last, F.T. and Fleming. L.V. (1990) Identifica-

tion of Ectomycorrhizas. *ITE Research Publication No 5*, Her Majesty's Stationery Office, London.
Ingold, C.T. (1966) The tetraradiate aquatic fungal spore. *Mycologia*, **58**, 43–56.
Ingold, C.T. (1971) *Fungal Spores, their Liberation and Dispersal*, Clarendon Press, Oxford.
Ingold, C.T. (1974) Growth and death of a fairy ring. *Bulletin of the British Mycological Society*, **8**, 74–5.
Ingold, C.T. (1975a) Hooker Lecture 1974. Convergent evolution in aquatic fungi: the tetraradiate spore. *Biological Journal of the Linnean Society*, **7**, 1–25.
Ingold, C.T. (1975b) *An Illustrated Guide to Aquatic and Water-borne Hyphomycetes (Fungi Imperfecti) with notes on their Biology*. Freshwater Biological Association Scientific Publication no. 30, The Ferry House, Ambleside, Cumbria, UK.
Ingold, C.T. (1976) The morphology and biology of freshwater fungi excluding phycomycetes, in *Recent Advances in Aquatic Mycology*, (ed. E.B.G. Jones), Elek Science, London, pp. 335–57.
Ingold, C.T. (1980) *Flammulina velutipes*. *Bulletin of the British Mycological Society*, **14**, 112–18.
Ingold, C.T. (1981) *Flammulina velutipes* in relation to drying and freezing. *Transactions of the British Mycological Society*, **76**, 150–2.
Ingold, C.T. (1982) Resistance of certain basidiomycetes to freezing. *Transactions of the British Mycological Society*, **79**, 554–6.
Ingold, C.T. and Zoberi, M.H. (1963) The asexual apparatus of Mucorales in relation to spore liberation. *Transactions of the British Mycological Society*, **46**, 115–34.
Ingraham, J.L. (1962) Factors affecting the lower limits of temperature for growth, in *Recent Progress in Microbiology*, vol. VIII, International Congress for Microbiology, University of Toronto Press, Montreal, pp. 201–12.
Iqbal, S.H. and Webster, J. (1973a) The trapping of aquatic hypomycete spores by air bubbles. *Transactions of the British Mycological Society*, **60**, 37–48.
Iqbal, S.H. and Webster, J. (1973b) Aquatic hypomycete spora of the River Exe and its tributaries. *Transactions of the British Mycological Society*, **61**, 331–46.
Irvine, J.A., Dix, N.J. and Warren, R.C. (1978) Inhibitory substances in *Acer platanoides* leaves. Seasonal activity and effects on growth of phylloplane fungi. *Transactions of the British Mycological Society*, **70**, 363–71.
Ishihara, T. (1980) The role of laccase in lignin biodegradation, in *Lignin Biodegradation: Microbiology, Chemistry and Potential applications*, vol. 2, (eds T.K. Kirk, T. Higuchi and Hou-min Chang), CRC Press, Boca Raton, FL, pp. 17–32.

Ishikawa, H., Schubert, W.J. and Nord, F.F. (1963a) Investigations on lignins and lignification 27. The enzymatic degradation of soft wood lignin by white-rot fungi. *Archives of Biochemistry and Biophysics*, **100**, 131–9.
Ishikawa, H., Schubert, W.J. and Nord, F.F. (1963b) Investigations on lignins and lignification 28. Degradation by *Polyporus versicolor* and *Fomes fomentarius* of aromatic compounds structurally related to soft wood lignin. *Archives of Biochemistry and Biophysics*, **100**, 140–9.
Iwahara, S. (1980) Microbial degradation of DHP, in *Lignin Biodegradation: Microbiology, Chemistry and Potential Applications*, vol. 1, (eds T.K. Kirk, T. Higuchi and Houmin Chang), CRC Press, Boca Raton, FL, pp. 151–70.
Jackson, R.M. (1957) Fungistasis as a factor in the rhizosphere phenomenon. *Nature*, **180**, 96–7.
Jackson, R.M. (1960) Soil fungistasis and the rhizosphere, in *The Ecology of Soil Fungi*, (eds D. Parkinson and J.S. Waid), Liverpool University Press, Liverpool, pp. 168–76.
Jain, M.K., Kapoor, K.K. and Mishra, M.M. (1979) Cellulase activity, degradation of cellulose and lignin, and humus formation by thermophilic fungi. *Transactions of the British Mycological Society*, **73**, 85–9.
Jalaluddin, M. (1967a) Studies on *Rhizina undulata*. I. Mycelial growth and ascospore germination. *Transactions of the British Mycological Society*, **50**, 449–59.
Jalaluddin, M. (1967b) Studies on *Rhizina undulata*. II. Observations and experiments in East Anglian plantations. *Transactions of the British Mycological Society*, **50**, 461–72.
Jamalainen, E.A. (1974) Resistance in winter cereal and grasses to low temperature parasitic fungi. *Annual Review of Phytopathology*, **12**, 281–302.
Jansson, H-B. (1982) Attraction of nematodes to endoparasitic nematophagous fungi. *Transactions of the British Mycological Society*, **79**, 25–9.
Jansson, H-B., and Nordbring-Hertz, B. (1979) Attraction of nematodes to living mycelium of nematophagous fungi. *Journal of General Microbiology*, **112**, 89–93.
Jansson, H-B. and Nordbring-Hertz, B. (1983) The endoparasitic fungus *Meria coniospora* infects nematodes specifically at the chemosensory organs. *Journal of General Microbiology*, **129**, 1121–6.
Jansson, H-B. and Nordbring-Hertz, B. (1984) Involvement of sialic acid in nematode chemotaxis and infection by an endoparasitic nematophagous fungus. *Journal of General Microbiology*, **130**, 39–43.
Jansson, H-B., Jeyaprakash, A. and Zuckerman, B.M. (1985) Control of root-knot nematodes on tomato by the endoparasitic fungus *Meria*

coniospora. Journal of Nematology, **17**, 327–9.

Jay, J.M. (1987) Meats, poultry and seafoods, in *Food and Beverage Mycology*, (ed. L.R. Beuchat), Van Nostrand Reinhold, New York, pp. 155–73.

Jenkinson, D.S. and Oades, J.M. (1979) A method for measuring adenosine triphosphate in soil. *Soil Biology and Biochemistry*, **11**, 193–9.

Jenkinson, D.S. and Powlson, D.S. (1976) The effects of biocidal treatments on metabolism in soil. I. Fumigation with chloroform. *Soil Biology and Biochemistry*, **8**, 167–77.

Jennings, D.H. (1983) Some aspects of the physiology and biochemistry of marine fungi. *Biological Reviews, Cambridge*, **58**, 423–59.

Jennings, D.H. (1984) Water flow through mycelia, in *The Ecology and Physiology of the Fungal Mycelium*, (eds D.H. Jennings and A.D.M. Rayner) *British Mycological Symposium 8*, Cambridge University Press, Cambridge, pp. 143–64.

Jennings, D.H. (1986) Fungal growth in the sea, in *The Biology of Marine Fungi*, (ed. S.T. Moss), Cambridge University Press, Cambridge, pp. 1–18.

Jenny, H. and Grossenbacher, K. (1963) Root surface boundary zones as seen in the electron microscope. *Soil Science Society of America Proceedings*, **27**, 273–7.

Jensen, K.F. (1969) Effect of constant and fluctuating temperature on growth of four wood-decaying fungi. *Phytopathology*, **5**, 645–7.

Jensen, V. (1974) Decomposition of angiosperm tree leaf litter, in *Biology of Plant Litter Decomposition*, vol. 1, (eds C.H. Dickinson and G.F.J. Pugh), Academic Press, London, pp. 69–104.

Jinks, J.L. (1952) Heterokaryosis: a system of adaptation in wild fungi. *Proceedings of the Royal Society London Series B*, **140**, 83–99.

Johnen, B.G. (1978) Rhizosphere microorganisms and roots stained with europium chelate and fluorescent brightener. *Soil Biology and Biochemistry*, **10**, 495–502.

Johnson, C.L. and Preece, T.F. (1979) Natural history of *Pilobolus kleinii*: experiments with sporangia in the cow digestive tract and the early stages of growth. *Transactions of the British Mycological Society*, **72**, 453–7.

Johnson, R.G. (1980) Ultrastructure of ascospore appendages of marine ascomycetes. *Botanica Marina*, **23**, 501–27.

Johnson, T.W. and Sparrow, F.K. (1961) *Fungi in Oceans and Estuaries*, Cramer, Weinheim.

Jones, D. and Watson, D. (1969) Parasitism and lysis by soil fungi of *Sclerotinia sclerotiorum* (Lib.) de Bary a phytopathogenic fungus. *Nature*, **224**, 287–8.

Jones, E.B.G. (1971) Aquatic fungi, in *Methods in Microbiology* 4, (ed. C. Booth), Academic Press, London and New York, pp. 335–65.

Jones, E.B.G. (1974) Aquatic fungi: freshwater and marine, in *Biology of Plant Litter Decomposition vol. 2*, (eds C.H. Dickinson and G.J.F. Pugh), Academic Press, London and New York, pp. 337–83.
Jones, E.B.G. (ed.) (1976a) *Recent Advances in Aquatic Mycology*, Elek Science, London.
Jones, E.B.G. (1976b) Lignicolous and algicolous fungi, in *Recent Advances in Aquatic Mycology*, (ed. E.B.G. Jones), Elek Science, London, pp. 1–49.
Jones, E.B.G. (1982) Decomposition by basidiomycetes in aquatic environments, in *Decomposer Basidiomycetes*, (eds J.C. Frankland, J.N. Hedger and M.J. Swift), *British Mycological Society Symposium 4*, Cambridge University Press, Cambridge, pp. 191–211.
Jones, E.B.G. and Bryne, P.J. (1976) Physiology of the higher marine fungi, in *Recent Advances in Aquatic Mycology*, (ed. E.B.G. Jones), Elek Science, London, pp. 135–75.
Jones, E.B.G. and Jennings, D.H. (1964) The effect of salinity on the growth of marine fungi in comparison with non-marine species. *Transactions of the British Mycological Society*, **47**, 619–25.
Jones, E.B.G. and Kuthubutheen, A.J. (1989) Malaysian mangrove fungi. *Sydowia*, **41**, 160–9.
Jones, E.B.G. and Moss, S.T. (1978) Ascospore appendages of marine ascomycetes: An evaluation of appendages as taxonomic criteria. *Marine Biology*, **49**, 11–26.
Jones, E.B.G. and Oliver, A.C. (1964) Occurrence of aquatic hyphomycetes on wood submerged in fresh and brackish water. *Transactions of the British Mycological Society*, **47**, 45–48.
Jones, E.B.G., Byrne, P.J. and Alderman, D.J. (1976) The response of fungi to salinity. *Vie et Milieu*, Supplement no. 22, 265–80.
Jones, E.B.G., Johnson, R.G. and Moss, S.T. (1983) Taxonomic studies of the Halosphaeriaceae: *Corollospora* Werdemann. *Botanical Journal of the Linnean Society*, **87**, 193–212.
Jones, E.B.G., Byrne, P. and Alderman, D.J. (1971) The response of fungi to salinity. *Vie et Milieu*, Supplement 22, 265–80.
Jones, P.C.T. and Mollison, J.E. (1948) A technique for the quantitative estimation of soil microorganisms. *Journal of General Microbiology*, **2**, 54–69.
Jones, R., Parkinson, S.M., Wainwright, M. and Kilham, K. (1991) Oxidation of thiosulphate by *Fusarium oxysporum* grown under oligotrophic conditions. *Mycological Research*, **95**, 1169–74.
Jorgensen, J.R. and Hodges, C.S. (1970) Microbial characteristics of a forest after twenty years of prescribed burning. *Mycologia*, **62**, 721–6.
Juniper, B.E. (1991) The leaf from the inside and outside: a microbe's perspective, in *The Microbial Ecology of Leaves*, (eds J.H. Andrews and S.S. Hirano), Springer-Verlag, New York, pp. 21–4.

Kadam, K.L. and Drew, D. (1986) Study of lignin biotransformation by *Aspergillus fumigatus* and white-rot fungi using ^{14}C-labelled and unlabelled Kraft lignins. *Biotechnology and Bioengineering*, **28**, 394–404.

Kane, B.E. and Mullins, J.T. (1973) Thermophilic fungi in a municipal waste compost system. *Mycologia*, **65**, 1087–100.

Karling, J.S. (1977) *Chytridiomycetarum Iconographia*, Lubrecht & Cramer, Vaduz.

Kates, M. and Baxter, R.M. (1962) Lipid composition of mesophilic and psychrophilic yeasts (*Candida* species) as influenced by environmental temperature. *Canadian Journal of Biochemistry and Physiology*, **40**, 1213–27.

Katznelson, H. (1960) Observations on the rhizosphere effect, in *Ecology of Soil Fungi*, (eds D. Parkinson and J.S. Waid), Liverpool University Press, Liverpool, pp. 192–20.

Katznelson, H., Rouatt, J.W. and Payne, T.M.B. (1954) Liberation of amino acids by plant roots in relation to desiccation. *Nature*, **174**, 1110–11.

Kaushik, N.K. and Hynes, H.B.N. (1971) The fate of dead leaves that fall into streams. *Archiv für Hydrobiologie*, **68**, 465–515.

Keilich, G., Bailey, P. and Liese, W. (1970) Enzymic degradation of cellulose, cellulose derivatives and hemicelluloses in relation to the fungal decay of wood. *Wood Science Technology*, **4**, 273–83.

Keller-Schierlein, W. and Diekmann, H. (1970) Stoffwechselprodukte von Mikroorganismen 85. Zür Konstitution des Coprogens. *Helvetica Chimica Acta*, **53**, 2035–44.

Kellock, L. and Dix, N.J. (1984a) Antagonism by *Hypomyces aurantius*. I. Toxins and hyphal interactions. *Transactions of the British Mycological Society*, **82**, 327–34.

Kellock, L. and Dix, N.J. (1984b) Antagonism by *Hypomyces aurantius*. II. Ultrastructure studies of hyphal disruptions. *Transactions of the British Mycological Society*, **82**, 335–8.

Kemp, R.F.O. (1975) Breeding biology of *Coprinus* species in the section Lanatuli. *Transactions of the British Mycological Society*, **65**, 375–88.

Kendrick, W.B. (1959) The time factor in the decomposition of coniferous leaf litter. *Canadian Journal of Botany*, **37**, 907–12.

Kendrick, W.B. and Burges, N.A. (1962) Biological aspects of the decay of *Pinus sylvestris* leaf litter. *Nova Hedwigia*, **4**, 313–42.

Kerekes, R. and Nagy, G. (1980) Membrane lipid composition of a mesophilic and psychrophilic yeast. *Acta Alimentaria*, **9**, 93–8.

Kerling, L.C.P. (1958) De microflora op het blad van *Beta vulgaris* L. *Tijdschriftt Plantenziekten*, **64**, 402–10.

Kerling, L.C.P. (1964) Fungi in the phyllosphere of rye and strawberry. *Mededelingen Landbhoogesch Opzoekstations Gent*, **29**, 885–95.

Kerry, B.R. (1984) Nematophagous fungi and the regulation of nematode populations in soil. *Helminthological Abstracts*, **B53**, 1–14.

Kerry, B.R. and Crump, D.H. (1980) Two fungi parasitic on females of cyst-nematodes. (*Heterodera* spp.). *Transactions of the British Mycological Society*, **74**, 119–25.

Keyser, P., Kirk, T.K. and Zeikus, J.G. (1978) Ligninolytic enzyme systems of *Phanerochaete chrysosporium* synthesized in relation to the fungal decay of wood. *Wood Science Technology*, **4**, 273–83.

Khan, M.A. (1987) Interspecies interactions in aquatic hyphomycetes. *Botanical Magazine, Tokyo*, **100**, 295–303.

Kilbertus, G. (1968) Vitesse de décomposition de *Pseudoscleropodium purum* (Hedw.) Fleisch dans la nature. *Revue d'Ecologie et de Biologie du Sol*, **5**, 237–44.

Kingham, D.L. and Evans, L.V. (1986) The *Pelvetia–Mycosphaerella* interrelationship, in *The Biology of Marine Fungi*, (ed. S.T. Moss), Cambridge University Press; Cambridge, pp. 177–87.

Kinkel, L.L., Andrews, J.H., Berbee, F.M. and Nordheim, E.V. (1987) Leaves as islands for microbes. *Oecologia*, **71**, 405–8.

Kirby, J.J.H. (1984) Microbial aspects of aquatic macrophyte decomposition. PhD thesis, University of Exeter.

Kirby, J.J.H., Webster, J. and Baker, J.H. (1990) A particle plating method for analysis of fungal community composition and structure. *Mycological Research*, **94**, 621–6.

Kirk, P.W. (1976) Cytochemistry of marine fungal ascospores, in *Recent Advances in Aquatic Mycology* (ed. E.B.G. Jones), Elek Science, London, pp. 177–92.

Kirk, P.W. (1980) The mycostatic effect of seawater on spores of terrestrial and marine higher fungi. *Botanica Marina*, **23**, 233–8.

Kirk, T.K. (1983) Degradation and conversion of lignocelluloses, in *The Filamentous Fungi*, (eds J.E. Smith, D.R. Berry and B. Kristiansen), Edward Arnold, London, pp. 266–95.

Kirk, T.K. and Farrell, R.L. (1987) Enzymatic 'combustion': the microbial degradation of lignin. *Annual Review of Microbiology*, **41**, 465–505.

Kirk, T.K. and Fenn, P. (1982) Formation and action of the ligninolytic system in basidiomycetes, in *Decomposer Basidiomycetes: their Biology and Ecology*, (eds J.C. Frankland, J.N. Hedger and M.J. Swift), *British Mycological Symposium 4*, Cambridge University Press, Cambridge, pp. 67–90.

Kirk, T.K. and Kelman, A. (1965) Lignin degradation as related to the phenoloxidases of selected wood-decaying basidiomycetes. *Phytopathology*, **55**, 739–45.

Kirk, T.K. and Moore, W.E. (1973) Removing lignin from wood with white-rot fungi and the digestibility of resulting wood. *Wood and Fiber*, 72–9.

Kirk, T.K., Higuchi, T. and Chang, H. (1980) *Lignin Biodegradation:*

Microbiology, Chemistry and Potential Application, 2 vols, CRC Press, Boca Raton, FL.

Klich, M.A. and Tiffany, L.H. (1985) Distribution and seasonal occurrence of aquatic Saprolegniaceae in Northwest Iowa. *Mycologia*, **77**, 373–80.

Kliejunas, J.T. and Ko, W.H. (1975) A technique for direct observation of nematode trapping by fungi in soil. *Mycologia*, **67**, 420–3.

Klotz, L.J., Stolzy, L.H. and Dewolf, T.A. (1963) Oxygen requirements of three root-rotting fungi in liquid medium. *Phytopathology*, **53**, 302–5.

Knudson, L. (1913) Tannic acid fermentation. *Journal of Biological Chemistry*, **14**, 159–84.

Ko, W.H. and Hora, F.K. (1971) Fungitoxicity in certain Hawaiian soils. *Soil Science*, **112**, 276–9.

Ko, W.H. and Hora, F.K. (1972) Identification of an Al ion as a soil fungitoxin. *Soil Science*, **113**, 42–5.

Ko, W.H. and Lockwood, J.L. (1967) Soil fungistasis: relation to fungal spore nutrition. *Phytopathology*, **57**, 894–901.

Ko, W.H. and Lockwood, J.L. (1970) Mechanism of lysis of fungal mycelia in soil. *Phytopathology*, **60**, 148–54.

Ko, W.H. Hora, F.K. and Herlicska, E. (1974) Isolation and identification of a volatile fungistatic factor from alkaline soil. *Phytopathology*, **64**, 1398–400.

Koch, J. (1974) Marine fungi on driftwood from the West coast of Jutland. *Friesia*, **10**, 208–50.

Koenigs, J.W. (1974a) Hydrogen peroxide and iron: a proposed system for decomposition of wood by brown-rot basidiomycetes. *Wood and Fiber*, **6**, 66–79.

Koenigs, J.W. (1974b) Production of hydrogen peroxide by wood-decaying fungi in wood and its correlation with weight loss, depolymerization and pH changes. *Archives of Microbiology*, **99**, 129–45.

Kohlmeyer, J. (1966) Ecological observations on arenicolous marine fungi. *Zeitschrift für Allgemeine Mikrobiologie*, **6**, 95–106.

Kohlmeyer, J. (1968) Marine fungi from the tropics. *Mycologia*, **60**, 252–70.

Kohlmeyer, J. (1969). Ecological notes on fungi in mangrove forests. *Transactions of the British Mycological Society*, **53**, 237–50.

Kohlmeyer, J. (1977) New genera and species of higher fungi from the deep sea (1615–5315 m). *Revue de Mycologie*, **41**, 189–206.

Kohlmeyer, J. (1981a) Distribution and ecology of conidial fungi in marine habitats, in *Biology of Conidial Fungi*, vol. I, (eds G.T. Cole and B. Kendrick), Academic Press, New York and London, pp. 357–72.

Kohlmeyer, J. (1981b) Marine fungi from Martinique. *Canadian Journal of Botany*, **59**, 1314–21.

Kohlmeyer, J. (1983) Geography of marine fungi. *Australian Journal of Botany*, Supplement Series 10, 67–76.

Kohlmeyer, J. (1984) Tropical marine fungi. *P.S.Z.N.I. Marine Ecology*, **5**, 329–78.

Kohlmeyer, J. (1986) *Ascocratera manglicola* gen. et sp. nov., and key to the marine Loculoascomycetes on mangroves. *Canadian Journal of Botany*, **64**, 3036–42.

Kohlmeyer, J. and Charles, T.M. (1981) Sclerocarps: undescribed propagules in a sand-inhabiting marine fungus. *Canadian Journal of Botany*, **9**, 1787–91.

Kohlmeyer, J. and Kohlmeyer, E. (1979) *Marine Mycology: The Higher Fungi*. Academic Press, London.

Kohlmeyer, J. and Volkmann-Kohlmeyer, B. (1989) New species of *Koralionastes* (Ascomycotina) from the Caribbean and Australia. *Canadian Journal of Botany*, **68**, 1554–9.

Kohlmeyer, J. and Volkmann-Kohlmeyer, B. (1991) Illustrated key to the filamentous higher marine fungi. *Botanica Marina*, **34**, 1–61.

Koske, R.E. and Duncan, I.W. (1974) Temperature effects on growth, sporulation and germination of some 'aquatic' hyphomycetes. *Canadian Journal of Botany*, **52**, 1387–91.

Koske, R.E. and Tessier, B. (1986) Growth of some wood and litter decay basidiomycetes at reduced water potential. *Transactions of the British Mycological Society*, **86**, 156–8.

Kouyeas, V. (1964) An approach to the study of moisture relations of soil fungi. *Plant and Soil*, **20**, 351–63.

Kozlowski, T.T. and Ahlgren, C.E. (eds) (1974) *Fire and Ecosystems*, Academic Press, London.

Kramer, C.L. (1982) Production, release and dispersal of basidiospores, in *Decomposer Basidiomycetes: their Biology and Ecology*, (eds J.C. Frankland, J.N. Hedger and M.J. Swift) *British Mycological Soceity Symposium 4*, Cambridge University Press, Cambridge, pp. 33–49.

Kuan, Ta-Li. and Erwin, D.C. (1981) Effect of soil matric potential of alfalfa. *Phytopathology*, **72**, 543–8.

Kuc, J. and Shain, L. (1977) Antifungal compounds associated with disease resistance in plants, in *Antifungal Compounds*, vol. 2, (eds M.R. Siegel and H.D. Sisler), Marcel Dekker; New York, pp. 497–535.

Kuo, M.J. and Alexander, M. (1967) Inhibition of the lysis of fungi by melanins. *Journal of Bacteriology*, **94**, 624–9.

Kuthubutheen, A.J. and Webster, J. (1986a) Water availability and the coprophilous fungus succession. *Transactions of the British Mycological Society*, **86**, 63–76.

Kuthubutheen, A.J. and Webster, J. (1986b) Effects of water availability on germination, growth and sporulation of corprophilous fungi. *Transactions of the British Mycological Society*, **86**, 77–91.

Lacey, J. (1971) The microbiology of moist barley stored in unsealed silos. *Annals of Applied Biology*, **69**, 187–212.

Lacey, J. (1975) Potential hazards to animals and man from microorganisms in fodder and grain. *Transactions of the British Mycological Society*, **65**, 171–84.

Laiho, O. (1970) *Paxillus involutus* as a mycorrhizal symbiont of forest trees. *Acta Forestalia Fennica*, **106**, 1–72.

Laing, S.A.K. and Deacon, J.W. (1991) Videomicroscopical comparison of mycoparasitism by *Pythium oligandrum*, *P. nunn* and an unnamed *Pythium* species. *Mycological Research*, **95**, 469–79.

Lal, S.P. and Yadav, A.S. (1964) A preliminary list of microfungi associated with the decaying stems of *Triticum vulgare* and *Andropogon sorghum*. *Indian Phytopathology*, **17**, 208–18.

Lambourne, L.J. and Reardon, T.F. (1962) The use of 'seasonal' regressions in measuring feed intake of grazing animals. *Nature, London*, **196**, 961–2.

Langvad, F. (1980) A simple and rapid method for qualitative and quantitative study of the fungal flora of leaves. *Canadian Journal of Microbiology*, **26**, 666–70.

Larsen, K. (1971) Danish endocoprophilous fungi, and their sequence of occurrence. *Botanisk Tidsskrift*, **66**, 1–32.

Larsen, M.J., Jurgensen, M.F., Harvey, A.E. and Ward, J.C. (1978) Dinitrogen fixation associated with sporophores of *Fomitopsis pinicola*, *Fomes fomentarius* and *Echinodontium tinctorium*. *Mycologia*, **70**, 1217–21.

Last, F.T. (1955) Seasonal incidence of *Sporobolomyces* on cereal leaves. *Transactions of the British Mycological Society*, **38**, 221–39.

Last, F.T. (1970) Factors associated with the distribution of some phylloplane microbes. *Netherlands Journal of Plant Pathology*, **76**, 140–3.

Last, F.T. and Fleming, L.V. (1985) Factors affecting the occurrence of fruit bodies forming sheathing (ecto-) mycorrhizas with roots of trees. *Proceedings of the Indian Academy of Sciences (Plant Section)*, **94**, 111–27.

Last, F.T. and Price, D. (1969) Yeast associated with living plants and their environs, in *The Yeasts*, vol. 1, (eds A.H. Rose and J.S. Harrison), Academic Press, London, pp. 181–218.

Last, F.T., Dighton, J. and Mason, P.A. (1987) Successions of sheathing mycorrhizal fungi. *Trends in Ecology and Evolution*, **2**, 159–61.

Last, F.T., Mason, P.A., Wilson, J. and Deacon, J.W. (1983) Fine roots and sheathing mycorrhizas: their formation, function and dynamics. *Plant and Soil*, **71**, 9–21.

Last, F.T., Pelham, J., Mason, P.A. and Ingelby, K. (1979) Influence of leaves on sporophore production by fungi forming sheathing mycorrhizas with *Betula* spp. *Nature*, **280**, 168–9.

Latter, P.M. and Heal, O.W. (1971) A preliminary study of the growth of fungi and bacteria from temperate and Antartic soils in relation to temperature. *Soil Biology and Biochemistry*, **3**, 365–79.

Lebau, J.B. and Hawn, E.J. (1963) Formation of hydrogen cyanide by the mycelial stage of a fairy ring fungus. *Phytopathology*, **53**, 1395–6.

Ledingham, R.J. and Chim, S.H.F. (1955) A flotation method for obtaining spores of *Helminthosporium sativum* from soil. *Canadian Journal of Botany*, **33**, 298–303.

Lehmann, P.F. (1976) Unusual fungi on pine leaf litter induced by urea and urine. *Transactions of the British Mycological Society*, **67**, 251–3.

Lehmann, P.F. and Hudson, H.J. (1977) The fungal succession on normal and urea-treated pine needles. *Transactions of the British Mycological Society*, **68**, 221–8.

Leightley, L.E. and Eaton, R.A. (1977) Mechanisms of decay of timber by aquatic micro-organisms. *British Wood Preservers Association Annual Convention*, 1–26.

Leightley, L.E. and Eaton, R.A. (1979) *Nia vibrissa* – a marine white rot fungus. *Transactions of the British Mycological Society*, **73**, 35–40.

Leisola, M.S.A. and Fiechter, A. (1985) New trends in lignin biodegradation. *Advances in Biotechnology Processes*, **5**, 59–89.

Leuchtmann, A. and Clay, K. (1988) Experimental infection of host grasses and sedges with *Atkinsonella hypoxylon* and *Balansia cyperi* (Balansiae, Clavicipitaceae). *Mycologia*, **80**, 291–7.

Levi, M.P. and Cowling, E.B. (1969) Role of nitrogen in wood deterioration. VII. Physiological adaptation of wood-destroying and other fungi to substrates deficient in nitrogen. *Phytopathology*, **59**, 460–8.

Levi, M.P. and Preston, R.D. (1965) A chemical and microscopic examination of the action of the soft-rot fungus *Chaetomium globosum* on beechwood (*Fagus sylvatica*). *Holzforschung*, **19**, 183–90.

Levi, M.P., Merill, W. and Cowling, E.B. (1968) Role of nitrogen in wood deterioration. VI. Mycelial fractions and model nitrogen compounds as substrates for growth of *Polyporus versicolor* and other wood destroying and wood inhabiting fungi. *Phytopathology*, **58**, 626–34.

Levy, J.P. (1975) Colonization of wood by fungi, in *Biological Transformation of Wood by Microorganisms*, (ed. W. Liese), Springer-Verlag, Berlin, pp. 16–23.

Levy, J.P. (1982) The place of basidiomycetes in the decay of wood in contact with the ground, in *Decomposer Basidiomycetes*, (eds T.C. Frankland, J.N. Hedger and M.J. Swift), *British Mycological Symposium 4*, Cambridge University Press, Cambridge, pp. 161–78.

Lewis, D.H. and Smith, D.C. (1967) Sugar alcohols (polyols) in fungi and green plants. I. Distribution, physiology and metabolism. *New Phytologist*, **66**, 143–84.

Lewis, J.A. and Papavizas, G.C. (1969) Effect of sulphur-containing

volatiles present in cabbage on *Aphanomyces euteiches. Phytopathology*, **59**, 1558.
Lewis, J.A. and Papavizas, G.C. (1970) Evolution of volatile sulphur-containing compounds from decomposition of crucifers in soil. *Soil Biology and Biochemistry*, **2**, 239–46.
Lewis, J.A. and Starkey, R.L. (1969) Decomposition of plant tannins by some microorganisms. *Soil Science*, **107**, 235–41.
Lewis, P.F. (1975) The possible significance of the hemicelluloses in wood decay. *Material und Organismen*, **3**, 113–19.
Liese, W. (1964) Über den Abbau verholzter Zellwände durch Moderfäulepilze. *Holz als Roh-und Werkstoff*, **24**, 289–95.
Liese, W. (1970) Ultrastructural aspects of woody tissue disintegration. *Annual Review of Phytopathology*, **8**, 231–58.
Liese, W. and Ammer, U. (1964) Über den Befall von Buchenholz durch Moderfäulepilze in Abhängigkeit von der Holzfeuchtigkeit. *Holzforschung*, **18**, 97–102.
Lilly, V.G. and Barnett, H.L. (1951) *Physiology of Fungi*, McGraw-Hill, London.
Lindeberg, G. (1944) Über die physiologie lignin abbauender Bodenhymenomyzeten. *Symbolae Botanicae Upsalienses*, **8**, 1–183.
Lindeberg, G. (1947) On the decomposition of lignin and cellulose in litter caused by soil-inhabiting hymenomycetes. *Arkiv för Botanik*, **33a**(10), 1–15.
Lindeberg, G. (1948) On the occurrence of polyphenol oxidases in soil-inhabiting basidiomycetes. *Physiologia Plantarum*, **1**, 196–205.
Lindeberg, G. (1949) Influence of enzymatically oxidised gallic acid on the growth of some hymenomycetes. *Svensk Botanisk Tidskrift*, **43**, 438–47.
Lindeberg, G. and Holm, G. (1952) Occurrence of tyrosinase and laccase in fruit bodies and mycelia of some hymenomycetes. *Physiologia Plantarum*, **5**, 100–12.
Lindeberg, G., Lindeberg, M., Lundgren, L. *et al.* (1980) Stimulation of litter-decomposing basidiomycetes by flavonoids. *Transactions of the British Mycological Society*, **75**, 455–9.
Lindenfelser, L.A. and Cieger, A. (1969) Production of antibodies by *Alternaria* species. *Developments in Industrial Microbiology*, **10**, 271–8.
Lindsey, B.I. and Pugh, G.J.F. (1976) Distribution of microfungi over the surfaces of attached leaves of *Hippophae rhamnoides*. *Transactions of the British Mycological Society*, **67**, 427–33.
Linford, M.B. and Yap, F. (1939) Root knot injury restricted by a fungus. *Phytopathology*, **29**, 596–609.
Linford, M.B., Yap, F. and Oliveira, J.M. (1938) Reduction of soil populations of the root-knot nematode during decomposition of organic matter. *Soil Science*, **45**, 127–41.
Lingappa, B.T.L. and Lockwood, J.L. (1963) Direct assay of soils for

fungistasis. *Phytopathology*, **53**, 529–31.
Lingappa, B.T. and Lockwood, J.L. (1964) Activation of soil microflora by fungus spores in relation to soil fungistasis. *Journal of General Microbiology*, **35**, 215–27.
Liou, J.Y. and Tzean, S.S. (1992) Stephanocysts as nematode-trapping and infecting propagules. *Mycologia*, **84**, 786–90.
Liwicki, R., Paterson, A., Macdonald, M.J. and Broda, P. (1985) Phenotypic classes of phenoloxidase – negative mutants of the lignin degrading fungus *Phanerochaete chrysosporium*. *Journal of Bacteriology*, **162**, 641–4.
Ljungdahl, L.G. and Eriksson, K.E. (1985) Ecology of microbial cellulose degradation. *Advances in Microbial Ecology*, **8**, 237–99.
Lochhead, A.G. and Thexton, R.H. (1947) Qualitative studies of soil microorganisms. VII. The rhizosphere effect in relation to the amino acid nutrition of bacteria. *Canadian Journal of Research*, **25**, 20–6.
Lockwood, J.L. (1964) Soil fungistasis. *Annual Review of Phytopathology*, **2**, 341–62.
Lockwood, J.L. (1977) Fungistasis in soils. *Biological Reviews, Cambridge*, **52**, 1–43.
Lockwood, J.L. (1981) Exploitation competition, in *The Fungal Community: Its Organisation and Role in the Ecosystem*, (eds D.T. Wicklow and G.C. Carroll), Marcel Dekker, New York, pp. 319–49.
Lockwood, J.L. (1992) Exploitation competition, in *The Fungal Community: Its Organisation and Role in the Ecosystem*, 2nd edn, (eds G.C. Carroll and D.T. Wicklow), Marcel Dekker, New York, pp. 243–63.
Lockwood, J.L. and Filonow, A.B. (1981) Responses of fungi to nutrient-limiting conditions and to inhibitory substances in natural habitats. *Advances in Microbial Ecology*, **5**, 1–61.
Lodha, B.C. 91974) Decomposition of digested litter, in *Biology of Plant Litter Decomposition*, (eds C.H. Dickinson and G.J.F. Pugh), Academic Press, London, pp. 213–41.
Loginova, L.G. and Tashpulatov, Z. (1967) Multicomponent cellulolytic enzymes of thermotolerant and mesophilic fungi closely related to *Aspergillus fumigatus*. *Microbiology (Washington)*, **36**, 828–31.
Loman, A.A. (1962) The influence of temperature on the location and development of decay fungi in lodgepole pine logging slash. *Canadian Journal of Botany*, **40**, 1545–59.
Loman, A.A. (1965) The lethal effects of periodic high temperature on certain lodgepole pine slash decaying basidiomycetes. *Canadian Journal of Botany*, **43**, 334–8.
Lord, K.A., Cayley, G.R. and Lacey, J. (1981) Laboratory application of preservatives to hay and effects of irregular distribution on mould development. *Animal Feed Science and Technology*, **6**, 73–82.
Lord, K.A., Lacey, J., Cayley, G.R. and Manlove, R. (1981) Fatty acids

as substrates and inhibitors of fungi from propionic acid-treated hay. *Transactions of the British Mycological Society*, **77**, 41–5.

Luard, E.J. (1982a) Accumulation of intracellular solutes by two filamentous fungi in response to growth at low steady state osmotic potential. *Journal of General Microbiology*, **128**, 2563–74.

Luard, E.J. (1982b) Growth and accumulation of solutes by *Phytophthora cinnamomi* and other lower fungi in response to changes in external osmotic potential. *Journal of General Microbiology*, **128**, 2583–90.

Luard, E.J. (1982c) Effect of osmotic shock on some intra-cellular solutes in two filamentous fungi. *Journal of General Microbiology*, **128**, 2575–81.

Luard, E.J. and Griffin, D.M. (1981) Effect of water potential on fungal growth and turgor. *Transactions of the British Mycological Society*, **76**, 33–40.

Luff, M.L. (1965) The morphology and microclimate of *Dactylis glomerata* tussocks. *Journal of Ecology*, **53**, 771–87.

Lumsden, R.D. (1981) Ecology of mycoparasitism, in *The Fungal Community*, (eds D.T. Wicklow and G.C. Carroll), Marcel Dekker, New York, pp. 295–318.

Lumsden, R.D., Byer, W.A. and Dow, R.L. (1975) Differential isolation of *Pythium* species from soil by means of selective media, temperature and pH. *Canadian Journal of Microbiology*, **21**, 606–12.

Lumsden, R.D., Ayers, W.A., Adams, P.B. *et al.* (1976) Ecology and epidemiology of *Pythium* species in field soil. *Phytopathology*, **66**, 1203–9.

Lund, A. (1934) Studies on Danish freshwater phycomycetes and notes on their occurrence. *Kongelige Danske Videnskabernes Selskabs Skrifter*, **9**, 1–98.

Lundborg, A. and Unestam, T. (1980) Antagonism against *Fomes annosus*. Comparison between different test methods *in vitro* and *in vivo*. *Mycopathologia*, **70**, 107–5.

Lussenhop, J. (1981) Analysis of microfungal component communities, in *The Fungal Community: its Organisation and Role in the Ecosystem*, (eds D.T. Wicklow and G.C. Carroll), Marcel Dekker, New York, pp. 37–45.

Lussenhop, J., Kumar, R., Wicklow, D.T. and Lloyd, J.E. (1980) Insect effects on bacteria and fungi in cattle dung. *Oikos*, **34**, 54–8.

Lynch, J.M. and Bragg, E. (1985) Microorganisms and soil aggregate stability. *Advances in Soil Science*, **2**, 133–71.

Lyr, H. (1960) Formation of ecto-enzymes by wood-destroying and wood-inhabiting fungi on various culture media. V. Complex medium as carbon source. *Archives of Microbiology*, **35**, 258–78.

Lyr, H. (1961) Hemmungsanalytische Untersuchungen an einigen Ektoenzymen holzzerstörender Pilze. *Enzymologia*, **23**, 231–48.

Lyr, H. (1962) Detoxification of heartwood toxins and chlorophenols by higher fungi. *Nature*, **195**, 289–90.

Lyr, H. (1963) Enzymatische Detoxifikation chlorierter phenole. *Phytopathologische Zeitschrift*, **47**, 73–83.

Lyr, H. (1965) On the toxicity of oxidized polyphenols. *Phytopathologische Zeitschrift*, **52**, 227–40.

Macarthur, R.H. and Wilson, E.D. (1967) *The Theory of Island Biogeography*, Princeton University Press, Princeton, NJ.

Macauley, B.J. and Griffin, D.M. (1969a) Effects of carbon dioxide and oxygen on the activity of some soil fungi. *Transactions of the British Mycological Society*, **53**, 53–62.

Macauley, B.J. and Griffin, D.M. (1969b) Effects of carbon dioxide and bicarbonate on the growth of some soil fungi. *Transactions of the British Mycological Society*, **53**, 223–8.

Macdougall, B.M. and Rovira, A.D. (1970) Sites of exudation of ^{14}C labelled compounds from wheat roots. *New Phytologist*, **69**, 999–1003.

Madelin, M.F. (1968) Fungi parasitic on other fungi and lichens, in *The Fungi. An Advanced Treatise*, vol. III, (eds G.C. Ainsworth and A.S. Sussman), Academic Press, London, pp. 253–66.

Magan, N. (1993) Tolerance of fungi to sulphur dioxide, in *Stress Tolerance of Fungi* (ed. D.H. Jennings), Marcel Dekker, New York, pp. 173–87.

Magan, N. and Lacey, J. (1984a) Effect of temperature and pH on water relations of field and storage fungi. *Transactions of the British Mycological Society*, **82**, 71–81.

Magan, N. and Lacey, J. (1984b) Effect of water activity, temperature and substrate interactions between field and storage fungi. *Transactions of the British Mycological Society*, **82**, 83–93.

Magan, N. and Lacey, J. (1985) Interactions between field and storage fungi on wheat grain. *Transactions of the British Mycological Society*, **85**, 29–37.

Magan, N. and Lynch, J.M. (1986) Water potential, growth and cellulolysis of fungi involved in decomposition of cereal residues. *Journal of General Microbiology*, **132**, 1181–7.

Magan, N. and McLeod, A.R. (1991) Effects of atmospheric pollutants on phyllosphere microbial communities, in *Microbial Ecology of Leaves*, (eds J.H. Andrews and S.S. Hirano), Springer-Verlag, New York, pp. 379–400.

Mallett, K.I. and Harrison, L.M. (1988) The mating system of the fairy ring fungus *Marasmius oreades* and the genetic relationship of fairy rings. *Canadian Journal of Botany*, **66**, 1111–16.

Mankau, R. (1962) Soil fungistasis and nematophagous fungi. *Phytopathology*, **52**, 611–5.

Mankau, R. (1975) A semiquantitative method for enumerating and observing parasites and predators of soil nematodes. *Journal of Nematology*, **7**, 119–22.
Mankau, R. (1980) Biological control of nematode pests by natural enemies. *Annual Review of Phytopathology*, **18**, 415–40.
Mankau, R. (1981) Microbial control of nematodes. In *Plant Parasitic Nematodes* vol. 3, (eds B.M. Zuckerman and R.A. Rhode), Academic Press, New York, pp. 475–94.
Manocha, M.S. (1975) Host-parasite relations in a mycoparasite. III. Morphological and biochemical differences in the parasitic- and axenic-culture spores of *Piptocephalis virginiana*. *Mycologia*, **67**, 382–91.
Manocha, M.S. and Campbell, C.D. (1978) The effect of growth temperature on the fatty acid composition of *Thamnidium elegans* Link. *Canadian Journal of Microbiology*, **24**, 670–4.
Manocha, M.S. and Deven, J.M. (1975) Host–parasite relations in a mycoparasite. IV. A correlation between levels of α – linolenic acid and parasitism of *Piptocephalis virginiana*. *Mycologia*, **67**, 1148–57.
Mansfield, J.W. (1983) Antimicrobial compounds, in *Biochemical Plant Pathology*, (ed. J.A. Callow), J. Wiley & Sons, New York, pp. 237–65.
Mansfield, J.W. and Deverall, B.J. (1974) Changes in wyerone acid concentrations in leaves of *Vicia faba* after infection by *Botrytis cinerea* or *B. fabae*. *Annals of Applied Biology*, **77**, 227–35.
Mansfield, J.W., Dix, N.J. and Perkins, A.M. (1975) Role of the phytoalexin pisatin in controlling saprophytic fungal growth on pea leaves. *Transactions of the British Mycological Society*, **64**, 507–11.
Mansfield, J.W., Hargreaves, J.A. and Boyle, F.C. (1974) Phytoalexin production by live cells in broad bean leaves infected with *Botrytis cinerea*. *Nature*, **252**, 316–17.
Marchant, R. (1970) The root surface of *Ammophila arenaria* as a substrate for microorganisms. *Transactions of the British Mycological Society*, **54**, 479–82.
Martin, J.P. and Haider, K. (1969) Phenolic polymers of *Stachybotrys atra*, *Stachybotrys chartarum* and *Epicoccum nigrum* in relation to humic acid formation. *Soil Science*, **107**, 260–70.
Martin, J.P., Haider, K. and Wolf, D. (1972) Synthesis of phenols and phenolic polymers by *Hendersonula toruloidea* in relation to humic acid formation. *Soil Science Society of America Proceedings*, **36**, 311–15.
Martin, J.T., Bant, R.F. and Burchill, R.T. (1957) Fungistatic properties of apple leaf wax. *Nature*, **180**, 796–7.
Marx, D.H. (1977) Tree host range and world distribution of the ectomycorrhizal fungus *Pisolithus tinctorius*. *Canadian Journal of Microbiology*, **23**, 217–23.

Marx, D.H. and Davey, C.B. (1970) Influence of temperature on aseptic synthesis of ectomycorrhizae by *Thelephora terrestris* and *Pisolithus tinctorius* on loblolly pine. *Forest Science*, **16**, 424–31.

Mason, P.A., Last, F.T., Pelham, J. and Ingleby, K. (1982) Ecology of some fungi associated with an ageing stand of birches (*Betula pendula* and *B. pubescens*). *Forest Ecology and Management*, **4**, 19–39.

Mason, P.A., Last, F.T., Wilson, J. et al. (1987) Fruiting and successions of ectomycorrhizal fungi, in *Fungal Infection of Plants*, (eds G.F. Pegg and P.G. Ayres), *British Mycological Society Symposium 13*, Cambridge University Press, Cambridge, pp. 253–68.

Masters, M.J. (1976) Freshwater phycomycetes on algae, in *Recent Advances in Aquatic Mycology*, (ed. E.B.G. Jones), Elek Science, London, pp. 489–512.

McBride, R.P. (1971) Microorganism interaction in the phyllosphere of larch, in *Ecology of Leaf Surface Microorganisms*, (eds T.F. Preece and C.H. Dickinson), Academic Press, London, pp. 545–55.

McBride, R.P. (1972) Larch leaf waxes utilized by *Sporobolomyces roseus* in situ. *Transactions of the British Mycological Society*, **58**, 329–31.

McBride, R.P. and Hayes, A.J. (1977) Phylloplane of European larch. *Transactions of the British Mycological Society*, **69**, 39–46.

McKenzie, E.H.C. and Hudson, H.J. (1976) Mycoflora of rust-infected and non-infected plant material during decay. *Transactions of the British Mycological Society*, **66**, 223–38.

McMurrough, A.H. and Rose, A.H. (1973) Effects of temperature variation on the fatty acid composition of a psychrophilic *Candida* sp. *Journal of Bacteriology*, **114**, 451–2.

Mehan, V.K. and Chohan, J.S. (1981) Effect of fungicides on leaf spot pathogens and the phylloplane mycoflora of ground-nut. *Transactions of the British Mycological Society*, **76**, 361–6.

Meier, H. (1962) Chemical and morphological aspects of the fine structure of wood. *Pure and Applied Chemistry*, **5**, 37–52.

Melin, E. (1946) Der Einfluss von Waldstreuextrakten auf das Wachstum von Bodenpilzen mit besonderer Berucksichtigung der Wurzelpilze von Bäumen. *Symbolae Botanicae Upsalienses*, **8**, 1–16.

Menge, J.A. and Grand, L.F. (1978) Effect of fertilization on production of epigeous basidiocarps by mycorrhizal fungi in loblolly pine plantations. *Canadian Journal of Botany*, **56**, 2357–62.

Meredith, D.S. (1962) Some fungi on decaying banana leaves in Jamaica. *Transactions of the British Mycological Society*, **45**, 335–47.

Merrill, W., French, D.W. and Wood, F.A. (1964) Decay of wood by species of the Xylariaceae. *Phytopathology*, **54**, 56–8.

Metwalli, A.A. and Shearer, C.A. (1989) Aquatic hypomycete communities in clear-cut and wooded areas of an Illinois stream. *Transactions of the Illinois Academy of Science*, **82**, 5–16.

Meyer, J.L. (1980) Dynamics of phosphorus and organic matter during leaf decomposition in a forest stream. *Oikos*, **34**, 44–53.

Meyers, S.P. and Hoyo, L. (1966) Observations on the growth of the marine hyphomycete *Varicosporina ramulosa*. *Canadian Journal of Botany*, **44**, 1133–40.

Meyers, S.P. and Reynolds, E.S. (1960) Occurrence of lignicolous fungi in Northern Atlantic and Pacific marine localities. *Canadian Journal of Botany*, **38**, 217–26.

Meyers, S.P. and Simms, J. (1960) Thalassiomycetes VI. *Comparative growth studies of Lindra thalassiae and lignicolous ascomycete species. Canadian Journal of Botany*, **43**, 379–92.

Meyers, S.P. and Simms, J. (1965) Thalassiomycetes VI. Comparative growth studies of *Lindra thalassiae* and lignicolous ascomycete species. *Canadian Journal of Botany*, **43**, 379–92.

Michaelides, J. and Kendrick, B. (1978) An investigation of factors retarding colonization of conifer needles by amphibious fungi in streams. *Mycologia*, **70**, 419–30.

Mikola, P. (1954) Experiments on the rate of decomposition of forest litter. *Communicationes Instituti Forestalis Fenniae*, **43**, 1–50.

Miller, C.E. (1976) Substrate-influenced morphological variations and taxonomic problems in freshwater, posteriorly uniflagellate phycomycetes, in *Recent Advances in Aquatic Mycology*, (ed E.B.G. Jones) Elek Science, London, pp. 469–87.

Miller, J.D. (1986) Toxic metabolites of epiphytic and endophytic fungi of conifer needles, in *Microbiology of the Phyllosphere*, (Eds N.J. Fokkema and J. van den Heuvel), Cambridge University Press, Cambridge, pp. 223–31.

Minter, D.W. (1981) Possible biological control of *Lophodermium seditiosum*, in *Current Research on Conifer Needle Diseases* (ed. C.S. Millar), Aberdeen University Press, Aberdeen, pp. 67–74.

Mishra, M.M., Kapoor, K.K., Jain, M.K. and Singh, C.P. (1981) Cellulose degradation by *Humicola lanuginosus*. *Transactions of the British Mycological Society*, **76**, 159–60.

Mishra, R.R. and Dickinson, C.H. (1981) Phylloplane and litter fungi of *Ilex aquifolium*. *Transactions of the British Mycological Society*, **77**, 329–37.

Mislivec, P.B. and Tuite, J. (1970) Temperature and relative humidity requirements of species of *Penicillium*, isolated from yellow dent corn kernels. *Mycologia*, **62**, 75–88.

Mitchell, C.P. and Dix, N.J. (1975) Growth and germination of *Trichoderma* spp. under the influence of soil fungistasis. *Transactions of the British Mycological Society*, **64**, 235–41.

Mitchell, C.P. and Millar, C.S. (1978) Mycoflora succession on Corsican pine needles colonized on the tree by three different fungi. *Transactions*

of the *British Mycological Society*, **71**, 303–17.
Mitchell, D.J. and Zentmyer, G.A. (1971) Effects of oxygen and carbon dioxide tensions on growth of several species of *Phytophthora*. *Phytopathology*, **61**, 787–91.
Mitchell, D.T. (1970) Fungus succession on dung on South African ostrich and Angora goat. *Journal of South African Botany*, **36**, 191–8.
Mitchell, R. and Alexander, M. (1963) Lysis of soil fungi by bacteria. *Canadian Journal of Microbiology*, **9**, 169–77.
Mohamed, S.H. (1986) The ecology of some agarics from disturbed soils and litter. MSc thesis, University of Stirling, UK.
Mohamed, S.H. and Dix, N.J. (1988) Resource utilization and distribution of *Coprinus comatus*, *Coprinus atramentarius*, *Lacrymaria velutina* and *Melanoleuca grammopodia*. *Transactions of the British Mycological Society*, **90**, 255–63.
Molina, R. and Trappe, J.M. (1982) Patterns of ectomycorrhizal specificity and potential among Pacific Northwest conifers and fungi. *Forest Science*, **28**, 423–58.
Molina, R., Massicotte, H.B. and Trappe, J.M. (1992a) Ecological role of specificity phenomena in ectomycorrhizal plant communities: potentials for interplant linkages and guild development, in *Mycorrhizas in Ecosystems*, (eds D.J. Read, D.H. Lewis, A.H. Fitter and I.J. Alexander) CAB International, Wallingford, UK, pp. 106–12.
Molina, R., Massicotte, H.B. and Trappe, J.M. (1992b) Specificity phenomena in mycorrhizal symbioses: community ecological consequences and practical implications, in: *Mycorrhizal Functioning: An Integrative Plant–Fungal Process*, (ed. M.F. Allen), CAB International, Wallingford, UK, pp. 106–12.
Mommaerts-Billiet, F. (1971) Recherches sur l'écosystème forêt serie B: La chênaie mélangée calcicole de Virelles-Blaimont. *Bulletin de la Société Royale de Botanique de Belgique*, **104**, 181–95.
Montgomery, R.A.P. (1982) The role of polysaccharidase enzymes in the decay of wood by basidiomycetes, in *Decomposer Basidiomycetes: their Biology and Ecology*, (eds J.C. Frankland, J.N. Hedger and M.J. Swift), *British Mycological Symposium 4*, Cambridge University Press, Cambridge, pp. 51–65.
Morton, A.G., England, D.J.F. and Towler, D.A. (1958) The physiology of sporulation in *Penicillium griseofulvum* Dierckx. *Transactions of the British Mycological Society*, **41**, 39–51.
Morton, L.H.G. and Eggins, H.O.W. (1976) Studies of interactions between wood-inhabiting microfungi. *Material und Organismen*, **11**, 197–214.
Morton, L.H.G. and Eggins, H.O.W. (1977) The effect of constant, alternating and fluctuating temperatures on the growth of some wood inhabiting fungi. *International Biodeterioration Bulletin*, **13**, 116–22.

Moser, M. (1949) Untersuchungen über den Einfluss von Waldbranden auf die Pilzvegetation I. *Sydowia*, **3**, 336–83.

Moser, M. (1958) Der Einfluss tiefer Temperaturen auf das Wachstum und die Lebensfahigkeit hoherer Pilze mit spezieller Berucksichtigung von Mykorrhiza-Pilzen. *Sydowia*, **12**, 386–99.

Moser, M. (1962) Die Rolle des Wassers im Leben der hoheren Pilze. *Schweizerische Zeitschrift fur Pilzkunde*, **9**, 129–41.

Moser, M. (1965) Der Wasserhaushalt hohere Pilze in Beziehung zu ihrem Standort. *Schweizerische Zeitschrift für Pilzkunde*, **11**, 161–72.

Moss, S.T. (ed.) (1986a) *The Biology of Marine Fungi*, Cambridge University Press, Cambridge, p. 382.

Moss, S.T. (1986b) Biology and phylogeny of the Labyrinthulales and Thraustochytriales, in *The Biology of Marine Fungi*, (ed. S.T. Moss), Cambridge University Press, Cambridge, pp. 105–29.

Mosse, B. (1975) A microbiologist's view of root anatomy, in *Soil Microbiology*, (ed. N. Walker), Butterworths, London, pp. 39–66.

Mowe, G., King, B. and Senn, S.J. (1983) Tropic responses of fungi to wood volatiles. *Journal of General Microbiology*, **129**, 779–84.

Mowll, J.L. and Gadd, G.M. (1985) Effect of vehicular lead pollution on phylloplane mycoflora. *Transactions of the British Mycological Society*, **84**, 685–9.

Mulinge, S.K. and Apinis, A.E. (1969) Occurrence of thermophilous fungi in stored moist barley grain. *Transactions of the British Mycological Society*, **53**, 361–70.

Mumma, R.O., Sekura, R.D. and Fergus, C.L. (1971) Thermophilic fungi: II. Fatty acid composition of polar and neutral lipids of thermophilic and mesophilic fungi. *Lipids*, **6**, 584–8.

Murakami, Y. (1987) Spatial distribution of *Russula* species in *Castanopsis cuspidata* forest. *Transactions of the British Mycological Society*, **89**, 187–93.

Mutch, R.W. (1970) Wildland fires and ecosystems: an hypothesis. *Ecology*, **51**, 1046–51.

Myers, J.A. and Cook, R.J. (1972) Induction of chlamydospore formation in *Fusarium solani* by abrupt removal of the organic carbon substrate. *Phytopathology*, **62**, 1148–53.

Mylyk, O.M. (1975) Heterokaryon incompatibility genes in *Neurospora crassa* detected using duplication producing chromosome rearrangements. *Genetics*, **80**, 107–24.

Myrold, D.D., Elliott, L.F., Papendick, R.I. and Campbell, G.S. (1981) Water potential–water content characteristics of wheat straw. *Soil Science Society of America Journal*, **45**, 329–33.

Nagel-de Boois, H.M. and Jansen, E. (1971) The growth of fungal mycelium in forest soil layers. *Revue d'Ecologie et de Biologie du Sol*, **8**, 509–20.

Nagy, L.A. and Harrower, K.M. (1979) Analysis of two Southern Hemisphere coprophilous fungus successions. *Transactions of the British Mycological Society*, **72**, 69–74.

Nakagiri, A., Tokumasu, S., Araki, H. *et al.* (1989) Succession of fungi in decomposing mangrove leaves in Japan. *Proceedings of the International Symposium on Microbial Ecology*, **5**, 297–301.

Nawawi, A. (1985) Aquatic hyphomycetes and other water-borne fungi from Malaysia. *Malayan Nature Journal*, **39**, 75–134.

Neergaard, P. (1977) *Seed Pathology*, vol. 1, Macmillan, London.

Newell, K. (1984a) Interactions between two decomposer basidiomycetes and a collembolan under Sitka spruce: distribution, abundance and selective grazing. *Soil Biology and Biochemistry*, **16**, 227–33.

Newell, K. (1984b) Interactions between two decomposer basidiomycetes and a collembolan under Sitka spruce: grazing and its potential effects on fungal distribution and litter decomposition. *Soil Biology and Biochemistry*, **16**, 235–9.

Newell, S.Y. (1976) Mangrove fungi: The succession in the mycoflora of red mangrove (*Rhizophora mangle* L.) seedlings, in *Recent Advances in Aquatic Mycology*, (ed. E.B.G. Jones), Elek Science, London, pp. 51–91.

Newell, S.Y. and Fell, J.W. (1992) Distribution and experimental responses to substrate of marine Oomycetes (*Halophytophthora* spp.) in mangrove ecosystems. *Mycological Research*, **96**, 851–6.

Newell, S.Y., Miller, J.D. and Fell, J.W. (1992) Rapid and pervasive occupation of fallen mangrove leaves by a marine zoosporic fungus. *Applied and Environmental Microbiology*, **53**, 2464–9.

Newhook, F.J. (1957) The relationship of saprophytic antagonism to control of *Botrytis cinerea* Pers. on tomatoes. *New Zealand Journal of Science and Technology*, **38**, 473–81.

Newman, E.I. (1969) Resistance to water flow in soil and plant. *Journal of Applied Ecology*, **6**, 1–12.

Newman, E.I. (1978) Root microorganisms: their significance in the ecosystem. *Biological Reviews*, **55**, 511–44.

Newman, E.I. (1985) The rhizosphere: carbon sources and microbial populations, in *Ecological Interactions in Soil*, (eds A.H. Fitter, D. Atkison, D.J. Read and M.B. Usher), Blackwell Publications, Oxford, pp. 107–22.

Newsham, K.K., Frankland, J.C., Boddy, L. and Ineson, P. (1992a) Effects of dry-deposited sulphur dioxide on fungal decomposition of angiosperm tree leaf litter. I. Changes in communities of fungal saprotrophs. *New Phytologist*, **122**, 97–110.

Newsham, K.K., Boddy, L., Frankland, J.C., and Ineson, P. (1992b) Effects of dry-deposited sulphur dioxide on fungal decomposition of angiosperm tree leaf litter. III. Decomposition rates and fungal respiration. *New Phytologist*, **122**, 127–40.

Nilsson, S. (1964) Freshwater hyphomycetes. Taxonomy, morphology and ecology. *Symbolae Botanicae Upsalienses*, **18**(2), 1–130.
Nilsson, T. (1975) Soft-rot fungi – decay patterns and enzyme production. *Material und Organismen*, **3**, 103–12.
Nilsson, T. and Daniel, G. (1989) Chemistry and microscopy of wood decay by some higher ascomycetes. *Holzforschung*, **43**, 11–18.
Nordbring-Hertz, B. (1973) Peptide-induced morphogenesis in the nematode-trapping fungus *Arthrobotrys oligospora*. *Physiologia Plantarum*, **29**, 223–33.
Nordbring-Hertz, B. (1984) Mycelial development and lectin–carbohydrate interactions in nematode-trapping fungi, in *The Ecology and Physiology of the Fungal Mycelium*, (eds D.H. Jennings and A.D.M. Rayner), *British Mycological Society Symposium* **8**, Cambridge University Press, Cambridge, pp. 419–32.
Nordbring-Hertz, B. and Jansson, H-B. (1984) Fungal development, predacity, and recognition of prey in nematode-destroying fungi, in *Current Perspectives in Microbial Ecology*, (eds M.J. Klug and C.A. Reddy), American Society for Microbiology, Washington, pp. 327–33.
Nordbring-Hertz, B., Friman, E. and Mattiason, B. (1982) A recognition mechanism in the adhesion of nematodes to nematode-trapping fungi, in *Lectins, Biology, Biochemistry and Clinical Biochemistry*, vol. 2, (ed. T.C. Bog-Hansen), Walter de Gruyter & Co., Berlin, pp. 83–90.
North, M.J. (1982) Comparative biochemistry of the proteinases of eucaryotic microorganisms. *Microbiological Reviews*, **46**, 308–40.
Nuss, I. (1975) Zur Ökologie der Porlinge. *Bibliotheca Mycologica*, Cramer, Vaduz, p. 45.
Nuss, I. (1986) Zur Ökologie der Porlinge. II. *Bibliotheca Mycologica*, Cramer, Berlin and Stuttgart, p. 105.
O'Donnell, J. and Dickinson, C.H. (1980) Pathogenicity of *Alternaria* and *Cladosporium* isolates on *Phaseolus*. *Transactions of the British Mycological Society*, **74**, 335–42.
Odunfa, V.S.A. and Oso, B.A. (1979) Fungal populations in the rhizosphere and rhizoplane of cowpea. *Transactions of the British Mycological Society*, **73**, 21–6.
Ogawa, M. (1985) Ecological characters of ectomycorrhizal fungi and their mycorrhizae. *JARQ*, **18**, 305–14.
Ogundero, V.W. (1980a) Fungal flora of poultry feeds. *Mycologia*, **72**, 200–2.
Ogundero, V.W. (1980b) Lipase activities of thermophilic fungi from mouldy groundnuts in Nigeria. *Mycologia*, **72**, 118–26.
Ogundero, V.W. (1981) Degradation of Nigerian palm products by thermophilic fungi. *Transactions of the British Mycological Society*, **77**, 267–71.
Okane, K. (1978) The seasonal change of aquatic phycomycetes in the Yokote river system in Japan. *Journal of Japanese Botany*, **53**, 245–52.

Okane, K. (1981) The distribution of aquatic phycomycetes and the abundance of their zoospores in strongly acidic rivers. *Japanese Journal of Ecology*, **31**, 405–12.

Old, K.M. (1967) Effects of natural soils on survival of *Cochliobolus sativus*. *Transactions of the British Mycological Society*, **50**, 615–24.

Old, K.M. and Nicholson, T.H. (1975) Electron microscopical studies of the microflora of roots of sand dune grasses. *New Phytologist*, **74**, 51–8.

Old, K.M. and Patrick, Z.A. (1979) Giant soil ameobae, potential biocontrol agents, in *Soil Borne Plant Pathogens*, (eds B. Schippers and W. Gams), Academic Press, London, pp. 617–28.

Old, K.M. and Robertson, W.M. (1970) Effects of lytic enzymes and natural soil on the fine structure of conidia of *Cochliobolus sativus*. *Transactions of the British Mycological Society*, **54**, 343–50.

Old, K.M. and Wong, J.N.F. (1976) Helically-lobed soil bacteria from fungal spores. *Soil Biology and Biochemistry*, **8**, 285–92.

Olsen, R., Odham, G. and Lindeberg, G. (1971) Aromatic substances in leaves of *Populus tremula* as inhibitors of mycorrhizal fungi. *Physiologia Plantarum*, **25**, 122–9.

Olthoff, T.H.A. and Estey, R.H. (1963) A nematotoxin produced by the nematophagous fungus *Arthrobotrys oligospora* Fresenius. *Nature, London*, **197**, 514–15.

Omar, M. and Heather, W.A. (1979) Effect of saprophytic phylloplane fungi on germination and development of *Melampsora larici-populina*. Transactions of the *British Mycological Society*, **72**, 225–31.

O'Neill, T.M. (1981) Narcissus smoulder: cause, epidemiology and host resistance. PhD thesis, University of Stirling.

O'Neill, T.M. and Mansfield, J.W. (1982) Antifungal activity of hydroxyflavans and other flavonoids. *Transactions of the British Mycological Society*, **79**, 229–37.

Onishi, H. (1963). Osmophilic yeasts. *Journal of Advanced Food Research*, **12**, 53–94.

Ooka, T. and Takeda, I. (1972) Studies on the peptide antibiotic suzukacillin. *Agricultural Biological Chemistry*, **36**, 112–19.

Oso, B.A. (1972) Conidial germination in *Cercospora arachidicola* Hori. *Transactions of the British Mycological Society*, **59**, 169–72.

Otjen, L. and Blanchette, R.A. (1985) Selective delignification of aspen wood blocks *in vitro* by three white-rot basidiomycetes. *Applied Environmental Microbiology*, **50**, 568–72.

Owens, L.D., Gilbert, R.G., Griebel, G.E. and Menzies, J.D. (1969) Identification of plant volatiles that stimulate microbial respiration and growth in soil. *Phytopathology*, **59**, 1468–72.

Pace, M.A. and Campbell, R. (1974) The effect of saprophytes on infection of leaves of *Brassica* sp. by *Alternaria brassicicola*. *Transactions of the British Mycological Society*, **63**, 193–6.

Pachenari, A. (1978) Necrotrophism in *Gliocladium roseum* (Bainier). MSc thesis, University of Stirling.

Pachenari, A. and Dix, N.J. (1980) Production of toxins and wall degradation enzymes by *Gliocladium roseum*. *Transactions of the British Mycological Society*, **74**, 561–6.

Padgett, D.E. (1978a) Observations on estuarine distribution of Saprolegniaceae. *Transactions of the British Mycological Society*, **71**, 141–3.

Padgett, D.E. (1978b) Salinity tolerance of an isolate of *Saprolegnia australis*. *Mycologia*, **70**, 1288–93.

Padgett, D.E. (1984) Evidence for extreme salinity tolerance by Saprolegniaceous fungi. *Mycologia*, **77**, 372–5.

Padgett, D.E., Kendrick, A.S., Hearth, J.H. and Webster, W.D. (1988) Influence of salinity, temperature and nutrient availability on the respiration of Saprolegniaceous fungi (Oomycetes). *Holarctic Ecology*, **11**, 119–26.

Pady, S.M. (1973) Ballistospore discharge in *Tilletiopsis minor*. *Canadian Journal of Botany*, **51**, 589–93.

Pady, S.M. (1974) Sporobolomycetaceae in Kansas. *Mycologia*, **66**, 333–8.

Page, R.M. (1959) Stimulation of asexual reproduction of *Pilobolus* by *Mucor plumbeus*. *American Journal of Botany*, **46**, 579–85.

Page, R.M. (1960) The effect of ammonia on growth and reproduction of *Pilobolus Kleinii*. *Mycologia*, **52**, 480–9.

Paine, R.L. (1968) Germination of *Polyporus betulinus* basidiospores on non-host species. *Phytopathology*, **58**, 1062.

Palmer, F.E., Emery, D.R., Stemmler, J. and Staley, J.T. (1987) Survival and growth of microcolonial rock fungi as affected by temperature and humidity. *New Phytologist*, **107**, 155–62.

Palmer, J.G., Murmanis, L. and Highley, T.L. (1983) Visualization of hyphal sheaths in wood decay Hymenomycetes. II. White-rotters. *Mycologia*, **75**, 1005–10.

Panasenko, V.T. (1967) Ecology of microfungi. *Botanical Review*, **33**, 189–215.

Panasenko, V.T. and Tatarenko, K.S. (1940) Psychrotolerant fungus flora of food products. *Mikrobiologia*, **9**, 579–84 (in Russian).

Papavizas, G.C. (1977a) Some factors affecting survival of sclerotia of *Macrophomina phaseolina* in soil. *Soil Biology and Biochemistry*, **9**, 337–41.

Papavizas, G.C. (1977b) Survival of sclerotia of *Macrophomina phaseolina* and *Sclerotium cepivorum* after drying and wetting treatments. *Soil Biology and Biochemistry*, **8**, 343–8.

Papavizas, G.C. (1985) *Trichoderma* and *Gliocladium*: biology, ecology and potential for biocontrol. *Annual Review of Phytopathology*, **23**, 23–54.

Papavizas, G.C. and Christensen, C.M. (1958) Grain storage studies 26. Fungus invasion and deterioration of wheats stored at low temperatures and moisture contents of 15 to 18 per cent. *Cereal Chemistry*, **35**, 27–34.

Papavizas, G.C. and Davey, C.B. (1961) Extent and nature of the rhizosphere of lupins. *Plant and Soil*, **14**, 215–36.

Papendick, R.I. and Campbell, G.S. (1981) Theory and measurement of water potential. *Soil Science Society of America Special Publication* no. 9, 1–22.

Papendick, R.I. and Mulla, D.J. (1986) Basic principles of cell and tissue water relations, in *Water, Fungi and Plants*, (eds P.G. Ayres and L. Boddy) *British Mycological Symposium 11*, Cambridge University Press, Cambridge, pp. 1–25.

Park, D. (1955) Experimental studies on the ecology of fungi in soil. *Transactions of the British Mycological Society*, **38**, 130–42.

Park, D. (1963) Evidence for a common fungal growth regulator. *Transactions of the British Mycological Society*, **46**, 541–8.

Park, D. (1965) Survival of microorganisms in soil, in *Ecology of Soil Borne Pathogens*, (eds K.F. Baker and W.C. Snyder) Murray, London, pp. 82–97.

Park, D. (1972a) Methods of detecting fungi in organic detritus in water. *Transactions of the British Mycological Society*, **58**, 281–90.

Park, D. (1972b) On the ecology of heterotrophic microorganisms in fresh-water. *Transactions of the British Mycological Society*, **58**, 291–9.

Park, D. (1975) A cellulolytic *Pythium* species. *Transactions of the British Mycological Society*, **65**, 249–57.

Park, D. (1976a) Carbon and nitrogen levels as factors influencing fungal decomposers, in *The Role of Terrestrial and Aquatic Organisms in Decomposition Processes*, (eds J.M. Anderson and J. Macfadyen), Blackwell, Oxford, pp. 41–59.

Park, D. (1976b) Nitrogen level and cellulose decomposition by fungi. *International Biodeterioration Bulletin*, **12**, 95–9.

Park, D. (1976c). Cellulose decomposition by a pythiaceous fungus. *Transactions of the British Mycological Society*, **66**, 65–70.

Park, D. (1977). *Pythium fluminum* sp. nov. with one variety and *P. uladhum* sp. nov. from cellulose in fresh-water habitats. *Transactions of the British Mycological Society* **69**, 225–31.

Park, D. (1980a) A method for isolating pigmented cellulolytic *Pythium* from soil. *Transactions of the British Mycological Society*, **75**, 491–517.

Park, D. (1980b) A two year study of numbers of cellulolytic *Pythium* in river water. *Transactions of the British Mycological Society*, **74**, 253.

Park, D. (1982a) Phylloplane fungi: tolerance of hyphal tips to drying. *Transactions of the British Mycological Society*, **79**, 174–8.

Park, D. (1982b) *Varicosporium* as a competitive soil saprophyte. *Transactions of the British Mycological Society*, **78**, 33–41.

Park, D. and Mckee, W. (1978) Cellulolytic *Pythium* as a component of the river mycoflora. *Transactions of the British Mycological Society*, **71**, 251–9.

Parker, A.D. (1979) Association between coprophilous ascomycetes and fecal substrates in Illinois. *Mycologia*, **71**, 1206–14.

Parker-Rhodes, A.F. (1955) Fairy ring kinetics. *Transactions of the British Mycological Society*, **38**, 5972.

Parkinson, D. and Clarke, J.H. (1964) Studies on fungi in the root region. III. Root surface fungi of three species of *Allium*. *Plant and Soil*, **20**, 166–74.

Parkinson, D. and Crouch, R. (1969) Studies on fungi in pinewood soil. V. Root mycofloras of seedlings of *Pinus nigra* var *laricio*. *Revue d'Ecologie et de Biologie du Sol*, **6**, 263–75.

Parkinson, D. and Pearson, R. (1967) Studies on fungi in the root region. VI. The occurrence of sterile dark fungi on the root surface. *Plant and Soil*, **27**, 113–19.

Parkinson, D., Gray, T.R.G and Williams, S.T. (1971) Methods for studying the ecology of soil microorganisms. *I.B.P. Handbook no. 19*, Blackwell, Oxford.

Parkinson, D., Taylor, G.S. and Pearson, R. (1963) Studies on fungi in the root region. I. Development of fungi on young roots. *Plant and Soil*, **19**, 332–49.

Parkinson, D. and Thomas, A. (1965) A comparison of methods for isolation of fungi from rhizospheres. *Canadian Journal of Microbiology*, **11**, 1001–8.

Parkinson, D., Visser, S. and Whittaker, J.B. (1979) Effects of collembolan grazing on fungal colonization of leaf litter. *Soil Biology and Biochemistry*, **11**, 529–35.

Pavlica, D.A., Hora, T.S., Breashaw, J.J. *et al.* (1978) Volatiles from soil influencing activities of soil fungi. *Phytopathology*, **68**, 758–65.

Pearce, R.B. (1987) Antimicrobial defences in secondary tissues of woody plants, in *Fungal Infection of Plants*, (eds G.F. Pegg and P.G. Ayres), *British Mycological Society Symposium 13*, Cambridge University Press, Cambridge, pp. 219–38.

Pearson, R. and Parkinson, D. (1961) The sites of excretion of ninhydrin-positive substances by broad bean seedlings. *Plant and Soil*, **13**, 391–6.

Pegler, D.N., Spooner, B.M. and Lewis Smith, R.I. (1980) Higher fungi of Antarctica, the subantarctic zone and Falkland Islands. *Kew Bulletin*, **35**, 500–62.

Perrott, P.E. (1960) Ecology of some aquatic phycomycetes. *Transactions of the British Mycological Society*, **43**, 19–30.

Persson, Y., Veenhuis, M. and Nordbring-Hertz, B. (1985) Morphogenesis and significance of hyphal coiling by nematode-trapping fungi in mycoparasitic relationships. *FEMS Microbiology Ecology*, **31**, 283–91.

Peters, B.G. (1955) A note on simple methods of recovering nematodes from the soil, in *Proceedings of the University of Nottigham 2nd Easter School in Agricultural Science*, (ed. D.K. Mck. Kevan), Butterworth, London, pp. 313–14.
Petersen, P.M. (1970a) Danish fireplace fungi. An ecological investigation on fungi on burns. *Dansk Botanisk Arkiv*, **27**, (3), 1–97.
Petersen, P.M. (1970b) Changes of the fungus flora after treatment with various chemicals. *Botanisk Tiddskrift*, **65**, 264–80.
Petersen, P.M. (1971) The macromycetes in a burnt forest area in Denmark. *Botanisk Tiddskrift*, **66**, 228–48.
Petersen, P.M. (1985) The ecology of Danish soil-inhabiting Pezizales with emphasis on edaphic conditions. *Opera Botanica*, **77**, 38.
Peterson, E.A. (1958) Observations on fungi associated with plant roots. *Canadian Journal of Microbiology*, **4**, 257–65.
Peterson, E.A. (1959) Seed-borne fungi in relation to colonization of roots. *Canadian Journal of Microbiology*, **5**, 579–82.
Petrini, L. and Petrini, O. (1985) Xylariaceous fungi as endophytes. *Sydowia*, **38**, 216–34.
Petrini, O. (1984) Endophytic fungi in British Ericaceae: a preliminary study. *Transactions of the British Mycological Society*, **83**, 510–12.
Petrini, O. (1986) Taxonomy of endophytic fungi of aerial plant tissues, in *Microbiology of the Phylloplane*, (eds N.J. Fokkema and J. van den Heuvel) Cambridge University Press, Cambridge, pp. 179–87.
Petrini, O. (1992) Fungal endophytes of tree leaves, in *Microbial Ecology of Leaves*, (eds J.H. Andrews and S.S. Hirano), Springer-Verlag, New York, pp. 179–97.
Petrini, O. Müller, E. and Luginbühl, M. (1979) Pilze als Endophyten von grünen Pflanzen. *Naturwissenschaften*, **66**, 262–3.
Petrini, O., Stone, J. and Carroll, F.E. (1982) Endophytic fungi in evergreen shrubs in Western Oregon: a preliminary study. *Canadian Journal of Botany*, **60**, 789–96.
Pianka, E.R. (1970) On r- and K-selection. *American Naturalist*, **104**, 592–7.
Pidacks, C., Whitehill, A.R., Pruess, L.M. *et al.* (1953) Coprogen, the isolation of a new growth factor required by *Pilobolus* species. *Journal of the American Chemical Society*, **75**, 6064–5.
Pirozynski, K.A. and Dalpé, Y. (1989) Geological history of the Glomaceae with particular reference to mycorrhizal symbiosis. *Symbiosis*, **7**, 1–36.
Pirozynski, K.A. and Hawksworth, D.L. (eds) (1988) *Coevolution of Fungi with Plants and Animals*, Academic Press, London.
Pirozynski, K.A. and Malloch, D.W. (1975) The origin of land plants: a matter of mycotrophism. *Biosystems*, **6**, 153–64.
Pitt, J.I. (1975) Food spoilage by xerophilic fungi, in *Water Relations of Foods*, (ed. R.B. Duckworth), Academic Press, London, pp. 273–307.

Pitt, J.I. and Christian, J.H.B. (1968) Water relations of xerophilic fungi isolated from prunes. *Applied Microbiology*, **16**, 1853–8.

Pitt, J.I. and Hocking, A.D. (1977) Influence of solute and hydrogen ion concentration on the water relations of some xerophilic fungi. *Journal of General Microbiology*, **101**, 35–40.

Pixton, S.W. and Warburton, S. (1971) Moisture content relative humidity equilibrium, at different temperatures, of some oilseeds of economic importance. *Journal of Stored Products Research*, **7**, 261–9.

Platt, M.W., Hadar, Y. and Chet, H. (1983) Fungal activities involved in lignocellulose degradation by *Pleurotus*. *Applied Microbiology and Biotechnology*, **20**, 150–4.

Plesofsky-Vig, N. and Brambl, R. (1993) Heat shock proteins in fungi, in *Stress Tolerance of Fungi*, (ed. D.H. Jennings), Marcel Dekker, New York, pp. 45–68.

Porter, D. (1989) Phylum Labyrinthomycota, in *Handbook of Protoctista*, (eds L. Margulis, J.O. Corliss, M. Melkonian and D.J. Chapman), Jones and Bartlett, Boston, pp. 388–98.

Potgieter, H.J. and Alexander, M. (1966) Susceptibility and resistance of several fungi to microbial lysis. *Journal of Bacteriology*, **91**, 1526–33.

Premdas, P.D. and Kendrick, B. (1991) Colonization of autumn-shed leaves by four aero-aquatic fungi. *Mycologia*, **83**, 317–21.

Prescott, C.E. and Parkinson, D. (1985) Effects of sulphur pollution on rates of litter decomposition in a pine forest. *Canadian Journal of Botany*, **63**, 1436–43.

Pugh, G.J.F. (1958) Leaf litter fungi found in *Carex paniculata* L. *Transactions of the British Mycological Society*, **41**, 185–95.

Pugh, G.J.F. (1967) Root colonization by fungi, in *Progress in Soil Biology*, (eds O. Graff and J.E. Satchell), North Holland, Amsterdam, pp. 21–6.

Pugh, G.J.F. (1980) Strategies in fungal ecology. *Transactions of the British Mycological Society*, **75**, 1–14.

Pugh, G.J.F. and Allsopp, D. (1982) Microfungi on Signy Island, South Orkney Islands. *British Antarctic Survey Bulletin*, **57**, 55–67.

Pugh, G.J.F. and Boddy, L. (1988) A view of disturbance and life strategies in fungi, in *Fungi and Ecological Disturbance*, (eds L. Boddy, R. Watling and A.J.E. Lyon), *Proceedings of the Royal Society of Edinburgh*, **94B**, 3–11.

Pugh, G.J.F. and Buckley, N.G. (1971a) *Aureobasidium pullulans*: an endophyte in sycamore and other trees. *Transactions of the British Mycological Society*, **57**, 227–31.

Pugh, G.J.F. and Buckley, N.G. (1971b) The leaf surface as a substrate for colonization by fungi, in *Ecology of Leaf Surface Microorganisms*, (eds T.F. Preece and C.H. Dickinson), Academic Press, London, pp. 431–46.

Pugh, G.J.F. and Evans, M.D. (1970) Keratinophilic fungi associated with birds. II. Physiological studies. *Transactions of the British Mycological Society*, **54**, 241–50.

Pugh, G.J.F. and Mathison, G.E. (1962) Studies on fungi in coastal soils. III. An ecological survey of keratinophilic fungi. *Transactions of the British Mycological Society*, **45**, 567–72.

Pugh, G.J.F. and Mulder, J.L. (1971) Mycoflora associated with *Typha latifolia*. *Transactions of the British Mycological Society*, **57**, 273–82.

Qasem, S.A. and Christensen, C.M. (1958) Influence of moisture content, temperature and time on the deterioration of stored corn by fungi. *Phytopathology*, **48**, 544–9.

Rai, B. and Singh, D.B. (1980) Antagonistic activity of some leaf surface microfungi against *Alternaria brassicae* and *Drechslera graminea*. *Transactions of the British Mycological Society*, **75**, 363–9.

Ramsbottom, J. (1953) *Mushrooms and Toadstools. A Study of the Activities of Fungi*, Collins, London.

Raper, J.R. (1966) Life cycles, basic patterns of sexuality, and sexual mechanisms, in *The Fungi: An Advanced Treatise*, vol. II, (eds G.C. Ainsworth and A.S. Sussman), Academic Press, New York and London, pp. 473–511.

Rayner, A.D.M. (1976) Dematiaceous hyphomycetes and narrow dark zones in decaying wood. *Transactions of the British Mycological Society*, **67**, 546–9.

Rayner, A.D.M. (1977a) Fungal colonization of hardwood stumps from natural sources. I. Non-basidiomycetes. *Transactions of the British Mycological Society*, **69**, 291–302.

Rayner, A.D.M. (1977b) Fungal colonization of hardwood stumps from natural sources. II. Basidiomycetes. *Transactions of the British Mycological Society*, **69**, 303–12.

Rayner, A.D.M. (1978) Interactions between fungi colonizing hardwood stumps and their possible role in determining patterns of colonization and succession. *Annals of Applied Biology*, **89**, 131–4.

Rayner, A.D.M. (1986) Water and the origins of decay in trees, in *Water Fungi and Plants*, (eds P.G. Ayres and L. Boddy), *British Mycological Symposium 11*, Cambridge University Press, Cambridge, pp. 321–41.

Rayner, A.D.M. (1991) The challenge of the individualistic mycelium. *Mycologia*, **83**, 48–71.

Rayner, A.D.M. and Boddy, L. (1986) Population structure and the infection biology of wood-decay in living trees. *Advances in Plant Pathology*, **5**, 119–60.

Rayner, A.D.M. and Boddy, L. (1988) *Fungal Decomposition of Wood: Its Biology and Ecology*, John Wiley & Sons, Chichester.

Rayner, A.D.M. and Coates, D. (1987) Regulation of mycelial organisation and responses, in *Evolutionary Biology of the Fungi*, (eds A.D.M.

Rayner, C.M. Brasier and D. Moore), *British Mycological Society Symposium 12*, Cambridge University Press, Cambridge, pp. 115–36.

Rayner, A.D.M. and Hedges, M.J. (1982) Observations on the specificity and ecological role of basidiomycetes colonizing dead elm wood. *Transactions of the British Mycological Society*, **78**, 370–3.

Rayner, A.D.M. and Todd, N.K. (1977) Intraspecific antagonism in natural populations of wood-decaying basidiomycetes. *Journal of General Microbiology*, **103**, 85–90.

Rayner, A.D.M. and Todd, N.K. (1979) Population and community structure and dynamics of fungi in decaying wood. *Advances in Botanical Research*, **7**, 333–420.

Rayner, A.D.M. and Webber, J.F. (1984) Interspecific mycelial interactions – an overview, in *The Ecology and Physiology of the Fungal Mycelium*, (eds D.H. Jennings and A.D.M. Rayner), *British Mycological Society Symposium 8*, Cambridge University Press, Cambridge, pp. 383–417.

Rayner, A.D.M., Boddy, L. and Dowson, C.G. (1987a) Temporary parasitism of *Coriolus* spp. by *Lenzites betulina*: a strategy for domain capture in wood decay fungi. *Microbial Ecology*, **45**, 53–8.

Rayner, A.D.M., Boddy, L. and Dowson, C.G. (1987b) Genetical interactions and developmental versatility during establishment of decomposer basidiomycetes in wood and tree litter, in *Ecology of Microbial Communities*, (eds M. Fletcher, Gray, T.R.G. and Jones, J.G.), *Society for General Microbiology Symposium 41*, Cambridge University Press, Cambridge, pp. 83–123.

Rayner, A.D.M., Watling, R. and Frankland, J.C. (1985) Resource relations – an overview, in *Developmental Biology of Higher Fungi*, (eds D. Moore, L.A. Casselton. D.A. Wood and J.C. Frankland) *British Mycological Society Symposium 10*, Cambridge University Press, Cambridge, pp. 1–40.

Rayner, A.D.M., Powell, K.A. Thompson, W. and Jennings, D.H. (1985) Morphogenesis of vegetative organs, in *Developmental Biology of the Higher Fungi*, (eds D. Moore, L.A. Casselton, D.A. Wood and J.C. Frankland), *British Mycological Symposium 10*, Cambridge University Press, Cambridge, pp. 249–79.

Read, D.J. (1984) The structure and function of the vegetative mycelium of mycorrhizal roots, in *The Ecology and Physiology of the Fungal Mycelium*, (eds D.H. Jennings and A.D.M. Rayner), *British Mycological Society Symposium 8*, Cambridge University Press, Cambridge, pp. 215–40.

Read, D.J. (1991) Mycorrhizas in ecosystems – Nature's response to the 'Law of the Minimum', in *Frontiers in Mycology* (ed. D.L. Hawksworth), CAB International, Wellington, UK, pp. 101–30.

Read, D.J. (1992) The mycorrhizal fungal community with special reference to nutrient mobilization, in *The Fungal Community: Its Organisation and Role in the Ecosystem*, 2nd edn, (eds G.C. Carroll and D.T. Wicklow), Marcel Dekker, New York, pp. 631–52.

Read, D.J. and Boyd, R. (1986) Water relations of mycorrhizal fungi and their host plants, in *Water, Fungi and Plants*, (eds P.G. Ayres and L. Boddy), *British Mycological Symposium 11*, Cambridge University Press, Cambridge, pp. 287–303.

Read, D.J., Leake, J.R. and Langdale, A.R. (1989) The nitrogen nutrition of mycorrhizal fungi and their host plants, in *Nitrogen, Phosphorus and Sulphur Utilization by Fungi* (eds L. Boddy, R. Marchant and D.J. Read), *British Mycological Symposium*, Cambridge University Press, Cambridge, pp. 181–204.

Read, S.J., Moss, S.T. and Jones, E.B.G. (1991) Attachment studies of aquatic hyphomycetes. *Philosophical Transactions of the Royal Society of London B*, **334**, 449–57.

Read, S.J., Moss, S.T. and Jones, E.B.G. (1992) Attachment and germination of conidia, in *The Ecology of Aquatic Hyphomycetes* (ed. F. Bärlocher) Springer-Verlag, Berlin, pp. 135–51.

Reddell, P. and Malajczuk, N. (1984) Formation of mycorrhizae by Jarrah (*Eucalyptus marginata* Donn ex Smith) in litter and soil. *Australian Journal of Botany*, **32**, 511–20.

Rees, G. (1980) Factors affecting the sedimentation of marine fungal spores. *Botanica Marina*, **23**, 375-85.

Rees, G., Johnson, R.G. and Jones, E.B.G. (1979) Lignicolous marine fungi from Danish sand dunes. *Transactions of the British Mycological Society*, **72**, 99–106.

Reese, E.T. and Levinson, H.S. (1952) A comparative study of the breakdown of cellulose by microorganisms. *Physiologia Plantarum*, **5**, 345–66.

Reese, E.T., Siu, R.G.H. and Levinson, H.S. (1950) The biological degradation of soluble cellulose derivatives and its relationship to the mechanism of cellulose hydrolysis. *Journal of Bacteriology*, **59**, 458–97.

Reilly, P.J. (1981) Xylanases: structure and function, in *Trends in the Biology of Fermentations for Fuels and Chemicals*, (ed. A. Hollaender) Plenum, New York, pp. 111–29.

Revay, A and Gönczol, J. (1990) Longitudinal distribution and colonization patterns of wood-inhabiting fungi in a mountain stream in Hungary. *Nova Hedwigia*, **51**, 505–20.

Reyes, A.A. and Mitchell, J.E. (1962) Growth response of several isolates of *Fusarium* in rhizospheres of host and non-host plants. *Phytopathology*, **52**, 1196–200.

Rice, P.F. (1970) Some biological effects of volatiles emanating from wood. *Canadian Journal of Botany*, **48**, 710–35.

Richardson, M.J. (1972) Coprophilous ascomycetes on different dung types. *Transactions of the British Mycological Society*, **58**, 37–48.

Richardson, M.J. and Leadbeater, G. (1972) *Piptocephalis fimbriata* sp. nov. and observations on the occurrence of *Piptocephalis* and *Syncephalis*. *Transactions of the British Mycological Society*, **58**, 205–15.

Richardson, M.J. and Watling, R. (1968) Key to fungi on dung. *Bulletin of the British Mycological Society*, **2**, 1843.

Richardson, M.J. and Watling, R. (1969) Key to fungi on dung. *Bulletin of the British Mycological Society*, **3**, 86–8, 121–4.

Richter, M., Wilms, W. and Scheffer, F. (1968) Determination of root exudates in a sterile continuous flow culture. II. Short term and long term variations of exudation intensity. *Plant Physiology*, **43**, 1747–54.

Ride, J.P. and Drysdale, R.B. (1972) A rapid method for the chemical estimations of filamentous fungi in plant tissue. *Physiological Plant Pathology*, **2**, 7–15.

Rishbeth, J. (1963) Stump protection against *Fomes annosus*. III. Inoculation with *Peniophora gigantea*. *Annals of Applied Biology*, **52**, 63–77.

Rishbeth, J. (1968) The growth rate of *Amillaria mellea*. *Transactions of the British Mycological Society*, **51**, 575–86.

Rishbeth, J. (1978) Effects of soil temperature and atmosphere on growth of *Armillaria* rhizomorphs. *Transactions of the British Mycological Society*, **70**, 213–20.

Ritchie, D. (1957) Salinity optima for marine fungi affected by temperature. *American Journal of Botany*, **44**, 870–4.

Ritchie, D. and Jacobsohn, M.K. (1963) The effects of osmotic and nutritional variation on the growth of a salt-tolerant fungus, *Zalerion eistla*, in *Symposium on Marine Microbiology*, (ed. C.H. Openheimer), Thomas, Springfield, IL, pp. 286–9.

Roberts, R.E. (1963) A study of the distribution of certain members of the Saprolegniaceae. *Transactions of the British Mycological Society*, **46**, 213–24.

Robinson, P.M. and Park, D. (1966) Volatile inhibitors of spore germination produced by fungi. *Transactions of the British Mycological Society*, **49**, 639–49.

Robinson, R.K. (1972) The production by roots of *Calluna vulgaris* of a factor inhibitory to growth of some mycorrhizal fungi. *Journal of Ecology*, **60**, 219–24.

Rogers, J.D. (1979) The Xylariaceae: systematic, biological and evolutionary aspects. *Mycologia*, **71**, 1–42.

Rogers, W.S. (1968) Amount of cortical and epidermal tissue shed from roots of apple. *The Journal of Horticultural Science*, **43**, 527–8.

Rohrmann, S. and Molitoris, H.P. (1986) Morphological and physiological adaptations of the Cyphellaceous fungus *Halocyphina villosa* (Aphyllophorales) to its marine habitat. *Botanica Marina*, **29**, 539–47.

Romell, L.-G. (1938) A trenching experiment in spruce forest and its

bearing on problems of mycotrophy. *Svensk Botanisk Tidskrift*, **32**, 89–99.
Rose, A.H. (1962) Biochemistry of the psychrophilic habit: Studies on the low maximum temperature, in *Recent Progress in Microbiology*, VIIIth International Congress for Microbiology, University of Toronto Press, Montreal, pp. 193–200.
Rosenberg, S.L. (1975) Temperature and pH optima for 21 species of thermophilic and thermotolerant fungi. *Canadian Journal of Microbiology*, **21**, 1535–40.
Rosenberg, S.L. (1978) Cellulose and lignocellulose degradation by thermophilic and thermotolerant fungi. *Mycologia*, **70**, 1–13.
Ross, D.J. (1987) Soil microbial biomass estimated by the fumigation-incubation procedure: seasonal fluctuations and influence of soil moisture content. *Soil Biology and Biochemistry*, **19**, 397–404.
Ross, D.J. (1988) Modifications to the fumigation procedure to measure microbial biomass C in wet soils under pasture: influence on estimates of seasonal fluctuations in the soil biomass. *Soil Biology and Biochemistry*, **20**, 377–83.
Ross, D.J. (1989) Estimation of soil microbial C by a fumigation extraction procedure: influence of soil moisture content. *Soil Biology and Biochemistry*, **21**, 767–72.
Rossall, S. and Mansfield, J.W. (1980) Investigation of the causes of poor germination of *Botrytis* spp. on broad bean leaf (*Vicia faba*). *Physiological Plant Pathology*, **16**, 369–82.
Rouatt, J.W. and Katznelson, H. (1961) A study of the bacteria on the root surface and in the rhizosphere soil of crop plants. *Journal of Applied Bacteriology*, **24**, 164–71.
Rovira, A.D. (1956) The nature of exudates from oats and peas. *Plant and Soil*, **7**, 178–94.
Rovira, A.D. (1959) Plant root excretions in relation to the rhizosphere effect. IV. Influence of plant species, age of plant, light, temperature and calcium nutrition on exudation. *Plant and Soil*, **11**, 53–64.
Rovira, A.D. (1973) Zones of exudation along plant roots and spatial distribution of microorganisms in the rhizosphere. *Pesticide Science*, **4**, 361–6.
Rovira, A.D. and McDougall, B.M. (1967) Microbiological and biochemical aspects of the rhizosphere, in *Soil Biochemistry*, (eds A.D. Mclaren and G.H. Peterson), Arnold, London, pp. 417–63.
Rovira, A.D. and Ridge, E.H. (1973) Exudation of C^{14} labelled compounds from wheat roots: influence of nutrients, microorgansisms and added organic compounds. *New Phytologist*, **72**, 1081–7.
Rovira, A.D., Newman, E.I., Bowen, H.J. and Campbell, R. (1974) Quantitative assessment of the rhizoplane by direct microscopy. *Soil Biology and Biochemistry*, **6**, 211–16.

Ruinen, J. (1956) Occurrence of *Beijerinckia* species in the phyllosphere. *Nature*, **177**, 220–1.
Ruinen, J. (1966) The phyllosphere. IV. Cuticle decomposition by microorganisms in the phyllosphere. *Annales de l'Institut Pasteur*, **3**, 342–6.
Ruscoe, Q.W. (1971) Mycoflora of living and dead leaves of *Nothofagus truncata*. *Transactions of the British Mycological Society*, **56**, 463–74.
Russell, E.W. (1961) *Soil Conditions and Plant Growth*, Longmans, London.
Safar, M.H. and Cooke, R.C. (1988) Exploitation of faecal resource units by coprophilous ascomycetes. *Transactions of the British Mycological Society*, **90**, 593–9.
Sagar, B.F. (1988) Microbial cellulases and their action on cotton fibres, in *Cotton Strip Assay: An Index of Decomposition in Soils*, (eds A.F. Harrison, A.F. Latter and D.W.H. Walton), Institute of Terrestrial Ecology, Grange-over-Sands, UK, pp. 17–20.
Sagara, N. (1992) Experimental disturbances and epigaeous fungi, in *The Fungal Community: Its Organisation and Role in the Ecosystem*, (eds G.C. Carroll and D.T. Wicklow), Marcel Dekker, New York, pp. 265–74.
Salt, G.A. (1979) The increasing interest in minor pathogens, in *Soil Borne Plant Pathogens*, (eds B. Schipper and W. Gams), Academic Press, London, pp. 289–312.
Sanders, P.F. and Anderson, J.M. (1979) Colonization of wood blocks by aquatic hyphomycetes. *Transactions of the British Mycological Society*, **73**, 103–7.
Sanders, P.F. and Webster, J. (1978) Survival of aquatic hyphomycetes in terrestrial situations. *Transactions of the British Mycological Society*, **71**, 231–7.
Sanders, P.F. and Webster, J. (1980) Sporulation responses of some aquatic hyphomycetes to flowing water. *Transactions of the British Mycological Society*, **74**, 601–5.
Satchell, J.E. and Lowe, D.G. (1967) Selection of leaf litter by *Lumbricus terrestris*, in *Progress in Soil Biology*, (eds O. Graff and J.E. Satchell), North-Holland, Amsterdam, pp. 102–19.
Saunders, G.F. and Campbell, L.L. (1966) Ribonucleic acid ribosomes of *Bacillus stearothermophilus*. *Journal of Bacteriology*, **91**, 332–9.
Savory, J.G. (1954) Breakdown of timber by ascomycetes and fungi imperfecti. *Annals of Applied Biology*, **41**, 336–47.
Savory, J.G. (1964) Dry rot: a re-appraisal. *Record of the Fourteenth Annual Convention of the British Wood Preservers Association*, 69–76.
Savory, J.G. and Pinion, L.C. (1958) Chemical aspects of decay of beech wood by *Chaetomium globosum*. *Holzforschung*, **12**, 99–103.
Saxena, G. and Mukerji, K.G. (1991) Distribution of nematophagous

fungi in Varanasi, India. *Nova Hedwigia*, **52**, 487–95.
Schanel, L. and Esser, K. (1971) The phenoloxidases of the ascomycete *Podospora anserina*. VII. Substrate specificity of laccases with different molecular structures. *Archives of Microbiology*, **77**, 111–17.
Schaumann, K. (1968) Marine höhere Pilze (ascomycetes und fungi imperfecti) aus dem Weser-Ästuar. *Veröffentlichen des Instituts Meeresforschung in Bremerhaven*, **11**, 93–117.
Schaumann, K. (1969) Über marine höhere Pilze von Holzsubstraten der Nordsee-insel Helgoland. *Berichte der Deutschen Botanischen Gesellschaft*, **82**, 307–27.
Schaumann, K. (1975) Ökologische Untersuchungen über höhere Pilze in Meer- und Brackwasser der Deutschen Bucht unter besonderer Berucksichtigung der holzbesiedelungen Arten. *Veröffentlichen des Instituts Meeresforschung in Bremerhaven*, **15**, 79–182.
Scheffer, T.C. (1986) O_2 requirements for growth and survival of wood-decaying and sapwood-staining fungi. *Canadian Journal of Botany*, **64**, 1957–63.
Scheffer, T.C. and Cowling, E.B. (1966) Natural resistance of wood to microbial deterioration. *Annual Review of Phytopathology*, **4**, 147–70.
Schippers, B., Boerwinkel, D.J. and Konings, H. (1978) Ethylene not responsible for inhibition of conidium germination by soil volatiles. *Netherlands Journal of Plant Pathology*, **84**, 101–7.
Schippers, B., Meijer, J.W. and Liem, J.I. (1982) The effect of ammonia from alkaline soils on germination and growth of several soil fungi. *Transactions of the British Mycological Society*, **79**, 253–9.
Schitzer, M. and Negroud, J.A. (1975) Further investigation on the chemistry of fungal humic acids. *Soil Biology and Biochemistry*, **7**, 365–71.
Schmiedeknecht, M. (1960) Feuchtigkeit als standort Faktor für mikroskopische Pilze. *Zeitschrift für Pilzkunde*, **25**, 69–77.
Schnürer, J. and Roswall, T. (1982) Fluorescein diacetate hydrolysis as a measure of total microbial activity in soil and litter. *Applied and Environmental Microbiology*, **43**, 1256–61.
Schobert, B. (1977) Is there an osmotic regulatory mechanism in algae and higher plants? *Journal of Theoretical Biology*, **68**, 17–26.
Schobert, B. and Tschesche, H. (1978) Unusual solution properties of proline and its interaction with proteins. *Biochimica et Biophysica Acta*, **541**, 270–7.
Schramm, J.E. (1966) Plant colonization studies on black wastes from anthracite mining in Pennsylvaania. *Transactions of the American Philosophical Society*, **56**, 1–194.
Schroth, M.N. and Snyder, W.C. (1961) The effect of host exudates on chlamydospore germination of the bean root fungus *Fusarium solani f phaseoli*. *Phytopathology*, **51**, 389–93.

Seal, K.J. and Eggins, H.O.W. (1972) The role of microorganisms in biodegradation of farm animal waste with particular reference to intensively produced wastes. A review. *International Biodeterioration Bulletin*, **8**, 95–100.

Seehann, G., Leise, W. and Kess, B. (1975) Lists of fungi in soft-rot tests. *International Research Group*. Wood Preservation, Princes Risborough, UK, Document no. IRG/WP/105.

Seifert, K. (1968) On the systematics of wood-rots: their chemical and physical characteristics. *Holz als Roh und Werkstoff*, **26**, 208–15.

Seitz, L.M., Sauer, D.B., Burrough, S.R. et al. (1979) Ergosterol as a measure of fungal growth. *Phytopathology*, **69**, 1202–3.

Setliff, E.C. and Eudy, W.W. (1980) Screening white-rot fungi for their capacity to delignify wood, in *Lignin Biodegradation: Microbiology Chemistry and Potential Applications*, vol. 1, (eds T. Kirk, T. Higuchi and Hou-min Chang), CRC Press, Boca Raton, FL, pp. 135–49.

Sewell, G.W.F. (1959) Studies of fungi in a *Calluna* heathland soil. I. Vertical distribution in soil on root surfaces. *Transactions of the British Mycological Society*, **42**, 343–53.

Sewell, G.W.F. and Brown, J.C. (1959) Ecology of *Mucor ramannianus* Moller. *Nature*, **183**, 1344–5.

Shantz, H.L. and Piemeisel, R.L. (1917) Fungus fairy rings in Eastern Colorado and their effect on vegetation. *Journal of Agricultural Research*, **11**, 191–245.

Sharma, P.D. (1973) Succession of fungi on decaying *Setaria glauca*. *Annals of Botany*, **37**, 203–8.

Sharma, P.D., Fisher, P.J. and Webster, J. (1977) Critique of the chitin assay technique for estimation of fungal biomass. *Transactions of the British Mycological Society*, **69**, 479–83.

Sharp, R.F. (1974) Nitrogen fixation in deteriorating wood: the incorporation of $^{15}N_2$ and the effect of environmental conditions on acetylene reduction. *Soil Biology and Biochemistry*, **7**, 9–14.

Sharp, R.F. and Millbank, J.W. (1973) Nitrogen fixation in deteriorating wood. *Experimenta*, **29**, 895–6.

Shaw, M. (1963) The physiology and host parasite relations of the rusts. *Annual Review of Phytopathology*, **1**, 259–94.

Shaw, P.J.A. (1992) Fungi, fungivores and fungal food webs, in *The Fungal Community: Its Organisation and Role in the Ecosystem*, 2nd edn, (eds G.C. Carroll and D.T. Wicklow), Marcel Dekker, New York, pp. 295–310.

Shaw, P.J.A., Dighton, J., Poskitt, J. and McLeod, A.R. (1992) The effects of sulphur dioxide and ozone on the mycorrhizas of Scots pine and Norway spruce in a field fumigation system. *Mycological Research*, **96**, 785–91.

Shearer, C.A. (1972) Fungi of the Chesapeake bay and its tributaries. III.

The distribution of wood-inhabiting ascomycetes and fungi imperfecti of the Patuxent river. *American Journal of Botany*, **59**, 961–9.
Shearer, C.A. (1992) The role of woody debris in the life cycles of aquatic hyphomycetes, in *The Ecology of Aquatic Hyphomycetes*, (ed. F. Bärlocher), Springer-Verlag, Berlin, pp. 77–98.
Shearer, C.A. (1993) The freshwater ascomycetes. *Nova Hedwigia*, **55**, 1–33.
Shearer, C.A. and Lane, L. (1983). Comparison of three techniques for the study of aquatic hyphomycete communities. *Mycologia*, **75**, 498–508.
Shearer, C.A. and Webster, J. (1985a) Aquatic hyphomycete community structure in the River Teign. I. Longitudinal distribution patterns. *Transactions of the British Mycological Society*, **84**, 489–501.
Shearer, C.A. and Webster, J. (1985b) Aquatic hyphomycete community structure in the River Teign. II. Temporal distribution patterns. *Transactions of the British Mycological Society*, **84**, 503–7.
Shearer, C.A. and Webster, J. (1985c) Aquatic hyphomycete community structure in the River Teign. III. Comparison of sampling techniques. *Transactions of the British Mycological Society* **84**, 509–18.
Shearer, C.A. and Webster, J. (1991) Aquatic hyphomycetes in the River Teign. IV. Twig colonization. *Mycological Research*, **95**, 413–20.
Shearer, C.A. and Zare-Maivan, H. (1988) *In vitro* hyphal interactions among wood- and leaf-inhabiting ascomycetes and fungi-imperfecti from freshwater habitats. *Mycologia*, **80**, 31–7.
Sherriff, D.W. (1973) An infra-red psychrometer for detecting changes in the humidity of leaf boundary layers. *Journal of Experimental Botany*, **24**, 641–7.
Sherwood, M. and Carroll, G.C. (1974) Fungal succession on needles and young twigs of old growth Douglas fir. *Mycologia*, **66**, 499–506.
Shigo, A.L. (1958) Fungi isolated from oak-wilt trees and their effects on *Ceratocystis fagacearum*. *Mycologia*, **50**, 757–69.
Shigo, A.L. (1964) Organism interactions in beech bark disease. *Phytopathology*, **54**, 263–9.
Shipton, A. and Brown, J.F. (1962) A whole leaf clearing and staining technique to demonstrate host pathogen relationships of wheat stem rust. *Phytopathology*, **52**, 1313.
Shortle, W.C. and Cowling, E.B. (1978) Interaction of live sapwood and fungi commonly found in discolored and decayed wood. *Phytopathology*, **68**, 617–23.
Shubin, V.I., Ronkonen, N.I. and Saukonen, A.V. (1977) The effect of fertilizers on the fructification of macromycetes on young birch trees. (*in Russian*) *Mikologiyai Fitopatologia*, **11**, 294–303.
Siegel, M.R. and Schardl, C.L. (1992) Fungal endophytes of grasses: detrimental and beneficial associations, *Microbial Ecology of Leaves*

(eds J.H. Andrews and S.S. Hirano), Springer-Verlag, New York, pp. 198–221.

Simon, E.W. (1974) Phospholipids and plant membrane permeability. *New Phytologist*, **73**, 377–420.

Sinensky, M. (1974) Homeoviscous adaptation – a homeostatic process that regulates the viscosity of membrane lipids in *Escherichia coli*. *Proceedings of the National Academy of Science*, **71**, 522–5.

Singh, D.B, Singh, S.P. and Gupta, R.C. (1979) Antifungal effect of volatiles from seeds of some Umbelliferae. *Transactions of the British Mycological Societyh*, **73**, 349–50.

Singh, N. and Webster, J. (1972) Effect of coprophilous species of *Mucor* and bacteria on sporangial production of *Pilobolus*. *Transactions of the British Mycological Society*, **59**, 43–9.

Singh, N. and Webster, J. (1973) Antagonism between *Stilbella erythrocephala* and other coprophilous fungi. *Transactions of the British Mycological Society* **61**, 487–95.

Singh, N. and Webster, J. (1976) Effect of dung extracts on the fruiting of *Pilobolus* species. *Transactions of the British Mycological Society*, **67**, 377–9.

Siu, R.G.H. (1951) *Microbial Decomposition of Cellulose*, Reinhold, New York.

Skidmore, A.M. (1976) Secondary spore production amongst phylloplane fungi. *Transactions of the British Mycological Society*, **66**, 161–3.

Skidmore, A.M. and Dickinson, C.H. (1976) Colony interactions and hyphal interference between *Septoria nodorum* and phylloplane fungi. *Transactions of the British Mycological Society*, **66**, 57–64.

Skujins, J.J., Potgieter, H.J. and Alexander, M. (1965) Dissolution of fungal cell walls by a streptomycete chitinase and $\beta(1-3)$ glucanase. *Archives of Biochemistry and Biophysics*, **3**, 358–64.

Smedegaard-Petersen, V. and Tolstrup, K. (1986) Yield reducing effect of saprophytic leaf fungi in barley crops, in *Microbiology of the Phyllosphere*, (eds N.J. Fokkema and J. van den Heuvel), Cambridge University Press, Cambridge, pp. 160–71.

Smith, A.M. (1973) Ethylene as the cause of soil fungistasis. *Nature*, **246**, 311–13.

Smith, A.M. (1976) Ethylene in soil biology. *Annual Review of Phytopathology*, **14**, 53–73.

Smith, J.D. (1957) Fungi and turf diseases. 7. Fairy rings. *Journal of the Sports Turf Research Institute*, **33**, 324–52.

Smith, J.D. (1980) Is biological control of *Marasmius oreades* fairy ring possible? *Plant Disease*, **64**, 348–55.

Smith, J.D. and Rus, R. (1978) Antagonism in *Marasmius oreades* fairy rings. *Journal of the Sports Turf Research Institute*, **54**, 97–105.

Smith, J.D. and Rupps, R. (1978) Antagonism in *Marasmius oreades* fairy

rings. *Journal of the Sports Turf Research Institute*, **54**, 97–105.
Smith, J.D., Jackson, N. and Woolhouse, A.R. (1989) Fairy rings, in *Fungal Diseases of Amenity Turf Grasses*, E. and F.N. Spon, London, pp. 341–52.
Smith, K.A. (1978) Inefficiency of ethylene as a regulator of soil microbial activity. *Soil Biology and Biochemistry*, **10**, 269–72.
Smith, M.L., Bruhn, J.N. and Anderson, J.B. (1992) The fungus *Armillaria bulbosa* is among the largest and oldest living organisms. *Nature*, **356**, 428–31.
Smith, R.T., Blanchard, R.O. and Shortle, W.C. (1981) Postulated mechanisms of biological control of decay fungi in red maple wounds treated with *Trichoderma harzianum*. *Phytopathology*, **71**, 496–8.
Smith, S.N., Armstrong, R.A. and Rimmer, J.J. (1984) Influence of environmental factors on zoospores of *Saprolegnia diclina*. *Transactions of the British Mycological Society*, **82**, 413–22.
Smith, S.N., Ince, E. and Armstrong, R.A. (1990) Effect of osmotic and matric potential on *Saprolegnia diclina* and *S. ferax*. *Mycological Research*, **94**, 71–7.
Smith, W.H. (1970) Root exudates of seedling and mature sugar maple. *Phytopathology*, **60**, 701–3.
Smith, W.H. (1972) The influence of artificial defoliation on exudates of sugar maple. *Soil Biology and Biochemistry*, **4**, 111–13.
Smith, W.L., Moline, H.E. and Johnson, K.S. (1979) Studies with *Mucor* species causing post harvest decay of fresh produce. *Phytopathology*, **69**, 865–9.
Sneh, B., Humble, S.J. and Lockwood, J.L. (1977) Parasitism of oospores of *Phytophthora megasperma* var. *sojae*, *P. cactorum*, *Pythium* spp. and *Aphanomyces euteiches* in soil by Oomycetes, Chytridiomycetes, Hyphomycetes, Actinomycetes and bacteria. *Phytopathology*, **67**, 622–8.
Söderstrom, B.E. (1975) Vertical distribution of microfungi in a spruce forest soil in the south of Sweden. *Transactions of the British Mycological Society*, **65**, 419–25.
Söderstrom, B.E. (1977) Vital staining of fungi in pure cultures and in soil with fluorescein diacetate. *Soil Biology and Biochemistry*, **9**, 59–63.
Somkuti, G.A. and Babel, F.J. (1968) Purification and properties of *Mucor pusillus* acid protease. *Journal of Bacteriology*, **95**, 1407–14.
Somkuti, G.A. and Somkuti, A.C. (1969) Lipase of *Mucor pusillus*. *Applied Microbiology*, **17**, 606–10.
Somkuti, G.A., Babel, F.J. and Somkuti, A.C. (1969) Cellulolysis by *Mucor pusillus*. *Applied Microbiology*, **17**, 888–92.
Sommers, L.E., Harris, R.F., Dalton, F.N. and Gardner, W.R. (1970) Water potential relations of three root-infecting *Phytophthora* species. *Phytopathology*, **60**, 932–4.

Soprunov, F.F. (1958) *Predacious Hyphomycetes and their Application in the Control of Pathogenic Nematodes*. Academy of Sciences Turkmenistan. English translation 1968, Israel Program for Scientific Translations.

Spano, S.D., Jurgensen, M.F., Larsen, M.J. and Harvey, A.E. (1982) Nitrogen fixing bacteria in Douglas fir residue decayed by *Fomitopsis pinicola*. *Plant and Soil*, **68**, 117–23.

Sparrow, F.K. (1960) *Aquatic Phycomycetes*, University of Michigan Press, Ann Arbor.

Sparrow, F.K. (1968) Ecology of freshwater fungi, in *The Fungi: An Advanced Treatise*, vol. III (eds G.C. Ainsworth and A.S. Sussman), Academic Press, London, pp. 41–93.

Sparrow, F.K. (1973) Chytridiomycetes. Hyphochytridiomycetes, in *The Fungi: An Advanced Treatise*, vol. IVB, (eds G.C. Ainsworth, F.K. Sparrow and A.S. Sussman), Academic Press, London, pp. 61–73.

Spensley, P.C. (1963) Aflatoxin, the active principle in turkey 'X' disease. *Endeavour* **22**, 75–9.

Sridhar, K.R. and Bärlocher, F. (1992a) Endophytic aquatic hyphomycetes of roots of spruce, birch and maple. *Mycological Research*, **96**, 305–8.

Sridhar, K.R. and Bärlocher, F. (1992b) Aquatic hyphomycetes in spruce roots. *Mycologia*, **84**, 580–4.

Sridhar, K.R., Chandrashekar, K.R. and Kaveriappa, K.M. (1992) Research on the Indian subcontinent, in *The Ecology of Aquatic Hyphomycetes*, (ed. F. Bärlocher), Springer-Verlag, Berlin, pp. 182–225.

Staley, J.T., Palmer, F. and Adams, J.B. (1982) Microcolonial fungi common inhabitants on desert rocks. *Science*, **215**, 1093–5.

Stalpers, J.A. (1978) Identification of wood-inhabiting Aphyllophorales in pure culture. *Studies in Mycology*, Baarn, **16**, 1–248.

Stanghellini, M.E. and Hancock, J.G. (1971) The sporangium of *Pythium ultimum* as a survival structure in soil. *Phytopathology*, **61**, 157–64.

Steiner, G.W. and Lockwood, J.L. (1969) Soil fungistasis: sensitivity of spores in relation to germination time and size. *Phytopathology*, **59**, 1084–92.

Stenesh, J. and Yang, C. (1967) Characterization and stability of ribosomes from mesophilic and thermophilic bacteria. *Journal of Bacteriology*, **93**, 930–6.

Stenlid, J. and Rayner, A.D.M. (1989) Environmental and endogenous controls of developmental pathways: variation and its significance in the forest pathogen *Heterobasidion annosum*. *New Phytologist*, **113**, 245–58.

Stenton, H. (1958) Colonization of roots of *Pisum sativum* by fungi. *Transactions of the British Mycological Society*, **41**, 74–80.

Sterne, R.E. and McCarver, T.H. (1978) Osmotic effects on radial growth

rate and specific growth rate of three soil fungi. *Canadian Journal of Microbiology*, **24**, 1434–7.
Sterne, R.E., Zentmyer, G.A. and Bingham, F.T. (1976) The effect of osmotic potential and specific ions on growth of *Phytophthora cinnamomi*. *Phytopathology*, **66**, 1398–402.
Stevens, J. (1987) Interaction of the soft-rot fungus *Phialophora richardsiae* and a bacterium. *Mycologia*, **79**, 794–7.
Stevens, L., Dix, N.J. and Thompstone, A. (1983) Effects of high water activity on growth and metabolism in *Aspergillus sejunctus*. *Transactions of the British Mycological Society*, **80**, 527–71.
Stevenson, D.R. and Thompson, C.J. (1976) Fairy ring kinetics. *Journal of Theoretical Biology*, **58**, 143–63.
Stirling, G.R. (1988) Biological control of plant parasitic nematodes, in *Diseases of Nematodes*, vol. II (eds G.O. Poinar and H-B. Jansson), Press, Boca Raton, FL.
Stirling, G.R. and Mankau, R. (1978) *Dactylella oviparasitica*, a new fungal parasite of *Meloidogyne* eggs. *Mycologia*, **70**, 774–83.
Stirling, G.R. and Mankau, R. (1979) Mode of parasitism of of *Meloidogyne* and other nematode eggs by *Dactylella oviparasitica*. *Journal of Nematology*, **11**, 282–8.
Stirling, G.R., McHenry, M.V. and Mankau, R. (1978) Biological control of root-knot nematode on peach. *California Agriculture*, **32**, 6–7.
Stirling, G.R., McHenry, M.V. and Mankau, R. (1979) Biological control of root-knot nematode (*Meloidogyne* spp.) on peach. *Phytopathology*, **69**, 806–9.
Stone, J.K. (1987) Initiation and development of latent infections by *Rhabdocline parkeri* on Douglas fir. *Canadian Journal of Botany*, **65**, 2614–21.
Stutzenberger, F.J., Kaufman, A.J. and Lossin, R.D. (1970) Cellulolytic activity in municipal solid waste composting. *Canadian Journal of Microbiology*, **16**, 553–60.
Suberkropp, K. (1984) Effect of temperature on seasonal occurrence of aquatic hyphomycetes. *Transactions of the British Mycological Society*, **82**, 53–62.
Suberkropp, K. (1991) Relationship between growth and sporulation of aquatic hyphomycetes on decomposing leaf litter. *Mycological Research*, **95**, 843–50.
Suberkropp, K. (1992) Interactions with invertebrates, in *The Ecology of Aquatic Hyphomycetes*, (ed. F. Bärlocher), Springer-Verlag, Berlin, pp. 118–34.
Suberkropp, K. and Klug, M.J. (1976) Fungi and bacteria associated with leaves during processing in a woodland stream. *Ecology*, **57**, 707–19.
Suberkropp, K. and Klug, M.J. (1980) Maceration of deciduous leaf litter by aquatic hyphomycetes. *Canadian Journal of Botany*, **58**, 1025–31.

Suberkropp, K. and Klug, M.J. (1981) Degradation of leaf litter by aquatic hyphomycetes, in *The Fungal Community: its Organisation and Role in the Ecosystem*, (eds D.T. Wicklow and G.C. Carroll), Marcel Dekker, New York, pp. 761–76.

Suberkropp, K., Arsuffi, T.L. and Anderson, J.P. (1983) Comparison of degradative ability, enzymatic activity and palatability of aquatic hyphomycetes grown on leaf litter. *Applied and Environmental Microbiology*, **46**, 237–44.

Suberkropp, K., Godshalk, G.L. and Klug, M.J. (1976) Changes in the chemical composition of leaves during processing in a woodland stream. *Ecology*, **57**, 720–7.

Subramanian, C.V. (1983) Hyphomycetes in the marine habitat, in *Hyphomycetes: Taxonomy and Biology*, Academic Press, London, pp. 283–94.

Sumner, J.L. and Morgan, E.D. (1969) The fatty acid composition of sporangiospores and vegetative mycelium of temperature-adapted fungi in the order Mucorales. *Journal of General Microbiology*, **59**, 215–21.

Sumner, J.L., Morgan, E.D. and Evans, H.C. (1969) The effect of growth temperature on the fatty acid composition of fungi in the order Mucorales. *Canadian Journal of Microbiology*, **15**, 515–20.

Sundman, V. and Nase, L. (1972) The synergistic ability of some wood-degrading fungi to transform lignins and lignosulfonates on various media. *Archiv für Mikrobiologie*, **86**, 339–48.

Sussman, A.S. (1973) Longevity and survivability of fungi, in *The Fungi: An Advanced Treatise*, vol. III, (eds G.C. Ainsworth and A.S. Sussman), Academic Press, London, pp. 447–76.

Sussman, A.S. and Halvorson, H.O. (1966) *Spores: their Dormancy and Germination*, Harper & Row, New York.

Sutherland, J.B. and Crawford, D.L. (1981) Lignin and glucan degradation by species of the Xylariaceae. *Transactions of the British Mycological Society*, **76**, 335–7.

Suzuki, S. (1961a) The diurnal migration of zoospores of aquatic fungi in a shallow lake. *Botanical Magazine, Tokyo*, **74**, 138–41.

Suzuki, S. (1961b) The vertical distributions of the zoospores of aquatic fungi during the circulation and stagnation periods. *Botanical Magazine, Tokyo*, **74**, 254–8.

Swift, M.J. (1976) Species diversity and the structure of microbial communities in terrestrial habitats, in *The Role of Terrestrial and Aquatic Organisms in Decomposition Processes*, (eds J.M. Anderson and A. MacFadyen), Blackwell, Oxford, pp. 185–222.

Swift, M.J. (1977) The ecology of wood decomposition. *Science Progress Oxford*, **64**, 175–99.

Swift, M.J. (1982) Basidiomycetes as components of forest ecosystems, in *Decomposer Basidiomycetes: their Biology and Ecology*, (eds J.C.

Frankland, J.N. Hedger and M.J. Swift), *British Mycological Symposium 4*, Cambridge University Press, Cambridge, pp. 307–37.

Swift, M.J. and Boddy, L. (1984) Animal-microbial interactions in wood decomposition, in *Invertebrate–Microbial Interactions*, (eds J.M. Anderson, A.D.M. Rayner and D.W.H. Walton), *British Mycological Symposium 6*, Cambridge University Press, Cambridge, pp. 89–131.

Swift, M.J., Healey, I.N., Hibberd, J.K. *et al.* (1976) The decomposition of branch-wood in the canopy and floor of mixed deciduous woodland. *Oecologia*, **26**, 139–49.

Swisher, R. and Carroll, G.C. (1980) Fluorescein diacetate hydrolysis as an estimator of microbial biomass on coniferous needle surfaces. *Microbial Ecology*, **6**, 217–26.

Tabak, H.H. and Cooke, W.B. (1968) Growth and metabolism of fungi in an atmosphere of nitrogen. *Mycologia*, **60**, 115–40.

Tansey, M.R. and Brock, T.D. (1971) Isolation of thermophilic and thermotolerant fungi from hot spring effluent and thermal soils of Yellowstone National Park. *Bacteriological Proceedings*, Abstract 36.

Tansey, M.R. and Jack, M.A. (1976) Thermophilic fungi in sun-heated soils. *Mycologia*, **68**, 1061–75.

Tansey, M.R., Murrmann, D.N., Behnke, B.K. and Behnke, E.R. (1977) Enrichment, isolation and assay of growth of thermophilic and thermotolerant fungi in lignin-containing media. *Mycologia*, **69**, 463–76.

Tariq, V.N. and Magee, A. (1990) Effects of volatiles from garlic bulb extract on *Fusarium oxysporum* f. sp. *lycopersici*. *Mycological Research*, **94**, 617–20.

Tate, K.R. and Jenkinson, D.S. (1982) Adenosine triphosphate measurement in soil: an improved method. *Soil Biology and Biochemistry*, **14**, 331–5.

Taylor, E.E. and Marsh, P.B. (1963) Cellulose decomposition by *Pythium*. *Canadian Journal of Microbiology*, **9**, 353–8.

Taylor, G.S. and Parkinson, D. (1961) The growth of saprophytic fungi on root surfaces. *Plant and Soil*, **15**, 261–7.

Taylor, G.S. and Parkinson, D. (1964) Studies on fungi in the root region II. The effects of certain environmental conditions on the development of root surface mycofloras of dwarf bean seedlings. *Plant and Soil*, **20**, 34–42.

Taylor, G.S. and Parkinson, D. (1965) Studies on fungi in the root region IV. Fungi associated with the roots of *Phaseolus vulgaris* L. *Plant and Soil*, **22**, 1–20.

Te Strake, D. (1959) Estuarine distribution and saline tolerance of some Saprolegniaceae. *Phyton. International Journal of Experimental Botany*, **12**, 147–52.

Thacker, D.G. and Good, H.M. (1952) The composition of air in trunks of

sugar maple in relation to decay. *Canadian Journal of Botany*, **30**; 475–85.
Theander, O. (1978) Leaf litter of some forest trees. Chemical compositon and microbiological activity. *Tappi*, **61**, 69–72.
Theden, G. (1961) Untersuchungen über die Fähigkeit holzzerstorender Pilze zur Trockenstarre. *Angewandte Botanik*, **35**, 131–45.
Theodorou, M.K., Lowe, S.E. and Trinci, A.P.J. (1992) Anaerobic fungi and the rumen ecosystem, in *The Fungal Community*, 2nd ed, (eds G.C. Carroll and D.T. Wicklow), Marcel Dekker, New York, pp. 43–72.
Theodoru, C. and Bowen, G.D. (1987) Germination of basidiospores of mycorrhizal fungi in the rhizosphere of *Pinus radiata* D. Don. *New Phytologist*, **106**, 217–23.
Thomas, F. (1905) Die Wachstumgeschwindikeit eines Pilzkreises von *Hydnum suaveolens* Scop. *Berichte der Deutschen Botanischen Gesellschaft*, **23**, 476–8.
Thomas, K., Chilvers, G.A. and Norris, R.H. (1989) Seasonal occurrence of conidia of aquatic hyphomycetes (fungi) in Lees Creek, Australian Capital Territory. *Australian Journal of Marine and Freshwater Research*, **40**, 11–23.
Thomas, K., Chilvers, G.A. and Norris, R.H. (1991a) Changes in the concentration of aquatic hyphomycetes spores in Lees Creek, ACT, Australia. *Mycological Research*, **95**, 178–83.
Thomas, K., Chilvers, G.A. and Norris, R.H. (1991b) A dynamic model of fungal spora in a freshwater stream. *Mycological Research*, **95**, 184–8.
Thomas, K., Chilvers, G.A. and Norris, R.H. (1992a) Aquatic hyphomycetes from different substrates: Substrate preference and seasonal occurrence. *Australian Journal of Marine and Freshwater Research*, **43**, 491–509.
Thomas, K., Chilvers, G.A. and Norris, R.H. (1992b) Litterfall in riparian and adjacent forest zones near a perennial upland stream in the Australian Capital Territory. *Australian Journal of Marine and Freshwater Research*, **43**, 511–16.
Thompson, W. (1984) Distribution, development and functioning of mycelial cord system of decomposer basidiomycetes of the deciduous woodland floor, in *The Ecology and Physiology of the Fungal Mycelium*, (eds D.H. Jennings and A.D.M. Rayner), *British Mycological Society Symposium 8*, Cambridge University Press, Cambridge, pp. 185–214.
Thompson, W. and Boddy, L. (1983) Decomposition of suppressed oak trees in even-aged plantations. II. Colonisation of trees roots by cord- and rhizomorph-producing basidiomycetes. *New Phytologist*, **93**, 277–91.
Thompson, W. and Rayner, A.D.M. (1982a) Structure and development of mycelium cord systems of *Phanerochaete laevis* in soil. *Transactions of the British Mycological Society*, **78**, 193–200.

Thompson, W. and Rayner, A.D.M. (1982b) Spatial structure of a population of *Tricholomopsis platyphylla* in a woodland site. *New Phytologist*, **92**, 103–14.

Thompson, W. and Rayner, A.D.M. (1983) Extent, development and function of mycelial cord systems in soil. *Transactions of the British Mycological Society*, **81**, 333–45.

Thompstone, A. and Dix, N.J. (1985) Cellulase activity in the Saprolegniaceae. *Transactions of the British Mycological Society*, **85**, 361–6.

Thorn, R.G. and Barron, G.L. (1984) Carnivorous mushrooms. *Science*, **224**, 76–8.

Thorn, R.G. and Barron, G.L. (1986) *Nematoctonus* and the tribe Resupinateae in Ontario, Canada. *Mycotaxon*, **25**, 231–453.

Thornton, D.R. (1963) The physiology and nutrition of some aquatic hyphomycetes. *Journal of General Microbiology*, **33**, 23–31.

Thornton, D.R. (1965) Amino acid analysis of fresh leaf litter and the nitrogen nutrition of some aquatic hyphomycetes. *Canadian Journal of Microbiology*, **11**, 657–62.

Timell, T.E. (1967) Recent progress in the chemistry of wood hemicelluloses. *Wood Science and Technology*, **1**, 45–70.

Timonin, M.I. (1939) The interactions of higher plants and soil microorganisms I. Microbial populations of the rhizosphere of seedlings of certain cultivated plants. *Canadian Journal of Research*, **18**, 307–17.

Timonin, M.I. and Lochhead, A.G. (1948) Distribution of microorganisms in the rhizosphere root system. *Transactions of the Royal Society of Canada*, **42**, 175–81.

Tiunova, N.A., Pirieva, D.A., Feniksova, R.V. and Kuznetsov, V.D. (1976) Formation of chitinase by actinomycetes in submerged cultures. *Mikrobiologiya*, **45**, 280–3.

Todd, N. and Rayner, A.D.M. (1978) Genetic structure of a natural population of *Coriolus versicolor*. *Genetical Research, Cambridge*, **32**, 55–6.

Todd, N.K. and Rayner, A.D.M. (1980) Fungal individualism. *Science Progress, Oxford*, **66**, 331–54.

Topps, J.H. and Wain, R.L. (1957) Fungistatic properties of leaf exudates. *Nature*, **179**, 652–3.

Trappe, J.M. (1962) Fungus associates of ectotrophic mycorrhizae. *Botanical Review*, **28**, 538–606.

Traquair, J.A. and Hawn, E.J. (1982) Pathogenicity of *Coprinus psychromorbidus* on alfalfa. *Canadian Journal of Plant Pathology*, **4**, 106–8.

Traquair, J.A. and Mckeen, W.E. (1986) Fine structure of root tip cells of winter wheat exposed to cultural filtrates of *Coprinus psychromorbidus* and *Marasmius oreades*. *Canadian Journal of Plant Pathology*, **8**, 59–64.

Tresner, H.D. and Hayes, J.A. (1971) Sodium chloride tolerance of terrestrial fungi. *Applied Microbiology*, **22**, 210–13.

Tribe, H.T. (1957) Ecology of micro-organisms in soil observed during their development upon buried cellulose film, in *Microbial Ecology*, (eds R.E.O. Williams and C.C. Spicer), *7th Symposium, Society for General Microbiology*, pp. 287–98.

Tribe, H.T. (1966) Interactions of soil fungi on cellulose film. *Transactions of the British Mycological Society*, **49**, 427–57.

Tribe, H.T. (1980) Prospects for the biological control of plant-parasitic nematodes. *Parasitology*, **81**, 619–39.

Tribe, H.T. and Mabadeje, S.A. (1972) Growth of moulds on media prepared without organic nutrients. *Transactions of the British Mycological Society*, **58**, 127–37.

Trojanowski, J., Haider, K. and Hutterman, A. (1984) Decomposition of ^{14}C-labelled lignin, holocellulose and lignocellulose by mycorrhizal fungi. *Archives of Microbiology*, **139**, 202–6.

Trolldenier, G. (1972) L'influence de nutrition potassique de haricots nains (*Phaseolus vulgaris* var. *nanus*) sur l'exsudation de substances organiques marquées au ^{14}C le nombres de bactéries rhizosphériques et la respiration des racines. *Revue de l'Ecologie et Biologie du Sol*, **9**, 595–603.

Tsuneda, I. and Kennedy, L.L. (1980) Basidiomycete spore germination on wood. *Mycologia*, **72**, 204–8.

Tubaki, K. (1961) Notes on some fungi and yeasts from Antarctica. *Antarctic Record*, **11**, 161–2.

Tukey, H.B. (1971). Leaching of substances from plants, in *Ecology of Leaf Surface Microorganisms*, (eds T.F. Preece and C.H. Dickinson), Academic Press, London, pp. 67–80.

Tunlid, A., Jansson, H-J. and Nordbring-Hertz, B. (1992) Fungal attachment to nematodes. *Mycological Research*, **96**, 401–12.

Turnau, K. (1984a) Post-fire cup fungi of Turbacz and Stare Wierchy mountains in the Gorce Range (Polish Western Carpathians). *Keszyty Naukowe Uniwersytetu Jagiellonskiego Prace Botaniczne*, **12**, 145–70.

Turnau, K. (1984b) Interactions between organisms isolated from burns. *Keszyty Naukowe Uniwersytetu Jagiellonskiego Prace Botaniczne*, **12**, 171–80.

Turner, S.M. and Newman, E.I. (1984) Fungal abundance on *Lolium perenne* roots: influence of nitrogen and phosphorus. *Transactions of the British Mycological Society*, **82**, 315–22.

Tyler, G. (1985) Macrofungal flora of Swedish beech forest related to soil organic matter and acidity characteristics. *Forest Ecology and Management*, **10**, 13–29.

Tyler, G. (1989) Edaphial distribution patterns of macrofungal species in deciduous forest in South Sweden. *Acta Oecologia, Oecologia Generalis*, **10**, 309–26.

Tyler, G. (1991) Effects of litter treatments on the sporophore production of beech forest macrofungi. *Mycological Research*, **95**, 1137–9.
Tzean, S.S. and Estey, R.H. (1978) Nematode-trapping fungi as mycopathogens. *Phytopathology*, **68**, 1266–70.
Ulken, A. and Sparrow, F.K. (1968) Estimation of chytrid propagules in Douglas Lake by the MPN = Pollen grain method. *Veröffentlichen des Instituts für Meeresforschung, Bremerhaven*, **11**, 83–8.
Unestam, T. (1966) Chitinolytic, cellulolytic and pectinolytic activity *in vitro* of some parasitic and saprophytic Oomycetes. *Physiologia Plantarum*, **19**, 15–30.
Valiela, I. (1969) The arthropod fauna of bovine dung in central New York and sources on its natural history. *Journal of the New York Entomological Society*, **77**, 210–20.
Valiela, I. (1974) Composition, food webs and population limitation in dung arthropod communities during invasion and succession. *American Midland Naturalist*, **92**, 370–85.
Vancura, V. (1964) Root exudates of plants I. Analysis of root exudates in barley and wheat in their natural phases of growth. *Plant and Soil*, **21**, 231–48.
Vancura, V. (1967) Root exudates of plants III. Effect of temperature and 'cold shock' on the exudation of various compounds from seeds and seedlings of maize and cucumber. *Plant and Soil*, **27**, 319–28.
Vancura, V. and Garcia, J.L. (1969) Root exudates of reversibly wilted millet plants (*Panicum miliaceum* L). *Oecologia Plantarum*, **4**, 93–8.
Vancura, V. and Hanzlikova, A. (1972) Root exudates of plants IV. Differences in chemical composition of seed and seedling exudates. *Plant and Soil*, **36**, 271–82.
Van den Heuvel, J. (1969) Effects of *Aureobasidium pullulans* on numbers of lesions on dwarf beans of *Alternaria zinniae*. *Netherlands Journal of Plant Pathology*, **75**, 300–7.
Van den Heuvel, J. (1971) Antagonism between pathogenic and saprophytic *Alternaria* species on bean leaves, in *Ecology of Leaf Surface Microorganisms*, (eds T.F. Preece and C.H. Dickinson), Academic Press, London, pp. 537–44.
Van Den Heuvel, J., Verheus, A.H. and Kruyswijk, C.J. (1978) Lack of phytoalexin involvement in the antagonism of *Alternaria tenuissima* and *Alternaria zinniae* on dwarf bean leaves. *Netherlands Journal of Plant Pathology*, **84**, 81–3.
Van der Drift, J. and Jansen, E. (1977) The grazing of springtails on hyphal mats and its influence on fungal growth and respiration, in *Soil Organisms as Components of Ecosystems*, (eds V. Lohm and T. Perssont), Ecological Bulletin 25, Stockholm, pp. 302–9.
Venkata Ram, C.S. (1956) Studies on the cellulolytic activity of fusaria with reference to bacterial and other cellulose substrates. *Proceedings of*

the *National Institute of Sciences of India*, **22B**, 204–11.
Visser, S. (1985) Role of the soil invertebrates in determining the composition of soil microbial communities, in *Ecological Interactions in Soil, Plants, Microbes and Animals*, (ed. A.H. Fitter), Blackwell Scientific, Oxford, pp. 297–317.
Vogt, K.A., Bloomfield, J., Ammirati, J.F. and Ammirati, S.R. (1992) Sporocarp production by basidiomycetes, with special emphasis on forest ecosystems, in *The Fungal Community: Its Organisation and Role in the Ecosystem*, (eds G.C. Carroll and D.T. Wicklow), Marcel Dekker, New York, pp. 563–81.
Wagner-Merner, D.T. (1972) Arenicolous fungi from the south and central Gulf Coast of Florida. *Nova Hedwigia*, **23**, 915–22.
Waid, J.S. (1957) Distribution of fungi within decomposing tissues of rye grass roots. *Transactions of the British Mycological Society*, **40**, 391–406.
Wainwright, M. (1988) Metabolic diversity of fungi in relation to growth and mineral cycling in soil – a review. *Transactions of the British Mycological Society*, **90**, 159–70.
Wainwright, M. (1992) The impact of fungi on environmental biogeochemistry, in *The Fungal Community: Its Organisation and Role in the Ecosystem*, 2nd ed, (eds G.C. Carroll and D.T. Wicklow), Marcel Dekker, New York, pp. 601–18.
Wainwright, M. (1993) Oligotrophic growth of fungi–stress or natural state? in *Stress Tolerance of Fungi*, (ed. D.H. Jennings), Marcel Dekker, New York, pp. 127–44.
Waksman, S.A., Cordon, T.C. and Hulpoi, N. (1939) Influence of temperature upon microbiological population and decomposition processes in composts of stable manure. *Soil Science*, **47**, 83–114.
Walker, J.A. and Maude, R.B. (1975) Natural occurrence and growth of *Gliocladium roseum* on the mycelium and sclerotia of *Botrytis allii*. *Transactions of the British Mycological Society*, **65**, 335–8.
Wall, C.J. and Lewis, B.G. (1978) Survival of *Mycocentrospora acerina* conidia. *Transactions of the British Mycological Society*, **70**, 157–60.
Wallwork, J.A. (1976) *The Distribution and Diversity of Soil Fauna*, Academic Press, London.
Warcup, J.H. (1951a) The ecology of soil fungi. *Transactions of the British Mycological Society*, **34**, 376–99.
Warcup, J.H. (1951b) Studies on the growth of basidiomycetes in soil. *Annals of Botany*, **15**, 305–17.
Warcup, J.H. (1955) On the origin of colonies of fungi developing on soil dilution plates. *Transactions of the British Mycological Society*, **38**, 298–301.
Warcup, J.H. (1959) Studies on basidiomycetes in soil. *Transactions of the British Mycological Society*, **42**, 45–52.
Warcup, J.H. (1981) Effect of fire on the soil microflora and other

non-vascular plants, in *Fire and the Australian Biota*, (eds A.M. Gill, R.H. Groves and I.R. Noble), Australian Academy of Science, Canberra, pp. 203–14.

Warcup, J.H. (1990) Occurrence of ectomycorrhizal and saprophytic Discomycetes after a wild fire in a eucalyptus forest. *Mycological Research*, **94**, 1065–9.

Warnock, D.W. (1971) Assay of fungal mycelium in grains of barley including the use of fluorescent antibody technique for individual fungal species. *Journal of General Microbiology*, **67**, 197–205.

Warren, R.C. (1972) Interference of common leaf saprophytic fungi with the development of *Phoma betae* lesions on sugar beet leaves. *Annals of Applied Biology*, **72**, 137–44.

Warren, R.C. (1976) Microbes associated with buds and leaves: some recent investigations on deciduous trees, in *Microbiology of Aerial Plant Surfaces*, (eds C.H. Dickinson and T.F. Preece), Academic Press, London, pp. 361–74.

Waterhouse, G.M. (1973) Peronosporales, in *The Fungi: An Advanced Treatise*, vol. IVB, (eds G.C. Ainsworth, F.K. Sparrow and A.S. Sussman), Academic Press, London, pp. 165–83.

Watkinson, S.C. (1975) The relation between nitrogen nutrition and formation of mycelial strands in *Serpula lacrimans*. *Transactions of the British Mycological Society*, **64**, 195–200.

Watkinson, S.C. (1984) Morphogenesis of the *Serpula lacrimans* colony in relation to its function in nature, in *The Ecology and Physiology of the Fungal Mycelium*, (eds D.H. Jennings and A.D.M. Rayner), *British Mycological Symposium 8*, Cambridge University Press, Cambridge, pp. 165–84.

Watling, R. (1982) Taxonomic status and ecological identity in the basidiomycetes, in *Decomposer Basidiomycetes: their Biology and Ecology*, (eds J.C. Frankland, J.N. Hedger and M.J. Swift), *British Mycological Society Symposium 4*, Cambridge University Press, Cambridge, pp. 1–32.

Watling, R. (1984) Macrofungi of birchwoods. *Proceedings of the Royal Society of Edinburgh*, **85B**, 129–40.

Watling, R. (1988) Larger fungi and some of earth's major catastrophies, in *Fungi and Ecological Disturbance*, (eds L. Boddy, R. Watling and A.J.E. Lyon), *Proceedings of the Royal Society of Edinburgh*, **94B**, 49–59.

Watling, R. and Lyon A.J.E. (1988) Field Excursion, in *Fungi and Ecological Disturbance*, (eds L. Boddy, R. Watling and A.J.E. Lyon), *Proceedings of the Royal Society of Edinburgh*, **94B**, 183–4.

Watson, R.D. (1960) Soil washing improves the value of the soil dilution and the plate count method of estimating populations of soil fungi. *Phytopathology*, **50**, 792–4.

Weaver, T. (1975) Fairy ring fungus as decomposers. *Proceedings of the*

Montana Academy of Sciences, **35**, 34–8.

Webber, J.F. and Hedger, J.N. (1986) Comparison of interaction between *Ceratocystis ulmi* and elm bark saprobes *in vitro* and *in vivo*. *Transactions of the British Mycological Society*, **86**, 93–101.

Webster, J. (1956) Succession of fungi on decaying cocksfoot culms I. *Journal of Ecology*, **44**, 517–44.

Webster, J. (1957) Succession of fungi on decaying cocksfoot culms II. *Journal of Ecology*, 45, 1–30.

Webster, J. (1959) Experiments with spores of aquatic hyphomycetes I. Sedimentation and impaction on smooth surfaces. *Annals of Botany, London*, NS, **23**, 595–611.

Webster, J. (1970). Presidential Address. Coprophilous Fungi. *Transactions of the British Mycological Society*, **54**, 161–80.

Webster, J. (1975) Further studies of sporulation of aquatic hyphomycetes in relation to aeration. *Transactions of the British Mycological Society*, **64**, 119–27.

Webster, J. (1987) Convergent evolution and the functional significance of spore shape in aquatic and semi-aquatic fungi, in *Evolutionary Biology of the Fungi* (eds A.D.M. Rayner, G.M. Brasier and D. Moore), *British Mycological Symposium, 12*, Cambridge University Press, Cambridge, pp. 191–201.

Webster, J. (1992) Anamorph–teleomorph relationships, in *The Ecology of Aquatic Hyphomycetes* (ed. F. Bärlocher) Springer-Verlag, Berlin, pp. 99–117.

Webster, J. and Davey, R.A. (1984) Sigmoid conidial shape in aquatic fungi. *Transactions of the British Mycological Society*, **82**, 43–52.

Webster, J. and Descals, E. (1981) Morphology, distribution and ecology of conidial fungi in freshwater habitats, in *Biology of Conidial Fungi*, vol. I; (eds G.T. Cole and B. Kendrick), Academic Press, New York, pp. 295–355.

Webster, J. and Dix, N.J. (1960) Succession of fungi on decaying cocksfoot culms III. A comparison of the sporulation and growth of some primary saprophytes on stem, leaf, blade and sheath. *Transactions of the British Mycological Society*, **43**, 85–99.

Webster, J. and Towfik, F.H. (1972) Sporulation of aquatic hyphomycetes in relation to aeration. *Transactions of the British Mycological Society*, **59**, 353–64.

Webster, J. and Benfield, E.F. (1986) Vascular plant breakdown in freshwater ecosystems. *Annual Review of Ecology and Systematics*, **17**, 567–74.

Webster, J., Moran, S.T. and Davey, R.A. (1976) Growth and sporulation of *Tricladium chaetocladium* and *Lunulospora curvula* in relation to temperature. *Transactions of the British Mycological Society*, **67**, 491–549.

Weete, J.D. (1980) *Lipid Biochemistry of Fungi and Other Organisms*, Plenum, New York.

Weidensaul, T.C. and Wood, F.A. (1974) Response of *Fusarium solani* to constant and fluctuating temperatures and its relationship to *Fusarium* canker of sugar maple. *Phytopathology*, **64**, 1018–24.

West, P.M. and Lochhead, A.G. (1940) The nutritional requirements of soil bacteria: a basis for determining the bacterial equilibrium of soils. *Soil Science*, **50**, 409–20.

Westermark, U. and Eriksson, K.E. (1974) Cellobiose: quinone oxidoreductase, a new wood-degrading enzyme from white rot fungi. *Acta Chemica Scandinavica B*, **28**, 209.

Wethered, J.M. and Jennings, D.H. (1985) Major solutes contributing to solute potential of *Thraustochytrium aureum* and *T. roseum* after growth in media of different salinities. *Transactions of the British Mycological Society*, **85**, 439–46.

Wethered, J.M., Metcalf, E. and Jennings, D.H. (1985) Carbohydrate metabolism in the fungus *Dendryphiella salina* VIII. The contribution of polyols and ions to the mycelial solute potential in relation to the external osmoticum. *New Phytologist*, **101**, 631–50.

Whalley, A.J.S. (1985) The Xylariaceae: some ecological considerations. *Sydowia Annales Mycologici Ser. II*, **38**, 369–82.

Whalley, A.J.S. and Watling, R. (1982) Distribution of *Daldinea concentrica* in the British Isles. *Transactions of the British Mycological Society*, **78**, 47–53.

Whipps, J.M., Lewis, K. and Cooke, R.C. (1988) Mycoparasitism and plant disease control, in *Fungi in Biological Control Systems* (ed. M.N. Burge) Manchester University Press, Manchester, pp. 161–87

Whittle, A.M. (1977) Mycoflora of cones and seeds of *Pinus sylvestris*. *Transactions of the British Mycological Society*, **69**, 47–57.

Wicklow, D.T. (1973) Microfungal populations in surface soils of manipulated prairie stands. *Ecology*, **54**, 1302–10.

Wicklow, D.T. (1975) Fire as an environmental cue initiating ascomycete development in a tall grass prairie. *Mycologie*, **67**, 852–62.

Wicklow, D.T. (1981) The coprophilous fungal community: a mycological system for examining ecological ideas, in *The Fungal Community: Its Organisation and Role in the Ecosystem*, (eds D.T. Wicklow and G.C. Carroll), Marcel Dekker, New York, pp. 47–75.

Wicklow, D.T. (1989) Parallels in the development of post-fire fungal and herb communities, in *Fungi and Ecological Disturbance*, (eds L. Boddy, R. Watling and A.J.E. Lyon), *Proceedings of the Royal Society of Edinburgh*, **94B**, 87–95.

Wicklow, D.T. (1992) The coprophilous fungal community: and experimental system, in *The Fungal Community: Its Organisation and Role in the Ecosystem*, 2nd ed, (eds G.C. Carroll and D.T. Wicklow) Marcel Dekker, New York, pp. 715–28.

Wicklow, D.T. and Hirschfield, B.J. (1979a) Evidence of a competitive

hierarchy among coprophilous populations. *Canadian Journal of Microbiology*, **25**, 855–8.
Wicklow, D.T. and Hirschfield, B.J. (1979b) Competitive hierarchy in post-fire ascomycetes. *Mycologia*, **71**, 47–54.
Wicklow, D.T. and Malloch, D. (1971) Studies in the genus *Thelebolus*. Temperature optima for growth and ascocarp development. *Mycologia*, **63**, 118–31.
Wicklow, D.T. and Moore, V. (1974) Effect of incubation temperature on the coprophilous fungus succession. *Transactions of the British Mycological Society*, **62**, 411–15.
Wicklow, D.T. and Yocom, D.H. (1981) Fungal species number and decomposition of rabbit faeces. *Transactions of the British Mycological Society*, **76**, 29–32.
Wicklow, D.T. and Yocom, D.H. (1982) Effect of larval grazing by *Lycoriella mali* (Diptera: Sciaridae) on species abundance of coprophilous fungi. *Transactions of the British Mycological Society*, **78**, 29–32.
Wicklow, D.T., and Zak, J.C. (1979) Ascospore germination of carbonicolous ascomycetes in fungistatic soils: an ecological interpretation. *Mycologia*, **71**, 238–42.
Wicklow, D.T., Angel, C.D.P. and Lussenhop, J. (1980) Fungal community expression in lagomorph versus ruminant feces. *Mycologia*, **72**, 1015–21.
Widden, P. (1979) Fungal populations from forest soils in southern Quebec. *Canadian Journal of Botany*, **57**, 1324–31.
Widden, P. (1984) The effects of temperature on competition for spruce needles among sympatric species of *Trichoderma*. *Mycologia*, **76**, 873–83.
Widden, P. (1986a) Microfungal community structure from forest soils in southern Quebec using discriminant function and factor analysis. *Canadian Journal of Botany*, **64**, 1402–12.
Widden, P. (1986b) Seasonality of forest soil microfungi in southern Quebec. *Canadian Journal of Botany*, **64**, 1413–23.
Widden, P. (1987) Fungal communities in soils along an elevation gradient in Northern England. *Mycologia*, **79**, 298–309.
Widden, P. and Abitol, J.J. (1980) Seasonality of *Trichoderma* species in a spruce forest soil. *Mycologia*, **72**, 775–84.
Widden, P. and HSU, D. (1987) The effects of temperature and litter types on competition between *Trichoderma* species. *Soil Biology and Biochemistry*, **19**, 89–93.
Widden, P. and Scattolin, V. (1988) Competitive interactions and ecological strategies of *Trichoderma* species colonizing spruce litter. *Mycologia*, **80**, 795–803.
Wilcox, W.W. (1970) Tolerance of *Polyporus amarus* to extractives from incense cedar heartwood. *Phytopathology*, **60**, 919–23.

Wildman, H.G. (1987) Fungal colonization of resources in soil – an island biogeographical approach. *Transactions of the British Mycological Society*, **88**, 291–7.
Wilkins, W.H. and Harris, G.C.M. (1946) The ecology of the larger fungi V. An investigation into the influence of rainfall and temperature on the seasonal production of fungi in a beechwood and in a pinewood. *Annals of Applied Biology*, **33**, 179–90.
Williams, E.N.D., Tood, N.K. and Rayner, A.D.M. (1971) Spatial development of populations of *Coriolus versicolor*. *New Phytologist*, **89**, 307–19.
Williams, E.N.D., Tood, N.K. and Rayner, A.D.M. (1981) Propagation and development of fruit bodies of *Coriolus versicolor*. *Transactions of the British Mycological Society*, **77**, 409–14.
Williams, S.T., Parkinson, D. and Burges, N.A. (1965) An examination of the soil washing technique by its application to several soils. *Plants and Soil*, **22**, 167–86.
Willoughby, L.G. (1962) The occurrence and distribution of reproductive spores of Saprolegniaceae in freshwater. *Journal of Ecology*, **50**, 733–59.
Willoughby, L.G. (1978). *Leptomitus lacteus*. CMI Descriptions of Pathogenic Fungi and Bacteria 597, Commonwealth Mycological Institute, Kew; UK.
Willoughby, L.G. and Archer, J.F. (1973) The fungal spora of a freshwater stream and its colonization pattern on wood. *Freshwater Biology*, **3**, 219–39.
Willoughby, L.G. and Collins, V.G. (1966) A study of the distribution of fungal spora and bacteria in Blelham Tarn and its associated streams. *Nova Hedwigia*, **12**, 150–71.
Willoughby, L.G. and Roberts, R.J. (1991) Occurrence of the sewage fungus *Leptomitus lacteus*, a nectotroph on perch (*Perca fluviatilis*), in Windermere. *Mycological Research*, **95**, 755–68.
Willoughby, L.G. and Redhead, K. (1973) Observations on the utilization of soluble nitrogen by aquatic fungi in nature. *Transactions of the British Mycological Society*, **60**, 598–601.
Willoughby, L.G., Pickering, A.D. and Johnson, H.G. (1984) Polycell-gel assay of water for spores of Saprolegniaceae (fungi), especially those of the *Saprolegnia* pathogen of fish. *Hydrobiologia*, **114**, 237–48.
Wilson, J.M. and Griffin, D.M. (1975) Respiration and radial growth of soil fungi at two osmotic potentials. *Soil Biology and Biochemistry*, **7**, 269–74.
Wilson, J.M. and Griffin, D.M. (1979) The effect of water potential on the growth of some soil basidiomycetes. *Soil Biology and Biochemistry*, **11**, 211–12.
Wise, L.E. and Jahn, E.C. (1952) *Wood Chemistry*, 2nd edn, Reinhold New York.

Wolf, F.T. (1971) An unusual occurrence of 'fairy rings'. *Mycologia*, **63**, 671–2.

Wood, R.M. (1969) Relation between cellulolytic and pseudocellulolytic microorganisms. *Biochimica et Biophysica Acta*, **192**, 531–4.

Wood, S.E. (1988) The monitoring and identification of *Saprolegnia parasitica* and its infection of salmonid fish. PhD thesis, University of Newcastle-upon-Tyne.

Wood, S.N. and Cooke, R.C. (1984) Use of semi-natural resource units in experimental studies on coprophilous fungi. *Transactions of the British Mycological Society*, **83**, 337–9.

Wood, S.N. and Cooke, R.C. (1986) Effect of *Piptocephalis* species on growth and sporulation of *Pilaira anomala*. *Transactions of the British Mycological Society*, **83**, 337–9.

Wood, S.N. and Cooke, R.C. (1987) Nutritional competence of *Pilaira anomala* in relation to exploitation of faecal resource units. *Transactions of the British Mycological Society*, **88**, 247–55.

Wood, T.M. (1969) Relation between cellulolytic and pseudocellulolytic microorganisms. *Biochimica et Biophysica Acta*, **192**, 531–4.

Wood, T.M. (1981) Co-operative action between enzymes involved in the degradation of crystalline cellulose, in *Colloque Cellulolyse Microbienne*, Marseille, CNRS, pp. 167–76.

Wood, T.M. and McCrae, S.I. (1986) The cellulase of *Penicillium pinophilum*. *Biochemical Journal*, **234**, 93–9.

Wood-Eggenschwiler, S. and Barlocher, F. (1983) Aquatic hyphomycetes in sixteen streams in France, Germany and Switzerland. *Transactions of the British Mycological Society*, **81**, 371–9.

Woodwell, G.M., Whittaker, R.H. and Houghton, R.A. (1975) Nutrient concentrations in plants in the Brookhaven oak–pine forest. *Ecology*, **56**, 318–32.

Wright, E. and Tarrant, R.F. (1957) Microbiological soil properties after logging and slash burning. *US Forest Service Pacific North West Forest and Range Experimental Station Research Note 157*.

Wyborn, C.H.E., Priest, D. and Duddington, C.L. (1969). Selective technique for the determination of nematophagous fungi in soils. *Soil Biology and Biochemistry*, **1**, 101–2.

Wynn-Williams, D.D. (1980) Seasonal fluctuations in microbial activity in Antarctic moss peat. *Biological Journal of the Linnean Society*, **14**, 11–28.

Yadav, A.S. (1966) The ecology of microfungi on decaying stems of *Heracleum sphondylium*. *Transactions of the British Mycological Society*, **49**, 471–5.

Yadav, A.S. and Madelin, M.F. (1968) The ecology of microfungi on decaying stems of *Urtica dioica*. *Transactions of the British Mycological Society*, **51**, 249–59.

Yamaski, I., Satomura, Y. and Yamamoto, T.T. (1951) Studies on *Sporobolomyces* red yeast (in Japanese). *Journal of the Agricultural Chemical Society Japan*, **24**, 399–402.

Yocom, D.H. and Wicklow, D.T. (1980) Community differentiation along a dune succession: an experimental approach with coprophilous fungi. *Ecology*, **61**, 868–80.

Yodzis, P. (1978) Competition for space and the structure of ecological communities. *Lecture notes in Biomathematics*, 25, Springer-Verlag, Berlin.

Zadrazil, F. (1980) Conversion of different plant waste into feed by basidiomycetes. *European Journal of Applied Microbiology and Biotechnology*, **9**, 243–8.

Zak, J.C. (1992) Responses of soil fungal communities to disturbance, in *The Fungal Community: Its Organization and Role in the Ecosystem*, 2nd ed, (eds D.T. Wicklow and G.C. Carroll), Marcel Dekker, New York, pp. 403–25.

Zak, J.C. and Wicklow, D.T. (1978a) Response of carbonicolous ascomycetes to aerated steam temperatures and treatment intervals. *Canadian Journal of Botany*, **56**, 2313–18.

Zak, J.C. and Wicklow, D.T. (1978b) Factors influencing patterns of ascomycete sporulation following simulated burning of prairie soils. *Soil Biology and Biochemistry*, **10**, 533–5.

Zak, J.C. and Wicklow, D.T. (1980) Structure and composition of a post-fire ascomycete community: Role of abiotic and biotic factors. *Canadian Journal of Botany*, **58**, 1915–22.

Zemek, J., Marvanova, Kuniak, L. and Kadlecikova, B. (1985) Hydrolytic enzymes in aquatic hyphomycetes. *Folia Microbiologica*, **30**, 363–72.

Ziegler, A.W. (1958) The Saprolegniaceae of Florida. *Mycologia*, **50**, 693–6.

INDEX

Page numbers in bold indicate the position of figures, and those in italic, tables.

Abies balsamea, heartwood fungi of 350
Absidia spp.
  optimal pH 58
  in rhizosphere 186
Absidia spinosa, rhizosphere fungus *199*
Acer platanoides
  antifungal compounds 49, *109*
  colonization of phylloplane by fungi 93–4, *108*, *109*
Acer pseudoplatanus
  deposition of spores on leaves of 98–9
  rate of decomposition of litter 124
Acer saccharum, leaf mapping of aquatic hyphomycetes **247**
Acid rain, effect on mycorrhizal fungi 394
Acremonium spp., soft-rot fungi 163
Acremonium griseoviride, distribution in soil *176*
Acrothecium spp.
  growth on *Dactylis glomerata* 133, **135**
  relative humidity requirement *139*
Actinomycetes
  invasion of fungal spores 20
  production of antifungal volatiles 182
Aegerita candida, aero-aquatic hyphomycete 262
  propagules **263**
Aeration, effects on fungal growth 54–7, 146
  see also Anaerobic conditions
Aflatoxin 70
  production by storage fungi 337
Agarics
  oxidation of phenolics 119
  leaf litter-decomposing 116, *118*, 119, 120–1, *365*
  lignin degradation 118
  phoenicoid forest fungi 305, 307
  polysaccharide hydrolysis 118
  wood-rotting 39
Agaricus spp.
  fairy ring fungi 371
  linear organs of 22
Agaricus campestris, spore size 87
Agaricus langei, growth on conifer litter 121
Agaricus tabularis
  growth rates of fairy rings *374*

Agaricus tabularis contd
  linear organs of 374
Agrocybe praecox, growth in
  disturbed environments 43
Agropyron repens, colonization by
  fungi 140–1
Aigialus spp., growth on
  mangroves 277
Alatospora acuminata, growth on
  submerged leaves 249, 252
Alder, see *Alnus glutinosa*
Aleuria aurantia, growth following
  volcanic eruption 320
Allescheria spp., heat-tolerant
  fungi 324
Allescheriella bathygena, depth of
  survival 279
Allium spp., antifungal substances
  produced by roots 187
Alnus spp.
  ectomycorrhizas 378
  endophyte communities 150
Alnus glutinosa
  colonization of submerged leaves
    by aquatic hyphomycetes
    249, 252
  rate of decomposition of litter
    124
Alpova diplophloeus,
  ectomycorrhizal fungus 378
Alternaria spp.
  antifungal antibiotic production
    107
  growth on phylloplane 92, 94, 97
  pigmented spores 114
  number of spores in the
    atmosphere 85, **86**
Alternaria alternata **132**
  antibiosis 107
  cellulose hydrolysis 115
  germ tube survival 113
  growth on
    herbs and grasses 143
    leaf litter 115

  phylloplane 92, 94–5
  relative humidity requirement *139*
Alternaria tenuis
  effect of matric potential on
    growth 64
  growth on
    Dactylis glomerata 131, **133**
    herbaceous plants 141
    sugar cane 142
Alternaria tenuissima, competition
  with *A. zinniae* 105–6
Alternaria zinniae, inhibition of
  spore germination by *A.
  tenuissima* 105–6
Amanita citrina, ectomycorrhizal
  fungus 394
Amanita muscaria
  ectomycorrhizal fungus 378, 385,
    393
  pH limits for growth *58*
Amanita pantherina, pH limits for
  growth *58*
Amanita rubescens, late-stage
  ectomycorrhizal fungus 385
  fruit-bodies **388**
Amino acids, requirements of fungi
  48
  antifungal volatile 182
  effect on sporulation of
    Pilobolus 221
Ammophila arenaria, invasion of
  roots by fungi 190
Amoebae, fungivorous **19**, 20
Amphibious fungi 260
Amylocarpus encephaloides,
  ascospores **268**
Anaerobes 55
  facultative 54–5
Anaerobic conditions, survival of
  fungi under 260, 264–5
Anamorph 1
Anastomosis **15**, 16
Angiosperms
  hemicellulose content of wood 30

leaf litter, fungal succession on 121–2
leaves, fungal succession on 92–5, 103
  rate of soft-rot decay 165–6
  see also named species
Anguillospora crassa **244**
  colonization of submerged wood 250
Anguillospora filiformis, spore production 254
Anguillospora cf. *gigantea*, inhibition of fungal growth by 253
Anguillospora rosea, survival in anaerobic conditions 260
Animals
  effects on
    fungal growth 78–80
    fungal microhabitats 80, 84
    fungal species diversity 79–80
  feeding on fungi 19–20
  role in
    litter decomposition 125–6
    spore dispersal 80
    substrates for saprophytic fungi 26
  see also Arthropods; Crustacea; Dung; Invertebrates
Antagonism 160
  direct 67, 71–7
    hyphal interference 76–7
    mycoparasitism 72–6
    overgrowth 71
  indirect 67–71
    antibiosis 69–71
    competition 67–9
  interspecific 213–15
  intraspecific 356–7, **358**, 359
  see also Competition
Antennospora quadricornuta, geographical distribution 282

Anthostomella minima, growth on sugar cane 142
Anthracobia spp., phoenicoid fungi 311–12
Anthracobia maurilabra 311
Anthracobia melaloma
  growth following volcanic eruption 320
  phenology 308
  seasonal variation **308**
Anthracophilous fungi, see Phoenicoid fungi
Antibiosis 69–71, 107
  role of *in vivo* 77–8
  species-specific 69
Antibiotics 6, 9–10, 253, 315
  chemical composition of 70
  effects on
    fungal growth 70–1
    fungal succession 82, 215
  production by
    coprophilous fungi 215
    nematophagous fungi 299
  in sclerotia 21
Antifungal substances 80, 215, 299
  detoxification of 49–51
  production by
    fungi 107, 182, 187
    higher plants 49, 52, 108–10, 149, 187
  see also Phenolics; Tannins
Aphanomyces spp., non-cellulolytic 45
Apiognomonia errabunda, growth on leaf litter **117**
Aplanospore 2, 7
Apodachlya spp., freshwater fungi 232
Aqualinderella spp.
  facultative anaerobes 55
  freshwater fungi 232
Aquatic fungi 225–83
  classification 225, *226*

Aquatic fungi *contd*
  habitats 226–7
  indwellers 225–6
  techniques of study 227
  transients 226
  see also Brackish water species;
    Freshwater fungi;
    Hyphomycetes; Marine
    fungi
Arabinoglucuronoxylan 30
Arachis hypogaea, rhizosphere 186,
  191
Arenariomyces trifurcata
  ascospore **268**
  growth on driftwood 281
  spores in sea foam 280
Arenicolous fungi 280
Armillaria spp.
  distribution of 53, 361
  growth on fallen timbers 354
  root-rot fungus 154
  strand-former 156
Armillaria bulbosa 9
  growth on fallen timbers 352,
    353, 354, **355**
  mycelial size 345
Armillaria mellea
  cords of 23
  distribution of **393**
  growth on fallen timbers 351
Arthrobotrys spp.
  biological control agents 301
  mycoparasites 296, 299
  predatory nematophagous fungi
    207, 285
Arthrobotrys dactyloides,
  nematode trap **287**
Arthrobotrys musiformis 292
  nematode attraction **298**
Arthrobotrys oligospora 292–3,
  296–8
  fungistasis of 181
  nematode attraction **298**
  nematode traps **287**

Arthrobotrys robusta 292
Arthrobotrys superba 301
  nematode attraction **297**
Arthropods
  colonization of dung 219
  detritivorous, effects on
    fungal growth 78–9
    fungal species diversity 79–80
  effects on fungal succession 84,
    219
  fungicide production by 80
  see also Animals; Crustacea;
    Invertebrates
Articulospora tetracladia **243**
Aschocytula obiones, phylloplane
  fungus of saltmarshes 197
Ascobolus spp., growth on dung
  207
Ascobolus albidus, growth on dung
  224
Ascobolus carbonarius, phoenicoid
  fungus 311–12, 314
  inhibition of seedling growth 312
Ascobolus crenulatus
  fruiting time *214*
  growth on
    copromes 214–15
    dung 216
  hyphal interference by *Coprinus*
    heptemerus **216, 217**
Ascobolus denudatus, growth on
  Pinus litter 122
Ascobolus furfuraceus, growth on
  dung 224
Ascobolus glaber, fruiting time *214*
Ascobolus immersus
  fruit-body **205**
  growth on dung 224
Ascobolus stictoideus, fruiting time
  214
Ascochyta pinodes, growth on
  phylloplane 93
Ascocoryne sarcoides, growth on
  dead wood 156

Ascodesmis nigricans, growth on dung 209
Ascomycetes
  ascospore appendages 267, **268, 269**
  conidia 2
  effect of heat and ash on 317, *318, 319*
  euryhaline species 277
  hemicellulose degradation 30–1
  lignin degradation 33–4
  marine **268, 269**
    ascospores 2
    effects of salinity on *273*, **275**
  nematophagous fungi 284
  phoenicoid fungi 317, *318, 319*, **320**
  pioneer colonizers of standing trees 154, 156
  terrestrial, effects of salinity on **275**
Ascomycotina
  cellulose hydrolysis 45
  degradation of animal structural protein by 38
  growth on dung 204, 212
  hemicellulose hydrolysis 45
  lignin degradation 145
  marine 267
    characteristics 267
  sclerotia 20
  sexual reproduction 15
  soft-rot fungi 163
Ascophanus microsporus, growth on dung 224
Ascospores 2
  appendages 267, **268, 269**
Aseptate species 70
Ash, chemical composition of **303**, *304*
Ash, see *Fraxinus excelsior*
Aspergillosis 337
Aspergillus spp.
  desert soil fungi 178
  inability to degrade plant lignins 34
  production of antibiotics 70
  xerotolerant fungi 64, *333*, 334–5
Aspergillus amstelodami, xerophilic fungus 339
Aspergillus candidus
  pH limits for growth 58
  storage fungus 335
Aspergillus clavatus, allergic disease due to 337
Aspergillus flavipes, pH limits for growth 58
Aspergillus flavus
  aflatoxin production 337
  antibiotic production 70
  storage fungus 335
Aspergillus fumigatus
  animal pathogen 325, 337
  heat-tolerant fungus *323*, 324
  pH limits for growth 58
  storage fungus 335
Aspergillus glaucus
  enzymic activity 337
  storage fungus 334
Aspergillus halophilicus, storage fungus 334
Aspergillus niger, pH limits for growth 58
Aspergillus ochraceus, ochratoxin production 337
Aspergillus repens
  pH limits for growth 58
  xerophilic fungus 65, 339
Aspergillus restrictus, storage fungus 334
Aspergillus sejunctus, xerophilic fungus 340
Aspergillus ustus, in rhizosphere soil 186
Asteromyces cruciatus
  conidia **270**
  in sand dunes 281

504 INDEX

Asteromyces cruciatus contd
  spore germination 272
Atkinsonella spp., endophytes 90
Aureobasidium spp., growth on wood 157
Aureobasidium pullulans 26, **91**
  antifungal antibiotic production 107
  endophyte *151*
  competition with parasitic fungi 106
  growth on
    Agropyron repens 141
    buds 97
    conifer needles 96
    Dactylis glomerata 133
    hogweed 141
    leaf litter 115–16
    phylloplane 92, 97–8, 103, 105, 114
  non-cellulolytic 115
  overgrowth by *Collybia peronata* 71
  polysaccharide slime production 114
  resting structures 114
  wood sapstainer 159
Auricularia spp.
  basidiocarp survival 364
  sporulation 364
Auricularia auricula-judae, survival at low water potential *170*
Auriscalpium vulgare, growth on *Pinus* cones **366**
Autolysis 14
Avena sativa, rhizosphere mycoflora 191

Bacteria, fungivorous 20
Balansia spp., endophytes 90
Balsam fir, *see Abies balsamea*
Barley, *see Hordeum vulgare*

Basidiomycetes
  bipolar species 343
  branched basidiospores 267, 269, **270**
  brown-rot fungi 166
  combative fungi 8
  conidia 2
  detoxification of antifungal substances by 51
  ectomycorrhizal fungi 377–8
    distinguishing features 382–3
    fruiting of 388–9
  fairy ring fungi 371
  genetic control of mating 343
  growth
    carbon dixide concentration and 146
    at low nitrogen levels 146–7
    optimal pH for 57
    oxygen partial pressures and 146
    stimulation of 51
  heart-rot 145
  hemicellulose degradation 30
  heterothallic 343
  homothallic 343
  hyphal interference by 76
  interactions in attached oak branches *348*
  life cycle 341, 343
  lignin degradation 31, 33–4, 36
  linear organs of 22–3, 25
  litter decomposers 122, *365*
  marine **270**
  mycelia 344–5
  nematophagous fungi 284
  overgrowth by 71
  pioneer colonizers of standing trees 154
  soil biomass 365
  substrate degradation by 45
  terrestrial macrofungi 341
    methods of study 343
    nutritional relationships *342*

tetrapolar species 343
tolerance of water stress 63–4, 169, *170*, 171
vegetative incompatibility mechanisms 16–17
wood-rotting *153*, 346
    basidiocarps **349**, **361**, **362**, **364**
    see also Macrofungi
Basidiomycotina
    antibiotic production 70
    degradation of lignin 145
    growth on dung 204, 212
    marine 267
    'non-weed' species 43, *44*
    sclerotia 20
    sexual reproduction 15
Basidiospores 2
    branched 267, 269, **270**
    production by terrestrial macrofungi 341
Basswood, see *Tilia americana*
Bathyascus vermisporus, depth of survival 279
Bavendamm test 36
Beech, see *Fagus sylvatica*
Benlate 107–8
Betula spp.
    antifungal substances 110
    colonization by microfungi *158*, *159*
    ectomycorrhizas 378
        succession *384*
    growth on coal spoil, mycorrhizal associations 391
    heart-rot basidiomycetes 148
    infection by macrofungi 359
    intraspecific antagonism of macrofungi growing on 357
    leaf litter, succession on 118
    soft-rot 165
Betula pendula
    antifungal compounds 49
    mycorrhizal fruiting on 389

Betula pubescens, ectomycorrhizal succession **386**
Bicarbonate ion concentration, effect on fungal growth 57
Biocontrol mechanisms
    interspecific competition 105–6
    necrotrophic fungi 74
Biological control agents
    Dactylella oviparasitica 301
    Drechmeria coniospora 301
    endophytes 90
    nematophagous fungi 300
        problems 300–1
    Paecilomyces lilacinus 301
    Peniophora gigantea 10
Biomass
    basidiomycetes in soil 365
    recycling of mycelium 14–15
Biotrophs 72
Birch, see *Betula* spp.
Bjerkandera spp., growth on fallen timbers 354
Bjerkandera adusta 10
    growth on
        beech logs **355**
        fallen timbers 352, *353*, *354*
    interaction with *Pseudotrametes gibbosa* 356
Blastocladia spp., facultative anaerobes 55
Boletus chrysenteron, ectomycorrhizal fungus 394
Boletus edulis
    ectomycorrhizal fungus 378
    pH limits for growth 58
Boletus subtometosus, pH limits for growth 58
Botrytis spp., parasitized by *Gliocladium roseum* 72
Botrytis cinerea
    cold-tolerance 327
    germination of spores 101–2
    growth on
        Dactylis glomerata 133

Botrytis cinerea contd
  growth on *contd*
    flowers 98
    herbaceous plants 141
    phylloplane 92, 95
  inhibition by plant antifungal substances 110
  pH limits for growth *58*
  rhizosphere fungus *199*
Bracken litter, succession on 122
Brackish water species 277
Brassica oleracea, changes in rhizosphere mycoflora 189
Broad bean, see *Vicia faba*
Brown-rot fungi 36, 154, 157, 161, 166–8
  degradation of substrate 166
  spread of 166
Bryosphaeria spp., psychrotolerant 327
Buds, colonization by fungi 97–8
Butt rot 152

Cabbage, see *Brassica oleracea*
Calluna vulgaris
  antifungal substances produced by roots 187
  rhizosphere 186
Calocera cornea
  growth on dead wood 157
  survival at low water potential *170*
Calvatia cyathiformis, fairy ring fungus *374*
Calvatia sculpta, cord formation in 25
Cancellidium applanatum, aero-aquatic hyphomycete 262
Candida spp.
  distribution in soil *176*
  psychrotolerant 327
Candida scottii, temperature range for growth *323*

Cantharellus tubaeformis, ectomycorrhizal fungus 391
Carbon dioxide concentration, effect on fungal growth 55, **56**, 57, 146
Carbonicolous fungi, *see* Phoenicoid fungi
Carbosphaerella leptosphaerioides, growth on driftwood 281
Castanea spp., heart-rot basidiomycetes of 148
Castanea sativa, antifungal substances 109, 149
Catechin 51
Catecholase reaction 34–6
Catenaria spp., endoparasitic nematophagous fungi 285
Catenaria anguillulae, infection of nematodes 294, **295**
Catinella olivacea, growth on dead wood 157
Cattle dung, growth of fungi on 211, 215, 222
Cellulolysis, rate of 47
Cellulolysis adequacy index (CAI) 30
Cellulolytic activity, factors determining 29
Cellulose
  chemical structure of 27, **28**
  content of lignified cell walls 162
  crystalline 27–8
    hydrolysis of 28
  degradation by fungi 29, 45–6, 82–3, 115, 164, 324
  hydrolysis of 28–30
    classification of fungi based on 29–30, 81
Cell wall structure of plants 161, **162**
Cephalosporium spp.
  growth on roots 192
  in rhizosphere 189
  soft-rots 163

Cephalosporium gramineum,
    antibiotic production 77
Ceratocystis spp., wood sapstainer
    159
Cereals
    root colonization 202
    see also named cereals
Ceriosporopsis spp., colonization
    of basswood by 276
Ceriosporopsis calyptrata,
    ascospores **269**
Ceriosporopsis cambrensis,
    estuarine distribution 278
Ceriosporopsis circumvestita,
    colonization of pine wood
    276
Ceriosporopsis halima
    ascospore **268**
    estuarine distribution 278
    geographical distribution 282
Ceuthospora pinastri, growth
    in litter 120
    on phylloplane 92
    on pine needles 96
Chaetocladium spp.
    growth on dung 207
    mycoparasites 218
Chaetocladium brefeldii,
    fruit-bodies **205**
Chaetomium spp.
    fruit-body **206**
    growth on
        dung 207
        worked timbers 159
    phoenicoid fungi 312
Chaetomium bostrychodes, growth
    on copromes 215
Chaetomium caprinum, fruiting
    time *214*
Chaetomium globosum
    rhizosphere fungus *199*
    soft-rot fungus 163, 165
Chaetomium insolens, cellulolytic
    fungus 46

Chaetomium thermophile,
    heat-tolerant fungus *323*, 324
Chamaecyparis spp., antifungal
    compounds 149
Cheilymenia spp., growth on dung
    224
Chestnut, see *Castanea sativa*
Chitin 31, **32**
Chlamydospores 5, 7, 17, **18**, 114
    formation of 17–18
Chondrostereum purpureum
    growth on
        dead wood 156
        fallen timbers 352, *353*, **355**,
            356
Chrysanthemum spp., production
    of antifungal substances 110
Chrysosporium spp., marker fungi
    of tundra soils 177
Chrysosporium pannorum
    cellulolytic fungus 46
    distribution in soil *176*
    psychrotolerant fungus 327
    temperature range for growth
        323
Chytridiales
    freshwater fungi 227–31
    marine 266
    parasites of phytoplankton 228–9
    saprophytic 229–31
        seasonal variation **232**
        techniques of measurement
            229
Chytridiomycetes
    freshwater fungi 227–31
    nematophagous fungi 284
Cirrenalia macrocephala,
    holeuryhaline species 277
Cladosporium spp. 26
    cellulose hydrolysis 115
    endophytes *151*
    growth on
        phylloplane 93–5, 97, 105,
            110

Cladosporium contd
  growth on *contd*
    Pinus litter 121
    wood 157
  overgrowth by *Collybia peronata* 71
  pigmented spores 114
  psychrotolerant 327
Cladosporium cladosporioides **91**
  competition with parasitic fungi 106
  growth on phylloplane 92, 94
    germ tube survival 113
  relative humidity requirement *139*
Cladosporium herbarum
  antibiosis 107
  distribution 40
  growth on
    angiosperm phylloplane 92, 103
    buds 97
    conifer needles 96
    Dactylis glomerata 131, 133
    herbs and grasses 141, 143
    leaf litter 115–16
    sugar cane 142
  pioneer colonizer 82, 113
    chlamydospore formation 114
  pH limits for growth *58*
  reduction of *Botrytis* rots by 106
  relative humidity requirement *139*
  xylan hydrolysis 82
Cladosporium macrocarpum, antibiosis 107
Classification of
  aquatic fungi 225, *226*, 242
  fungi, by ability to decompose cellulose 29–30, 81
  nematophagous fungi 295, *296*
  phoenicoid fungi 305–8
Clathrospaerina spp., aero-aquatic hyphomycetes 262

Clathrospaerina zalewskii
  colonization of submerged leaves by 264
  propagules **263**
Clavariopsis aquatica **244**
  inhibition of fungal growth by 253
  spore production 254
Clavatospora longibranchiata, growth on submerged leaves 252
Clavicipitaceae, endophytes 90
Clitocybe spp.
  fairy ring fungi
  leaf litter decomposing 118
Clitocybe alexandri, litter decomposing fungus *118*
Clitocybe clavipes, litter decomposing fungus *118*
Clitocybe flaccida 26
  growth on litter 366
Clitocybe geotropa, litter decomposing fungus *118*
Clitocybe infundibuliformis, litter decomposing fungus *118*
Clitocybe nebularis
  basidiocarps **371**
  competition with other fungi 372
  fairy ring fungus 372, 375
  growth on litter *118*, 366
    of conifers 120
  overgrowth by *Collybia peronata* 71
  ruderal fungus 372
Clitocybe odora, growth on beech leaf litter *118*, 119
C:N ratio 47–8
  influence on cord morphogenesis 25
Coal spoil, mycorrhizal associations of trees growing on 391
Cochliobolus sativus
  attack by fungivorous amoebae **19**

INDEX 509

inhibited by yeast 106
Coemansia spp., growth on dung 204
Coenocytes 13
Cold stores, growth of psychrotolerant fungi in 327
Cold-tolerant fungi, *see* Psychrotolerant fungi
Collagen, hydrolysis of 38
Collybia spp.
  fairy ring fungi 371
  leaf litter decomposing *118*
Collybia butyracea 26, 397
  growth on litter *118*, 366
  inhibition of mycorrhizal fungi 397
  overgrowth by *Collybia peronata* 71
Collybia confluens, overgrowth by 71
Collybia dryophila
  growth on beech leaf litter *118*, 119
  overgrowth by 71, **73**
Collybia maculata, growth on conifer litter 120
Collybia peronata
  cord formation in 26
  growth on leaf litter 119
  mutual inhibition with *Penicillium janthinellum* **69**
  overgrowth by 71, **72**
Colonization 3
  of attached branches 346, **347**, 348
  of buds 97–8
  of dung 211–12
  flare-up phase 68
  of flowers 98
  of leaf litter 115–16
  of leaves 88–114
    Agropyron repens 140–1
    Dactylis glomerata 129–40
    cereal 102–3

    development of communities 89–99
    factors affecting 99–114
    phylloplane 89–114
    shed, in water 248–9, 252
    submerged 262, 264
    techniques for studying 88–9
  pattern of 42
    species number 42
    species turnover 42
  of radicles 197
  of roots 197–202
    bean 198–201
    cereal 202
  of timber
    fallen 351–6
    standing 345–51
    worked 157, 159
  of wood 145–71, 249–52, 275–83, 345–56
  see also named fungal groups, plant species and substrates
Colonizers
  pioneer
    replacement of 81–2
    weak parasites 81
    'weed' species 81
  primary 83
    of *Dactylis glomerata* **132**, 133–4, 137
    of grasses 143
    of the phylloplane 113–14
    wood-rotting microfungi 157–8
  secondary 83
    of *Dactylis glomerata* **132**, 133
    of grasses 143
    of leaf litter 115
    macrofungi 346
    wood-rotting microfungi 159
Combat, definition 6
Combative fungi 8–10, 44, 82
  definition 8
  examples of 9–10

510 INDEX

Commensalism 46, 48, 82, 159
Communities
  on *Alnus* spp. 150
  on bean plants 41
  climax 83–4
  development of 45–80
    influence of C and N 45–8
    on living leaves 89–114
  dominant species 41
  on *Fagus sylvatica* 41
  forest soil 177
  on grasses 127–41
  grassland soil 177
  mature 80
  pioneer 80–1
    replacement of 81–2
    weak parasites in 81
    'weed' species in 81
  rhizosphere 184, 186–7, 189, *199*
  soil 175–9
  structure of 39–84
Competition
  coprophilous fungi 213–15
  definition 6
  Ingoldian aquatic hyphomycetes 252–3
  interspecific *105*, 106–8, 214–15, 397
  macrofungi 350
  mycorrhizal fungi 397
  pathogenic fungi 106–7
  phoenicoid fungi 315
  phylloplane fungi *105*, 106–8
  see also Antagonism
Compost, formation of 322–3, 325–6
  fungal succession 323–4
  thermotolerant fungi in 322–3, 325
Conidia 2, 94, 269, **270**, 274
  of aero-aquatic hyphomycetes 261–2
  of Ingoldian aquatic hyphomycetes 241–2

Conidial traps 285
Coniferous forest, phoenicoid fungi 311
Conifers
  ectomycorrhizas 378
  endophytes of 90
  growth of macrofungi on 350
  fungal growth on 120–2, 397
  pH 121
  phenolics in 121
  needles
    colonization by aquatic hyphomycetes 249
    fungal growth on 96
    spore deposition on 98
    resistance to soft-rots 165
  see also named species; Gymnosperms
Coniochaeta spp.
  growth on dung 224
  phoenicoid fungi 312
Coniophora puteana, growth on
  fallen timber 351
  worked timbers 157
Coniothyrium spp., endophytes *151*
Conolus versicolor, growth on beech logs *354*
Contact inhibition 6
  between phylloplane fungi and pathogens 107
  method of studying 107
Coprinus spp.
  combative fungi 82
  growth on dung 207
  pH optimum for growth 57
Coprinus angulatus, phoenicoid fungus 308
Coprinus atramentarius
  growth in disturbed environments 43
  survival at low water potential *170*
Coprinus cinereus, heat-tolerant fungus *323*

INDEX 511

Coprinus comatus, growth in disturbed environments 43
Coprinus disseminatus, distribution of 39
Coprinus heptemerus
  fruiting-body **206**
  fruiting time *214*
  hyphal interference by 76, 215, **216**, **217**, 218
Coprinus miser, growth on dung 209
Coprinus patouillardii, fruiting time *214*
Coprinus plicatilis, growth following volcanic eruptions 320
Coprinus psychromorbidus, cold tolerance 328
Coprinus radiatus, fruiting time *214*
Coprinus stercoreus, growth on dung 209
Coprobia granulata, growth on dung 222, 224
Coprogen 220
Copromes 214
Coprophagy 222
Coprophilous fungi 203–24
  antibiotic production 215
  autecological studies 224
  competition 213–15
  flora on dung from different animals 222, *223*, 224
  fruit-bodies **205**, **206**
  hyphal interference 215–18
  methods of studying 204
  mycoparasitism 218
  predation 219–20
  stimulation of germination by heat 314
  succession of 207–22
  synergism 220–2
  see also Dung
Coprotus granuliformis, growth on dung 209

Cord-forming fungi
  distribution of 359, 361
  growth on fallen timbers 352–3, 355
Cords **22, 24**
  advantages of 23
  differentiation in 23
  evolution of 23, 25
  formation, control of 25–6
  function of 23
  morphogenesis, influence of C:N ratio on 25
  structure 23
  transport of water in 23
Coriolus spp., growth on fallen timbers 354
Coriolus versicolor 9–10
  basidiocarps **349**
  growth on
    attached branches 346, **360**
    beech logs *353*, **355**
    dead wood 156
    fallen timbers 352–3
  hyphal barrage with *Stereum hirsutum* **67**
  interaction with *Daldinia* 350
  intraspecific antagonism **358**, 359
  mycoparasitized by
    Hypomyces aurantius 75
    Lenzites betulina 356
  white-rot fungus 166–7
Corollospora maritima
  ascospore **268**
  depth of survival 279
  distribution
    geographical 282
    vertical 277
  growth on driftwood 281
  perithecium **268**
  spores
    germination 272
    in sea foam 280
  stenohaline species 277
Corsican pine, *see Pinus maritima*

Corticium spp., growth on dead wood 156
Corticium evolvens, growth on beech wood *353*, 355
Cortinarius spp., late-stage mycorrhizal fungi 385, 393
Cortinarius pistorius, ectomycorrhizal fungus 378
Corylus avellana
  antifungal substances 109
  succession on leaf litter of 118
Coryne spp., growth on fallen timbers 352
Coryne sarcoides, growth on fallen timbers 352, *353*, *354*, **355**, 356
Cremasteria cymatilis, effect of salinity on vegetative growth 272
Cresolase reaction 34–5
Cristella sulphurea, growth on beech logs *354*
Crustacea, interactions with marine fungi 280
Cryptocline spp., endophytes *151*
Cryptococcus spp.
  growth on phylloplane 92
  psychrotolerant fungi 327
  temperature range for growth *323*
Cryptococcus terreus, soil inhabiting 184
Cryptosporiopsis spp., endophytes 89, *151*
C-selection 6–7
  see also Combative fungi
Ctenomyces serratus, growth on feathers 8
Cupressus spp., antifungal compounds 149
Curvularia spp., effects of CO_2 concentration on growth of 57
Curvularia lunata
  growth on sugar cane 142

inhibition by antifungal volatiles 182
Cuticle, protection against fungal growth 100
Cylindrocarpon spp.
  colonization of submerged wood 250
  growth on
    roots 193
    wood 157
Cylindrocarpon destructans, growth on roots 192, 199, **200**, 201–2
Cylindrocarpon didymum, growth on roots 192
Cylindrocarpon magnusianum, temperature range for growth *323*
Cylindrocladium spp., growth on wood 157
Cystoderma amianthinum, growth on conifer litter 121
Cytoplasmic streaming 14

Dacrymyces stillatus
  growth on dead wood 157
  survival at low water potential *170*
Dactylaria spp., predatory nematophagous fungi 285
Dactylaria candida, predatory nematophagous fungus 296, 299
  nematode attraction **298**
  trapping structures **286**
Dactylaria gracilis 296
Dactylella oviparasitica
  biological control agent 301
  nematode egg parasite 285
Dactylis glomerata **130**
  fungal colonization of 129–30, **131**, *132*, 133, **134**, **135**, **136**, 137–40
  effects of humidity on 137–9
  effects of moisture on *137*, 138–9

effects of nutritional status on 137
effects of physical conditions on 136
effects of senescence on 137–8
primary saprophytes **132**, 133–4, 137
secondary saprophytes **132**, 133
growth of 128–9
Daedaleopsis confradosa, sporulation 362–3
Daldinia spp., endophytes 150, *151*
Daldinia concentrica
  competition with other macrofungi 350
  distribution 40
  fruit-bodies **155**
  growth on attached branches 348
  xylariaceous rot fungus 168–9
Dasyscyphus grevillei, growth on hogweed 142
Dasyscyphus niveus, growth on dead wood 157
Dasyscyphus sulphureus, growth on nettle 142
Debaryomyces spp., osmotolerant fungi *333*
Dedikaryotization 359
Dendryphiella salina
  conidia **270**
  phylloplane fungus of saltmarshes 197
  sand dune fungus 281
  spore germination 272
  inhibition by sea water 271
Dendryphion comosum, growth on herbaceous plants 141
Desert, fungal species characteristic of 178
Desmazierella acicola, growth in conifer litter 121
Deuteromycetes
  conidia 2, 269
  euryhaline species 277
  marine 269, **270**
  nematophagous fungi 284
Deuteromycotina
  antibiotic production 70
  cellulose degradation 45
  facultative anaerobes 55
  growth on
    bracken litter 122
    dung 204, 209
  hemicellulose degradation 30, 31, 45
  marine 267
  reproduction 2
  sclerotia 20
  soft-rots 163
Dew
  analysis of 99–100
  source of water 65, 112
  see also Honey dew
Diatoms, parasitized by chytrids 228–9
Dictyosporium pelagicum, vertical distribution 277
Dictyosporium toruloides, growth on herbaceous plants 142
Die-back 154
Digitatispora marina 267
  basidiospores **270**
  Phoma pattern of vegetative growth 273
  sporulation 274
Dikaryons 16
Di–mon mating 359
Discomycetes, phoenicoid forest fungi 305–6
Discosia artocreas, growth on leaf litter 116–17
Discula quercina, growth on leaf litter 116
Dispersal, of spores 2–3, 17, 80, 86–7
Distribution
  component restricted species 39

Distribution *contd*
  factors limiting 40
    aeration 54–7
    animals 78–80
    hydrogen ion concentration 57–9
    inhibitors 48–9
    interspecific interactions 66–78
    nitrogen 48
    temperature 53–4
    water availability 59–66
  non-component restricted species 39
  taxon-selective 39
Disturbance, definition 5
Disturbed environments, colonization by fungi 43
Dithane 107
Ditiola radicata, survival at low water potential *170*
Ditylenchus myceliophagus, biological control of 301
Doratomyces spp.
  growth on worked timbers 159
  soft rots 163
  soil inhabiting fungi 115
Dothichiza pythiophila, growth on phylloplane 92
  stimulation of 51
Drechmeria spp., endoparasitic nematophagous fungi 285, **295**
  antibiotic activity 299
Drechmeria coniospora
  endoparasitic nematophagous fungus **289**, 294–6
  biological control agent 301
  nematode attraction 297–8
Drechmeria psychrophila, nematophagous fungus 293
Drechslera sorokiniana, inhibition by phylloplane fungi 106–7
Driftwood, fungal growth on 280
Dry-rot fungi 157

Dung
  adaptations of fungi growing on 203
  colonization by fungi 211–12
  from different animals, comparison of fungal flora 222, *223*, 224
  fungal growth on 203–24
  fungal succession on 207–22
    analysis of 212–23, **214**, 215–22
    appearance of fruit-bodies **208**, 213
    environmental effects 207–12
    microhabitat effects 211
    rainfall effects 211
    temperature effects 207–9
    water regime effects 209, **210**, 211
  mycelial growth rate on 212
  spore germination on 212
  substrate for fungal growth 203
  see also Coprophagous fungi
Dung beetles 219

Earthworms, leaf preference of 124
Ecology, fungal
  autecological approach 3
  techniques 4
Ectomycorrhizae 312, 376–88, 391, 393–5
  see also Mycorrhizal fungi
Elder, *see Sambucus nigra*
Endophragmia spp., growth on leaf litter 117
Endophytes
  as biological control agents 90
  of conifers 90
  of grasses 90
    effects on plant growth 90
    insecticide production by 90, 152
  phylloplane fungi as 151
  soil fungi as 150
  stem 150
  symbiotic 89

wood-decay fungi as 150, *151*
  on non-hosts 151–2
  in woody plants 89–90
Endozoic fungi 285
Entomophthorales, conidia 2
Enzymes, extracellular 13
Epichloe spp., endophytes 90
Epicoccum spp., growth on phylloplane 94
Epicoccum nigrum **132**
  epiphyte *151*
Epicoccum purpurascens
  cellulolytic fungus 115
  growth on
    buds 97
    Dactylis glomerata 133
    herbaceous plants 141
    Pinus litter 121
  phylloplane fungus 92, 105
    chlamydospore formation 114
    pigmented spores 114
    relative humidity requirement *139*
Ethylene, regulator of microbial growth in soil 182, 187–8
Euglypha denticulata 20
Eumycota spp., psychrotolerant fungi 327
Eurotium repens, synergism with *Viennatidia fimicola* 220
Euryhaline species 277
Exidia glandulosa
  basidiocarps **349**
  decay of attached branches by 346, 348
  sporulation 364
  white-rot fungus 154
Extractives
  content in wood 148–9
  effects on distribution of wood-rot fungi 148–9
  stimulators of fungal growth 149
Extreme environments, fungi of 322–40

see also Osmotolerant fungi;
  Psychrotolerant fungi;
  Thermotolerant fungi;
  Xerotolerant fungi

Fagus spp.
  heart-rot basidiomycetes of 148
  macrofungi, growth on
    attached branches 350
    litter 366
  stem endophytes of 150
Fagus sylvatica 9–10
  antifungal compounds 49, 109
  colonization
    by marine fungi 276
    of submerged leaves by aquatic hyphomycetes 249
  fungal communities on 41
  litter
    rate of decomposition *118*, 124
    succession on 116, *117*, 118–19
  macrofungi, growth on
    fallen timbers 352–6
    logs *353*, *354*, **355**, *357*, **360**
  mycorrhizae 381, 390
  pathogens of 156
  root mycoflora of 193, *194*
Fairy ring fungi 371–6
  age 374
  basidiocarps **371**, **373**
  bilateral extinction 375
  free rings 371
  indifference phenomenon 375
  mycelial zone, fungal species in 376
  polarity of development 372
  radial growth rate *374*
  somatic incompatibility 374–5
  tethered rings 371
  toxins produced by 375
  unilateral extinction 375
Fairy rings 14, **371**, **372**, **373**
  grassland 373
    death of grass 375

Fairy rings *contd*
  grassland *contd*
    stimulation of grass growth 375–6
    kinetics 375
Farmers' lung disease 337
Fertilizers, effects on
  cell leakage by leaves 100
  fruiting of mycorrhizal fungi 391–4
Ferulic acid 51
Fibroporia vaillantii, growth on worked timbers 157
Filamentous fungi, growth on dung 204
Fireplace fungi 308
Fires
  bonfires 302
  effects on
    mycorrhizal fungi 391
    soil 303–5
  fungi growing at the site of, *see* Phoenicoid fungi
  natural 302
Fistulina hepatica, selectivity of 148
Flagellospora curvula
  growth on submerged leaves 252
  temperature characteristics of 256
Flagellospora penicillioides,
  temperature characteristics of 256
Flammulina velutipes
  basidiocarps **364**
  sporulation 363
  survival at low water potential *170*
Flavanonols 149
Flowers, colonization of by fungi 98
Fomes spp., perennial sporophytes 362
Fomes fomentarius, distribution 40
Forest soils, typical fungal communities 177

Fraxinus spp., stem endophytes 150
Fraxinus excelsior
  antifungal substances 109
  attached branches, decay by macrofungi 348, 350
  litter
    rate of decomposition 124
    succession on 118
French bean, *see Phaseolus vulgaris*
Freshwater fungi 227–65
  Chytridiomycetes 227–31
  conidia 2
  Oomycetes 231–41
  Peronosporales 239–41
  see also Hyphomycetes, aero-aquatic; Hyphomycetes, Ingoldian aquatic
Fruiting time, minimal 213, *214*
Fungicides 69–71, 80
  effect on phylloplane colonization 107–8
Fungicoles 74–6
Fungi imperfecti, effects of salinity on **275**
  see also Deuteromycotina
Fungistasis 18, 69–71, 80, 179–80, **181**, 182–3, 201
  inhibitory volatiles 181–2
  nutritional hypothesis 180
  overcome by roots 183
  phoenicoid fungi and 315
  residual 182
Fungivores 19–20
Fusarium spp.
  cellulose and carboxymethylcellulose hydrolysis 82
  chlamydospores **18**
    formation 18
  colonization of submerged wood 250

ecosystem distribution *178*
grassland soil fungi 177
growth in carbon-free media 8
inability to degrade plant lignins 34
inhibition by tannins 37
leaf litter fungi 116
pathogenic *192*
pioneer colonizers 82
rhizosphere fungi 189, 191, *192*, *199*
root
  colonizers *199*
  invasion by 81
soil fungi 115, 175, 184
spore dispersal 2
wood-rotting fungi 157
Fusarium aqueductuum, sewage fungus 232
Fusarium culmorum, growth on roots 192, 202
Fusarium lateritum, endophyte *151*
Fusarium nivale, cold tolerance 328
Fusarium oxysporum
  facultative anaerobe 54
  pH limits for growth *58*
  root
    colonizer 192, 199, **200**
    invasion by 201
Fusarium sambucinum, growth on roots 192
Fusarium solani
  facultative anaerobe 54
  growth on roots 192, 201
Fusidic acid 71
Fusidium spp., antibiotic production 71
Fusticeps bullatus, aero-aquatic hyphomycete 262
  spores 262

Galactoglucomannan 30

Galerina spp., psychrotolerant fungi 327
Galleropsis desertorum, water stress tolerance 64
Gallic acid 51, *109*, 115–16
Ganoderma spp., perennial sporophytes 362
Ganoderma adspersum
  spore growth on attached branches 350
  production 362
  size 87
Ganoderma applanatum
  spore production 362
  white-rot fungus 168
Ganoderma tsugae, white-rot fungus 167
Gäumannomyces graminis, growth on roots 202
  salinity and 272
Gelasinospora reticulospora, growth following volcanic eruption 320
Geniculosporium spp., endophytes *151*
Geococcus vulgaris 20
Geopetalum carbonarium, phoenicoid fungus 308
Geopyxis carbonaria, phoenicoid fungus 308, 311–12
  growth on lime-rich substrates 315
Geotrichum candidum, sewage fungus 232
Gliocladium spp.
  in carbon-free media 8
  on roots 192
  soil inhabiting 115, 184
Gliocladium roseum
  growth on conifer litter 121
  necrotroph 72
  rhizosphere fungus *199*
  root
    colonizer *199*, **200**

518   INDEX

Gliocladium roseum contd
   root *contd*
      invasion by 201
Gliomastix spp.
   growth on roots 193
   rhizosphere fungi 189
Gliomastix convoluta, effects of
      CO_2 concentration on growth
      of **56**, 57
Glucomannan 30
Glucuronoxylan **30**
Gnomonia gnomon, growth of
   teleomorphs on leaf litter
   118
Gnomonia setacea, growth of
   teleomorphs on leaf litter 118
Gonatobotrys simplex,
   mycoparasite 95
Goniopila monticola, growth on
   submerged leaves 252
Grasses
   characteristics of fungi growing
      on *139*, 140
   endophytes of 90
   fungal communities on 128–41
      primary saprophytes 143
      secondary colonizers 143
      soil fungi 143–4, 177
   root colonization 202
   see also *Agropyron repens*;
      Dactylis glomerata;
      Grassland
Grassland
   burnt, phoenicoid fungal flora
      312–13
   fairy rings 373, 375–6
   soil fungal communities 177
Guignardia citricarpa, growth on
   sugar cane 142
Gymnosperms, hemicellulose
   content of wood 30
   see also named species; Conifers
Gyrophragmium dunaii, water
   stress tolerance 64

Habitat exploitation by fungi 39
Hadacin 71
Halocyphina villosa
   growth on mangroves 277
   sporulation 274
   vegetative growth
      effect of salinity on 272
      Phoma pattern of 273
Halosphaeria spp., colonization of
   basswood by 276
Halosphaeria appendiculata,
   colonization of beech wood 276
Halosphaeria hamata
   colonization of beech wood 276
   geographical distribution 282
Halosphaeria mediosetigera
   ascospore **268**
   marine soft-rot fungus 279
   spore germination, inhibition by
      sea water 271
   vegetative growth 272
Halosphaeria quadriremis,
   ascospore **268**
Hansenula spp., osmotolerant
   fungi 333
Harposporium spp., endoparasitic
   nematophagous fungi 285
   antibiotic activity 299
Harposporium anguillulae,
   endoparasitic nematophagous
   fungus **289**, 292, 296, 298
   nematode attraction **297**, 298
Hartig net 376
   formation of 380–1
Hazel, see *Corylus avellana*
Heart-rot fungi 146
   selectivity of 147–8, 154
   S-selected 149
Heartwood, macrofungi colonizing
   350–1
Heather, see *Calluna vulgaris*
Heat-shock proteins 329
Hebeloma spp., early-stage
   mycorrhizal fungi 385

INDEX 519

Hebeloma crustuliniforme, ectomycorrhizal fungus 380, 385, 394
Hebeloma mesophaeum, ectomycorrhizal fungus 380, 391
Hedera spp., stem endophytes 150
Helicodendron spp.
 aero-aquatic hyphomycetes 262
 propagule 262
 survival under anaerobic conditions 264
Helicodendron giganteum
 colonization of submerged leaves by 264
 propagules **263**
Helicodendron triglitziense, survival of desiccation 265
Helicoon spp., aero-aquatic hyphomycetes 262
 propagules 262
Heliscella stellata, growth on submerged leaves 252
Heliscus lugdunensis, **244**
Heliscus tentaculus, temperature characteristics of 256
Helminthosporium sativum
 effect of pollen on infection by 104
 germ tube survival 113
Helotium caudatum, growth on leaf litter 118
Hemicellulose
 chemical structure of 30
 hydrolysis of 30–1
 quality in wood 148
Hendersonia acicola, growth on pine needles 96–7, 120
Heracleum sphondylium, fungal colonization of 141–2
Herbaceous plants, fungal succession
 primary saprophytes 143
 role of soil fungi 143–4
 secondary colonizers 143

 see also Heracleum sphondylium; *Saccharum officinarum*; *Urtica dioica*
Herodera marioni, biological control of 300
Heterobasidion annosum
 basidiospore production 362
 colonization by 10
 competition with *Peniophora gigantea* 10
 K-selection 10
 mycelium versatility 10
 selectivity of 148
 sporulation 363
 survival at low water potential *170*
Heterokaryon formation 15, 16
Hiliscus lugdunensis, colonization of submerged wood 250
Hirst spore trap 85
Hogweed, *see Heracleum sphondylium*
Hohenbuehelia spp., nematodes as nitrogen supply for 147
Holeuryhaline species 277
Holly, *see Ilex aquifolium*
Holomorph 1
Honey dew, effect on phylloplane colonization 100
Hordeum vulgare
 pollen effects on fungal colonization of leaves *102*, 103
 rhizosphere mycoflora 189
 root exudates *195*
Hormiscium sp., growth on hogweed 142
Hormonema spp., endophytes *151*
Humicola spp.
 grassland soil fungi 177
 growth on worked timbers 159
 rhizosphere fungi *199*
 soil inhabiting 115
Humicola alopallonella
 brackish water species 277

Humicola alopallonella contd
   growth on mangroves 277
Humicola fuscoatra, distribution in
   soil *176*
Humicola insolens, heat-tolerant
   fungus *323*, 324
Humicola thermophile, cellulolytic
   fungus 46
Humus
   phenolics in 123–4
   rate of decay 123
   structure 123
Hyaline cells 114
Hydnum suaveolens, fairy ring
   fungus *374*
Hydrogen ion concentration,
   effects on fungal growth 57–9
   see also pH
Hygrophoropsis aurantiaca, growth
   on conifer litter 121, 397
Hymenomycetes
   inability to utilize nitrates 48
   inhibition by antifungal volatiles
      52
Hyphae
   barriers **67**
   branching of 12
   cavities, formed by soft-rot fungi
      164–5
   fusion 15
      lethal reactions 16
   growth of 12
   interference 9, 76–7, 82
      coprophilous fungi 215, **216**,
         217, 218
   intermingling of different species
      67
   lysis of in overgrowth 71, **73**
   non-self fusion 15
   pigmentation of 21, **67**
   primary branches 12–13
   productivity, measurement of
      172–4
   self-to-self fusion 15

skeletal 23
space occupied by 68
Hyphoderma spp., wood-rotting,
   nematophagous fungi 299
Hypholoma capnoides, basidiocarp
   production 39
Hypholoma fasciculare 9
   growth on fallen timber 351, *354*
   mycelial growth 344
   strand-former 156
Hyphomycetes **243**, **244**
   aero-aquatic 261–5
   colonization of submerged
      leaves by 262, 264
   effects of salinity on 265
   growth and survival under
      anaerobic conditions
      264–5
   survival of desiccation 265
   techniques for studying 262
   effects of salinity on **273**, *273*,
      275
Ingoldian aquatic 241–61
   classification 242
   colonization of shed leaves
      248–9, 252
   colonization of wood 249, *250*,
      251, 252
   competition 253
   conidia 241–2
   distribution, along river
      courses 257–8
   ecological studies 248–61
   effect of altitude on
      distribution 256–7
   effect of dissolved oxygen on
      distribution 259–60
   effect of pH on distribution
      257, **258**, 259
   effects of competition on
      colonization 252
   effects of salinity on 259
   effects of temperature on 256
   identification 242

interactions with invertebrates 261
resource preference 249
seasonal variation in 253–4, **255**, 256
spore production 253, 255
spores **242**
succession on submerged leaves 252
techniques for study 245–6, **247**, 248
in terrestrial habitats 260
Hypolimnion, Saprolegniaceae in 235
Hypomyces aurantius
  antibiotic production by 9
  necrotroph **75, 76**
Hypoxylon spp.
  endophytes 90, 150, *151*
  growth on fallen timbers 352
  xylariaceous rot 168–9
Hypoxylon fragiforme, growth on fallen timbers *353*, 354
Hypoxylon mammatum, tree pathogen 156
Hypoxylon multiforme, fruit-bodies **155**
Hypoxylon nummularium, growth on fallen timbers *353*, 354, **355**
Hypoxylon punctulatum, tree pathogen 156
Hypoxylon rubiginosum
  growth on attached branches 348
  tree pathogen 156
Hysterangium crassum, ectomycorrhizal fungus 381

Identification of fungi 4, 242
Ilex spp., stem endophytes 150
Ilex aquifolium, colonization of phylloplane 93
Incense cedar, *see Libocedrus decurrens*

Incompatibility 343
Inhibition, mutual **69**
Inhibitors 48, **50**
  effects on fungal distribution 48–9
  preformed 108–9
  see also Antifungal substances
Inonotus dryadeus, growth on attached branches 350
Inonotus hispidus, growth on attached branches 350
Insecticides, production by endophytes 90, 152
Interactions, interspecific 66–78
  direct antagonism 71–7
  indirect antagonism 67–71
  in vivo–in vitro correlation 77–8
Interference, hyphal 9, 76–7, 82
  coprophilous fungi 215–18
Invertebrates, interactions with
  aquatic hyphomycetes 261
  wood-rotting fungi 160–1
  see also Animals; Arthropods; Crustacea
In vivo–in vitro correlation 77–8
Iodophanus carneus, growth on dung 207
Island theory 42

Juniperus spp., stem endophytes 150

Keratin, hydrolysis of 38
Kickxella spp., growth on dung 204
Kriegeriella mirabilis, growth on pine litter 120
K-selection strategies, definition 6

Laboulbeniales, chitin degradation by 8
Labyrinthulales, obligate marine fungi 266
Laccaria spp., early-stage ectomycorrhizal fungi 385
Laccaria amethystea 380, 397

522 INDEX

Laccaria bicolor 380
Laccaria laccata, early-stage ectomycorrhizal fungus 378, 391, 394, 397
  fruit-bodies **387**
Laccaria proxima 385, 391
Lacrymaria velutina, growth in disturbed environments 43
Lactarius spp., ectomycorrhizal fungi 393
Lactarius camphoratus 395, 397
Lactarius delicious
  ectomycorrhizal fungus 378
  pH limits for growth *58*
Lactarius piperatus 394
Lactarius pubescens 385, 391
Lactarius rufus 391
Lactarius subdulcis 391, 394
Lactarius tabidus 397
Lactarius turpis 391
Laetiporus sulphureus
  basidiospore formation 362
  selectivity of 148
Lagenidiales, marine 266
Lagenocystis spp., growth on roots 202
Larix europea, fungal growth on needles of 96
Lasiobolus ciliatus, growth on dung 224
Latent invasion strategy 348
Lead pollution, effects on phylloplane colonization 97
Leaf glands, production of antifungal substances 110
Leaf hairs
  effect on microclimate 112–13
  protection against fungal colonization 87
Leaf-inhabiting fungi, *see* Phylloplane fungi
Leaf litter
  bracken, succession on 122
  colonization by soil fungi 115–16

conifer
  pH of 121
  succession on 120–2
decomposition of 122–7
  Ao horizon 123, 125
  Aoo horizon 123, 125
  by macrofungi 365, **366**, 367–76
  role of agarics 116, *118*, 119, *365*
  role of soil animals 125–6
  effects of sulphur dioxide on fungal communities of 127
  production 122–3
  pH of 59
  rate of decay 124, 126
  factors affecting 126
  secondary mycoflora 115
  succession on 114–16, *117*, 118–22
  comparison of angiosperm and conifer litter 121–2
  water-holding capacity 119
  water potential of 66, 126
Leaf shape, effect on microclimate 112
Leaf surface, *see* Phylloplane; Phylloplane fungi
Leaves
  cell leakage, effects on colonization by fungi 99–100
  colonization of
    abaxial surface 98
    adaxial surface 98
    development of communities 89–99
  effects of climate 111–14
  effects of plant inhibitors 108–11
  factors affecting 99–114
  in freshwater 248, 252
  interspecific competition 105–8
  pollen effect 100–4

secondary saprophytic phase 95
techniques for studying 88–9
decomposition in freshwater 260–1
fungal succession 252
leaching 248
deposition of spores on 85, **86**, 87–8
factors affecting 86–7
rain wash-out 86
sedimentation 86
splash dispersal 86
wind impaction 86
germination of spores on 100–2
growth of fungal parasites 90, 92, 95–6
growth of symbionts on 89
gymnosperm, fungal succession on 96
senescence, effects on fungal growth 100
see also Phylloplane; Phylloplane fungi
Leccinum spp., late-stage mycorrhizal fungus 385
Leccinum aurantiacum, ectomycorrhizal fungus 380
Leccinum scabrum, ectomycorrhizal fungus, fruit-bodies **388**
Lemonniera aquatica **244**
growth on submerged leaves 252
temperature characteristics of 256
Lemonniera centrosphaera, growth on submerged leaves 252
Lemonniera terrestris, growth on submerged leaves 252
Lentinus lepideus, growth on worked timbers 157
Lenzites betulina 9–10
mycoparasite of *Coriolus versicolor* 356

Lenzites saepiaria
growth on fallen timber 351
inhibition by antifungal volatiles 52
Lepiota amianthina, litter decomposing fungus *118*
Lepiota procera, litter decomposing fungus *118*
Lepiota sordida, fairy ring fungus *374*
Leptodontium, growth on wood 158
Leptomitales
facultative anaerobes 55
freshwater fungi 231–3
Leptomitus lacteus 55
freshwater fungus 232–3
sewage fungus 232
Leptosphaeria acuta, growth on nettle 141
Leptosphaeria doliolum, growth on nettle 142
Leptosphaeria microscopica **132**
growth on *Dactylis glomerata* 133
relative humidity requirement *139*
Leptosphaeria nigrans
growth on *Dactylis glomerata* 133
relative humidity requirement *139*
Leptosphaeria sacchari, growth on sugar cane 142
Leptostroma spp., endophytes 89, 90
protection of plants against insects 152
Leptostroma pinastri, growth in litter 120
Libocedrus decurrens, heartwood fungi of 350
Lichens, marine 267
Light, supply to plants, effect on rhizosphere mycoflora 193, *194*

Ligniera spp., growth on roots 202
Lignin 32, 145, 148, 164, 166
  chemical structure of **33**
  degradation of 29, 33–7, 167–9
    by agarics 118
    by ascomycetes 33–4, 145
    by basidiomycetes 31, 33–4, 36, 145
    by heat-tolerant fungi 324, 326
  distribution in plant cell walls 161–2
Lignincola laevis, vertical distribution 277
Lignocellulose 145
Ligustrum vulgare, antifungal substances 109
Lime, effects on fruiting of
  mycorrhizal fungi 391
  phoenicoid fungi 315, *316*
Lime, *see Tilia platyphylla*
Lindra thallassiae
  spore germination 272
  vegetative growth 272
Linear organs 22–6
  structure of 22
  transport in 22–3
Liquidambar styraciflua, attack by soft-rot fungi 165
Litter, *see* Leaf litter
Littoral muds, Saprolegniaceae in 234–5
Lodgepole pine, *see Pinus contorta* var. *latifolia*
Lopadostoma turgidum 9
Lophodemium arundinaceum, growth on sugar cane 142
Lophodermella conjuncta, growth on pine needles 96
Lophodermella sulcigena, growth on pine needles 96
Lophodermium conigenum, endophyte 90

Lophodermium pinastri, growth on
  phylloplane 92
  pine bud scales 97
  pine needles 96
Lulworthia spp.
  ascospores **269**
  colonization of basswood by 276
  growth on mangroves 277
Lulworthia floridana
  effect of salinity on vegetative growth 272, *273*
  estuarine distribution 278
Lulworthia medusa, vegetative growth 272
Lunulospora curvula **243**
  spore production 254, **255**
  temperature characteristics 256
Lupinus nootkatensis, rhizosphere 186
Lycoperdon spp., fairy ring fungi 371
Lycoperdon pyriforme, growth on dead wood 157

Macrofungi
  basidiocarp survival 364
  basidiospore production 341
  colonization of
    fallen timber, effect of temperature on 351
    wood 152–4, 156–7, 345–52, *353*, *354*, **355**, 356, *357*, **358**, **360**, **361**, **362**
  competition between 350
  fruit-bodies
    of litter decomposing **366**
    succession 364
  intraspecific antagonism 356–9
  litter decomposers *342*, 365–76
  mycorrhizal *342*, 376–97
    see Mycorrhizal fungi
  nutritional relationships of *342*
  phenology 362–4

INDEX 525

pioneer colonizers of
  attached branches 346, **347**, 348
  fallen timbers 352
secondary colonizers of attached branches 346
spore production 362
  effect of temperature on 362–4
wood-rotting *342*, 345–64
  basidiocarps **349, 361, 362, 364**
  fallen timber 351–6
  interactions between colonies 356, *357*, **360**
  standing trees 345–51
Malthouse workers' lung disease 337
Maltose 27
Maneb 107–8
Mangroves, colonization by fungi 275, 277
Marasmius spp.
  fairy ring fungus 371
  leaf litter decomposing 118
Marasmius androsaceus
  competition with *Mycena galopus* 368, **369**, 370
  cords **22**
  grazing by arthropods 79, **370**
  growth on conifer litter 120–1
  stimulation of growth of 51
Marasmius oreades, fairy ring fungus **372, 373**, *374*, 375–6
Marine fungi 265–83
  brackish water species 277
  culture of 269–71
  environment 266–7
  euryhaline species 277
  facultative 266
  holeuryhaline species 277
  lignicolous 276–83
    depth of survival 279
    on driftwood 280
    effect of temperature on distribution 282–3
  estuarine distribution 278–9
  fruiting of 276–81
  geographical distribution 281–3
  growth on sea grasses 279
  host preference 276
  interactions with crustacea 280
  methods of studying 276
  succession 276
  vertical distribution of 277, **278**
  obligate 266
  reproduction 274–6
    effects of salinity on *273*, 274–5
  spore germination 271–2
    effect of mycostatic factors 271
    effect of salinity 271–2
  stenohaline species 277
  substrata 275
  vegetative growth 272–3
    effect of salinity 272
    effect of salinity and temperature 272, **273**
Massarina aquatica, inhibition of fungal growth by 253
Matric potential 60, 64–5
Melampsora laricipopulina, attacked by mycoparasites 95
Melanin, deposition in spore and hyphal walls 21
Melanomma spp., growth on fallen timbers 352
Melanomma pulvis-pyrius, growth on beech logs *353*
Membrane structure 329–30
Memnospores 2, 17
Mesophiles 53–4, 322
Microdochium bolleyi, growth on roots 202
Microfungi, *see* Wood-rotting fungi
Micromphale perforans
  growth on conifer litter 120
  stimulation of growth of 51

Microsclerotia 114
Microsporum spp., skin pathogen 38
Microthyrium microscopicum, growth on leaf litter **117**
Mildews, on leaves 95
Mindeniella spp., facultative anaerobes 55
Mineral cycling, role of fungi 179
Mites 20
Modifiers 49
Mollisia cinerea, growth on dead wood 157
Mollisia melaleuca, growth on dead wood 157
Mollisia palustris, growth on *Dactylis glomerata* 133–4
Mollisina acerina, growth on leaf litter 116–17
Monacrosporium spp., predatory nematophagous fungi 207, 285, 292
Monacrosporium cionopagum, nematophagous fungus 293, 296
  nematode attraction **298**
Monacrosporium doedycoides, nematophagous fungus, traps **287**
Monacrosporium ellipsosporum, nematophagous fungus 300
  nematode attraction **297, 298**
  trapping structures **286**
Monascus bisporus, osmotolerant fungus 332, *333*
Monascus bisporus, xerophilic fungus 339
Monodictys pelagica
  in sand dunes 281
  vertical distribution 277
Monokaryons 17
Montagnea arenaria, water stress tolerance 64

Mortierella spp.
  chitin degradation 32
  distribution *176, 178*
  growth on wood 157
  hemicellulose degradation 31
  inability to hydrolyse cellulose 29, 45
  pioneer colonizers 82, 199, 201
  psychrotolerant fungi 327
  rhizosphere fungi 186
  root fungi **200**
  soil fungi 116, 175, 184
  xylan hydrolysis 45, 82
Mortierella alpina, optimal pH 58
Mortierella isabellina, optimal pH 58
Mortierella minutissima, optimal pH 58
Mortierella ramanniana
  optimal pH 58
  temperature range for growth *323*
Moss
  association of phoenicoid fungi with 316–17
  succession on 122
Moulds, non-decay 157–8
  antagonists of wood-rotting fungi 160
  role in wood decay 160
Mount St Helens, volcanic eruption 319–20
Mucigel sheath 188, **189**
Mucor spp. 7–8
  facultative anaerobes 54
  growth
    on beech logs *354*, **355**
    in carbon-free media 8
    on dung 207
    on wood 157
  hemicellulose degradation 31
  inability to hydrolyse cellulose 29, 45
  inhibition by tannins 37

pioneer colonizers 82, 199, 201
psychrotolerant fungi 327
rhizosphere fungi 186
root fungi **200**
soil fungi 116, 175, 184
spore size 87
thermophilic, membrane structure 330
xylan hydrolysis 45, 82
Mucor circinelloides, cold tolerance 327
Mucor hiemalis
  distribution in soil *176*
  fruiting time *214*
  growth on bracken litter 122
  inhibited by gallic acid 116
  response to water stress 62
  'secondary sugar' fungus 46
  rhizosphere fungus *199*
  soil inhabiting 184
  temperature range for growth *323*
Mucor mucedo
  cold tolerance 327
  fruiting time *214*
  sporopollenin in zygospores of 21
  temperature range for growth *323*
Mucor oblongisporus
  psychrotolerant fungus 331
  'secondary sugar' fungus 46
  temperature range for growth *323*
Mucor piriformis, cold tolerance 327
Mucor plumbeus, rhizosphere fungus *199*
Mucor racemosus, pH limits for growth *58*
Mucor strictus
  psychrophile 331–2
  temperature range for growth *323*
Mucoraceae, effects of salinity on **275**

Mucorales
  asexual spores of 2
  conidia 2
  inability to utilize nitrates 48
  nitrogen requirements 47
  nutrition of 7
  reproduction 7
  R-selected fungi 7–8
Mud
  anaerobic conditions 264
  Saprolegniaceae in 234–5
  see also Hyphomycetes, aero-aquatic
Mutualism 46, 48, 90, 160
Mycelia sterilia, soil fungus 175
Mycelium 12–17
  ageing of 14
  cords 22–3, 25–6, 344
    see also Cord-forming fungi; Cords
  depth of penetration into solid substrata 13
  distribution in soil horizons *177*
  estimation of, importance in studies of succession 83
  extension of 12–13
  fans 344
  heterokaryotic 15
  modes of growth 344
    factors affecting 344
  nuclear migration, rates of 15
  primary homokaryotic 344
  secondary dikaryotic 344
Mycelium radicis atrovirens, distribution in soil *176*
Mycena spp., leaf litter decomposing fungi 118
Mycena epipterygia, litter decomposing fungus *118*
Mycena galericulata, growth on dead wood 156
Mycena galopus
  autecology 3, 366–71
  basidiocarps **367**

Mycena galopus contd
  competition with *Marasmius androsaceus* 368, **369**, **370**
  grazed by springtails 368, **370**
  growth on litter 366–7
    bracken 122, **367**
    conifer 120, 369
    leaf 119
  horizontal distribution 370–1
  mycelium detection 367
  seasonal variation in abundance 368
  white-rot fungus 367
Mycena pura, growth on beech leaf litter 119
Mycena rosella, litter decomposing fungus *118*
Mycena vulgaris, litter decomposing fungus *118*
Mycocentrospora acerina, chlamydospore formation 18
Mycogone perniciosa, nematode damage **19**
Mycoparasites, fungi 20, 95, 356
  coprophilous 218
Mycoparasitism 6, 9, 82
  mechanism of 72–6
Mycorrhiza 1, 312
  sheathing 343–4, 376–9
    fruit-bodies **387**, **388**
Mycorrhizal fungi
  acid rain, effect on 394
  arrival in the rhizosphere 379–80
  attachment to root surface 380
  distribution, effect of
    metal ion saturation on **390**, **392**, **393**
    organic matter on **390**, **392**, **393**
  early-stage 385
    fruit-bodies **387**
    R-selected 387
  fruiting, effects of
    fertilizer on 391–4
    lime on 391–2
    rainfall on 395, **396**, 397
    temperature on 395
  Hartig net formation 380–1
  host specificity 378–9
  infection by, factors affecting 383, 389–94
  interspecific competition 397
  late-stage 385–6
    fruit-bodies **388**
    S-selected 387
  mycelial mats 381
  mycelium distribution 381
  phenology 394–7
  seasonal abundance of sporophytes 394–5, 397
  somatic incompatibility *382*
  spore germination, stimulation by rhizosphere 379–80
  succession of **383**, *384*, 385, **386**, 387–8, **389**
Mycosphaerella ascophylli, growth on algae 267
Mycosphaerella punctiformis, growth on leaf litter 116, **117**, 118
Mycosphaerella recucita, growth on *Dactylis glomerata* 133
Mycostasis, *see* Fungistasis
Mycostatic factor, in sea water 271
Mycotoxins, accumulation in stored grain 337
Myxomphalia maura, phoenicoid fungus 308, 311
  basidiocarps **307**
  growth
    following volcanic eruption 320
    lime-treated plots 316
  seasonal variation **309**
Myxomycetes, growth on dung 204
Myzocytium spp., endoparasitic nematophagous fungi 285, 292

Naemacyclus niveus, growth on phylloplane 92
Nautosphaeria cristaminuta, colonization of beech wood 276
Necrotrophs 72, **74**, **75**, **76**
Nematoctonus spp.
  endoparasitic nematophagous fungi 285, 299
  nematotoxin production 299
  wood-rotting fungi 299
Nematoctonus leiosporus, endoparasitic nematophagous fungus **289**
Nematodes 20
  attracted by nematophagous fungi 296, **297**, **298**
  biological control of 300–1
  fungal parasites of 207
  fungivorous **19**
  habitat 284
  infection by endoparasitic nematophagous fungi 298–9
  plant-parasitic 284–5
  predator–prey relationships 294, **295**
  see also Nematophagous fungi
Nematophagous fungi 147, 284–301
  abundance 292–4
  attraction of nematodes by 296, **297**, **298**
  biological control using 300–1
  classification 295, *296*
  distribution 291–4
  ecological characteristics 294–300
  endoparasitic 285, **288**, **289**
    factors affecting distribution 292
    in sewage 294, **295**
  nematotoxin production 298–9
  parasites of eggs or cysts 285
  predator–prey relationships 294, **295**
  predatory 285
  effect of organic matter on activity **293**, **294**
  factors affecting distribution of 292–3
  succession of 293, **295**
  techniques for studying 285, 290–1
  trapping structures **286**, **287**
  stimulation of trap formation by nematodes 297
  survival structures 299–300
  techniques for studying 290–1
  also wood-rotting fungi 299
Nematophthora gynophila, nematode egg parasite 285
Neotiella hetieri, phoenicoid fungus 311
Nettle, see *Urtica dioica*
Neurospora spp., phoenicoid fungi 314
Neurospora crassa
  ascospores
    heat activation of germination 314
    sporopollenin in 21
  growth on charred vegetation 305
Neurospora sitophila, heat-tolerant fungus *323*
Nia vibrissa
  basidiospore **270**
  marine fungus 267
  white-rot fungus 279
Nicotiana tabacum, rhizosphere 186
Nigrospora sphaerica, growth on sugar cane 142
Nitrates, fungi unable to utilize 48
Nitrogen
  concentration of in freshwater substrata 240
  effects of availability on succession 84
  effects on rhizosphere mycoflora 194–5

Nitrogen *contd*
  fixation 147
    increased following burning 303
  requirements of fungi 47–8, 68
  sources of 147, 160
Nodulisporium spp.
  endophytes *151*
  growth on fallen timbers 352, *353*
'Non-weed' species
  characteristics 43–4
  ecological adaptations *44*
Norway maple, *see Acer platanoides*
Norway spruce, *see Picea abies*
Nutritional hypothesis 212

Oak, *see Quercus* spp.
Oceanitis scuticella, depth of survival 279
Ochratoxin 337
Octospora spp., phoenicoid fungi 311
Oidiodendron spp.
  distribution in soil *176*
  ecosystem distribution *178*
  forest soil fungi 177
Oidiodendron fuscum,
  soil-inhabiting fungus 175
Oligotrophs 1, 8, 179
Olpidium spp., growth on roots 202
Omphalina spp., psychrotolerant fungi 327
Onygena spp., growth on keratin-rich substrata 8
Onygena equina, keratin degradation by **9**, 38
Oomycetes
  freshwater fungi 231–41
  hyphae of 13
  inability to hydrolyse cellulose 29, 45

nematophagous fungi 284
response to water stress 62
Oospores 2, 18
Ophiobolus erythrosporus, growth on herbaceous plants 142
Opportunistic decomposers, *see* 'Weed' species
Opportunists 116
Orbimyces spectabilis, conidial development 274
Oryza sativa, root colonization 202
Osmophiles 1
Osmotic potential 60
Osmotolerant fungi 64, 332–3
  accumulation of polyols in 338–9
Oudemansiella mucida
  distribution of 39
  selectivity of 148
  soft-rot fungus 163
Overgrowth 71, **72**
Oxygen, effect on growth of fungi 54, 146
  see also Aeration
Ozone, effects on phylloplane colonization 97
Ozonium spp., growth on beech logs *354*

Paecilomyces spp.
  chitin degradation by 32
  heat-tolerant fungi 324
  soil fungi 175
Paecilomyces carneus, forest soil fungus 177
Paecilomyces lilacinus
  biological control agent 301
  nematode cyst parasite 285
Panaeolus sphinctrinus, hyphal interference 217
Panaeolus subbalteatus, growth in disturbed environments 43
Papulaspora spp.
  grassland soil fungi 177
  rhizosphere fungi *199*

INDEX 531

soil inhabiting 184
Parasites, symptomless 89
Parasitic fungi
  endoparasites 285
  growth on leaves 90, 92, 95–6
  nematode infection by 285, 292, 294, 298–9
Parthenolide 110
Pastures, fairy rings in 372
Pathogenic fungi
  of animals 38, 325, 337
  competition with
    phylloplane fungi 107
    yeast 106
  of plants 104, 106, 156
    cold tolerance 327–8
    resistance of plants 51
  in rhizosphere of host 191, *192*
Paxillus atrotomentosus, growth on dead wood 156
Paxillus involutus
  ectomycorrhizal fungus 378–9, 391, 393
  pH limits for growth *58*
Pea, *see Pisum sativum*
Peanut, *see Arachis hypogaea*
Penicillium spp.
  antibiotic production 70
  chitin degradation by 32
  endophytes 150
  facultative anaerobes 54
  forest soil fungi 177
  genetic variation in 16
  growth on
    dung 209
    leaf litter 115–16, 119, 121
    roots 192, 198, 201
    wood 157
  insensitivity to soil fungistasis 201
  optimal pH 58
  phenolic metabolism 115
  psychrotolerant fungi 327
  rhizosphere fungi 186

soil fungi 115, 175, 184
spore size 87
stimulation of growth of 51
tannin decomposition by 37
xerophiles 64
xerotolerant fungi 65, *333*, *334*, *336*
Penicillium brevicompactum
  distribution in soil *176*
  storage fungus 336
Penicillium capsulatum, storage fungus 336
Penicillium chrysogenum, temperature range for growth *323*
Penicillium citrinum, rhizosphere fungus *199*
Penicillium cyclopium, pH limits for growth *58*
Penicillium daleae, distribution in soil *176*
Penicillium expansum, rhizosphere fungus *199*
Penicillium hordei, storage fungus 336
Penicillium italicum, pH limits for growth *58*
Penicillium janthinellum
  mutual inhibition with *Collybia peronata* **69**
  rhizosphere fungus *199*
  root fungus *199*, **200**
Penicillium lilacinum, soil inhabiting 184
Penicillium luteum, pH limits for growth *58*
Penicillium nigricans
  effects of CO_2 concentration on growth of **56**, 57
  rhizosphere fungus *199*
Penicillium piceum, storage fungus 336
Penicillium pinophilum, cellobiohydrolases 28–9

532   INDEX

Penicillium rubrum
  antibiotic production 70
  rubratoxin production 337
Penicillium spinulosum, nitrogen requirement 48
Penicillium stoloniferum, antibiotic production 71
Penicillium verrucosum, storage fungus 336
Peniophora spp., endophytes 151
Peniophora gigantea, biological control agent 10
Peniophora limitata, growth on attached branches 348
Peniophora lycii, growth on attached branches 348
Peniophora phlebioides, growth on fallen timber 351
Peniophora quercina
  basidiocarps **349**
  growth on attached branches 346, **347**, 348, 350
Periconia abyssa, depth of survival 279
Periconia cookei, growth on herbaceous plants 141
Periconiella echinochloae, growth on sugar cane 142
Peronosporales, freshwater fungi 239–41
  colonization of submerged leaves 240, **241**
  effect of rainfall on numbers of 240
  techniques for the study of 239–40
Pestalotia spp., Phoma pattern of vegetative growth 272
Peziza spp., phoenicoid fungi 312
  growth following volcanic eruption 320
Peziza endocarpoides, phoenicoid fungus 311
  inhibition of seedling germination by 312

Peziza ostracoderma, phoenicoid fungus 314
Peziza petersii, fruit bodies **306**
Peziza praetervisa, phoenicoid fungus 308, 311, 314–15
  growth on lime-rich substrata 315
Peziza trachycarpa, phoenicoid fungus 311
  effects on distribution of Ingoldian aquatic hyphomycetes 257, **258**, 259
  limits for fungal growth 58
  optima for fungal growth, *see* named fungi; Hydrogen ion concentration
  of sea water 266
Phacidium infestans
  cold tolerance 328
  temperature range for growth 323
Phaeolus schweinitzii
  basidiospore formation 362
  growth on attached branches 350, 359
  selectivity of 148
Phallus impudicus 9
  growth on fallen timber 351, 352, *353*, *354*, **355**
  strand-former 156
Phanerochaete spp., growth on fallen timbers 354
Phanerochaete chrysosporium
  heat-tolerant fungus 324, 326
  lignin degradation 36–7, 326
  white-rot fungus **167**
Phanerochaete laevis, linear organs of 23
Phanerochaete velutina, 9
  combative fungus 82
  distribution 361
  growth on fallen timbers 352, *353*, *354*, **355**
  strand-former 156

survival at low water potential 170
Phaseolus vulgaris
  fungal communities on *41*
  rhizosphere communities 184, 189, *199*
  roots
    colonization of 198, *199*, **200**, 201
    exudates from 188
    surface mycoflora 192, *193*
Phellinus ferreus, growth on attached branches 346, 348, 350
Phellinus pini, selectivity of 148
Phenolics
  antifungal substances 49, **50**, 149
  action of 49
  in conifer litter 121
  in humus 123–4
  inhibition of aquatic hyphomycetes by 249
  levels in plants 126, 149
Phenols 115
  level in wood 145
  removal by soft-rot fungi 160
Phialophora spp.
  growth on
    bracken litter 122
    wood 158
    soft-rot fungi 163
    soil fungi 175
Phialophora radicicola, growth on roots 202
Phlebia merismoides, combative fungus 82
Phlebia radiata, growth on dead wood 156
Phlebia rufa, decay of attached branches by 346
Phoenicoid fungi 302–21
  cellulases 312
  classification 305–8
  community, effects of abiotic and biotic factors on **320**

competition, reduced by heat 314–15
competitive hierarchy 315
cumulative fruiting frequency 309, **310**
distribution 311
ecological characteristics 305–13
ectomycorrhizal formation 312
effect of
  ash on 317, *318*, *319*
  heat on 313–14, 317, *318*, *319*
  lime on 315, *316*
experimental studies 313–19
in forest sites 311–12
fruit-bodies **306**, **307**
fruiting
  following volcanic eruptions 319–21
  frequency 309, 311
  patterns 311
  stimulated by heat 314
  succession *313*
fungistasis and 315
germination, stimulated by heat 314
in grassland sites 312–13
growth
  factors affecting 305
  on litter 312
  on wood 312
lignin degradation by 312
phenology 308
phenol oxidase production 312
also plant parasitic 311–12
seasonal variation **308**, **309**
seedling growth inhibited by 312
trophic relationships 312
Pholiota carbonaria, phoenicoid fungus 311
  basidiocarps **307**
  growth following volcanic eruption 320

Phoma spp.
  endophytes 89, *151*
  vegetative growth 272
    response to salinity and temperature **273**, **274**
Phoma betae, inhibition by phylloplane fungi 106
Phoma pattern 272, 278
Phomopsis spp., endophytes 89, *151*
Phomopsis asteriscus, growth on hogweed 141
Phomopsis oblonga 152
Phomopsis perniciosa, growth on phylloplane 95
Phylloplane 88
  angiosperm 92
  colonization by fungi 89–114
    effect of fungal parasites on 95–6
    effects of cell leakage 99–100
    effects of climate 111–14
    effects of plant inhibitors on 108–11
    effects of pollutants on 97
    interspecific competition 105–8
    pollen effect 100–4
    primary colonizers, characteristics 113–14
    seasonal variation 93–4
  growth of weakly parasitic fungi on 92
  gymnosperm 92
  humidity 112–13
  microclimate 99, 111–13
  temperature 112–13
Phylloplane fungi 81, **91**
  colonization of radicles by 197
  persisting in leaf litter 115, 120
Phyllosphere 88
Phyllosticta spp., endophyte 89
Phytoalexins 108, 111
Phytophthora spp.
  freshwater fungi 239
  oospores as food for fungivores 20
  reproduction, water requirement for 63
Phytophthora cinnamomi, water stress 63–4
Phytophthora infestans, temperature range for growth *323*
Phytoplankton, parasitized by chytrids 228, **229**, **230**
Picea abies, litter production 122
Picea sitchensis
  colonization by macrofungi 359
  spore deposition on needles of 98
Pichia spp., osmotolerant fungi 333
Pigmentation, advantages of 21
Pilaira spp., growth on dung 207
Pilaira anomala
  fruit-bodies **205**
  fruiting time *214*
  growth on
    copromes 224
    dung 209, 211
  nutrition 8
Pilobolus spp.
  fruit-bodies **205**
  growth
    on dung 207
    effect of coprogen on 220–1
  sporangium 2
  sporulation, effect of ammonia on 221
Pilobolus crystallinus
  fruiting time *214*
  growth on dung 209
Pilobolus kleinii, effect of ammonium and acetate on fruiting 221, **222**
Pinus spp.
  antifungal compounds 149
  cones, macrofungal growth on 366

fungal succession on needles of 96
heart-rot basidiomycetes on 148
mycorrhizae 381–2
saprophytic chytrids on 230–1
soft-rot fungi on 165
stem endophytes 150
Pinus contorta var. *latifolia*, macrofungal growth on fallen timber, 351
Pinus maritima, fungal succession on litter 120
Pinus sylvestris
  antifungal volatiles produced by 52
  fungus stimulators produced by 51
  litter
    rate of decomposition of 124, 126
    succession on 121
  marine fungi, growth on 276
  soft-rot fungi, growth on 163–5
Pipe rot 348
Piptocephalis spp.
  growth on dung 207
  mycoparasite 218
Piptocephalis cylindrospora, fruit-body **205**
Piptoporus betulinus
  basidiocarp **361**
  infection of standing trees by 359
  selectivity of 148
  spore size 87
  sporulation 363
Piricauda pelagica, holeuryhaline species 277
Pisolithus arhizus, ectomycorrhizal fungus 378, 395
Pisum sativum, colonization of leaves 93
  radicles *197*

Plasmogamy 15
Pleomorphism 17
Pleospora vagans
  growth on *Dactylis glomerata* 133
  relative humidity requirement *139*
Pleurophragmium simplex, growth on nettle 142
Pleurotus spp., nematodes as nitrogen source for 147
Pleurotus ostreatus
  bacteria as nitrogen source for 147
  lignin degradation by 36
  nematotoxin production 299
  survival at low water potential *170*
  wood-rotting, nematophagous fungus 299
Pluteus cervinus
  distribution of 39
  growth on dead wood 156
Podocarpus ferrugineus, antifungal compounds 149
Podosphaera leucotricha, inhibition by plant antifungal substances 110
Podospora spp.
  growth on dung 207
  phoenicoid fungi 313
Podospora appendiculata, growth on dung 224
Podospora curvicolla, growth on dung 209
Podospora curvula, growth on dung 224
Podospora minuta, fruiting time *214*
Podospora pilosa, antibiotic production 315
Podospora setosa, growth on dung 224
Podospora tetraspora, growth on dung 209

Podospora vesticola
  fruit-body **206**
  growth on dung 224
Podsol, fungal distribution in 175
Point growth 25
Pollen
  effect 100–1, *102*, **103**, **104**
  growth of saprophytic chytrids on 229
  pathogen lesion development 106
  stimulation of
    fungal spore germination by 102
    pathogenic fungi 104
Pollutants, effects on
  ectomycorrhizal fungi 394
  leaf litter communities 127
  phylloplane colonization 97
Polyols, role in tolerance to water stress 338–9
Polyphenol oxidases 34, **35**, 36
  detoxification of antifungal substances 49–51
Polyporus amarus, growth on attached branches 350
Polyporus ciliatus, basidiospore formation 362
Polyporus squamosus
  basidiospore formation 362
  spore size 87
Polysaccharides, in plant cell wall *163*
  see also Cellulose; Hemicellulose; Starch
Polyscytalum fecundissimum, growth on leaf litter 117
Polyspores 156
Populus tremula
  antifungal compounds 49
  pathogens of 156
  production of fungus stimulators 51
Poronia punctata, antibiotic production 215

Primary resource capture, definition 6
Prior colonization effect 81
Privet, *see Ligustrum vulgare*
Propagules, of aero-aquatic hyphomycetes 262, **263**
Protascus subuliformis, endoparasitic nematophagous fungus **288**
Proteins, hydrolysis of 37–8
Psathyrella carbonicola, growth following volcanic eruptions 320
Psathyrella hydrophilum 9
  growth on dead wood 156
Pseudohiatula tenacella, growth on *Pinus* cones 366
Pseudoperithecia 140
Pseudotrametes gibbosa
  growth on dead wood 9, 156
  interaction with *Bjerkandera adusta* 356
Pseudotsuga spp., antifungal compounds 149
Pseudotsuga menziesii, mycorrhizae 381
Psilocybe spp., growth in disturbed environments 43
Psychrophiles 1, 54, 322, 331
Psychrotolerant fungi 54, 208, 322, 327–8
  adaptations 331
  membrane structure 331
  sterile forms of 54
Puffball, *see Lycoperdon pyriforme*
Pycnidia 140
Pyrenopeziza revincta, growth on hogweed 142
Pyrenopeziza urticola, growth on nettle 142
Pyronema spp., phoenicoid fungi 311, 314
Pyronema domesticum, phoenicoid fungus 314

Pyrophilous fungi, *see* Phoenicoid fungi
Pyropyxis rubra, phoenicoid fungus 312
  inhibition of seedling germination 312
Pythiaceae, tolerance of water stress 63
Pythium spp. 7
  cellulose hydrolysis by 29, 45, 82
  colonizers of leaf debris in streams 240, **241**
  food for amoebae 20
  freshwater fungi 239–40
  growth on roots 202
  invasion of roots by 81
  nutrition 7
  optimal pH for growth of 57
  pioneer colonizers 82
  reproduction 7
  water requirement of 63
Pythium fluminum var. *fluminum*, freshwater fungus 239–40
Pythium oligandrum, 'secondary sugar' fungus 46
Pythium ultimum, germination of sporangia 8
Pythium ultimum, rhizosphere fungus *199*

Quercus spp.
  antifungal compounds 109, 149
  colonization of
    attached branches **347**, 350
    submerged wood *250*, *251*, 252
  growth on coal spoil, mycorrhizal associations 391
  heart-rot basidiomycetes of 148
  macrofungal growth on **347**, 350
    intraspecific antagonism of 357
  microfungal colonization of *158*
  Mycena galopus on litter of 367–8
  pathogens of 156
  stem endophytes 150

Quercus petraea
  litter production 122
  macrofungal decay of attached branches 345–6, 348
Quercus robur
  litter
    rate of decomposition of 124
    succession on 118
  macrofungal decay of attached branches 345–6, 348

Rabbit dung, growth of fungi on 207–9, 211, 213, 215, 219, 223
Ramularia sp., mycoparasite 95
Remispora maritima, vertical distribution 277
Remispora pilleata, brackish water species 277
Reproductive cycles 1–2
  asexual 2
  parasexual 2, 16
  pleomorphic fungi 1
  sexual 2
Respiration, measurement of 174
Resupinates 156
Rhabdocline parkeri, endophyte 90
  protection of plants against insects 152
Rhacodium spp., psychrotolerant fungi 327
Rhinocladiella spp., growth on wood 158
Rhipidium spp., facultative anaerobes 55
Rhizina undulata, phoenicoid fungus 311–12, 314
  fruit-bodies **306**
  growth following volcanic eruptions 320
  inhibition of seedling growth 312
  plant parasitic fungus 311–12
Rhizoctonia spp.
  effects of CO_2 concentration on growth of 57

Rhizoctonia spp. *contd*
  growth on roots 193, 202
Rhizoctonia solani
  effects of CO_2 concentration on
    growth of 56
  inhibition by antifungal volatiles
    182
Rhizoctonia tuliparum, sclerotia of
  21
Rhizomorphs **22**-3, **24**, 25–6, 156
Rhizomucor miehei, heat-tolerant
  fungus *323*, 324
Rhizomucor pusillus
  animal pathogen 325
  heat-tolerant *323*, 324
Rhizophydium spp., saprophyte on
  pollen 229–30, **231**
Rhizophydium planktonicum,
  parasite of phytoplankton 228,
  229, **230**
Rhizoplane, fungal growth in 183–4
Rhizopogon cokeri,
  ectomycorrhizal fungus 378
Rhizopogon luteolus,
  ectomycorrhizal fungus 379
Rhizopus spp. 7
  hemicellulose degradation 31
  inability to hydrolyse cellulose
    29
  pioneer colonizer 82
  thermophilic fungi, membrane
    structure 330
  xylan hydrolysis 45, 82
Rhizopus sexualis, cold tolerance
  327
Rhizopus stolonifer, cold tolerance
  327
Rhizosphere 183–96
  antifungal substances in 187
  effect 183–4, 186–8
    on ectomycorrhizal fungi
      379–80
    increase with time 189–91
    with increasing depth 186

  ethylene production in the
    187–8
  extent of 186
  fungal
    communities 184, **185**, 186–7,
      199
    growth in 183–4
    spore germination in 183
  material sloughed off roots in
    190–1
  mycoflora
    development of 189
    differences between plants
      191
    effects of environmental
      conditions on 192
    effects of physiological
      changes in the plant on
      192, 193, *194*
    harmful effect on plant growth
      196
    pathogenic fungi in 191, *192*
    plant competition and 196
    stimulation of plant growth by
      196
  root exudates in 187–9, 193–4,
    195, 196
Rhodotorula spp., growth on
  phylloplane 92
Rhodotus palmatus, distribution of
  39
Rhopalomyces spp., nematode egg
  parasites 285
Rhopalomyces elegans, parasite of
  nematodes 207
Rice, *see Oryza sativa*
Rigidoporus spp., perennial
  sporophytes 362
Rigidoporus ulmarius, growth on
  attached branches 350
Ripartites tricholoma, phoenicoid
  fungus 308, 311
Root fungi, techniques for studying
  174

INDEX 539

Roots
  antifungal substances released into rhizosphere 187
  colonization by fungi 184, **185**, **190**, *197*, **198**, *199*, **200**, 201–2
    early *197*, 201
    mature systems 198, 201
  death of 191
  exudates into rhizosphere 187–9, 193–4, *195*, 196
  fungal growth stimulated by 183–4, 186–8, 379–80
  invasion by fungi 190
  mycorrhizal spore germination stimulated by 379–80
  shedding of material into rhizosphere 190–1
Root tips, source of organic matter for the soil 188
R-selected fungi, *see* Ruderals
R-selection 6–7
R:S ratio 184–6
Rubratoxin 70, 337
Ruderals 7–8, 44, 81, 157
  aquatic hyphomycetes 254
  Clitocybe nebularis 372
  early-stage mycorrhizal fungi 387
  Pilaira anomala 224
  Trichoderma spp. 10
Runner formation 71
Ruscus spp., stem endophytes 150
Russula spp.
  inhibition by *Collybia butyracea* 397
  late-stage mycorrhizal fungi 385, 393, 397
Russula fragilis, ectomycorrhizal fungus 395
Russula mairei, ectomycorrhizal fungus 391
Rusts, 95
Rye, *see Secale cereale*

Ryparobius spp., growth on dung 207
Ryparobius dubius, fruiting time *214*

Saccharomyces baillii, water requirement *333*
Saccharomyces rouxii, osmotolerant yeast *333*
  accumulation of polyols in 338
Saccharum officinarum, fungal succession on 142
Saccobolus spp., growth on dung 207
Saccobolus versicolor 209
Salix spp.
  growth on coal spoil, mycorrhizal associations 391
  heart-rot basidiomycetes 148
Saltmarshes, colonization of roots by phylloplane fungi 197
Sambucus nigra, antifungal substances 109
Sand dunes, fungi characteristic of 280–1
Saprolegnia spp.
  dispersal of 3
  inability to hydrolyse cellulose 45
  reproduction, water requirement of 63
Saprolegnia ferax/mixta, in sewage effluent 236
Saprolegnia parasitica, salinity tolerance 238
Saprolegniaceae
  in lake muds 234, **235**
  tolerance of water stress 63
Saprolegniales
  cellulose hydrolysis by 29
  freshwater fungi 233–9
  distribution 234–8
  effect of pH on distribution 236–7

Saprolegniales *contd*
freshwater fungi *contd*
effect of salinity on
distribution 237
periodicity 238–9
techniques for studying 233–4
inability to utilize nitrates 48
Saprolenia diclina, freshwater
fungus 238
Sapromyces spp.
facultative anaerobes 55
freshwater fungi 232
Saprophytes, growth on
phylloplane 90, 92
Saprotrophs, substrates for 26–38
Sapstainers 159
Sapwood, limitations to fungal
growth in 154
Sapwood-rot fungi 146
Schizophyllum commune
basidiocarp survival 364
growth following volcanic
eruptions 320
inhibition by antifungal volatiles
52
survival at low water potential
170
Schizopora paradoxa
growth on attached branches 346
survival at low water potential
171
Sciarid fly 79
Sclerocarp 281–2
Sclerophoma pythiophila, growth
on
litter 120
pine needles 96
pine strobili 97
Sclerotia 5, 20–1, 140
antibiotics in 21
formation of 20–1
survival of 21
Sclerotinia sclerotiorum, pH limits
for growth 58

Sclerotium rolfsii, sclerotia
formation 21
Scytalidium album, antagonist of
wood-rotting fungi 160
Sea foam, fungal spores in **242**, 280
mycostatic factor in 271
pH of 266
salt content 266
temperature range 266
Secale cereale, colonization of
leaves, effects of pollen **103**,
104
Secondary resource capture 6, 71
Secondary strategies 10–11
'Secondary sugar' fungi 46
Seedlings
infection by mycorrhizal fungi
385
inhibition by phoenicoid fungi
312
Selenophoma donacis, growth on
Dactylis glomerata 133
Septa 13–14
Septoria nodorum, inhibition by
saprophytic fungi 106
Sequoia sempervirens, inhibition of
soft-rot fungi 165
Serpula lacrimans
brown-rot fungus 168
cord initiation in 25
growth on worked timbers 157
linear organs of 23, 25
Sewage
nematophagous fungi in 294, **295**
Saprolegniaceae in 236
Sewage fungus 232
Silver maple, *see Acer saccharum*
Sistotrema brinkmannii 9
Sitka spruce, *see Picea sitchensis*
Slime, polysaccharide 114
Slime moulds, growth on dung 204
Soft-rot fungi 158, 160–1, 163, **164**,
165–6, 324
cellulose degradation by 164

establishment of 163
growth conditions 163
hemicellulose degradation by 164
lignin degradation by 164
marine fungi 279
spread of 164–5
Soil
  basidiomycete biomass 365
  carbon dioxide concentration in 55–6
  cellulolytic activity in 29
  changes in after burning 303–5
  fungal activity in 179–83
  fungal communities 177
    seasonal variation 178–9
  fungal distribution in 175–9
    see also Rhizosphere
  fungi, see Soil fungi
  fungistasis 68, 179–83, 201
    inhibitory volatiles 181–2
    nutritional hypothesis 180
    overcome by roots 183
    residual 182
  horizons 123, 125, 175
    distribution of fungi in 175, *176, 177*
  mineral, fungal activity in 179–83
  mor 126, 365, 391
  mull 127, 365, 391
  mycelial networks of 179
  mycorrhizae in 377
  mycostasis 179
  nutrient availability, effect of heat on 313
  oxygen levels in 54
  pH, increase after burning 303
  physical properties, effects of burning on 304
  R:S values 184–6
  type, effect on mycorrhizal associations 390–1
  water potential of 61
  water sorption isotherms **62**
Soil fungi
  antibiotic production by 70
  autochthonous species 115–16
  colonization of roots by 197
  effects of burning on 305
  techniques for studying 172–5
  zygmogenous species 116
Sordaria spp.
  growth on dung 207
  phoenicoid fungi 313
Sordaria fimicola
  anastomosing hyphae **15**
  fruiting time *214*
  growth on dung 212–13
Sordaria humana, growth on dung 211
Sordaria macrospora, growth on copromes 215
Spaerobolus stellatus, growth on dead wood 157
Species turnover 42
Spegazzinia tessarthra, growth on sugar cane 142
Sphaerosporella brunnea
  ectomycorrhizal formation 312
  phoenicoid fungus 312
Spirodactylon spp., growth on dung 204
Spores 17–20
  atmospheric, numbers of 85
  autolysis of 18–19
  deposition on leaves
    abaxial and adaxial leaf surfaces 98
    effects of leaf structure on 99
    factors affecting 87
    on veins 99
  deposition on plants
    rain wash-out 86
    sedimentation 86
    splash dispersal 86
    wind impaction 86
  desiccation-resistant 340
  discharge, mechanisms 203

Spores *contd*
  dispersal of 2–3, 17
  functions 17
  germination
    inhibition by bacteria 101–2
    on leaf surfaces 100–2
    stimulation by pollen 102, 104
  leakage from **101**
  pigmentation of 21, 114
  secondary 94
  size, effect on deposition and dispersal 87
  survival following digestion 203
  survival time 19
  wastage of 18–19
    consumption by animals 19–20
Sporidesmium salinum, effect of salinity on vegetative growth 272
Sporobolomyces spp., growth on phylloplane 92, 95
Sporobolomyces pararoseus, competition with parasitic fungi 106
Sporobolomyces roseus **91**
  effects of lead pollution on colonization by 97
  growth on conifer needles 96
Sporobolomyces ruberrimus, antifungal antibiotic production 107
Sporocybe spp., soft-rot fungi 163
Sporomiella spp., growth on dung 207
Sporopollenin 21
Sporormia bipartis, growth on dung 224
Sporormiella spp., phoenicoid fungi 313
Sporormiella intermedia
  fruiting time *214*
  growth on dung 208
Sporormiella pilosella, phoenicoid fungus 315

Sporotrichum pulverulentum, β1–4 glucanase complex 28
Springtails 20
  effects of grazing on fungal growth 79, 368–70
S-selected fungi, *see* Stress-tolerant fungi
S-selection 7
Stachybotrys atra, growth on hogweed 142
Staling 70, 82
Starch 27
Steccherinum fimbriatum, cords **24**
Stemphylium spp., growth on phylloplane 92, 94
Stemphylium botryosum
  antibiosis 107
  germ tube survival 114
Stemphylium solani, pH limits for growth *58*
Stems, microclimate 113
Stenohaline species 277
Stereum gausapatum
  decay of attached branches by 346, 348
  white-rot fungus 154
Stereum hirsutum 9
  basidiocarps **349**
  growth on
    attached branches 346
    beech logs *353*, *354*, **355**
    dead wood 156
    fallen timber 351–3
  hyphal barrage with *Coriolus versicolor* **67**
  sporulation 364
  survival at low water potential *170*
Stereum sanguinolentum
  growth on
    attached branches 350
    fallen timber 351
  inhibition by antifungal volatiles 52

white-rot fungus 168
Stilbella erythrocephala, growth on dung 207, 209–10
    antibiotic production 215
Stilbenes 149
Stimulators of fungal growth 51–2, 66–7
Stinkhorns, see Phallus impudicus
Storage
    conditions to prevent fungal growth 336
    of grain
        fungal colonization 334–6
        health hazards 337–8
Storage fungi 334–7
    deterioration of plant oils by 336
    interaction between water activity and temperature 336
    mycotoxin production 337
Strand-formers 156
Streptomyces griseus, production of antifungal volatiles 182
Streptomyces halstedii, production of antifungal volatiles 182
Stress 5, 8
Stress-tolerant fungi 8, 11, *44*, 149
    late-stage mycorrhizal fungi 387
Stysanus stemonitis, soil inhabiting fungus 184
Substrata for fungal growth 26–38
    water relations of 65–6
Successions 80–4
    cellulose hydrolysis during 83
    climax of 83–4
    coprophilous fungi 207–22
    factors influencing course of 84
    fungal 43
        on angiosperm leaves 92–5, 103
        in compost 323–4
        on dung 207–22
        on gymnosperm leaves 96
        on herbaceous plants 143–4
        on leaf litter 114–22

on leaves in freshwater 252
on moss litter 122
on submerged leaves 252
on sugar cane 142
on wood 83, 276
Ingoldian aquatic hyphomycetes 252
lignicolous marine fungi 276
mycorrhizal fungi 383–8
predatory nematophagous fungi 293–5
studies of, importance of mycelia estimations 83
wood-rotting microfungi 159
Sugar cane, see Saccharum officinarum
'Sugar fungi' 45, 81–2
Suillus spp., ectomycorrhizal fungi 378
Suillus aeruginascens 395
Suillus bovinus
    ectomycorrhizal fungus 381
    somatic incompatibility *382*
Suillus granulatus 379, 395
Suillus grevillei
    ectomycorrhizal fungus 380–1, 395
    pH limits for growth *58*
Suillus hirtellus 394
Suillus luteus 379
Suillus plorans 395
Suillus variegatus 395
Sulphur dioxide, effects on
    leaf litter communities 127
    phylloplane colonization 97
Sweetgum, see Liquidambar styraciflua
Sycamore, see Acer pseudoplatanus
Symbionts
    endophytes 89
    mutualistic 376
    neutral 89, 150
    study of 150

Sympodiella acicola, growth on pine litter 120
Synergism 47
  coprophilous fungi 220–2

Talaromyces thermophilus,
  heat-tolerant fungus 324
  temperature range for growth *323*
  thermostable proteins 328
Tannins
  antifungal properties 149
  complexes with polyphenol oxidases 51
  condensed 37
  degradation of 37, 115
  hydrolysable 37
  inhibitory activity of 49
  level in plant tissues 126, 145
Taxus spp., heart-rot basidiomycetes 148
Teleomorph 1
Temperature
  effects on fungal growth 53–4, 66
  physiology of adaptation to extremes of 328–32
  ranges for growth of fungi *323*
Tephrocybe carbonaria, phoenicoid fungus 308, 311
  seasonal variation **309**
Terpenes, antifungal agents 149
Terrestrial fungi, effect of salinity on 275
  see also Macrofungi
Tetrachaetum elegans **243**
  colonization of submerged leaves 249, 252
Tetracladium marchalianum **244**
  colonization of submerged leaves 249
  inhibition of fungal growth by 253
Tetraploa aristata **132**

growth on *Dactylis glomerata* 133–4, **136**, 142
growth on sugar cane 142
relative humidity requirement *139*
Thamnidium spp., psychrotolerant fungi 327, 331
Thelebolus spp.
  growth on dung 207–8
  psychrotolerance 208
Thelebolus nanus, growth on dung 224
Thelebolus stercoreus
  fruit-body **205**
  growth on dung 224
Thelephora terrestris, early-stage mycorrhizal fungus 385, 391
  fruit-bodies **387**
Thermomyces lanuginosus
  heat-tolerant fungus 324
  temperature range for growth *323*
  thermostable proteins of 328
  'secondary sugar' fungus 46
Thermophiles 1, 53–4, 322
  heat-shock proteins 329
  membrane structure 329–30
  thermostability of enzymes 328–9
Thermotolerant fungi 1, 53–4, 322–6
  deterioration of plant oils by 325, 336
  habitats 322, 325
  physiological behaviour 324
  potential uses 325–6
  substrate utilization 324
  see also Storage fungi
Thielavia angulata, growth on roots 202
Thielavia terrestris, heat-tolerant fungus 324
Thraustochytriales, obligate marine fungi 266
Thuja spp., antifungal compounds 149

Thysanophora penicilloides, stimulation of growth of 51
Tieghemiomyces spp., growth on dung 204
Tilia spp., rate of decomposition of litter 124
Tilia americana, colonization by marine fungi 276
Tilia playphylla, antifungal substances 109
Tilia vulgaris, production of volatiles affecting fungal growth 52
Tobacco, see *Nicotiana tabacum*
Tolypocladium spp., marker fungi of tundra soils 177
Tolypocladium geodes, distribution in soil *176*
Top rot 152
Torula herbarum **132**
   growth on herbaceous plants 141
   relative humidity requirement *139*
Torulopsis candida, competition with parasitic fungi 106
Trehalose, role in desiccation-resistant spores 340
Triadimefon 107
Trichocladium asperum
   rhizosphere fungus *199*
   soil inhabiting 184
Trichocladium opacum
   estuarine distribution 278
   growth on herbaceous plants 142
Trichoderma spp.
   antibiotic production by 70, 73
   antifungal volatiles produced by 70, 182
   chitin degradation by 32
   colonization by, effect of temperature on 10–11
   combined strategies 10
   endophytes 150
   in carbon-free media 8
   on roots 192–3, 201
   on wood 157, *354*
   insensitivity to soil fungistasis 201
   in litter
      conifer 121
      leaf 115
   necrotrophs 72–4
   optimal pH 58
   seasonal variation of 178–9
   soil fungi 115, 175
Trichoderma hamatum, C-selected fungus 11
Trichoderma harzianum
   growth on wood 160
   necrotroph 74
   nitrogen requirement 48
   oligotroph 8, 10
Trichoderma koningii
   C-selected fungus 11
   early root colonizer 199
   nitrogen requirement 48
   pH limits for growth *58*
   rhizosphere fungus *199*
Trichoderma polysporum
   distribution in soil *176*
   stress-tolerant strategy 11
Trichoderma viride
   β1–4 cellobiohydrolases 28
   distribution in soil *176*
   early root colonizer 199
   facultative anaerobe 54
   hyphal chlamydospore of **18**
   necrotroph **74**
   nitrogen requirement 48
   optimal pH 58
   rhizosphere fungi *199*
   root fungi *199*, **200**
   stress-tolerant strategy 11
Trichodermin 70
Tricholomopsis spp., growth on fallen timbers 354

Tricholomopsis platyphylla 9
  basidiocarps **362**
  distribution 39, 359, 361
    of mycelial types **363**
  growth on
    beech logs **355**
    fallen timbers 352, *353*, *354*, 355
  strand-former 156
Tricholomopsis rutilans
  basidiocarp production 39
  growth on dead wood 156
Trichophaea abundans, phoenicoid fungus 311–12, 314
Trichophaea hemisphaerioides,
  phoenicoid fungus 308, 311–12
  growth following volcanic eruptions 320
  growth on lime-rich substrates 315
Trichophyton spp., skin pathogens 38
Trichothecium cystosporium,
  nematophagous fungus 293
Trichothecium roseum
  rhizosphere fungus *199*
  soil inhabiting 184
Tricladium spp., growth on mangroves 277
Tricladium chaetocladium **243**
  spore production 254, **255**
Tricladium splendens **243**
  colonization of shed leaves 249
  soft-rot fungus **164**
  survival in anaerobic conditions 260
Triscelophorus monosporus,
  colonization of shed leaves 249
Triticum aestivum, roots
  exudates into rhizosphere 187, *195*
  material in rhizosphere 191
  surface fungi 189

  effects of pH on 192
Tropic responses, to volatiles produced by plants 52
Tropolones 149
Troposorella monospora, growth on pine litter 120
Tubaria autochthona, distribution 40
Tumularia aquatica **244**
Tundra soils, fungal marker populations 177–8
Typhula spp., cold tolerance 328

Ulmus spp.
  growth of macrofungi on attached branches 350
  rate of decomposition of litter 124
Unit community 40–1
Urtica dioica, fungal colonization of 141–2
Ustulina deusta
  tree pathogen 156
  xylariaceous rot 168

Varicosporina ramulosa
  conidial development 274
  effect of temperature on distribution 282
  sclerocarp 281
  spores in sea foam 280
Varicosporium elodeae **243**
  colonization of leaves by 260
Verticicladium trifidum, growth on pine litter 120
Verticillium spp.,
  chitin degradation by 32
  endoparasitic nematophagous fungi 285
  survival of sclerotia 21
Verticillium balanoides,
  endoparasitic nematophagous fungus 292
Verticillium bulbillosum,
  distribution in soil *176*

INDEX 547

Verticillium chlamydosporium,
  nematode cyst parasite 285
Verticillium lecanii, mycoparasite 95
Verticillium obovatum,
  endoparasitic nematophagous
  fungus **288**
Vicia faba
  antifungal substance in cuticular
    wax 110
  root exudates from 188
Viennotidia spp. growth on dung
  207
Viennotidia fimicola
  fruiting-body **206**
  growth on dung 220
  synergism with *Eurotium repens*
    220
Vitamins, requirements of fungi 48
Volcanic eruptions, fruiting of
  phoenicoid fungi following
  319–21
Vuilleminia comedens
  decay of attached branches by
    346, **347**, 348
  endophyte 151
  white-rot fungus 154

Water
  activity 59–60, *61*, *333*
  availability, effects on succession
    84
  potential 59–60, *61*, *333*
    optima for fungal growth 61,
      63–5
    relations of fungi 61–2
    sorption isotherms **62**
    stress 62
      tolerance to 62–3, 332, 338–40
      see also Osmotolerant fungi;
        Xerotolerant fungi
Waterlogging 55
Waxes, antifungal properties of 110
'Weed' species
  characteristics 43, 70

ecological adaptation *44*
  in pioneer communities 81
Wet-rot fungi 157
Wheat, *see Triticum aestivum*
White-rot fungi 34, 36, 154, 161,
  166, **167**, 168, 326, 353
  distribution 168
  effects on wood 167
  macrofungi 345–6, 367
  marine 279
  metabolism of carbohydrate and
    lignin 166–8
Wood
  aeration of 146
  antifungal properties of 149
  carbon dioxide levels in 145–6
  cellulose
    content of 27
    distribution in cell walls 162
  colonization of 9, 150–61
    by aquatic hyphomycetes 249,
      250, *251*, 252
    endophytes 150–2
    factors affecting 148, 159–60
    fallen timbers 156–7, 351–6
    fungal characteristics 146
    fungus–invertebrate
      interactions 160–1
    interspecific interactions 147
    by macrofungi 152–7, 345–56
    by marine fungi 275–83
    by microfungi 157–60
    standing trees 153–6, 345–51
    water relations 169–71
  composition of 145
  decay of
    fallen timber 351–6
    by macrofungi 345–64
    in the marine environment
      279–80
    patterns of 152
    standing trees 345–51
    worked timbers 157, 159
  hemicellulose content of 30, 148

Wood *contd*
  lignin
    content of 148
    distribution in cell walls 161–2
  medullary rays, influence on decay 152
  moisture content, change on decay of 157, 160, 170
  nitrogen content of 146
  oxygen levels in 55
  parenchyma, influence on rate of decay 152
  phenolic compounds in 148–9
  rot, types of 161–9
  succession on 83
  tannins, protective 51
  temperature fluctuation in 53
  tropical 149
Wood-rotting fungi 29, 39, 69
  axial colonization by 152
  colonization of attached branches 345–6, **347**, 348, 350, 359
  colonization of dead wood 156–7, 351–6
  colonization of standing trees 153–6
    basidiomycetes 154
    macrofungi 345–51
    pioneer species 153–4
    route of entry 154
  colonization of worked timbers 157, 159
  distribution, effect of chemical composition of wood on 148
  entry into host 152
  growth at low nitrogen levels 47
  heart-rot fungi 146
    selectivity of 147–8, 154
    S-selected 149
  hemicellulose degradation 31
  host switching 40
  interactions with invertebrates 160–1
  lignin degradation 34, 36
  microfungi 158–60
    primary colonizers 157–8
    secondary colonizers 159
    succession of 159
  radial spread 152
  sapwood-rot fungi 146
  sensitivity to water stress 169
  survival at low water potentials 170–1
  see also Macrofungi
Wood-stainers 146, 159

Xenospores 2, 17
Xerophilic fungi 64, 339
Xerotolerant fungi 62–5, 332–8
  accumulation of polyols in 338–9
  assessment of tolerance 332
  physiology of adaptation 338–40
Xylaria spp.
  endophytes 90, 150, *151*
  growth on fallen timbers 352
Xylaria carpophila
  distribution 40
  growth on beech cupules **366**
Xylaria hypoxylon 9
  fruit-bodies **155**
  growth on fallen timbers 352, *353*, *354*, **355**
  xylariaceous rot 168
Xylariaceae
  degradation of lignin by 45
  fruit-bodies **155**
Xylariaceous rots 168–9
  lignin oxidation by 169

Yeasts
  antibiosis 107
  bark-colonizing, degradation of tannins by 37
  competition with fungal pathogens 106
  growth on
    buds 97

flowers 98
phylloplane 92–6, 98, 105, 114
marine 267
production of polysaccharide
    slimes 114
psychrotolerant 327
soil inhabiting 184

Zalerion maritimum
conidium **270**
depth of survival 279
Phoma pattern of vegetative
    growth 273
spore germination 272
stenohaline species 277
vertical distribution 277
Zineb 107

Zone-lines 357
Zoospores 2
Zygomycetes
    inability to hydrolyse cellulose 29
    nematophagous fungi 284
Zygomycotina
    growth on dung 204, 212
    hyphae of 13
    non-cellulytic 45
Zygorhynchus vuillemini, effects of
    CO_2 concentration on growth
    of **56**, 57
Zygorrhizidium affluens, parasite
    of phytoplankton 228
Zygorrhizidium planktonicum,
    parasite of phytoplankton 228
Zygospores 2, 7, 18